AF478045

Cystic Fibrosis

A Trilogy of Biochemistry, Physiology, and Therapy

A subject collection from *Cold Spring Harbor Perspectives in Medicine*

Cystic Fibrosis
A Trilogy of Biochemistry, Physiology, and Therapy

A subject collection from *Cold Spring Harbor Perspectives in Medicine*

Edited by

John R. Riordan
University of North Carolina at Chapel Hill

Richard C. Boucher
University of North Carolina at Chapel Hill

Paul M. Quinton
University of California, San Diego

COLD SPRING HARBOR LABORATORY PRESS
Cold Spring Harbor, New York • www.cshlpress.org

Cystic Fibrosis: A Trilogy of Biochemistry, Physiology, and Therapy
A Subject Collection from *Cold Spring Harbor Perspectives in Medicine*
Articles online at www.perspectivesinmedicine.org

Executive Editor	Richard Sever
Managing Editor	Maria Smit
Project Manager	Barbara Acosta
Permissions Administrator	Carol Brown
Production Editor	Diane Schubach
Production Manager/Cover Designer	Denise Weiss
Publisher	John Inglis

Front cover artwork: The cover presents a trilogy of icons for areas of investigations into cystic fibrosis since its introduction into the medical literature 70 years ago: biochemical genetics, physiology, and therapy. The icons represent the progression of book topics in three parts from the genetics and biochemistry that identified and characterized the consequences of the missing phenylalanine (DF) in cystic fibrosis transmembrane conductance regulator (CTFR), to the physiological consequences of mutations on ion transport and mucus abnormalities in affected tissues, to the ultimate objective of effective therapy for the disease. Cover art created by Sean Needham.

Library of Congress Cataloging-in-Publication Data

Cystic fibrosis : a trilogy of biochemistry, physiology, and therapy / edited by John R. Riordan, University of North Carolina, Chapel Hill, Richard C. Boucher, University of North Carolina, Chapel Hill, Paul M. Quinton, University of California San Diego.
        page cm
"A subject collection from Cold Spring Harbor perspectives in medicine"--
Includes bibliographical references and index.
ISBN 978-1-936113-34-7 (hardcover : alk. paper)
 1. Cystic fibrosis--Pathophysiology. I. Riordan, John R., editor of compilation. II. Boucher, Richard C., editor of compilation. III. Quinton, Paul M., editor of compilation.

 RC858.C95C9336 2012
 616.3'72--dc23
                         2012032952

10 9 8 7 6 5 4 3 2 1

Contents

Contents

Preface

AN ALTERNATIVE TITLE FOR THIS BOOK could have been *Cystic Fibrosis: From Patient to Molecules and Back* as this bidirectional cycle has been, and continues to be, the focus of the international community that strives to understand and treat this disease. The journey certainly began with the clinical description of cystic fibrosis (CF), but, as methods of care and research discoveries have evolved, advances have been made from entry at all points in the cycle. The book, organized in three sections, somewhat arbitrarily begins with molecular studies, moves to studies of physiological dysfunction, and ends with therapy and treatment. The findings and interpretations presented in the chapters of each section are very much dependent on those in the other sections. We hope that this organization may contribute to a strong overall integration of the diverse range of approaches and their outcomes that may not be obvious from just their linear listing.

In another attempt at overall integration, nearly all of the chapters were written jointly by authors from two different laboratories or clinics. Even though this collaboration may have added to their burdens, the authors pursued it enthusiastically and produced very balanced coverage of their topics. Nevertheless, because of the finite size limitations, the book is far from comprehensive with many important issues and contributions from authorities in the field not included. It is hoped that these omissions will be remedied in revised editions of the book.

Despite the vacancies, we hope the contents do convey the excitement of the progress that has been made toward dissecting and conquering CF and the anticipation of the further breakthroughs on our doorsteps. We thank all of the authors for their labors of love and the exceptional editorial and publishing staff at Cold Spring Harbor Laboratory Press for their patience and persistence. The project manager, Barbara Acosta, is a saint. The editors are not so saintly!

JOHN R. RIORDAN
RICHARD C. BOUCHER
PAUL M. QUINTON

The Cystic Fibrosis Gene: A Molecular Genetic Perspective

Lap-Chee Tsui[1] and Ruslan Dorfman[2]

[1]The University of Hong Kong, Hong Kong, Special Administrative Region, China
[2]Geneyouin Inc., Maple, Ontario L6A 0H6, Canada

Correspondence: tsuilc@hku.hk; Ruslan.dorfman@geneyouin.ca

The positional cloning of the gene responsible for cystic fibrosis (CF) was the important first step in understanding the basic defect and pathophysiology of the disease. This study aims to provide a historical account of key developments as well as factors that contributed to the cystic fibrosis transmembrane conductance regulator (CFTR) gene identification work. A redefined gene structure based on the full sequence of the gene derived from the Human Genome Project is presented, along with brief reviews of the transcription regulatory sequences for the CFTR gene, the role of mRNA splicing in gene regulation and CF disease, and various related sequences in the human genome and other species. Because CF mutations and genotype–phenotype correlations are covered by our colleagues Ferec and Cutting in the next chapter, we only attempt to provide an introduction of the CF mutation database here for reference purposes.

The cloning of the gene responsible for cystic fibrosis (CF) is a classic example of disease gene identification based on genetic linkage analysis. The first genetic analysis for CF could be traced to 1946, when Andersen and Hodges proposed that the disease was caused by a recessive mutation (Andersen and Hodges 1946), but linkage analyses were only briefly applied by Morton and Allen for cystic fibrosis of the pancreas in 1956 (Allen et al. 1956; Morton and Steinberg 1956). In fact, genetic linkage for disease gene identification was not broadly attempted before the discovery of polymorphic DNA, termed "restriction fragment length polymorphisms" markers in the early 1980s, and the suggestion of their use as markers for genetic mapping (Botstein et al. 1980). Hence, early CF linkage and association studies were conducted with polymorphic biochemical markers, such as immunoglobulins and protein markers (Tsui et al. 1985b; Schmiegelow et al. 1986). Ironically, however, the first CF linkage was detected with a serum enzyme marker, paraoxonase (PON) (Eiberg et al. 1985).

IDENTIFICATION OF CFTR GENE

There have been previous reviews on the identification of the cystic fibrosis gene (Tsui and Buchwald 1991; Tsui 1995). To avoid repeating all that has been said, we have decided to provide only a historical account of key developments as well as factors that contributed to the CF gene identification.

1. CF was first found linked to PON, but this linkage did not directly result in identification of the CF gene because the chromosome location of PON was not known. Nevertheless, the finding confirmed CF to be a relatively homogeneous genetic disease and suggested that it would be possible to identify the CF gene by genetic linkage analysis with a large number of nuclear (two-generation) families.

2. The first DNA marker found linked to CF was an anonymous marker known as CRI-917 (Tsui et al. 1985a), which was subsequently designated D7S15.

3. Before the chromosome location of D7S15 was eventually published (Knowlton et al. 1985), rumor spread that the CF gene was localized to chromosome 7 (Newmark 1985). This information quickly led to the detection of linkage between CF and several other DNA markers on the long arm of chromosome 7 (Wainwright et al. 1985; White et al. 1985). In other words, the localization of CF to the long arm of chromosome 7 was almost instantly confirmed. However, further gene mapping was limited by the lack of human sequence data (Botstein and Risch 2003; Cardon and Abecasis 2003; Flint et al. 2005) because, essentially for each candidate region, a physical map had to be created, overlapping DNA segments isolated and analyzed for the presence of coding sequences (known as "chromosome walking"), and each gene assessed for its candidacy as the sought-after gene. This general approach was subsequently dubbed "positional cloning" (Collins 1990, 1992). The technique of gene hopping or jumping techniques (Collins et al. 1987) was also used in an attempt to reduce the number of DNA fragments required to be isolated while moving along the chromosome (Kerem et al. 1989a,b; Rommens et al. 1989a,b).

4. The identification of the CF gene was published in a series of three papers in 1989 (Kerem et al. 1989a; Riordan et al. 1989; Rommens et al. 1989a). The gene was named *cystic fibrosis transmembrane conductance regulator* (*CFTR* for short). The supporting evidence for the gene was the discovery of the most common mutation in CF, named ΔF508 (now renamed p.F508del). Because this 3-bp deletion was found on a relatively rare extended chromosome haplotype, it was possible to apply allelic association (commonly known as "linkage disequilibrium") in the fine mapping of the CF gene, and haplotype mapping (ancestral recombination) was proposed as a means to map disease genes by Cox and Chakravarti (Cox et al. 1989).

5. The strong allelic association led to the suggestion that the p.F508del mutation arose from one single event (Cutting et al. 1989; Kerem et al. 1989a). Furthermore, the occurrence of this common mutation in a region with extended linkage disequilibrium was used to form part of the argument for the initial identification of the CFTR gene (Kerem et al. 1989a).

6. It is of interest to note, however, that haplotype analysis was used to assess the genetic heterogeneity of the disease before the cloning of the gene (Tsui and Buchwald 1988). For example, it was used to show that pancreatic-sufficient CF patients had more variable and different haplotypes than pancreatic-insufficient patients (Kerem et al. 1989b).

7. It is also of interest to note that a great deal of effort was used to argue for the 3-bp p.F508del deletion as a bona fide mutation because it was not immediately apparent. Its status was somewhat questionable until the results of subsequent functional analysis of the p.F508del-CFTR protein (Drumm et al. 1991). The status of *CFTR* as the CF-causing gene was confirmed earlier, however, by the discovery of additional disease-casing mutations (Cutting et al. 1990).

8. Although CF is a relatively homogeneous, monogenic disease with distinct clinical

features, it is possible that a very small number of clinically diagnosed CF patients do not have mutations in the *CFTR* gene. In other words, some patients may be "phenocopies" of CF; despite their classical CF appearance, the disease in these patients may be caused by mutations in other genes. In support of this possibility, it was reported that mouse models with ENaC channel overexpression showed a phenotype reminiscent of CF (Donaldson and Boucher 2007). In fact, some patients with CF-like disease but mutations in only one copy of their *CFTR* genes were found to carry gain-of-function mutations in ENaC-encoding subunits (Sheridan et al. 2005; Azad et al. 2009; Fajac et al. 2009).

cDNA CLONING, MUTATION ANALYSIS, AND *CFTR* GENE STRUCTURE

The initial characterization of the *CFTR* gene was based on the isolation and alignment of genomic clones with the cDNA clones derived from the T84 human colonic adenocarcinoma cell line and those containing the p.Phe508del mutation isolated from a CF sweat gland cDNA library. Some of corresponding genomic DNA segments were missed because either the exons therein were small and thus refractory to isolation by hybridization or certain exons were skipped in the cDNA clones used for the construction of consensus sequence. Therefore, the *CFTR* gene was thought to contain 24 exons (Riordan et al. 1989). Subsequently, it became apparent that the gene contains 27 exons (Table 1); exons 6b, 14b, and 17b were missed in the original publication.

It is now well established that the full-length CFTR mRNA contains 6128 nucleotides. Alignment of this sequence with the genomic DNA sequence derived from the Human Genome Project has provided the precise exon–intron structure and intronic sequences of the *CFTR* gene (Table 1).

Based on the recent sequencing and functional data, the transcription module of *CFTR* was established to be >216 kb in size (see Fig. 1). The *CFTR* gene itself spans only 189.36 kb; however, the immediate promoter can be extended

as far as 20.9 kb upstream, where the CTCF-dependent insulator element is located—the expanded promoter region includes the regulatory binding element required for proper gene expression (Blackledge et al. 2007). An insulator element 6.8 kb downstream from *CFTR* includes a DHS site with a CTCF binding site followed by additional DHS sites, at +6.8 kb, +7.0 kb, and +15.6 kb, which appear to contribute to DNA looping, limiting the boundary of the *CFTR* transcriptional unit (Blackledge et al. 2009). Thus, the total size of the CFTR transcription unit between these two insulating elements is ∼216.7 kb.

SEGMENTAL DUPLICATION OF THE EXON 9 REGION

One of complications in the *CFTR* gene structure analysis stemmed from the segmental duplication of the ∼30-kb region including exon 9 (Rozmahel et al. 1997). These multiple segmental duplications containing exon 9 and its flanking intron sequences (Fig. 2) have been found in multiple locations across the human genome (Liu et al. 2004). Although how these sequences became amplified with two flanking LINE1 elements is not exactly clear, multiple copies of exon 9-related sequences become problematic when the exon 9 sequence is being examined for mutation analysis, especially when PCR amplification is used (El-Seedy et al. 2009). The pyrimidine track immediately upstream of exon 9 also appears to influence mRNA splicing, causing exon 9 skipping (see below) (Chu et al. 1991, 1993).

REGULATION OF CFTR TRANSCRIPTION

To understand the pathophysiology of CF, one of the approaches is to discover the CFTR expression pattern in different tissue surveys (Gregory et al. 1990; Trezise and Buchwald 1991; Ward et al. 1991). CFTR is found to be expressed in the epithelial cells of a variety of tissues and organs, whose functions are significantly affected in CF patients: lung and trachea, pancreas, liver, intestines, and sweat glands. Low levels of CFTR transcripts can be found in kidney, uterus, ovary,

Table 1. Summary of *CFTR* coding regions according to historical and current nomenclature

Exon	Historical exon name	Start	End	Length	Exon	Start	End	Length
Exon 1	Exon 1	117,120,017	117,120,201	185	Intron 1-2	117,120,202	117,144,306	24,105
Exon 2	Exon 2	117,144,307	117,144,417	111	Intron 2-3	117,144,418	117,149,087	4670
Exon 3	Exon 3	117,149,088	117,149,196	109	Intron 3-4	117,149,197	117,170,952	21,756
Exon 3	Exon 3	117,170,953	117,171,168	216	Intron 4-5	117,171,169	117,174,329	3161
Exon 5	Exon 5	117,174,330	117,174,419	90	Intron 5-6	117,174,420	117,175,301	882
Exon 6	Exon 6a	117,175,302	117,175,465	164	Intron 6-7	117,175,466	117,176,601	1136
Exon 7	Exon 6b	117,176,602	117,176,727	126	Intron 7-8	117,176,728	117,180,153	3426
Exon 8	Exon 7	117,180,154	117,180,400	247	Intron 8-9	117,180,401	117,182,069	1669
Exon 9	Exon 8	117,182,070	117,182,162	93	Intron 9-10	117,182,163	117,188,694	6532
Exon 10	Exon 9	117,188,695	117,188,877	183	Intron 10-11	117,188,878	117,199,517	10,640
Exon 11	Exon 10	117,199,518	117,199,709	192	Intron 11-12	117,199,710	117,227,792	28,083
Exon 12	Exon 11	117,227,793	117,227,887	95	Intron 12-13	117,227,888	117,230,406	2519
Exon 13	Exon 12	117,230,407	117,230,493	87	Intron 13-14	117,230,494	117,231,987	1494
Exon 14	Exon 13	117,231,988	117,232,711	724	Intron 14-15	117,232,712	117,234,983	2272
Exon 15	Exon 14a	117,234,984	117,235,112	129	Intron 15-16	117,235,113	117,242,879	7767
Exon 16	Exon 14b	117,242,880	117,242,917	38	Intron 16-17	117,242,918	117,243,585	668
Exon 17	Exon 15	117,243,586	117,243,836	251	Intron 17-18	117,243,837	117,246,727	2891
Exon 18	Exon 16	117,246,728	117,246,807	80	Intron 18-19	117,246,808	117,250,572	3765
Exon 19	Exon 17a	117,250,573	117,250,723	151	Intron 19-20	117,250,724	117,251,634	911
Exon 20	Exon 17b	117,251,635	117,251,862	228	Intron 20-21	117,251,863	117,254,666	2804
Exon 21	Exon 18	117,254,667	117,254,767	101	Intron 21-22	117,254,768	117,267,575	12,808
Exon 22	Exon 19	117,267,576	117,267,824	249	Intron 22-23	117,267,825	117,282,491	14,667
Exon 23	Exon 20	117,282,492	117,282,647	156	Intron 23-24	117,282,648	117,292,895	10,248
Exon 24	Exon 21	117,292,896	117,292,985	90	Intron 24-25	117,292,986	117,304,741	11,756
Exon 25	Exon 22	117,304,742	117,304,914	173	Intron 25-26	117,304,915	117,305,512	598
Exon 26	Exon 23	117,305,513	117,305,618	106	Intron 26-27	117,305,619	117,306,961	1343
Exon 27	Exon 24	117,306,962	117,308,715	1754				

Data are based on the Ensembl release 61–Feb 2011 (http://www.ensembl.org/Homo_sapiens/Transcript/Exons?db=core;g=ENSG00000001626;r=7:117119358-117308719;t=ENST00000003084).

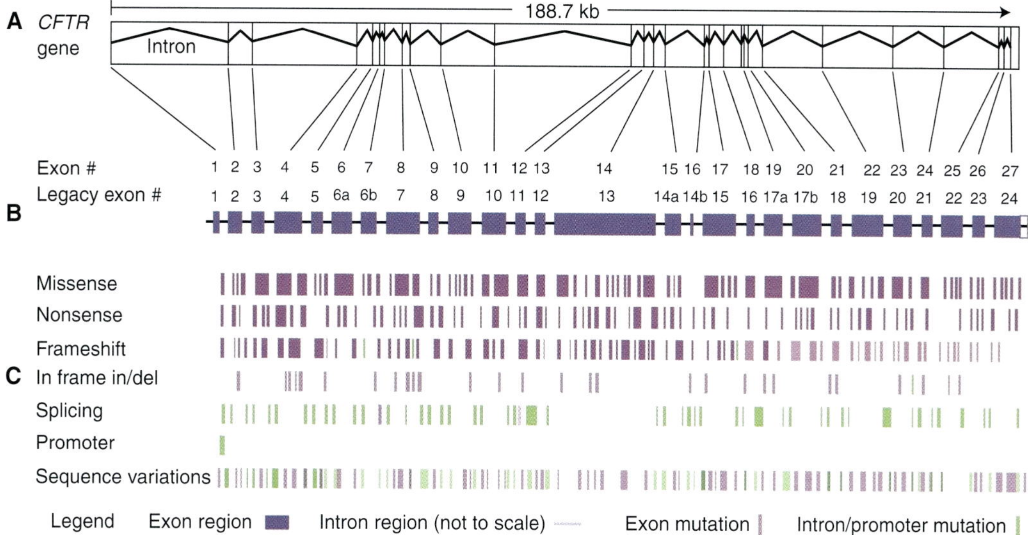

Figure 1. Scheme of *CFTR* gene, transcript, and mutation distribution. (*A*) Scaled schematic of exons and introns followed by current and historical exon numbering. (*B*) A rescaled exon scheme with minimized introns, which are drawn not to scale. (*C*) Distribution of known mutations and polymorphisms as vertical bars.

thyroid, and even higher levels in salivary gland and bladder, but the epithelial cell function is not seriously compromised in tissues and organs of CF patients. It is possible that there is sufficient compensation of the missing function by other ion transporters. It is of interest to note that the low levels of CFTR expression in these tissues are driven off an alternative promoter (McCarthy and Harris 2005). These studies have also led to the identification of several tissue-specific splicing isoforms and splicing elements (see below).

The promoter region of the *CFTR* gene lacks a TATA box and is quite GC rich, although generally it is not methylated.[3] As such, the CFTR promoter has not been well defined. The immediate promoter region contains an important *cis*-acting element within 632 nucleotides upstream of the translation start site (Lewandowska et al. 2010). The definition of the key promoter regions is also suggested by several mutations that appear to disrupt transcription initiation; for example, the c.-234T>A (-102T>A) muta-

tion was first reported by Claustres and coworkers (Romey et al. 1999). Mutations in more distant regions have been shown to affect the efficiency of transcription: The c.-1750A >G polymorphism reduces the transcription efficiency by >50% (Lewandowska et al. 2010).

The predominant CFTR transcription start site is as described in the original study by Rommens et al. (1989a), meaning that the majority of CFTR transcripts are driven from the key promoter, described in the previous section. However, although the canonical transcripts are found in cells with high CFTR expression, alternative transcription start sites are apparently used in cell lines with low expression levels (McCarthy and Harris 2005). CFTR transcripts from even more distant transcription start sites between −868 and −794 can be found in CFPAC and T84 cell lines (McCarthy and Harris 2005).

The immediate promoter region has also been characterized by consensus binding sites for several transcriptions factors: CTCF, AP-1, SP1, GRE, CRE, C/EBP, and Y-box proteins (McCarthy and Harris 2005). DNase I hypersensitive site (DHS) mapping has been used to map various putative enhancer sequences

---

[3]http://www.ensembl.org/Homo_sapiens/Location/View? g=ENSG00000001626;r=7:117078291-117109965#r=7:117 094129-117220824.

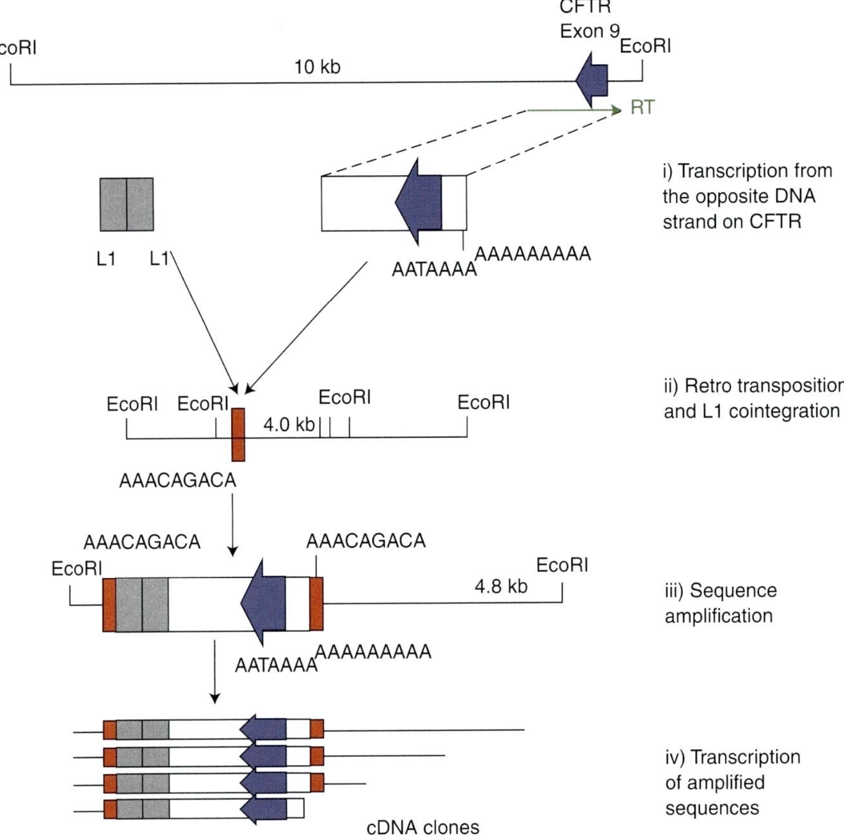

Figure 2. Scheme reconstructed based on data from the Rozmahel et al. (1997) publication showing the schematic diagram of proposed exon 9 duplication by retrotransposition.

within the CFTR intragenic regions (Table 2). Presumably, multiple transcription factors can bind to chromatin at these sequences, opening the DNA and extending the physical interactions with the promoter, thereby affecting transcription. HNF1α binding sites, indicative of putative enhancer elements, can be found in multiple locations inside introns 10, 17a, and 20 (Mouchel et al. 2004; McCarthy et al. 2009); it has been shown that RNAi-mediated inhibition of the HNF1α could lead to reduction of CFTR expression. Additional enhancer elements have been located in introns 1 and 11, and HNF1α and p300 are involved in the regulation of CFTR expression (Ott et al. 2009a). However, understanding of the transcriptional control of *CFTR* gene expression is far from comprehensive.

SPLICING

As described in the previous section, the majority of CFTR transcripts start from around the same site. There are, however, transcripts with alternative splicing, resulting in skipping of different exons. Some alternatively spliced species can produce truncated CFTR protein with partial function; for example, an alternatively spliced CFTR transcript missing exon 5 found in heart may produce an active channel albeit with significantly reduced function (Xie et al. 1995).

Many of the alternatively spliced transcripts are expected to introduce translation reading frameshifts, thereby introducing premature translation termination that can induce nonsense-mediated RNA decay, removing the aberrant mRNA species. For some mutations,

Table 2. List of genomic regulatory regions in the human *CFTR* gene

Regulatory region	Analysis	Location	Length (bp)
ENSR00000633238	Ensembl Regulatory Build	7:117110774−117111212	481
ENSR00000633239	Ensembl Regulatory Build	7:117116334−117117806	1617
ENSR00000633240	Ensembl Regulatory Build	7:117119693−117120138	488
ENSR00000633241	Ensembl Regulatory Build	7:117121119−117121448	360
ENSR00000633242	Ensembl Regulatory Build	7:117125053−117125742	756
ENSR00000633243	Ensembl Regulatory Build	7:117141802−117141975	186
ENSR00000633244	Ensembl Regulatory Build	7:117145373−117145667	319
ENSR00000633245	Ensembl Regulatory Build	7:117199627−117199834	226
ENSR00001050415	Ensembl Regulatory Build	7:117215123−117215415	317
ENSR00000633246	Ensembl Regulatory Build	7:117220696−117220951	280
ENSR00000633247	Ensembl Regulatory Build	7:117221137−117221424	312
ENSR00000633248	Ensembl Regulatory Build	7:117222043−117223014	1068
ENSR00000633249	Ensembl Regulatory Build	7:117227963−117229004	1144
ENSR00000683157	Ensembl Regulatory Build	7:117236602−117237299	764
ENSR00000633250	Ensembl Regulatory Build	7:117237932−117238146	233
ENSR00000633251	Ensembl Regulatory Build	7:117240434−117240790	387
ENSR00000633252	Ensembl Regulatory Build	7:117240865−117241512	708
ENSR00000633253	Ensembl Regulatory Build	7:117259273−117259612	370
ENSR00000633254	Ensembl Regulatory Build	7:117279990−117280443	496
ENSR00000633255	Ensembl Regulatory Build	7:117285455−117286319	949
ENSR00000633256	Ensembl Regulatory Build	7:117288545−117288840	320
ENSR00000633257	Ensembl Regulatory Build	7:117293467−117294015	603
ENSR00000633258	Ensembl Regulatory Build	7:117294880−117295179	330
ENSR00000633259	Ensembl Regulatory Build	7:117299551−117299849	323
ENSR00000633260	Ensembl Regulatory Build	7:117305402−117305885	532
ENSR00000633261	Ensembl Regulatory Build	7:117306006−117306648	703
ENST00000003084	miRanda microRNA target predictions: hsa-miR-28-3p	7:117307497−117307518	22
ENSR00001050416	Ensembl Regulatory Build	7:117307574−117307827	278
ENST00000003084	miRanda microRNA target predictions: hsa-miR-28-3p	7:117307750−117307771	22
ENST00000003084	miRanda microRNA target predictions: hsa-miR-433	7:117308205−117308229	25
ENSR00000633262	Ensembl Regulatory Build	7:117309937−117310180	268

however, the resulting mRNA species with exon skipping can persist and produce a dysfunctional protein (Hull et al. 1994). In the case of the c.2491G>T (p.Glu831X, also known as 2623G>T, E831X) variant, the presence of an alternative spliced mutant transcript may explain a mild disease presentation and pancreatic sufficiency in patients carrying this nonsense mutation (Hinzpeter et al. 2010).

Based on studies with minigene constructs, Hinzpeter et al. (2010) have proposed that the c.2491G>T mutation not only introduces a premature stop codon, but also creates a binding site (CAG*TA*G) for the splicing factor that recognizes the consensus sequence NAGNAG. The study shows that the mutation can lead to the production of three different mRNAs: CFTR_831-873del, CFTR_E831X, and CFTR_ E831del, and, unexpectedly, the latter transcript could bypass the premature termination codon and produce a functional p.E831del protein.

ROLE OF SPLICING REGULATION IN CF DISEASE

Mutations occurring at the intron−exon boundaries with highly conserved sequences, that is, splicing junctions at -1, -2, -3, and $+1$, $+2$, $+3$ positions are expected to affect splicing of the immediately adjacent exons. The consequence of such mutations, for example, c.1117-1G>A and c.1209+1G>A, are generally considered to be severe, whereas mutations occurring at more distant positions, for example, $+5$, $+6$, or -5 and -6, are mild, typically associated with mild CF disease. Although the sequences between $+6$ and $+30$ are to some degree conserved in evolution, suggestive of a possible role of them in splicing (Aznarez et al. 2008), there is not much information regarding the severity of mutations found beyond the ±6 positions, except for c.1117-18G>T, c.1209+ 18A>C, and c.2909-15T>G, which are frequently found in infertile males.

Some of the mutations are found to affect splicing efficiency, presumably because they alter the binding sites of splicing factors (Aznarez et al. 2003). The most noted example in the *CFTR* gene is the polypyrimidine track located in intron 8 in front of exon 9; the length of the T-track varies from five to nine Ts (Chu et al. 1993). Although the 5T variant is associated with low efficiency of splicing, the effect by itself is insufficient to cause any known disease. However, if the 5T allele is present in combination with another mild mutation, p.Arg117His (R117H) (Gervais et al. 1993), the combined effect of R117H-5T (reduced protein function and lower abundance of correctly spliced transcript) is apparently sufficient to cause CF, albeit the pancreatic sufficient form (Chu et al. 1993; Massie et al. 2001; Zuccato et al. 2004). Other examples of mutations affecting splicing efficiency include several missense (D648V and T665S) and nonsense (E664X) mutations, presumably due to the disruption of the ESE elements within exon 13 (Aznarez et al. 2003).

It is of interest to note that synonymous variants (or polymorphisms) are much less common in the *CFTR* gene than in other disease-related genes. The reason for this is unknown, but it may be a property of genomic regions with strong allelic association (or linkage disequilibrium). However, some of these rare synonymous variants, for example, c.2679G>T (p.Gly893Gly), c.2898G>A, c.3204C>T, c.3285

A>T, and c.3897A>G (rs1800131), could have an effect on splicing because they occur in the consensus sequences for ESEs (exon splicing enhancers) or ESRs (exon splicing repressors). The exon 12 variant c.2679G>T (p.Gly893Gly), for example, is associated with a mild form of CF; the mutation affects a highly conserved sequence and causes skipping of exon 12 (Pagani et al. 2005; Raponi et al. 2007). In fact, the sequence alteration in exon 12 has been found coupled with a new 5′ splice site in exon 15, resulting in a transcript lacking 76-amino-acid codons (Faa et al. 2010). This and other observations bring about a suggestion that some of the synonymous changes can affect CFTR mRNA folding and, in turn, protein synthesis (Bartoszewski et al. 2010). Supporting this view, synonymous mutations have been shown to affect posttranslational protein modification and protein stability and were found to be overrepresented among the "top hits" identified in several Genome Wide Association Studies of common diseases (Chen et al. 2010; Li et al. 2010; Plotkin and Kudla 2010), again showing a crucial, but frequently underestimated, role of synonymous variations in disease pathophysiology.

TRANS-SPLICING

Trans-splicing has been described for the *CFTR* gene, making the initial characterization of the gene structure difficult (Rommens et al. 1989a). It is unclear how often it happens for other genes, but *trans*-splicing has been suggested as a possible strategy for the correction of *CFTR* mutations (Wardle and Harris 1995; Mansfield et al. 2000; Buratti and Baralle 2001; Liu et al. 2002). For example, it has been feasible to reconstruct a full-length CFTR transcript with recombinant DNA containing exons 10–24, provided in *trans*, in transfected cells (Mansfield et al. 2000, 2003; Liu et al. 2002). It remains to be determined, however, whether it is possible to engineer gene replacement therapy by introducing a large CFTR cDNA construct into patient cells. Nevertheless, this approach may be useful for gene repair for rare missense and frameshift mutations, which are

     Cite this article as *Cold Spring Harb Perspect Med* doi: 10.1101/cshperspect.a009472

unlikely to benefit from drugs with molecular chaperone properties.

STUDY OF CFTR IN MICE AND OTHER SPECIES

Because of the difficulty in studying the *CFTR* gene in human tissues, extensive studies have been devoted to the isolation and characterization of genes homologous to *CFTR* in other species. CFTR (ABCC7) is a member of the C subfamily of ABC transporters, which includes sulfonylurea receptors (SURs) and multidrug resistance–associated proteins (MRPs). The ABCC subfamily proteins have two transmembrane membrane-spanning domains (MSDs), each consisting of six transmembrane (TM) segments, although some the MRP channels have an additional MSD_0 (Deeley et al. 2006; Toyoda et al. 2008). All ABC proteins also have two nucleotide-binding domains (NBDs), but only CFTR has the novel R domain (CFTR protein structure and function) (for review, see Hunt et al. 2013; Hwang and Kirk 2013). Better understanding of CFTR function may derive from studies of orthologous genes in other species. One line of such investigation has been attempts to search for naturally occurring *CFTR* gene mutations in various animal species, including sheep, guinea pigs, ferrets, dogs, and even chickens (Tebbutt 1995). Unfortunately, although some sequence variations have been identified, none of them appear to correlate with any CF-like disease.

Because mouse is the best-studied experimental animal for human diseases, several CFTR-deficient murine lines were created with the use of recombinant DNA technologies (Snouwaert et al. 1992). These include simple knockout (KO) and specific gene replacements on various genetic strains of mice, providing a spectrum of clinical phenotypes (Davidson and Rolfe 2001; Guilbault et al. 2007; Wilke et al. 2011). Despite undetectable CFTR transcripts in all the complete KO strains—*Cftr*^*tm1Unc*, *Cftr*^*tm1Cam*, *Cftr*^*tm1Hsc*, *Cftr*^*tm3Bay*, *Cftr*^*tm3Uth*— the survival varied from <5% to 40%. Furthermore, survival in the hypomorphic strains *Cftr*^*tm1Hgu* and *Cftr*^*tm1Bay* was largely unaffected, reminiscent of the residual CFTR function for some CF mutations and disease severity in humans, particularly with respect to pancreatic status and extent of CF lung disease (Wilke et al. 2011). The *Cftr*^*tm1Kth*, *Cftr*^*tm2Cam*, and *Cftr*^*tm1Eur* mice were created to study the p.Phe508del mutation, whereas other lines, such as *Cftr*^*tm1G551D*, for other missense mutations found in CF patients. Despite intestinal obstruction and reduced weight, the symptoms observed in CFTR-deficient mice appear to be very different from those of the human disease, particularly in terms of pulmonary involvement.

The availability of CFTR-deficient mice has, however, allowed the study of possible modifying genes for CF, because the same CFTR mutation(s) can result in different disease presentations among different CF patients. Variability in phenotypic manifestations was also observed in mice, and it appeared to be largely dependent on the genetic background of the knockout mice (Rozmahel et al. 1996). Indeed, these differences were used to map the *CFM1* and several other genetic modifier loci associated with varying "disease severity" in CFTR-deficient mice (Rozmahel et al. 1996; Haston et al. 2002a,b; Haston and Tsui 2003).

The CFTR-deficient pig (Rogers et al. 2008 a,b) appears to be a very promising animal model for CF because the mutant animals essentially recapitulate all of the CF-specific gastrointestinal abnormalities and the critical aspects of CF lung disease (Meyerholz et al. 2010). For example, meconium ileus is fully penetrant in the mutant animals: All of the newborn CFTR-deficient piglets suffer from intestinal obstruction and defects in gallbladder development. However, there are differences in disease severity between the completely null CF pigs (Rogers et al. 2008a,b) and the p.Phe508del pigs (Ostedgaard et al. 2011).

Ferrets with *CFTR* gene knockout may provide yet another promising animal model for CF. The CFTR-deficient animals have been found to show many of the characteristics of human CF disease, including defective airway chloride transport and submucosal gland fluid secretion;

meconium ileus with variable penetrance (75% MI); pancreatic, liver, and vas deferens diseases; and a predisposition to lung infection in early life (Sun et al. 2010).

CROSS-SPECIES STUDIES OF CFTR SEQUENCES

Study of sequences conserved across multiple species may identify potential regulatory sequences important for *CFTR* gene expression and splicing. Such information may be obtained from the ENCODE project (ENCyclopedia Of DNA Elements), which has been developed to identify conserved regions and putative gene expression regulatory regions; *CFTR* is one of the several candidate genes included in the study (ENCODE Project Consortium 2004; Birney et al. 2007). The ChIP-seq method that involves DNA digestion and chromatin immunoprecipitation followed by sequencing was used to identify the transcription factor and histone binding sites. RNA sequencing was used for more accurate gene and exon mapping. DNA methylation and other DNA modifications were assessed in multiple cell types. The conserved regions and functional DNA modification and protein–DNA interaction sites identified by ENCODE (http://encodeproject.org) are compared with the *CFTR* variants recorded in the Cystic Fibrosis Mutation Database in the PhenCode browser (http://globin.bx.psu.edu/phencode/) (Giardine et al. 2007).

However, not many CFTR mutations have been found associated with these putative regulatory regions thus far (Ott et al. 2009b). There are several possible explanations for the apparent paucity of mutations detected in these regions. First, the biological significance of these sequences has not been investigated. Second, mutation screening studies in CF patients have been primarily concentrated in the coding regions, including at most the immediate flanking intronic regions. Third, mutations in regulatory regions may only affect the expression levels of CFTR, resulting in very mild disease, not recognized as CF.

Identification of crucial regulatory sequences that can direct proper tissue specificity may be critically useful in the construction of CFTR expression vectors for CF gene therapy because, although CFTR deficiency could be rescued by overexpression of wild-type cDNA (Drumm et al. 1990), expression of the gene in atopic tissues could lead to adverse side effects. One study showed that atopic expression of CFTR in the heart could cause arrhythmia and sudden death of transgenic animals (Ye et al. 2010). Therefore, most of the CF gene therapy studies have only used heterologous promoters that are well characterized. Among these, the KRT18 promoter appears to be promising in targeting CFTR expression in the lung epithelial cells (Koehler et al. 2001).

CF MUTATION DATABASE AND MUTATION NOMENCLATURE

The Cystic Fibrosis Genetic Analysis Consortium was formed to facilitate CFTR mutation analysis, at a time before the gene structure was fully characterized. A set of general agreements and guidelines was agreed upon at the North American CF Conference in Florida on October 13, 1989. Briefly, members of the Consortium were encouraged to report data to the Consortium before submission for publication. Because it would usually take at least three months for any publication to come to print, early communication within the group would allow other members to use the information and screen their respective populations in a coordinated and concerted effort. The Consortium also helped disseminate information and ideas regarding novel methods and technical tips that would improve genetic testing. Members agreed that the group that reported the information would hold the priority to publish the primary data. Other members of the Consortium might use the information for their own research before the publication of the original report but might not publish any secondary observations without the consent of the original group.

In addition, a standardized mutation nomenclature system was proposed (Beaudet and Tsui 1993). This was the first time a systematic nomenclature was used to describe human disease mutations. These guidelines were adopted

Table 3. Summary statistics for classification of *CFTR* mutations recorded in CFMDB (January 2011)

Mutation type	Count	Frequency (%)
Missense	765	40.41
Frameshift	303	16.01
Splicing	225	11.89
Nonsense	160	8.45
In frame in/del	37	1.95
Large in/del	49	2.59
Promoter	15	0.79
Sequence variation	269	14.21
Unknown	70	3.7
Total	1893	

by the genetic community and applied in the naming of mutations in other genes, and laid the basis for subsequent versions of mutation nomenclature developed by the Human Genome Variation Society.[4]

After the launching of the Consortium in 1989, its membership quickly grew to include 130 groups of CF laboratories from more than 30 countries. The Consortium exchanged mutation reports and facilitated the discussions and research on genotype–phenotype correlations, confirmed clinical data, and enabled development of genetic screening tools. With the development of Internet technologies, because the data held by the Consortium were essentially open to the entire scientific community, a Web-based Cystic Fibrosis Mutation Database (CFMDB) was launched in 1995. Even to date, it is one of the most cited genetic databases and is visited by more than 500 unique users per day. The Database contains information for more than 1890 disease-causing mutations and variants of different types[5] (Tables 3 and 4: mutation distribution summary). Although the changes in HGV nomenclature rules have necessitated a revision (April 2010) of CFMDB, all of the previous (legacy) names for CF mutations are retained as a display option along with new maps and up-to-date numbering systems according to the current knowledge of the gene structure.

All new mutation submissions, however, would only appear in the revised HGV format.

CONCLUDING REMARKS

It has often been said that, had the full human genome been mapped and sequenced, the task of finding the cystic fibrosis gene would have taken only a fraction of the time and cost. The identification of the *CFTR* gene has, however, provided an entry point to understand the basic defect underlying this multisystemic disease, whereby better detection and treatment can be devised for the sake of all of the affected individuals.

In addition to providing the first example of positional cloning, the CF gene cloning work also provided the concept of haplotype mapping, which was illustrated on the cover of the September 8, 1989 issue of *Science* (see Fig. 3). Although it was not properly introduced in the report of Kerem et al. (1989), the solution for fine-mapping of human traits by haplotype analysis was initially attempted by Cox and Chakravarti (Cox et al. 1989) and later properly introduced by Lander and Kruglyak (Kruglyak 1997, 2005, 2008).

The spirit of sharing and cooperation was exemplified in CF genetic research as early as the CF gene mapping stage; the primary DNA marker data set used in the CF gene mapping study was shared with the human gene mapping community, so that different algorithms on multipoint mapping could be tested (Spence et al. 1989).

After identification of the *CFTR* gene, crucial information for the detection of additional mutations was shared among members of the CF Genetic Analysis Consortium before the gene structure was entirely elucidated and published (Zielenski et al. 1991). New mutation data were also shared before formal publication as describe in the preceding section.

Most remarkable, however, was the honorable competition and cooperative spirit that could be found even during the race to the cloning of the CF gene. In 1987, the group led by Robert Williamson had used a rather elegant technique and identified a DNA fragment very

---

[4]http://www.hgvs.org/mutnomen/history.html.

[5]http://www.genet.sickkids.on.ca/StatisticsPage.html.

Table 4. Distribution of mutations in CFTR protein domains

Domain	Exon number	Missense	Frameshift/stop	In frame in/del	Splicing	Total
MSD1	Exon 4	18	12	2	0	32
MSD2	Exon 4-5	13	5	2	1	21
MSD3	Exon 6a-6b	10	3	0	0	13
MSD4	Exon 6b	10	5	1	0	16
MSD5	Exon 7-8	13	7	2	0	22
MSD6	Exon 8	9	9	0	0	18
NBD1	Exon 10-13	147	37	3	1	188
R	Exon 13-14a	58	87	1	1	147
MSD7	Exon 15	13	6	0	0	19
MSD8	Exon 16-17a	12	10	2	1	25
MSD9	Exon 17b	25	2	0	0	27
MSD10	Exon 17b	12	9	0	0	21
MSD11	Exon 19	2	7	0	0	9
MSD12	Exon 20	6	7	0	0	13
NBD2	Exon 20-24	74	65	0	0	139
Total		422	271	13	4	710[a]

[a]Please note that the total number of mutations is smaller because synonymous variants are not shown in the summary; also, different cDNA mutations that lead to identical missense changes and mutations outside of the functionally defined domains are now shown in the summary. Therefore, the total number of missense mutations is substantially smaller than indicated in Table 3.

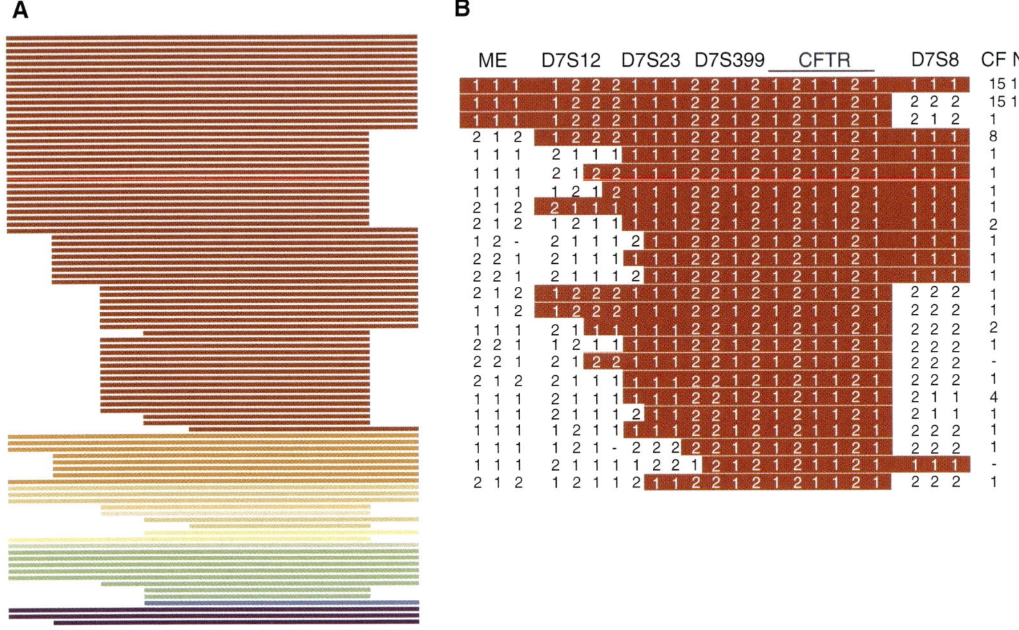

Figure 3. Schematic diagram illustrating the concept of haplotype mapping. (A) The color bars represent chromosomes from different patients with cystic fibrosis; this diagram was used as the basis for the cover design of the September 8, 1989 issue of *Science*. (B) The DNA marker haplotypes associated with the CF chromosomes carrying ΔF508; the "1s" and "2s" inside the colored bars on the *left* represent schematic alleles of DNA markers around the CFTR gene; the original data appeared in Table 3 on p. 1076 of Kerem et al. (1989); the numbers on the *right* denote the counts of CF and normal (N) chromosomes with the corresponding DNA marker haplotypes shown on the *left*.

close to the CF gene (Scambler et al. 1986; Estivill et al. 1987). In fact, their data misled them and many others to believe that they had cloned the gene. Due credit must be given to Williamson and his colleagues, however, because they were quick to admit to the scientific world that the candidate gene they had isolated was not the right gene (Wainwright et al. 1988). Their timely sharing of this valuable piece of information reversed the funding decision at the National Institutes of Health (of the United States), so that the CF gene cloning work in several laboratories could be continued without any interruption of funding.

We would like to end this work with a quote from a previous review article (Tsui and Estivill 1991):

> Finally, there has been tremendous competition in the field of disease gene cloning. . . . It is difficult to describe the feelings of performing repeated searches for clone after clone without an open reading frame, constructing and screening genomic and cDNA libraries one after another, watching the come and go of a candidate gene, and working under the fear that the gene has perhaps been identified by another group . . . although being the first to report appears to be the only consolation of this hard work, one should keep in mind that gene identification is in fact only the first step . . . in understanding a disease. . . .

REFERENCES

Reference is also in this collection.

Allen FH Jr, Dooley RR, Shwachman H, Steinberg AG. 1956. Linkage studies with cystic fibrosis of the pancreas. *Am J Hum Genet* **8:** 162–176.

Andersen DH, Hodges RG. 1946. Celiac syndrome; genetics of cystic fibrosis of the pancreas, with a consideration of etiology. *Am J Dis Child* **72:** 62–80.

Azad AK, Rauh R, Vermeulen F, Jaspers M, Korbmacher J, Boissier B, Bassinet L, Fichou Y, des Georges M, Stanke F, et al. 2009. Mutations in the amiloride-sensitive epithelial sodium channel in patients with cystic fibrosis-like disease. *Hum Mutat* **30:** 1093–1103.

Aznarez I, Chan EM, Zielenski J, Blencowe BJ, Tsui LC. 2003. Characterization of disease-associated mutations affecting an exonic splicing enhancer and two cryptic splice sites in exon 13 of the cystic fibrosis transmembrane conductance regulator gene. *Hum Mol Genet* **12:** 2031–2040.

Aznarez I, Barash Y, Shai O, He D, Zielenski J, Tsui LC, Parkinson J, Frey BJ, Rommens JM, Blencowe BJ. 2008. A systematic analysis of intronic sequences downstream of 5′ splice sites reveals a widespread role for U-rich motifs and TIA1/TIAL1 proteins in alternative splicing regulation. *Genome Res* **18:** 1247–1258.

Bartoszewski RA, Jablonsky M, Bartoszewska S, Stevenson L, Dai Q, Kappes J, Collawn JF, Bebok Z. 2010. A synonymous single nucleotide polymorphism in ΔF508 *CFTR* alters the secondary structure of the mRNA and the expression of the mutant protein. *J Biol Chem* **285:** 28741–28748.

Beaudet AL, Tsui LC. 1993. A suggested nomenclature for designating mutations. *Hum Mutat* **2:** 245–248.

Birney E, Stamatoyannopoulos JA, Dutta A, Guigo R, Gingeras TR, Margulies EH, Weng Z, Snyder M, Dermitzakis ET, Thurman RE, et al. 2007. Identification and analysis of functional elements in 1% of the human genome by the ENCODE pilot project. *Nature* **447:** 799–816.

Blackledge NP, Carter EJ, Evans JR, Lawson V, Rowntree RK, Harris A. 2007. CTCF mediates insulator function at the CFTR locus. *Biochem J* **408:** 267–275.

Blackledge NP, Ott CJ, Gillen AE, Harris A. 2009. An insulator element 3′ to the *CFTR* gene binds CTCF and reveals an active chromatin hub in primary cells. *Nucleic Acids Res* **37:** 1086–1094.

Botstein D, Risch N. 2003. Discovering genotypes underlying human phenotypes: Past successes for Mendelian disease, future approaches for complex disease. *Nat Genet* **33:** 228–237.

Botstein D, White RL, Skolnick M, Davis RW. 1980. Construction of a genetic linkage map in man using restriction fragment length polymorphisms. *Am J Hum Genet* **32:** 314–331.

Buratti E, Baralle FE. 2001. Characterization and functional implications of the RNA binding properties of nuclear factor TDP-43, a novel splicing regulator of CFTR exon 9. *J Biol Chem* **276:** 36337–36343.

Cardon LR, Abecasis GR. 2003. Using haplotype blocks to map human complex trait loci. *Trends Genet* **19:** 135–140.

Chen R, Davydov EV, Sirota M, Butte AJ. 2010. Non-synonymous and synonymous coding SNPs show similar likelihood and effect size of human disease association. *PLoS ONE* **5:** e13574.

Chu CS, Trapnell BC, Murtagh JJ Jr, Moss J, Dalemans W, Jallat S, Mercenier A, Pavirani A, Lecocq JP, Cutting GR, et al. 1991. Variable deletion of exon 9 coding sequences in cystic fibrosis transmembrane conductance regulator gene mRNA transcripts in normal bronchial epithelium. *EMBO J* **10:** 1355–1363.

Chu CS, Trapnell BC, Curristin S, Cutting GR, Crystal RG. 1993. Genetic basis of variable exon 9 skipping in cystic fibrosis transmembrane conductance regulator mRNA. *Nat Genet* **3:** 151–156.

Collins FS. 1990. Identifying human disease genes by positional cloning. *Harvey Lect* **86:** 149–164.

Collins FS. 1992. Positional cloning: Let's not call it reverse anymore. *Nat Genet* **1:** 3–6.

Collins FS, Drumm ML, Cole JL, Lockwood WK, Vande Woude GF, Iannuzzi MC. 1987. Construction of a general human chromosome jumping library, with application to cystic fibrosis. *Science* **235:** 1046–1049.

Cox TK, Kerem B, Rommens J, Iannuzzi MC, Drumm M, Collins FS, Dean M, Tsui L-C, Chakravarti A. 1989. Mapping of the cystic fibrosis gene using putative ancestral recombinants. *Am J Hum Genet* **45**: A136.

Cutting GR, Antonarakis SE, Buetow KH, Kasch LM, Rosenstein BJ, Kazazian HH Jr. 1989. Analysis of DNA polymorphism haplotypes linked to the cystic fibrosis locus in North American black and Caucasian families supports the existence of multiple mutations of the cystic fibrosis gene. *Am J Hum Genet* **44**: 307–318.

Cutting GR, Kasch LM, Rosenstein BJ, Zielenski J, Tsui LC, Antonarakis SE, Kazazian HH Jr. 1990. A cluster of cystic fibrosis mutations in the first nucleotide-binding fold of the cystic fibrosis conductance regulator protein. *Nature* **346**: 366–369.

Davidson DJ, Rolfe M. 2001. Mouse models of cystic fibrosis. *Trends Genet* **17**: S29–S37.

Deeley RG, Westlake C, Cole SP. 2006. Transmembrane transport of endo- and xenobiotics by mammalian ATP-binding cassette multidrug resistance proteins. *Physiol Rev* **86**: 849–899.

Donaldson SH, Boucher RC. 2007. Sodium channels and cystic fibrosis. *Chest* **132**: 1631–1636.

Drumm ML, Pope HA, Cliff WH, Rommens JM, Marvin SA, Tsui LC, Collins FS, Frizzell RA, Wilson JM. 1990. Correction of the cystic fibrosis defect in vitro by retrovirus-mediated gene transfer. *Cell* **62**: 1227–1233.

Drumm ML, Wilkinson DJ, Smit LS, Worrell RT, Strong TV, Frizzell RA, Dawson DC, Collins FS. 1991. Chloride conductance expressed by ΔF508 and other mutant CFTRs in *Xenopus* oocytes. *Science* **254**: 1797–1799.

Eiberg H, Mohr J, Schmiegelow K, Nielsen LS, Williamson R. 1985. Linkage relationships of paraoxonase (PON) with other markers: Indication of PON–cystic fibrosis synteny. *Clin Genet* **28**: 265–271.

El-Seedy A, Dudognon T, Bilan F, Pasquet MC, Reboul MP, Iron A, Kitzis A, Ladeveze V. 2009. Influence of the duplication of CFTR exon 9 and its flanking sequences on diagnosis of cystic fibrosis mutations. *J Mol Diagn* **11**: 488–493.

ENCODE Project Consortium. 2004. The ENCODE (ENCyclopedia Of DNA Elements.) Project. *Science* **306**: 636–640.

Estivill X, Farrall M, Scambler PJ, Bell GM, Hawley KM, Lench NJ, Bates GP, Kruyer HC, Frederick PA, Stanier P, et al. 1987. A candidate for the cystic fibrosis locus isolated by selection for methylation-free islands. *Nature* **326**: 840–845.

Faa V, Coiana A, Incani F, Costantino L, Cao A, Rosatelli MC. 2010. A synonymous mutation in the *CFTR* gene causes aberrant splicing in an Italian patient affected by a mild form of cystic fibrosis. *J Mol Diagn* **12**: 380–383.

Fajac I, Viel M, Gaitch N, Hubert D, Bienvenu T. 2009. Combination of ENaC and CFTR mutations may predispose to cystic fibrosis–like disease. *Eur Respir J* **34**: 772–773.

Flint J, Valdar W, Shifman S, Mott R. 2005. Strategies for mapping and cloning quantitative trait genes in rodents. *Nat Rev Genet* **6**: 271–286.

Gervais R, Dumur V, Rigot JM, Lafitte JJ, Roussel P, Claustres M, Demaille J. 1993. High frequency of the R117H cystic fibrosis mutation in patients with congenital absence of the vas deferens. *N Engl J Med* **328**: 446–447.

Giardine B, Riemer C, Hefferon T, Thomas D, Hsu F, Zielenski J, Sang Y, Elnitski L, Cutting G, Trumbower H, et al. 2007. PhenCode: Connecting ENCODE data with mutations and phenotype. *Hum Mutat* **28**: 554–562.

Gregory RJ, Cheng SH, Rich DP, Marshall J, Paul S, Hehir K, Ostedgaard L, Klinger KW, Welsh MJ, Smith AE. 1990. Expression and characterization of the cystic fibrosis transmembrane conductance regulator. *Nature* **347**: 382–386.

Guilbault C, Saeed Z, Downey GP, Radzioch D. 2007. Cystic fibrosis mouse models. *Am J Respir Cell Mol Biol* **36**: 1–7.

Haston CK, Tsui LC. 2003. Loci of intestinal distress in cystic fibrosis knockout mice. *Physiol Genomics* **12**: 79–84.

Haston CK, Corey M, Tsui LC. 2002a. Mapping of genetic factors influencing the weight of cystic fibrosis knockout mice. *Mamm Genome* **13**: 614–618.

Haston CK, McKerlie C, Newbigging S, Corey M, Rozmahel R, Tsui LC. 2002b. Detection of modifier loci influencing the lung phenotype of cystic fibrosis knockout mice. *Mamm Genome* **13**: 605–613.

Hinzpeter A, Aissat A, Sondo E, Costa C, Arous N, Gameiro C, Martin N, Tarze A, Weiss L, de Becdelievre A, et al. 2010. Alternative splicing at a NAGNAG acceptor site as a novel phenotype modifier. *PLoS Genet* **6**: e1001153.

Hull J, Shackleton S, Harris A. 1994. The stop mutation R553X in the *CFTR* gene results in exon skipping. *Genomics* **19**: 362–364.

* Hunt JF, Wang C, Ford RC. 2013. Cystic fibrosis transmembrane conductance regulator (ABCC7) structure. *Cold Spring Harb Perspect Med* **3**: a009514.

* Hwang T-C, Kirk KL. 2013. The CFTR ion channel: Gating, regulation, and anion permeation. *Cold Spring Harb Perspect Med* **3**: a009498.

Kerem B, Rommens JM, Buchanan JA, Markiewicz D, Cox TK, Chakravarti A, Buchwald M, Tsui LC. 1989a. Identification of the cystic fibrosis gene: Genetic analysis. *Science* **245**: 1073–1080.

Kerem BS, Buchanan JA, Durie P, Corey ML, Levison H, Rommens JM, Buchwald M, Tsui LC. 1989b. DNA marker haplotype association with pancreatic sufficiency in cystic fibrosis. *Am J Hum Genet* **44**: 827–834.

Knowlton RG, Cohen-Haguenauer O, Van Cong N, Frezal J, Brown VA, Barker D, Braman JC, Schumm JW, Tsui LC, Buchwald M, et al. 1985. A polymorphic DNA marker linked to cystic fibrosis is located on chromosome 7. *Nature* **318**: 380–382.

Koehler DR, Hannam V, Belcastro R, Steer B, Wen Y, Post M, Downey G, Tanswell AK, Hu J. 2001. Targeting transgene expression for cystic fibrosis gene therapy. *Mol Ther* **4**: 58–65.

Kruglyak L. 1997. What is significant in whole-genome linkage disequilibrium studies? *Am J Hum Genet* **61**: 810–812.

Kruglyak L. 2005. Power tools for human genetics. *Net Genet* **37**: 1299–1300.

Kruglyak L. 2008. The road to genome-wide association studies. *Nat Rev Genet* **9**: 314–318.

Lewandowska MA, Costa FF, Bischof JM, Williams SH, Soares MB, Harris A. 2010. Multiple mechanisms influence regulation of the cystic fibrosis transmembrane conductance regulator gene promoter. *Am J Respir Cell Mol Biol* **43:** 334–341.

Li Y, Vinckenbosch N, Tian G, Huerta-Sanchez E, Jiang T, Jiang H, Albrechtsen A, Andersen G, Cao H, Korneliussen T, et al. 2010. Resequencing of 200 human exomes identifies an excess of low-frequency non-synonymous coding variants. *Nat Genet* **42:** 969–972.

Liu X, Jiang Q, Mansfield SG, Puttaraju M, Zhang Y, Zhou W, Cohn JA, Garcia-Blanco MA, Mitchell LG, Engelhardt JF. 2002. Partial correction of endogenous ΔF508 CFTR in human cystic fibrosis airway epithelia by spliceosome-mediated RNA *trans*-splicing. *Nat Biotechnol* **20:** 47–52.

Liu X, Li X, Li M, Acimovic YJ, Li Z, Scherer SW, Estivill X, Tsui LC. 2004. Characterization of the segmental duplication LCR7–20 in the human genome. *Genomics* **83:** 262–269.

Mansfield SG, Kole J, Puttaraju M, Yang CC, Garcia-Blanco MA, Cohn JA, Mitchell LG. 2000. Repair of CFTR mRNA by spliceosome-mediated RNA *trans*-splicing. *Gene Ther* **7:** 1885–1895.

Mansfield SG, Clark RH, Puttaraju M, Kole J, Cohn JA, Mitchell LG, Garcia-Blanco MA. 2003. 5′ exon replacement and repair by spliceosome-mediated RNA *trans*-splicing. *RNA* **9:** 1290–1297.

Massie RJ, Poplawski N, Wilcken B, Goldblatt J, Byrnes C, Robertson C. 2001. Intron-8 polythymidine sequence in Australasian individuals with CF mutations R117H and R117C. *Eur Respir J* **17:** 1195–1200.

McCarthy VA, Harris A. 2005. The *CFTR* gene and regulation of its expression. *Pediatr Pulmonol* **40:** 1–8.

McCarthy VA, Ott CJ, Phylactides M, Harris A. 2009. Interaction of intestinal and pancreatic transcription factors in the regulation of *CFTR* gene expression. *Biochim Biophys Acta* **1789:** 709–718.

Meyerholz DK, Stoltz DA, Pezzulo AA, Welsh MJ. 2010. Pathology of gastrointestinal organs in a porcine model of cystic fibrosis. *Am J Pathol* **176:** 1377–1389.

Morton NE, Steinberg AG. 1956. Sequential test for linkage between cystic fibrosis of the pancreas and the MNS locus. *Am J Hum Genet* **8:** 177–189.

Mouchel N, Henstra SA, McCarthy VA, Williams SH, Phylactides M, Harris A. 2004. HNF1α is involved in tissue-specific regulation of CFTR gene expression. *Biochem J* **378:** 909–918.

Newmark P. 1985. Testing for cystic fibrosis. *Nature* **318:** 309.

Ostedgaard LS, Meyerholz DK, Chen JH, Pezzulo AA, Karp PH, Rokhlina T, Ernst SE, Hanfland RA, Reznikov LR, Ludwig PS, et al. 2011. The ΔF508 mutation causes CFTR misprocessing and cystic fibrosis-like disease in pigs. *Sci Transl Med* **3:** 74ra24.

Ott CJ, Blackledge NP, Kerschner JL, Leir SH, Crawford GE, Cotton CU, Harris A. 2009a. Intronic enhancers coordinate epithelial-specific looping of the active CFTR locus. *Proc Natl Acad Sci* **106:** 19934–19939.

Ott CJ, Suszko M, Blackledge NP, Wright JE, Crawford GE, Harris A. 2009b. A complex intronic enhancer regulates expression of the CFTR gene by direct interaction with the promoter. *J Cell Mol Med* **13:** 680–692.

Pagani F, Raponi M, Baralle FE. 2005. Synonymous mutations in CFTR exon 12 affect splicing and are not neutral in evolution. *Proc Natl Acad Sci* **102:** 6368–6372.

Plotkin JB, Kudla G. 2010. Synonymous but not the same: The causes and consequences of codon bias. *Nat Rev Genet* **12:** 32–42.

Raponi M, Baralle FE, Pagani F. 2007. Reduced splicing efficiency induced by synonymous substitutions may generate a substrate for natural selection of new splicing isoforms: The case of CFTR exon 12. *Nucleic Acids Res* **35:** 606–613.

Riordan JR, Rommens JM, Kerem B, Alon N, Rozmahel R, Grzelczak Z, Zielenski J, Lok S, Plavsic N, Chou JL, et al. 1989. Identification of the cystic fibrosis gene: Cloning and characterization of complementary DNA. *Science* **245:** 1066–1073.

Rogers CS, Hao Y, Rokhlina T, Samuel M, Stoltz DA, Li Y, Petroff E, Vermeer DW, Kabel AC, Yan Z, et al. 2008a. Production of CFTR-null and CFTR-ΔF508 heterozygous pigs by adeno-associated virus-mediated gene targeting and somatic cell nuclear transfer. *J Clin Invest* **118:** 1571–1577.

Rogers CS, Stoltz DA, Meyerholz DK, Ostedgaard LS, Rokhlina T, Taft PJ, Rogan MP, Pezzulo AA, Karp PH, Itani OA, et al. 2008b. Disruption of the *CFTR* gene produces a model of cystic fibrosis in newborn pigs. *Science* **321:** 1837–1841.

Romey MC, Guittard C, Carles S, Demaille J, Claustres M, Ramsay M. 1999. First putative sequence alterations in the minimal CFTR promoter region. *J Med Genet* **36:** 263–264.

Rommens JM, Iannuzzi MC, Kerem B, Drumm ML, Melmer G, Dean M, Rozmahel R, Cole JL, Kennedy D, Hidaka N, et al. 1989a. Identification of the cystic fibrosis gene: Chromosome walking and jumping. *Science* **245:** 1059–1065.

Rommens JM, Zengerling-Lentes S, Kerem B, Melmer G, Buchwald M, Tsui LC. 1989b. Physical localization of two DNA markers closely linked to the cystic fibrosis locus by pulsed-field gel electrophoresis. *Am J Hum Genet* **45:** 932–941.

Rozmahel R, Wilschanski M, Matin A, Plyte S, Oliver M, Auerbach W, Moore A, Forstner J, Durie P, Nadeau J, et al. 1996. Modulation of disease severity in cystic fibrosis transmembrane conductance regulator deficient mice by a secondary genetic factor. *Nat Genet* **12:** 280–287.

Rozmahel R, Heng HH, Duncan AM, Shi XM, Rommens JM, Tsui LC. 1997. Amplification of CFTR exon 9 sequences to multiple locations in the human genome. *Genomics* **45:** 554–561.

Scambler PJ, Law HY, Williamson R, Cooper CS. 1986. Chromosome mediated gene transfer of six DNA markers linked to the cystic fibrosis locus on human chromosome seven. *Nucleic Acids Res* **14:** 7159–7174.

Schmiegelow K, Eiberg H, Tsui LC, Buchwald M, Phelan PD, Williamson R, Warwick W, Niebuhr E, Mohr J, Schwartz M, et al. 1986. Linkage between the loci for cystic fibrosis and paraoxonase. *Clin Genet* **29:** 374–377.

Sheridan MB, Fong P, Groman JD, Conrad C, Flume P, Diaz R, Harris C, Knowles M, Cutting GR. 2005.

Mutations in the β-subunit of the epithelial Na$^+$ channel in patients with a cystic fibrosis-like syndrome. *Hum Mol Genet* **14:** 3493–3498.

Snouwaert JN, Brigman KK, Latour AM, Malouf NN, Boucher RC, Smithies O, Koller BH. 1992. An animal model for cystic fibrosis made by gene targeting. *Science* **257:** 1083–1088.

Spence MA, Tsui LC, Hodge SE, MacCluer JW, Elston RC. 1989. Genetic Analysis Workshop 6: Multipoint mapping of loci in the region of cystic fibrosis. *Prog Clin Biol Res* **329:** 1–9.

Sun X, Sui H, Fisher JT, Yan Z, Liu X, Cho HJ, Joo NS, Zhang Y, Zhou W, Yi Y, et al. 2010. Disease phenotype of a ferret CFTR-knockout model of cystic fibrosis. *J Clin Invest* **120:** 3149–3160.

Tebbutt SJ. 1995. Animal studies of cystic fibrosis. *Mol Med Today* **1:** 336–342.

Toyoda Y, Hagiya Y, Adachi T, Hoshijima K, Kuo MT, Ishikawa T. 2008. MRP class of human ATP binding cassette (ABC) transporters: Historical background and new research directions. *Xenobiotica* **38:** 833–862.

Trezise AE, Buchwald M. 1991. In vivo cell-specific expression of the cystic fibrosis transmembrane conductance regulator. *Nature* **353:** 434–437.

Tsui LC. 1995. The cystic fibrosis transmembrane conductance regulator gene. *Am J Respir Crit Care Med* **151:** S47–S53.

Tsui LC, Buchwald M. 1988. No evidence for genetic heterogeneity in cystic fibrosis. *Am J Hum Genet* **42:** 184.

Tsui LC, Buchwald M. 1991. Biochemical and molecular genetics of cystic fibrosis. *Adv Hum Genet* **20:** 311–152.

Tsui L-C, Estivill X. 1991. Identification of disease genes on the basis of chromosomal localization. In *Genome analysis. Vol. 3: Genes and phenotypes* (ed. Davies KE, Tilghman S), pp. 1–36. Cold Spring Harbor Laboratory Press, Cold Spring Harbor, NY.

Tsui LC, Buchwald M, Barker D, Braman JC, Knowlton R, Schumm JW, Eiberg H, Mohr J, Kennedy D, Plavsic N, et al. 1985a. Cystic fibrosis locus defined by a genetically linked polymorphic DNA marker. *Science* **230:** 1054–1057.

Tsui LC, Cox DW, McAlpine PJ, Buchwald M. 1985b. Cystic fibrosis: Analysis of linkage of the disease locus to red cell and plasma protein markers. *Cytogenet Cell Genet* **39:** 238–239.

Wainwright BJ, Scambler PJ, Schmidtke J, Watson EA, Law HY, Farrall M, Cooke HJ, Eiberg H, Williamson R. 1985. Localization of cystic fibrosis locus to human chromosome 7cen-q22. *Nature* **318:** 384–385.

Wainwright BJ, Scambler PJ, Stanier P, Watson EK, Bell G, Wicking C, Estivill X, Courtney M, Boue A, Pedersen PS. 1988. Isolation of a human gene with protein sequence similarity to human and murine int-1 and the *Drosophila* segment polarity mutant wingless. *EMBO J* **7:** 1743–1748.

Ward CL, Krouse ME, Gruenert DC, Kopito RR, Wine JJ. 1991. Cystic fibrosis gene expression is not correlated with rectifying Cl$^-$ channels. *Proc Natl Acad Sci* **88:** 5277–5281.

Wardle CJ, Harris A. 1995. Down-regulation of CFTR expression in an HT29-derived cell line by stable antisense RNA production. *Biochim Biophys Acta* **1261:** 417–423.

White R, Woodward S, Leppert M, O'Connell P, Hoff M, Herbst J, Lalouel JM, Dean M, Vande Woude G. 1985. A closely linked genetic marker for cystic fibrosis. *Nature* **318:** 382–384.

Wilke M, Buijs-Offerman RM, Aarbiou J, Colledge WH, Sheppard DN, Touqui L, Bot A, Jorna H, de Jonge HR, Scholte BJ. 2011. Mouse models of cystic fibrosis: Phenotypic analysis and research applications. *J Cyst Fibros* **10:** S152–S171.

Xie J, Drumm ML, Ma J, Davis PB. 1995. Intracellular loop between transmembrane segments IV and V of cystic fibrosis transmembrane conductance regulator is involved in regulation of chloride channel conductance state. *J Biol Chem* **270:** 28084–28091.

Ye L, Zhu W, Backx PH, Cortez MA, Wu J, Chow YH, McKerlie C, Wang A, Tsui LC, Gross GJ, et al. 2010. Arrhythmia and sudden death associated with elevated cardiac chloride channel activity. *J Cell Mol Med* **15:** 2307–2316.

Zielenski J, Rozmahel R, Bozon D, Kerem B, Grzelczak Z, Riordan JR, Rommens J, Tsui LC. 1991. Genomic DNA sequence of the cystic fibrosis transmembrane conductance regulator (CFTR) gene. *Genomics* **10:** 214–228.

Zuccato E, Buratti E, Stuani C, Baralle FE, Pagani F. 2004. An intronic polypyrimidine-rich element downstream of the donor site modulates cystic fibrosis transmembrane conductance regulator exon 9 alternative splicing. *J Biol Chem* **279:** 16980–16988.

Assessing the Disease-Liability of Mutations in *CFTR*

Claude Ferec[1] and Garry R. Cutting[2]

[1]Faculté de Médecine et des Sciences de la Santé, Université de Bretagne Occidentale; Centre Hospitalier Universitaire (CHU), Hôpital Morvan; INSERM, U1078 Brest, France

[2]McKusick-Nathans Institute of Genetic Medicine, Johns Hopkins University School of Medicine, Baltimore, Maryland 21204

Correspondence: gcutting@jhmi.edu

More than 1900 mutations have been reported in the cystic fibrosis transmembrane conductance regulator (CFTR), the gene defective in patients with cystic fibrosis. These mutations have been discovered primarily in individuals who have features consistent with the diagnosis of CF. In some cases, it has been recognized that the mutations are not causative of cystic fibrosis but are responsible for disorders with features similar to CF, and these conditions have been termed CFTR-related disorders or CFTR-RD. There are also mutations in CFTR that do not contribute to any known disease state. Distinguishing *CFTR* mutations according to their penetrance for an abnormal phenotype is important for clinical management, structure/function analysis of CFTR, and understanding the molecular and cellular mechanisms underlying CF.

The primary repository for cystic fibrosis transmembrane conductance regulator (CFTR) mutations is a database curated by investigators at the Hospital for Sick Children in Toronto (http://www.genet.sickkids.on.ca/cftr/Home .html). The CF Mutation Database was initiated by Lap-Chee Tsui soon after the cloning of the CFTR gene in 1989. In addition to cataloging DNA variation in *CFTR*, each entry includes the predicted effect of a mutation on gene function. Eighty-two percent of the ∼1900 reported mutations in the CF Mutation Database have putative deleterious effects, whereas 14% appear to be sequence variants of no functional consequence, and the remaining 4% are of unknown effect.

The vast majority of mutations (97.5%) reported in CFTR involves one or a few nucleotides. The remaining mutations (2.5%) involve a larger number of nucleotides such as deletions of entire exons or multiple exons of the gene. Missense mutations (i.e., those that change an amino acid) account for 40% of the reported mutations. They are the most common form of mutation involving one or a few nucleotides. Frameshift mutations (i.e., those that alter the reading frame) account for almost 16%, whereas splicing and nonsense mutations account for 12% and 8.5%, respectively. Mutations involving the insertion or deletion of nucleotides in multiples of three that leave the reading frame intact are uncommon (∼2%). However, among these in-frame deletions is p.Phe508del, a deletion of three nucleotides in exon 10 of CFTR that causes the loss of a single phenylalanine residue at codon 508. The F508del mutation is found in 70% of CF alleles in the Caucasian population (Bobadilla et al. 2002). The commonness of the F508del mutation is the primary reason that CF is one of the most prevalent lethal autosomal recessive disorders among Caucasians. The remainder of the mutations predicted to be deleterious (0.8%) lie within the promoter of the CFTR gene.

Most of the 1900 mutations reported in CFTR have not been experimentally evaluated for their effect on the function of CFTR. Mutations that have a frequency in CF patients exceeding 1% have been investigated in a variety of expression systems for their effect on the splicing of CFTR RNA or the function of the CFTR protein (Cheng et al. 1990; Gregory et al. 1991; Sheppard et al. 1993). Several dozen other mutations reported in patients have been investigated for their functional effects (e.g., Seibert et al. 1996a, 1997). The functional consequences of the remaining ~1850 CFTR mutations are unknown. Thus, the disease liability of most of the less common CFTR mutations is uncertain. The problem is further complicated as a number of mutations were reported in patients in whom the diagnosis of cystic fibrosis was not verified.

To rigorously ascertain the disease liability of the less common CFTR mutations, a project entitled Clinical and Functional Translation of CFTR (CFTR2; http://www.cftr2.org/) was initiated in 2009 (Sosnay et al. 2011). The project was supported by the U.S. CF Foundation to elucidate the medical implications of as many CFTR mutations as possible. By 2012, the CFTR2 Project had accrued clinical and CF mutations data on almost 40,000 patients from North America and Europe. Collection was based on data derived from CF Patient Registries and large CF clinics and, as such, all data were derived from patients diagnosed with CF by a medical professional. From these data, it was recognized that ~160 mutations account for 96%–97% of CF alleles. The CFTR2 Project has created a new website that provides clinical information associated with each of the 160 mutations when found in *trans* with a CF-causing mutation (as defined by the American College of Medical Genetics [Watson et al. 2004]). Disease liability of each of the 160 mutations is assessed using clinical, functional, and genetic evidence. The CFTR2 database provides a complement to the mutation repository in Toronto. Over the next few years, the CFTR2 Project will evaluate additional mutations beyond the current panel of ~160.

DISTRIBUTION WORLDWIDE

The hallmark of the mutational spectrum in the *CFTR* gene is the very high frequency of the F508del in the Caucasian population (Cystic Fibrosis Genetic Analysis Consortium 1990). Worldwide, the F508del mutation is responsible for about two-thirds of all CF chromosomes in the Caucasian population. However, mutations like the W1282X (p.Trp1282X) show a higher frequency than F508del in some populations such as Ashkenazi Jews (Abeliovich et al. 1992; Shoshani et al. 1992). In 2002, Bobadilla et al. published a large worldwide analysis of *CFTR* mutations showing that the F508del frequency varies from 100% in the Faroe Islands of Denmark to 24.5% in Turkey, and confirmed the northwest to southeast gradient in the F508del distribution in Europe (European Working Group on CF Genetics 1990; Bobadilla et al. 2002).

Worldwide, only a limited number of mutant alleles display a frequency higher than 1%, as for example the G542X (p.Gly542X), which is common in the Mediterranean area of Europe and Africa, which may be associated with the Phoenician migration around the Mediterranean Sea (Loirat et al. 1997). The G551D (p.Gly551Asp) arose in the Celtic population more recently and is still prevalent in Ireland and Brittany (Scotet et al. 2003b), and the N1303K (p.Asn1303Lys) is also rather frequent in populations from the center of Europe (Osborne et al. 1992). Besides these rather common mutations, which are largely distributed in the Caucasian population, there are also some founder effects of mutated alleles observed as, for example, in the Reunion Island, where the Y122X (p.Tyr122X) mutation displays a frequency of 24% (Duguéperoux et al. 2004), or the 394delTT (p.Leu88IlefsX22) referred to as a Nordic mutation found in countries bordering the Baltic sea (Schwartz et al. 1994). Another mutation that is common in central and eastern Europe is the deletion of exons 2 and 3 of the *CFTR*, which is the mark of the Slavic origin of this large deletion (Dork et al. 2000). In North America, the distribution of *CFTR* mutations reflects European descent (the five more common mutations in the U.S. with a frequency over 1% are F508del, G542X, G551D, W1282X, and N1303K) (Bobadilla et al. 2002). Racial admixture of *CFTR* mutations is also apparent. For example, the mutation 3120+1G→A (c.2988+1G→A) found in Native Africans with a high prevalence in South Africa is the second most prevalent allele in African-American populations of CF patients (Macek et al. 1997). Precise knowledge of the worldwide distribution of the *CFTR* mutations has facilitated the implementation of CF newborn screening in most of the countries of Europe and in most of the states in the U.S. and Australia (Farrell et al. 2005).

DIFFERENT CLASSES OF MUTATIONS

Understanding the effect of a mutation on the function of a gene is an essential step in resolving how a mutation causes disease. Mutations that introduce premature termination codons (PTC) such as frameshift, nonsense, and splicing mutations are expected to be deleterious. Other changes in the coding sequence that change a single amino acid (e.g., substitution) may or may not affect function of the encoded protein. Mutations that permit protein to be expressed invariably require experimental studies to assess their effect on the protein. CFTR is an integral membrane protein with multiple different functional and structural domains. Alteration in residue in one location may have local effects on domain function but consequences on other properties are also possible.

To provide context organizing the functional consequences of CFTR mutation, Welsh and Smith proposed a framework of five major classes of defect (Welsh and Smith 1993). Class I mutations involve changes that alter the stability of RNA. These are primarily mutations that introduce a PTC to the RNA transcript. PTC mutations generally cause the loss of messenger RNA because of nonsense-mediated RNA decay (NMRD) (Frischmeyer and Dietz 1999). Mutations in CFTR including G542X, R553X (p.Arg553X), and W1282X have been shown to lead to NMRD in primary airway cells (Hamosh et al. 1991, 1992b; Will et al. 1995). There are a few exceptions to the concept that PTC mutations cause NMRD. A notable example is the R1162X (p.Arg1162X) mutation in which the transcript containing a PTC is stable (Rolfini and Cabrini 1993; Will et al. 1995). Patients homozygous for R1162X showed severe pancreatic exocrine deficiency and variable lung disease severity that overlapped with measures observed in F508del homozygotes (Gasparini et al. 1992). Measurements of CFTR function in vivo in two R1162X homozygotes were similar to those recorded in 74 F508del homozygotes (Stanke et al. 2008). Another exception is the mutation E831X (p.Glu831X) that occurs at the 5′ splice site of exon 14a (Hinzpeter et al. 2010). The mutation alters splicing of CFTR such that three nucleotides encoding the mutant nonsense codon are omitted and a protein is synthesized missing the glutamic acid residue at codon 831. Although the resulting protein is stable, it has reduced function and is associated with CF, albeit a milder pancreatic sufficient form (Hinzpeter et al. 2010). Thus, PTC mutations either cause a loss of the RNA transcript and absence of protein or, in a few rare cases, cause stable RNA transcript but a modified protein that is either degraded or malfunctional. In almost all cases, these functional effects produce a CF phenotype. An important exception is when a PTC occurs in the last exon or the last third of a penultimate exon of a gene. In this situation, the RNA transcript is not recognized by the NMRD mechanism and the transcript is stable (Frischmeyer and Dietz 1999). PTC mutations occur in the last exon of CFTR, which are associated with stable transcript and a CFTR protein that is properly folded and present at the cell membrane. In a few cases, the resulting truncated protein is sufficiently functional to ameliorate life-shortening lung disease, although dysfunction is apparent in other organ systems (sweat gland [Mickle et al. 1998] or vas deferens [Claustres et al. 2000]).

Class II mutations cause a defect in biogenesis of CFTR. A prime example is the common mutation F508del in which the loss of a single amino acid leads to misfolding of the protein and redirection to degradative machinery in the cell (Kopito 1999). Lowering the temperature of the cell allows a portion of CFTR-bearing F508del to fold properly, and F508del is an excellent example of a naturally occurring temperature-sensitive mutant in humans (Denning et al. 1992). Mutations elsewhere in the gene that caused loss of stability include missense mutations such as G480C (p.Gly480Cys) and others that affect domain–domain interactions in CFTR such as R1070Q (p.Arg1070Gln) (Smit et al. 1995; Serohijos et al. 2008). The evaluation of mutations that alter protein stability are particularly useful in dissecting the structure of CFTR (Patrick et al. 2012).

Class III and class IV mutations are associated with a production of a CFTR molecule that is properly folded and trafficked to the apical membrane. Class III mutations affect the activation of the protein and a prime example is the G551D mutation that causes CFTR to be resistant to activation, even though normal levels of mutant protein are present at the cell surface (Drumm et al. 1991; Anderson and Welsh 1992; Schwiebert et al. 1995). Class IV mutations alter the conductivity of the chloride channel in CFTR. In some cases, these mutations are intimate to the ion conduction pore of the CFTR channel (Sheppard et al. 1993), whereas others appear to affect conductivity through allosteric mechanisms (Seibert et al. 1996b).

Class V mutations reduce the level of wild-type CFTR that is produced by splicing mutations. Although some wild-type CFTR transcript and protein is produced, the level is insufficiently high to avoid the development of the CF phenotype. A key example is the 3849+10kb (c.3717+12191C→T) mutation (Highsmith et al. 1994). This mutation activates a cryptic splice site in intron 19 of CFTR that causes mis-splicing of exon 19. However, the canonical splice sites are able to complete with the cryptic site resulting in the production of 5%–10% of wild-type transcript. Analysis of additional "leaky" splice mutations affirmed that ~10% of the levels of wild-type CFTR transcript are required to escape the pulmonary complications of CF (Highsmith et al. 1997; Kerem et al. 1997).

A variety of other mechanistic defects have been associated with CFTR mutations. These have not been formally incorporated into new classes beyond the original five. These include mutations that change the targeting of CFTR (Guggino and Stanton 2006), others that alter the interaction between CFTR and other proteins such as the epithelial sodium channel (ENaC) (Schreiber et al. 1999) and the outwardly rectifying chloride channel (Fulmer et al. 1995), and mutations that affect the membrane residence of CFTR (Silvis et al. 2003).

Understanding the implications of the mutations on CFTR function is critical for the design of therapeutics for cystic fibrosis. Mutations such as F508del that cause misfolding of the protein require therapeutics that resolve the misfolding to some extent so that a fraction of mutant protein can transit the Golgi and be directed to the cell membrane. On the other hand, mutations such as G551D require compounds that can activate a mutant protein that is already at the cell surface in levels comparable to what is seen with a normal protein.

GENOTYPE/PHENOTYPE CORRELATIONS

The degree of correlation between CFTR genotype and phenotype differs among organ systems (Mickle and Cutting 2000). During the hunt for the CF gene, it was shown that allelic variants of CFTR likely influence the pancreatic status in CF patients. Using markers linked to the CF locus, investigators in Toronto showed the robust association between the markers and pancreatic status (Kerem et al. 1989). On identification of the CFTR gene, it was shown that F508del and several other loss-of-function mutations including G551D and G542X are highly associated with pancreatic insufficiency. Other mutations that alter residual CFTR function such as R117H (p.Arg117His) and R347P (p.Arg347Pro) have been associated with the pancreatic sufficiency (Kristidis et al. 1992). The correlation with pancreatic status was confirmed in studies of a large collection of CF patients aggregated from different clinics in North America and Europe (The Cystic Fibrosis Genotype-Phenotype Consortium 1993). Correlation of CFTR genotype with other organ system disease has been more subtle. As patients with pancreatic sufficiency also have lower sweat chloride, mutations associated with pancreatic sufficiency show a rough correlation with sweat chloride concentration (Wilschanski et al. 1995). After accounting for pancreatic status, correlation with sweat chloride has not been evident. Furthermore, correlation with lung function has been difficult to detect (The Cystic Fibrosis Genotype-Phenotype Consortium 1993). Well-powered studies that reveal correlations between genotype and pancreatic status and sweat chloride concentrations have not shown correlation with lung function (as measured by the 1-s forced expiratory volume). However, when a sufficiently large number of patients with identical genotypes are found in a single geographic location such as Belgium and in French Canada, a correlation with decline in lung function becomes evident for the A455E (p.Ala455Glu) mutation (Gan et al. 1995; De Braekeleer et al. 1997).

A complementary approach to correlating genotype with phenotype is to aggregate patients according to the functional class of mutation. In general, mutations in class I to III are associated with complete or near complete loss of CFTR function (<3% of wild-type CFTR function) (Welsh and Smith 1993). Mutations in these classes are primarily associated with pancreatic insufficiency, higher rates of pulmonary infection, and higher mortality (Kubesch et al. 1993; Koch et al. 2001; Ahmed et al. 2003; McKone et al. 2003; Green et al. 2010). On the other hand, residual CFTR function tends to be higher (3%–10%) in class IV and V mutations. Consequently, patients with these mutations tend to have higher rates of pancreatic sufficiency and less elevated sweat chloride levels (Koch

et al. 2001; Ahmed et al. 2003; McKone et al. 2003; Green et al. 2010). As residual CFTR function is increased, such as in the case of the R117H mutation bearing the 7T variant, lung function is improved sufficiently such that a CF diagnosis may not be evident. Males bearing the combination of F508del mutation panel with R117H-7T are generally discovered because of infertility caused by absence of the vas deferens (see below). These men can have normal lung function (Colin et al. 1996), but some manifest pulmonary disease observed in patients with CF (Castellani et al. 1999).

Classifying mutations according to a single functional effect is problematic as a number of mutations alter several CFTR properties such as folding (class II) and ion conduction (class IV) (Champigny et al. 1995; Sheppard et al. 1995). Furthermore, there appears to be finer gradations of dysfunction among patients with pancreatic insufficiency. Meconium ileus (MI), an obstruction of the intestine at the ileo-colic junction at birth, affects ∼15% of CF patients. MI occurs almost exclusively in patients with pancreatic insufficiency (PI-CF); however, the rate of MI can differ by CFTR genotype. PI-CF patients bearing nonsense mutation G542X have shown the highest rate of MI, whereas those with F508del have an intermediate rate, and patients with the G551D mutation have a relatively low rate of MI (Hamosh et al. 1992a; Kristidis et al. 1992; Feingold and Guilloud-Bataille 1999).

One of the reasons that correlations among organ systems may vary is because of the presence of environmental and genetic modifiers. It was noted shortly after the cloning of CFTR that individuals homozygous for the common mutation F508del had a wide range of lung function (Kerem et al. 1990). This indicated that other factors such as modifier genes and environmental exposures must contribute to variation in patients with identical CFTR genotypes. The analysis of twins and siblings with cystic fibrosis confirmed that the genetic factors contributed substantially to variation of lung function (Mekus et al. 2000; Vanscoy et al. 2007). Studies of twins as they transit to different environments revealed that nongenetic (environmental and stochastic) factors contribute about equally to variation in lung function (Collaco et al. 2010). Among nongenetic factors, unique exposures accounted for 2/3 of the variability, whereas shared environment (i.e., living in the same home) accounted for the remaining third. These studies indicate that correlations between CFTR genotype and phenotype can be confounded by the differences in both genetic and nongenetic modifiers among patients. The effect of these confounders will likely differ among different samples of CF patients. One approach to address confounding by modifiers is to study large numbers of patients with different CFTR genotypes to tease out correlations conferred by different CFTR mutations. The CFTR2 project has a large collection of patients to enable this type of study. Indeed, functional studies, performed on over 50 different missense mutations found in at least nine patients in the CFTR2 database, revealed a correlation between CFTR genotype and both sweat chloride concentration and lung function.

CFTR-RELATED DISORDERS

After the discovery of the *CFTR* gene, it became evident that dysfunction of CFTR in a single organ could be associated with clinical phenotypes distinct from cystic fibrosis. This is illustrated in the paper by Dumur et al. (1990) who reported the presence of the F508del in 9/17 infertile males because of congenital bilateral absence of vas deferens (CBAVD). A role for CFTR in male infertility was extended by Anguiano et al. who provided evidence of compound heterozygosity for *CFTR* mutations (i.e., F508del and R117H) in CBAVD patients (Anguiano et al. 1992). Other groups confirmed the presence of two *CFTR* mutations in a subset of males with CBAVD (De Braekeleer and Ferec 1996; Wilschanski et al. 1996). However, the distribution of *CFTR* mutations differed between those with CF and those with CBAVD (Dörk et al. 1997; Claustres et al. 2000). The most frequent is a poly T tract polymorphism located in intron 8 (namely, the IVS8-5T) in the noncoding sequence of the gene, which is an exon 9 splice variant (Chillón et al. 1995). The poly T sequence had been studied by Chu et al. (1993) who have showed an inverse relationship between the length of the polythymidine tract and the proportion of exon 9–CFTR mRNA transcripts. The second one is a complex haplotype with the R117H missense mutation in exon 7 associated in *cis* with the IVS8-7T (Kiesewetter et al. 1993; Mercier et al. 1995; Cuppens et al. 1998). This was the beginning of a long debate and discussion in the community on the frontier between CF and CFTR-related disorders (CFTR-RD).

The criteria for the diagnosis of CF proposed by Rosenstein and Cutting (1998) was widely accepted worldwide and slightly modified by Farrell et al. (2008) more recently with the introduction of neonatal screening for CF. However, it was important to distinguish patients who do not meet the diagnosis criteria for CF but for whom there was evidence of CFTR dysfunction and finally, recently, the term CFTR-related disorders (CFTR-RD) has been accepted by the community (Bombieri et al. 2011). A CFTR-RD is defined as a clinical entity associated with CFTR dysfunction that does not fulfill the diagnostic criteria for CF. Three main entities illustrate this phenotype: congenital bilateral absence of the vas deferens with CFTR dysfunction, acute or recurrent chronic pancreatitis with CFTR dysfunction, and disseminated bronchiectasis associated with CFTR dysfunction.

CFTR Mutations Associated with CBAVD

Congenital bilateral absence of vas deferens is the most common cause of male sterility in otherwise healthy males with an incidence of about 1:1000 males (Holsclaw et al. 1971; Mak and Jarvi 1996). Many reports from different countries have identified that the disorder is associated with mutations in the *CFTR* gene, and the development of new technologies allowing the scanning and the rapid sequencing of the coding sequence of the gene shows that CBAVD patients are carrying in about 80% of cases two mutated alleles in the *CFTR* gene (Jarvi et al. 1998; Claustres et al. 2000). But it is also widely accepted that CF and CBAVD have complete separate clinical and prognostic characteristics and CBAVD is now considered as the paradigm of CFTR-RD.

As previously shown in this work, genotype/phenotype studies have clearly identified two categories of CF-causing disease mutations, severe mutations (with no functional CFTR protein) belonging to classes I–III, and mild mutations (more likely to have enough residual CFTR activity to sustain pancreatic function) and belonging to classes IV–V. The classical genotype found in CBAVD patients correspond to the association of a CF-causing mutation in *trans* with a mutated allele, which is generally a missense variation different from classical CFTR mild mutations. This has led us to the creation of a third group of CFTR mutations termed CFTR-RD alleles. The two most common CFTR-RD alleles are the IVS8-5T, a poly T tract in intron 8 of the *CFTR*, and the R117H associated with the 7T tract (Chillón et al. 1995). The IVS8-5T allele, which is present on at least 5% of "normal *CFTR* genes," is the most frequent CFTR-RD allele worldwide (Bombieri et al. 2011). Its frequency in CBAVD males is 5–8 times higher than in the general population (Chillón et al. 1995; Zielenski et al. 1995). About 35% of CBAVD males of European descent have inherited a common CF mutation in one gene and the IVS8-5T allele on the other chromosome. The IVS8 poly tract of T is also a genetic modifier of the R117H when associated in *cis*. The combination of R117H-7T is found in CBAVD, but R117H associated in *cis* with a 5T is a complex CF allele generally associated with a mild phenotype of CF (Kiesewetter et al. 1993).

The IVS8-5T is considered as a CBAVD mutation with incomplete penetrance. The polymorphic dinucleotide repeat IVS8–TG lying immediately upstream of the IVS8-TG tract was also found to influence exon 9 splicing (Cuppens et al. 1998). The longer the IVS8-TGn and the shorter the IVS8-Tn repeats, the greater is exon 9 skipping, leading to reduced levels of CFTR RNA transcript (Niksic et al. 1999; Hefferon et al. 2004). Consequently, CBAVD is more likely to occur in males carrying IVS8-5T and longer IVS8-Tn repeats of 12 or 13 (TG12 TG13) (Groman et al. 2004).

The vast majority of CFTR-RD mutations found in CBAVD patients are not present in mutation screening tests, as commercially available mutation-screening kits only identify the most common CF-causing mutations. Thus, the identification of CFTR-RD mutations require a full exploration of coding/flanking CFTR sequences by scanning (DHPLC or HRM techniques) and sequencing methods (Audrezet et al. 2008). Using these methods, it is possible to identify 85% of patients harboring two mutations, one of those two being a CFTR-RD mutation (Claustres et al. 2000).

CFTR Mutations in Idiopathic Chronic Pancreatitis

In the human exocrine pancreas, CFTR is predominantly expressed at the apical membrane of the ductal and centroacinar cells that line the small pancreatic ducts and controls cAMP-stimulated HCO3 secretion into the duct lumen. The major effect of CFTR in pancreatic ducts is to dilute and alkalinize the protein-rich acinar secretions thereby preventing the formation of protein plugs that predispose to pancreatic injury (Park et al. 2010).

In 1998, two groups simultaneously reported an association between *CFTR* mutations and idiopathic chronic pancreatitis (ICP) (Cohn et al. 1998; Sharer et al. 1998). Until that time, most published studies had only analyzed the most common CF-causing mutations and combined data from these earlier studies indicated that 18% of subjects with ICP had common CF-causing mutations, whereas 2% were compound heterozygotes (Audrezet et al. 2002). Subsequent studies analyzing all of the *CFTR* coding sequence corroborated the early findings and, moreover, showed that about 30% of ICP patients are carrying at least one CF mutation, and among those 10% are compound heterozygote, with a CF-causing mutation and in *trans* a CFTR-RD mutated allele (Audrezet et al. 2002). These studies showed that the risk of ICP increases 6.3 times with a CF-causing mutation and 37 times with a CF-causing mutation plus a milder allele in *trans* (Cohn 2005; Cohn et al. 2005).

Besides the association of *CFTR* alleles and ICP, mutations in *PRSS1* and in its inhibitor *SPINK1* have been correlated with hereditary pancreatitis (Whitcomb et al. 1996; Witt et al. 2000). These studies have paved the way to the identification of hereditary pancreatitis disease-causing mutations as well as susceptibility factors, the most common being the N34S located in *SPINK1* (Chen et al. 1999). *SPINK1* encodes a serine peptidase inhibitor (kazal type 1). Gene–gene interactions have been documented in individuals who inherit both low penetrance SPINK1 variations and *CFTR* mutations in ICP. A total of 4% of ICP patients are *trans*-heterozygotes of *SPINK1/CFTR* mutations (Audrezet et al. 2002). Whether the coinheritance of *SPINK1* and *CFTR* variants/mutations is a *bona fide* example of digenic inheritance or interaction between a disease-causing gene and a genetic modifier is unclear in most cases (Chen 2009).

CFTR Mutations and Disseminated Bronchiectasis

Disseminated bronchiectasis (DB) is the third documented CFTR-related disorder. DB is a pathological description of lung disease characterized by an abnormal and irreversible dilatation of thick-walled bronchi. Affected areas are inflamed and easily collapsible resulting in air flow obstruction and impaired clearance of secretions (Pasteur et al. 2000).

An increased incidence of *CFTR* mutation has been found in bronchiectasis and at least one *CFTR* mutation was reported in 10%–50% of a series of patients in different studies (Girodon et al. 1997; Bombieri et al. 1998; Casals et al. 2004; King et al. 2004). Often in these patients, only one mutation is CF-causing and no particular mutations have been associated only with bronchiectasis. IVS8-5T has been observed at higher than normal frequency in patients with bronchiectasis, but the frequency is not as high as observed in patients with CBAVD. The frequency and nature of *CFTR* gene mutations reported in bronchiectasis patients differs considerably among publications. This is likely caused by the more heterogeneous nature of bronchiectasis with respect to the other CFTR-RD and also possibly how exhaustive the search was for mutations in the *CFTR* gene. At the moment, *CFTR* gene mutation screening is mostly performed for research purposes and it is not generally advisable that it be used for bronchiectasis screening, but rather for CF exclusion.

DIAGNOSIS BASED ON THE KNOWLEDGE OF *CFTR* MUTATIONS

To date, more than 1900 mutations have been identified in the *CFTR* gene and the list is probably incomplete. *CFTR* mutations vary in their frequency and distribution all over the world in the Caucasian population. Extraordinary improvements in molecular analysis techniques enable the development of high sensitivity mutation panels for most populations and ethnic groups that can routinely be used for diagnosis applications. Among those applications are prenatal diagnosis, newborn screening, as well as cascade screening in affected families. All are based on the precise knowledge of the mutations in the *CFTR* gene.

Prenatal Diagnosis of CF

Prenatal diagnosis (PD) is today offered to couples whose offspring have a one-in-four risk of CF, by directly analyzing in the fetus *CFTR* mutations previously identified in both parents. PD is carried out from chorionic villus sampling extracted at 10–12 weeks of gestation or from amniotic fluid at 16–17 weeks of gestation. In addition to the parents of an already affected child, PD is offered to one-in-four risk couples who have been identified through cascade screening in families, or following the detection of an echogenic bowel on ultrasound examinations during the pregnancy (Scotet et al. 2000, 2002, 2008).

The reproductive decision making of CF couples varies largely among countries. For example, in Brittany, an 18-year experience of prenatal diagnosis of CF was recently reported, and although the life expectancy of patients with CF has considerably improved, the great majority of couples chose pregnancy termination when PD indicated that the fetus had CF (95.9%) (Scotet et al. 2003a, 2008). The situation is similar in Australia where Sawyer et al. (2006) ana-

     Cite this article as *Cold Spring Harb Perspect Med* doi: 10.1101/cshperspect.a009480

lyzed the reproductive attitudes of a cohort of parents with a CF child and 82% of those couples declared they would use PD in a subsequent pregnancy. The situation is different in the U.S., where parents of CF children choose PD only for only 20%–25% of subsequent pregnancies (Mischler et al. 1998).

Newborn Screening for CF

Newborn screening (NBS) for CF has been largely implemented in Australia and in many countries in Europe and, more recently, in most of the states in the U.S. (Comeau et al. 2004). Improvement in outcome for CF patients diagnosed by newborn screening (as opposed to diagnosis based on symptoms) is not as clearly beneficial as for other conditions, such as phenylketonuria. However, the general consensus in Europe and in the U.S. is that there is sufficient evidence to support NBS for CF (Farrell et al. 2005). NBS for CF is based on elevated levels of immunoreactive trypsinogen in the blood of the newborn. Trypsinogen is present at high levels in the CF newborn because of pancreatic exocrine disease. As the screening test is not specific for CF, most NBS programs perform DNA testing to identify known *CFTR* gene mutations; this strategy is called IRT/DNA strategy (Farrell et al. 2008). The sensitivity of the IRT test coupled with a highly specific test utilizing a panel of the most common CF-causing mutations provides a two-tier test with excellent sensitivity and specificity (Scotet et al. 2000; Castellani et al. 2009). NBS for CF is the first example of the successful introduction of a panel of mutations in a screening test for a hereditary disease.

CONCLUDING REMARKS

A tremendous amount of information about the mutations in CFTR has been accumulated since the discovery of the gene 23 years ago. Although the challenge of curing the disease remains, knowledge of the genetic basis of CF provided immense insight into the pathophysiology of the disease. In addition, identification and characterization of the mutations in *CFTR* has aided the diagnosis of CF, facilitated ge-

netic counseling for families, and screening of newborns and carriers in the general population. Last but not least, it becomes more evident that the precise knowledge of the *CFTR* mutations will be a prerequisite for the rationale basis for mutation-specific therapies that are now imminent (Accurso et al. 2010).

ACKNOWLEDGMENTS

This work is supported by grants from the National Institutes of Health (NIDDK DK44003 to G.R.C.) and the Cystic Fibrosis Foundation (CFF-CUTTIN08A0, 09A0 and 11A0 to G.R.C.).

REFERENCES

Abeliovich D, Lavon IP, Lerer I, Cohen T, Springer C, Avital A, Cutting GR. 1992. Screening for five mutations detects 97% of cystic fibrosis (CF) chromosomes and predicts a carrier frequency of 1:29 in the Jewish Ashkenazi population. *Am J Hum Genet* **51:** 951–956.

Accurso FJ, Rowe SM, Clancy JP, Boyle MP, Dunitz JM, Durie PR, Sagel SD, Hornick DB, Konstan MW, Donaldson SH, et al. 2010. Effect of VX-770 in persons with cystic fibrosis and the G551D-CFTR mutation. *N Engl J Med* **363:** 1991–2003.

Ahmed N, Corey M, Forstner G, Zielenski J, Tsui LC, Ellis L, Tullis E, Durie P. 2003. Molecular consequences of cystic fibrosis transmembrane regulator (CFTR) gene mutations in the exocrine pancreas. *Gut* **52:** 1159–1164.

Anderson MP, Welsh MJ. 1992. Regulation by ATP and ADP of CFTR chloride channels that contain mutant nucleotide-binding domains. *Science* **257:** 1701–1704.

Anguiano A, Oates RD, Amos JA, Dean M, Gerrard B, Stewart C, Maher TA, White MB, Milunsky A. 1992. Congenital bilateral absence of the vas deferens—a primarily genital form of cystic fibrosis. *JAMA* **267:** 1794–1797.

Audrezet MP, Chen JM, Le Marechal C, Ruszniewski P, Robaszkiewicz M, Raguenes O, Quere I, Scotet V, Ferec C. 2002. Determination of the relative contribution of three genes—the cystic fibrosis transmembrane conductance regulator gene, the cationic trypsinogen gene, and the pancreatic secretory trypsin inhibitor gene—to the etiology of idiopathic chronic pancreatitis. *Eur J Hum Genet* **10:** 100–106.

Audrezet MP, Dabricot A, Le Marechal C, Ferec C. 2008. Validation of high-resolution DNA melting analysis for mutation scanning of the cystic fibrosis transmembrane conductance regulator (CFTR) gene. *J Mol Diagn* **10:** 424–434.

Bobadilla JL, Macek M, Fine JP, Farrell PM. 2002. Cystic fibrosis: A worldwide analysis of CFTR mutations—Correlation with incidence data and application to screening. *Hum Mutat* **19:** 575–606.

Bombieri C, Benetazzo M, Saccomani A, Belpinta F, Gile LS, Luisetti M, Pignatti PF. 1998. Complete mutational screening of the CFTR gene in 120 patients with pulmonary disease. *Hum Genet* **103**: 718–722.

Bombieri C, Claustres M, De Boeck K, Derichs N, Dodge J, Girodon E, Sermet I, Schwarz M, Tzetis M, Wilschanski M, et al. 2011. Recommendations for the classification of diseases as CFTR-related disorders. *J Cyst Fibros* **10**: S86–S102.

Casals T, De Gracia J, Gallego M, Dorca J, Rodriguez-Sanchon B, Ramos MD, Gimenez J, Cistero-Bahima A, Olveira C, Estivill X. 2004. Bronchiectasis in adult patients: An expression of heterozygosity for CFTR gene mutations? *Clin Genet* **65**: 490–495.

Castellani C, Bonizzato A, Pradal U, Filicori M, Foresta C, La Sala GB, Mastella G. 1999. Evidence of mild respiratory disease in men with congenital absence of the vas deferens. *Respir Med* **93**: 869–875.

Castellani C, Southern KW, Brownlee K, Dankert RJ, Duff A, Farrell M, Mehta A, Munck A, Pollitt R, Sermet-Gaudelus I, et al. 2009. European best practice guidelines for cystic fibrosis neonatal screening. *J Cyst Fibros* **8**: 153–173.

Champigny G, Imler JL, Puchelle E, Dalemans W, Gribkoff V, Hinnrasky J, Dott K, Barbry P, Pavirani A, Lazdunski M. 1995. A change in gating mode leading to increased intrinsic Cl⁻ channel activity compensates for defective processing in a mild form of the disease. *EMBO J* **14**: 2417–2423.

Chen JM. 2009. Chronic pancreatitis: Genetics and pathogenesis. *Annu Rev Genomics Hum Genet* **10**: 63–87.

Chen JM, Audrezet MP, Mercier B, Quere I, Ferec C. 1999. Exclusion of anionic trypsinogen and mesotrypsinogen involvement in hereditary pancreatitis without cationic trypsinogen gene mutations. *Scand J Gastroenterol* **34**: 831–832.

Cheng SH, Gregory RJ, Marshall J, Paul S, Souza DW, White GA, O'Riordan CR, Smith AE. 1990. Defective intracellular transport and processing of CFTR is the molecular basis of most cystic fibrosis. *Cell* **63**: 827–834.

Chillón M, Casals T, Mercier B, Bassas L, Lissens W, Silber S, Romey MC, Ruiz-Romero BS, Verlingue C, Claustres M, et al. 1995. Mutations in the cystic fibrosis gene in patients with congenital absence of the vas deferens. *N Engl J Med* **332**: 1475–1480.

Chu CS, Trapnell BC, Curristin S, Cutting GR, Crystal RG. 1993. Genetic basis of variable exon 9 skipping in cystic fibrosis transmembrane conductance regulator mRNA. *Nat Genet* **3**: 151–156.

Claustres M, Guittard C, Bozon D, Chevalier F, Verlingue C, Ferec C, Girodon E, Cazeneuve C, Bienvenu T, Lalau G, et al. 2000. Spectrum of CFTR mutations in cystic fibrosis and in congenital absence of the vas deferens in France. *Hum Mutat* **16**: 143–156.

Cohn JA. 2005. Reduced CFTR function and the pathobiology of idiopathic pancreatitis. *J Clin Gastroenterol* **39**: S70–S77.

Cohn JA, Friedman KJ, Noone PG, Knowles MR, Silverman LM, Jowell PS. 1998. Relation between mutations of the cystic fibrosis gene and idiopathic pancreatitis. *N Engl J Med* **339**: 653–658.

Cohn JA, Neoptolemos JP, Feng J, Yan J, Jiang Z, Greenhalf W, McFaul C, Mountford R, Sommer SS. 2005. Increased risk of idiopathic chronic pancreatitis in cystic fibrosis carriers. *Hum Mutat* **26**: 303–307.

Colin AA, Sawyer SM, Mickle JE, Oates RD, Milunsky A, Amos JA. 1996. Pulmonary function and clinical observations in men with congenital bilateral absence of the vas deferens. *Chest* **110**: 440–445.

Collaco JM, Blackman SM, McGready J, Naughton KM, Cutting GR. 2010. Quantification of the relative contribution of environmental and genetic factors to variation in cystic fibrosis lung function. *J Pediatr* **157**: 802–807.

Comeau AM, Parad RB, Dorkin HL, Dovey M, Gerstle R, Haver K, Lapey A, O'Sullivan BP, Waltz DA, Zwerdling RG, et al. 2004. Population-based newborn screening for genetic disorders when multiple mutation DNA testing is incorporated: A cystic fibrosis newborn screening model demonstrating increased sensitivity but more carrier detections. *Pediatrics* **113**: 1573–1581.

Cuppens H, Lin W, Jaspers M, Costes B, Teng H, Vankeerberghen A, Jorissen M, Droogmans G, Reynaert I, Goossens M, et al. 1998. Polyvariant mutant cystic fibrosis transmembrane conductance regulator genes: The polymorphic (TG)m locus explains the partial penetrance of the 5T polymorphism as a disease mutation. *J Clin Invest* **101**: 487–496.

Cystic Fibrosis Genetic Analysis Consortium. 1990. Worldwide survey of the δF508 mutation—Report from the Cystic Fibrosis Genetic Analysis Consortium (CFGAC). *Am J Hum Genet* **47**: 354–359.

De Braekeleer M, Ferec C. 1996. Mutations in the cystic fibrosis gene in men with congenital bilateral absence of the vas deferens. *Mol Hum Reprod* **2**: 677.

De Braekeleer M, Allard C, Leblanc JP, Simard F, Aubin G. 1997. Genotype-phenotype correlation in cystic fibrosis patients compound heterozygous for the A455E mutation. *Hum Genet* **101**: 208–211.

Denning GM, Anderson MP, Amara JF, Marshall J, Smith AE, Welsh MJ. 1992. Processing of mutant cystic fibrosis transmembrane conductance regulator is temperature-sensitive. *Nature* **358**: 761–764.

Dörk T, Dworniczak B, Aulehia-Scholz C, Wieczorek D, Böhm I, Mayerova A, Seydewitz H, Nieschlag E, Meschede D, Horst J, et al. 1997. Distinct spectrum of CFTR gene mutations in congenital absence of vas deferens. *Hum Genet* **100**: 367–377.

Dörk T, Macek M Jr, Mekus F, Tummler B, Tzountzouris J, Casals T, Krebsova A, Koudova M, Sakmaryova I, Macek M Sr, et al. 2000. Characterization of a novel 21-kb deletion, CFTRdele2,3(21 kb), in the CFTR gene: A cystic fibrosis mutation of Slavic origin common in Central and East Europe. *Hum Genet* **106**: 259–268.

Drumm ML, Wilkinson DJ, Smit LS, Worrell RT, Strong TV, Frizzell RA, Dawson DC, Collins FS. 1991. Chloride conductance expressed by δF508 and other mutant CFTRs in *Xenopus* oocytes. *Science* **254**: 1797–1799.

Dugueperoux I, Bellis G, Lesure JF, Renouil M, Flodrops H, De Braekeleer M. 2004. Cystic fibrosis at the Reunion Island (France): Spectrum of mutations and genotype-phenotype for the Y122X mutation. *J Cyst Fibros* **3**: 185–188.

Dumur V, Gervais R, Rigot JM, Lafitte JJ, Manouvrier S, Biserte J, Mazeman E, Roussel P. 1990. Abnormal distribution of CF δ F508 allele in azoospermic men with congenital aplasia of epididymis and vas deferens [letter]. *Lancet* **336:** 512.

European Working Group on CF Genetics. 1990. Gradient of distribution in Europe of the major CF mutation and of its associated haplotype. *Hum Genet* **85:** 436–445.

Farrell PM, Lai HJ, Kosorok MR, Laxova A, Green CG, Collins J, Hoffman G, Laessig R, Rock MJ, Splaingard ML. 2005. Evidence on improved outcomes with early diagnosis of cystic fibrosis through neonatal screening: Enough is enough! *J Pediatr* **147:** S30–S36.

Farrell PM, Rosenstein BJ, White TB, Accurso FJ, Castellani C, Cutting GR, Durie PR, LeGrys VA, Massie J, Parad RB, et al. 2008. Guidelines for diagnosis of cystic fibrosis in newborns through older adults: Cystic Fibrosis Foundation consensus report. *J Pediatr* **153:** S4–S14.

Feingold J, Guilloud-Bataille M. 1999. Genetic comparisons of patients with cystic fibrosis with or without meconium ileus. Clinical Centers of the French CF Registry. *Ann Genet* **42:** 147–150.

Frischmeyer PA, Dietz HC. 1999. Nonsense-mediated mRNA decay in health and disease. *Hum Mol Genet* **8:** 1893–1900.

Fulmer SB, Schwiebert EM, Morales MM, Guggino WB, Cutting GR. 1995. Two cystic fibrosis transmembrane conductance regulator mutations have different effects on both pulmonary phenotype and regulation of outwardly rectified chloride currents. *Proc Natl Acad Sci* **92:** 6832–6836.

Gan KH, Veeze HJ, van den Ouweland AMW, Halley DJJ, Scheffer H, Van Der Hout A, Overbeek SE, deJongste JC, Bakker W, Heijerman HGM. 1995. A cystic fibrosis mutation associated with mild lung disease. *N Engl J Med* **333:** 95–99.

Gasparini P, Borgo G, Mastella G, Bonizzato A, Dognini M, Pignatti PF. 1992. Nine cystic fibrosis patients homozygous for the CFTR nonsense mutation R1162X have mild or moderate lung disease. *J Med Genet* **29:** 558–562.

Girodon E, Cazaneuve C, Lebargy F, Chinet TC, Costes B, Ghanem N, Martin J, Lemay S, Scheid P, Housset B, et al. 1997. CFTR gene mutations in adults with disseminated bronchiectasis. *Eur J Hum Genet* **5:** 149–155.

Green DM, McDougal KE, Blackman SM, Sosnay PR, Henderson LB, Naughton KM, Collaco JM, Cutting GR. 2010. Mutations that permit residual CFTR function delay acquisition of multiple respiratory pathogens in CF patients. *Respir Res* **11:** 140.

Gregory RJ, Rich DP, Cheng SH, Souza DW, Paul S, Manavalan P, Anderson MP, Welsh MJ, Smith AE. 1991. Maturation and function of cystic fibrosis transmembrane conductance regulator variants bearing mutations in putative nucleotide-binding domains 1 and 2. *Mol Cell Biol* **11:** 3886–3893.

Groman JD, Hefferon TW, Casals T, Bassas L, Estivill X, Des GM, Guittard C, Koudova M, Fallin MD, Nemeth K, et al. 2004. Variation in a repeat sequence determines whether a common variant of the cystic fibrosis transmembrane conductance regulator gene is pathogenic or benign. *Am J Hum Genet* **74:** 176–179.

Guggino WB, Stanton BA. 2006. New insights into cystic fibrosis: Molecular switches that regulate CFTR. *Nat Rev Mol Cell Biol* **7:** 426–436.

Hamosh A, Trapnell BC, Zeitlin PL, Montrose-Rafizadeh C, Rosenstein BJ, Crystal RG, Cutting GR. 1991. Severe deficiency of CFTR mRNA carrying nonsense mutations R553X and W1316X in respiratory epithelial cells of patients with cystic fibrosis. *J Clin Invest* **88:** 1880–1885.

Hamosh A, King TM, Rosenstein BJ, Corey M, Levison H, Durie P, Tsui LC, McIntosh I, Keston M, Brock DJH, et al. 1992a. Cystic fibrosis patients bearing the common missense mutation Gly→Asp at codon 551 and the δF508 are indistinguishable from δF508 homozygotes except for decreased risk of meconium ileus. *Am J Hum Genet* **51:** 245–250.

Hamosh A, Rosenstein BJ, Cutting GR. 1992b. CFTR nonsense mutations G542X and W1282X associated with severe reduction of CFTR mRNA in nasal epithelial cells. *Hum Mol Genet* **1:** 542–544.

Hefferon TW, Groman JD, Yurk CE, Cutting GR. 2004. A variable dinucleotide repeat in the *CFTR* gene contributes to phenotype diversity by forming RNA secondary structures that alter splicing. *Proc Natl Acad Sci* **101:** 3504–3509.

Highsmith WE Jr, Burch LH, Zhou Z, Olsen JC, Boat TE, Spock A, Gorvoy JD, Quittell L, Friedman KJ, Silverman LM, et al. 1994. A novel mutation in the cystic fibrosis gene in patients with pulmonary disease but normal sweat chloride concentrations. *N Engl J Med* **331:** 974–980.

Highsmith WE Jr, Lauranell BH, Zhaoqing Z, Olsen JC, Strong TV, Smith T, Friedman KJ, Silverman LM, Boucher RC, Collins FS, et al. 1997. Identification of a splice site mutation (2789+5G→A) associated with small amounts of normal CFTR mRNA and mild cystic fibrosis. *Hum Mutat* **9:** 332–338.

Hinzpeter A, Aissat A, Sondo E, Costa C, Arous N, Gameiro C, Martin N, Tarze A, Weiss L, de Becdelievre A, et al. 2010. Alternative splicing at a NAGNAG acceptor site as a novel phenotype modifier. *PLoS Genet* **6:** e1001153.

Holsclaw DS, Perlmutter AD, Jockin H, Shwachman H. 1971. Genital abnormalities in male patients with cystic fibrosis. *J Urol* **106:** 568–574.

Jarvi K, McCallum S, Zielenski J, Durie P, Tullis E, Wilchanski M, Margolis M, Asch M, Ginzburg B, Martin S, et al. 1998. Heterogeneity of reproductive tract abnormalities in men with absence of the vas deferens: Role of cystic fibrosis transmembrane conductance regulator gene mutations. *Fertil Steril* **70:** 724–728.

Kerem BS, Buchanan JA, Durie P, Corey ML, Levison H, Rommens JM, Buchwald M, Tsui LC. 1989. DNA marker haplotype association with pancreatic sufficiency in cystic fibrosis. *Am J Hum Genet* **44:** 827–834.

Kerem E, Corey M, Kerem BS, Rommens J, Markiewicz D, Levison H, Tsui LC, Durie P. 1990. The relation between genotype and phenotype in cystic fibrosis—Analysis of the most common mutation (δF508). *N Engl J Med* **323:** 1517–1522.

Kerem E, Rave-Harel N, Augarten A, Madgar I, Nissim-Rafinia M, Yahav Y, Goshen R, Bentur L, Rivlin J, Aviram M, et al. 1997. A cystic fibrosis transmembrane conductance regulator splice variant with partial penetrance associated with variable cystic fibrosis presentations. *Am J Resp Crit Care Med* **155:** 1914–1920.

Kiesewetter S, Macek M Jr, Davis C, Curristin SM, Chu CS, Graham C, Shrimpton AE, Cashman SM, Tsui LC, Mickle J, et al. 1993. A mutation in the cystic fibrosis transmembrane conductance regulator gene produces different phenotypes depending on chromosomal background. *Nature Genet* **5:** 274–278.

King PT, Freezer NJ, Holmes PW, Holdsworth SR, Forshaw K, Sart DD. 2004. Role of CFTR mutations in adult bronchiectasis. *Thorax* **59:** 357–358.

Koch C, Cuppens H, Rainisio M, Madessani U, Harms H, Hodson M, Mastella G, Navarro J, Strandvik B, McKenzie S. 2001. European Epidemiologic Registry of Cystic Fibrosis (ERCF): Comparison of major disease manifestations between patients with different classes of mutations. *Pediatr Pulmonol* **31:** 1–12.

Kopito RR. 1999. Biosynthesis and degradation of CFTR. *Physiol Rev* **79:** S167–S173.

Kristidis P, Bozon D, Corey M, Markiewicz D, Rommens J, Tsui LC, Durie P. 1992. Genetic determination of exocrine pancreatic function in cystic fibrosis. *Am J Hum Genet* **50:** 1178–1184.

Kubesch P, Dörk T, Wulbrand U, Kälin N, Neumann T, Wulf B, Geerlings H, Weibbrodt H, von der Hardt H, Tümmler B. 1993. Genetic determinants of airways' colonization with *pseudomonas aeruginosa* in cystic fibrosis. *Lancet* **341:** 189–193.

Loirat F, Hazout S, Lucotte G. 1997. G542X as a probable Phoenician cystic fibrosis mutation. *Hum Biol* **69:** 419–425.

Macek M Jr, Mackova A, Hamosh A, Hilman BC, Selden RF, Lucotte G, Friedman KJ, Knowles MR, Rosenstein BJ, Cutting GR. 1997. Identification of common cystic fibrosis mutations in African-Americans with cystic fibrosis increases the detection rate to 75%. *Am J Hum Genet* **60:** 1122–1127.

Mak V, Jarvi K. 1996. The genetics of male infertility. *J Urol* **156:** 1245–1256.

McKone EF, Emerson SS, Edwards KL, Aitken ML. 2003. Effect of genotype on phenotype and mortality in cystic fibrosis: A retrospective cohort study. *Lancet* **361:** 1671–1676.

Mekus F, Ballmann M, Bronsveld I, Bijman J, Veeze H, Tummler B. 2000. Categories of δF508 homozygous cystic fibrosis twin and sibling pairs with distinct phenotypic characteristics. *Twin Res* **3:** 277–293.

Mercier B, Verlingue C, Lissens W, Silber S, Novelli G, Bonduelle M, Audrézet MP, Férec C. 1995. Is congenital bilateral absence of vas deferens a primary form of cystic fibrosis? Analyses of the CFTR gene in 67 patients. *Am J Hum Genet* **56:** 272–277.

Mickle JE, Cutting GR. 2000. Genotype-phenotype relationships in cystic fibrosis. *Med Clin North Am* **84:** 597–607.

Mickle JE, Macek M Jr, Fulmer-Smentek SB, Egan MM, Schwiebert E, Guggino W, Moss R, Cutting GR. 1998. A mutation in the cystic fibrosis transmembrane conductance regulator gene associated with elevated sweat chloride concentrations in the absence of cystic fibrosis. *Hum Mol Genet* **7:** 729–735.

Mischler EH, Wilfond BS, Fost N, Laxova A, Reiser C, Sauer CM, Makholm LM, Shen G, Feenan L, McCarthy C, et al. 1998. Cystic fibrosis newborn screening: Impact on reproductive behavior and implications for genetic counseling. *Pediatrics* **102:** 44–52.

Niksic M, Romano M, Buratti E, Pagani F, Baralle FE. 1999. Functional analysis of *cis*-acting elements regulating the alternative splicing of human CFTR exon 9. *Hum Mol Genet* **8:** 2339–2349.

Osborne L, Santis G, Schwarz M, Klinger K, Dork T, McIntosh I, Schwartz M, Nunes V, Macek M Jr, Reiss J, et al. 1992. Incidence and expression of the N1303K mutation of the cystic fibrosis (CFTR) gene. *Hum Genet* **89:** 653–658.

Park HW, Nam JH, Kim JY, Namkung W, Yoon JS, Lee JS, Kim KS, Venglovecz V, Gray MA, Kim KH, et al. 2010. Dynamic regulation of CFTR bicarbonate permeability by $[Cl^-]_i$ and its role in pancreatic bicarbonate secretion. *Gastroenterology* **139:** 620–631.

Pasteur MC, Helliwell SM, Houghton SJ, Webb SC, Foweraker JE, Coulden RA, Flower CD, Bilton D, Keogan MT. 2000. An investigation into causative factors in patients with bronchiectasis. *Am J Respir Crit Care Med* **162:** 1277–1284.

Patrick AE, Karamyshev A, Millen L, Thomas PJ. 2012. Alteration of CFTR transmembrane span integration by disease-causing mutations. *Mol Biol Cell* **22:** 4461–4471.

Rolfini R, Cabrini G. 1993. Nonsense mutation R1162X of the cystic fibrosis transmembrane conductance regulator gene does not reduce messenger RNA expression in nasal epithelial tissue. *J Clin Invest* **92:** 2683–2687.

Rosenstein BJ, Cutting GR. 1998. The diagnosis of cystic fibrosis: A consensus statement. *J Pediatr* **132:** 589–595.

Sawyer SM, Cerritelli B, Carter LS, Cooke M, Glazner JA, Massie J. 2006. Changing their minds with time: A comparison of hypothetical and actual reproductive behaviors in parents of children with cystic fibrosis. *Pediatrics* **118:** e649–e656.

Schreiber R, Hopf A, Mall M, Greger R, Kunzelmann K. 1999. The first-nucleotide binding domain of the cystic fibrosis transmembrane conductance regulator is important for inhibition of the epithelial Na^+ channel. *Proc Natl Acad Sci* **96:** 5310–5315.

Schwartz M, Anvret M, Claustres M, Eiken HG, Eiklid K, Schaedel C, Stolpe L, Tranebjaerg L. 1994. 394delTT: A Nordic cystic fibrosis mutation. *Hum Genet* **93:** 157–161.

Schwiebert EM, Egan ME, Hwang T-H, Fulmer SB, Allen SS, Cutting GR, Guggino WB. 1995. CFTR regulates outwardly rectifying chloride channels through an autocrine mechanism involving ATP. *Cell* **81:** 1–20.

Scotet V, De Braekeleer M, Roussey M, Rault G, Parent P, Dagorne M, Journel H, Lemoigne A, Codet JP, Catheline M, et al. 2000. Neonatal screening for cystic fibrosis in Brittany, France: Assessment of 10 years' experience and impact on prenatal diagnosis. *Lancet* **356:** 789–794.

Scotet V, De Braekeleer M, Audrezet MP, Quere I, Mercier B, Dugueperoux I, Andrieux J, Blayau M, Ferec C. 2002. Prenatal detection of cystic fibrosis by ultrasonography: A retrospective study of more than 346,000 pregnancies. *J Med Genet* **39:** 443–448.

Scotet V, Audrezet MP, Roussey M, Rault G, Blayau M, DeBraekeleer M, Ferec C. 2003a. Impact of public health strategies on the birth prevalence of cystic fibrosis in Brittany, France. *Hum Genet* **113:** 280–285.

 Cite this article as *Cold Spring Harb Perspect Med* doi: 10.1101/cshperspect.a009480

Scotet V, Barton DE, Watson JB, Audrezet MP, McDevitt T, McQuaid S, Shortt C, De Braekeleer M, Ferec C, Le Marechal C. 2003b. Comparison of the CFTR mutation spectrum in three cohorts of patients of Celtic origin from Brittany (France) and Ireland. *Hum Mutat* **22**: 105.

Scotet V, Dugueperoux I, Audrezet MP, Blayau M, Boisseau P, Journel H, Parent P, Ferec C. 2008. Prenatal diagnosis of cystic fibrosis: The 18-year experience of Brittanny (western France). *Prenat Diagn* **28**: 197–202.

Seibert FS, Linsdell P, Loo TW, Hanrahan JW, Clarke DM, Riordan JR. 1996a. Disease-associated mutations in the fourth cytoplasmic loop of cystic fibrosis transmembrane conductance regulator compromise biosynthetic processing and chloride channel activity. *J Biol Chem* **271**: 15139–15145.

Seibert FS, Linsdell P, Loo TW, Hanrahan JW, Riordan JR, Clarke DM. 1996b. Cytoplasmic loop three of cystic fibrosis transmembrane conductance regulator contributes to regulation of chloride channel activity. *J Biol Chem* **271**: 27493–27499.

Seibert FS, Jia Y, Mathews CJ, Hanrahan JW, Riordan JR, Loo TW, Clarke DM. 1997. Disease-associated mutations in cytoplasmic loops 1 and 2 of cystic fibrosis transmembrane conductance regulator impede processing or opening of the channel. *Biochemistry* **36**: 11966–11974.

Serohijos AW, Hegedus T, Aleksandrov AA, He L, Cui L, Dokholyan NV, Riordan JR. 2008. Phenylalanine-508 mediates a cytoplasmic-membrane domain contact in the CFTR 3D structure crucial to assembly and channel function. *Proc Natl Acad Sci* **105**: 3256–3261.

Sharer N, Schwarz M, Malone G, Howarth A, Painter J, Super M, Braganza J. 1998. Mutations of the cystic fibrosis gene in patients with chronic pancreatitis. *N Engl J Med* **339**: 645–652.

Sheppard DN, Rich DP, Ostedgaard LS, Gregory RJ, Smith AE, Welsh MJ. 1993. Mutations in CFTR associated with mild-disease-form Cl⁻ channels with altered pore properties. *Nature* **362**: 160–164.

Sheppard DN, Ostedgaard LS, Winter MC, Welsh MJ. 1995. Mechanism of dysfunction of two nucleotide binding domain mutations in cystic fibrosis transmembrane conductance regulator that are associated with pancreatic sufficiency. *EMBO J* **14**: 876–883.

Shoshani T, Augarten A, Gazit E, Bashan N, Yahav Y, Rivlin Y, Tal A, Seret H, Yaar L, Kerem E, et al. 1992. Association of a nonsense mutation (W1282X), the most common mutation in the Ashkenazi Jewish cystic fibrosis patients in Israel, with severe disease presentation. *Am J Hum Genet* **50**: 222–228.

Silvis MR, Picciano JA, Bertrand C, Weixel K, Bridges RJ, Bradbury NA. 2003. A mutation in the cystic fibrosis transmembrane conductance regulator generates a novel internalization sequence and enhances endocytic rates. *J Biol Chem* **278**: 11554–11560.

Smit LS, Strong TV, Wilkinson DJ, Macek M Jr, Mansoura MK, Wood DL, Cole JL, Cutting GR, Cohn JA, Dawson DC, et al. 1995. Missense mutation (G480C) in the CFTR gene associated with protein mislocalization but normal chloride channel activity. *Hum Mol Genet* **4**: 269–273.

Sosnay PR, Castellani C, Corey M, Dorfman R, Zielenski J, Karchin R, Penland CM, Cutting GR. 2011. Evaluation of the disease liability of CFTR variants. *Methods Mol Biol* **742**: 355–372.

Stanke F, Ballmann M, Bronsveld I, Dork T, Gallati S, Laabs U, Derichs N, Ritzka M, Posselt HG, Harms HK, et al. 2008. Diversity of the basic defect of homozygous CFTR mutation genotypes in humans. *J Med Genet* **45**: 47–54.

The Cystic Fibrosis Genotype-Phenotype Consortium. 1993. Correlation between genotype and phenotype in patients with cystic fibrosis. *N Engl J Med* **329**: 1308.

Vanscoy LL, Blackman SM, Collaco JM, Bowers A, Lai T, Naughton K, Algire M, McWilliams R, Beck S, Hoover-Fong J, et al. 2007. Heritability of lung disease severity in cystic fibrosis. *Am J Respir Crit Care Med* **175**: 1036–1043.

Watson MS, Cutting GR, Desnick RJ, Driscoll DA, Klinger K, Mennuti M, Palomaki GE, Popovich BW, Pratt VM, Rohlfs EM, et al. 2004. Cystic fibrosis population carrier screening: 2004 revision of American College of Medical Genetics mutation panel. *Genet Med* **6**: 387–391.

Welsh MJ, Smith AE. 1993. Molecular mechanisms of CFTR chloride channel dysfunction in cystic fibrosis. *Cell* **73**: 1251–1254.

Whitcomb DC, Gorry MC, Preston RA, Furey W, Sossenheimer MJ, Ulrich CD, Martin SP, Gates LJ Jr, Amann ST, Toskes PP, et al. 1996. Hereditary pancreatitis is caused by a mutation in the cationic trypsinogen gene. *Nat Genet* **14**: 141–145.

Will K, Dörk T, Stuhrmann M, von der Hardt H, Ellemunter H, Tümmler B, Schmidtke J. 1995. Transcript analysis of CFTR nonsense mutations in lymphocytes and nasal epithelial cells from cystic fibrosis patients. *Hum Mutat* **5**: 210–220.

Wilschanski M, Zielenski J, Markiewicz D, Tsui LC, Corey M, Levison H, Durie PR. 1995. Correlation of sweat chloride concentration with classes of the cystic fibrosis transmembrane conductance regulator gene mutations. *J Pediatr* **127**: 705–710.

Wilschanski M, Corey M, Durie P, Tullis E, Bain J, Asch M, Ginzburg B, Jarvi K, Buckspan M, Hartwick W. 1996. Diversity of reproductive tract abnormalities in men with cystic fibrosis. *JAMA* **276**: 607–608.

Witt H, Luck W, Hennies HDC, Classen M, Kage A, Lass U, Landt O, Becker M. 2000. Mutations in the gene encoding the serine protease inhibitor, Kazal Type 1 are associated with chronic pancreatitis. *Nat Genet* **25**: 213–216.

Zielenski J, Patrizio P, Corey M, Handelin B, Markiewicz D, Asch R, Tsui LC. 1995. CFTR gene variant for patients with congenital absence of vas deferens. *Am J Hum Genet* **57**: 958–960.

The CFTR Ion Channel: Gating, Regulation, and Anion Permeation

Tzyh-Chang Hwang[1] and Kevin L. Kirk[2]

[1]Dalton Cardiovascular Research Center, University of Missouri-Columbia, Columbia, Missouri 65211

[2]Department of Cell, Developmental and Integrative Biology and Gregory Fleming James Cystic Fibrosis Research Center, University of Alabama at Birmingham, Birmingham, Alabama 35294

Correspondence: klkirk@uab.edu

Cystic fibrosis transmembrane conductance regulator (CFTR) is an ATP-gated anion channel with two remarkable distinctions. First, it is the only ATP-binding cassette (ABC) transporter that is known to be an ion channel—almost all others function as transport ATPases. Second, CFTR is the only ligand-gated channel that consumes its ligand (ATP) during the gating cycle—a consequence of its enzymatic activity as an ABC transporter. We discuss these special properties of CFTR in the context of its evolutionary history as an ABC transporter. Other topics include the mechanisms by which CFTR gating is regulated by phosphorylation of its unique regulatory domain and our current view of the CFTR permeation pathway (or pore). Understanding these basic operating principles of the CFTR channel is central to defining the mechanisms of action of prospective cystic fibrosis drugs and to the development of new, rational treatment strategies.

Being a member of the ABC (ATP-binding cassette) protein superfamily (Riordan et al. 1989), cystic fibrosis transmembrane conductance regulator (CFTR) shows the characteristic topology of two transmembrane domains (TMDs), each conjoined to a cytoplasmic nucleotide-binding domain (NBD1 and NBD2). Unique to CFTR, however, is a regulatory (R) domain connecting the two pseudo-symmetrical halves, each composed of a TMD and an NBD (Fig. 1A). With very few exceptions, members of the ABC transporter family harvest the energy of ATP hydrolysis to drive an uphill movement of a wide variety of substrates into (importers) or out of (exporters) the cell. ABC importers, present exclusively in prokaryotes, are important for up-

take of nutrients (Rees et al. 2009). ABC exporters, on the contrary, are found in all kingdoms of life. In humans, phytogenetic analysis classifies 48 ABC proteins into seven subfamilies (ABCA–ABCG) and CFTR belongs to the ABCC subfamily (Dean and Annilo 2005).

While other members of the ABCC subfamily use the free energy of ATP hydrolysis to export large hydrophobic anions (Leier et al. 1994; Loe et al. 1996), CFTR is a bona fide ATP-gated ion channel (Anderson et al. 1991a; Nagel et al. 1992), carrying out the function of passive diffusion of chloride ions (Bear et al. 1992). However, phosphorylation of the R domain by the cAMP-dependent protein kinase (PKA) is a prerequisite for channel gating by ATP (see

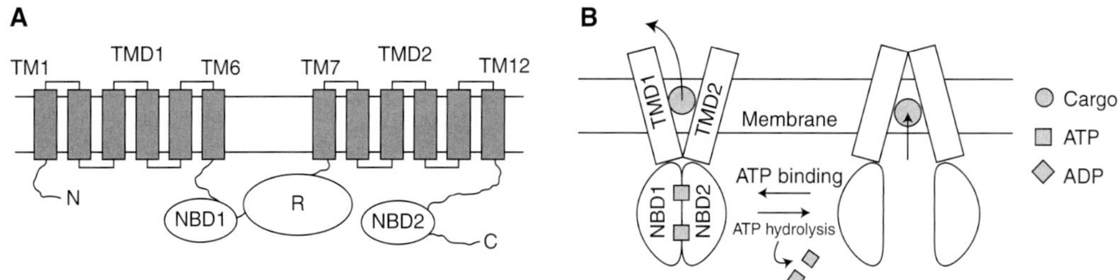

Figure 1. CFTR topology and alternating access model for ABC transporters. (*A*) Simplified topological model emphasizing the domain structure of a CFTR chloride channel, including cytoplasmic amino (N) and carboxyl (C) termini, two nucleotide-binding domains (NBD1, NBD2), the regulatory (R) domain, and the predicted transmembrane domains (TMD1 and TMD2). (*B*) Alternating access model for the ABC transporter that extrudes its substrate (i.e., exporter). In the apo state of exporters, the configuration of the TMDs is inward facing. The two NBDs are far apart. Once ATP binds to the Walker A and B motifs of the "head" subdomain, two NBDs dimerize in a head-to-tail configuration and this NBD dimerization provides the driving force to flip-flop TMDs so that the substrate binding site is now outward facing.

below for details). Crystallographic studies of several ABC proteins reveal at least two conformations of the protein: "inward-facing" TMDs with monomeric ATP-free NBDs and "outward-facing" TMDs with dimeric ATP-bound NBDs (Rees et al. 2009). A popular model has since been proposed (Fig. 1B). For ABC exporters, in the absence of ATP, the substrate-binding site in TMDs is accessible from the cytoplasmic side of the membrane; ATP binding triggers dimerization of the two NBDs, which is accompanied by conformational changes of TMDs so that the substrate-binding site now faces the extracellular environment.

Although the cargos for CFTR and ABC exporters differ immensely, a few pieces of evidence support the close evolutionary relationship between CFTR and ABC exporters. First, their TMDs conform to a "6 + 6" topological fold (Fig. 1A). Second, crystallographic studies of both prokaryotic and eukaryotic ABC exporters have revealed that TM6 and TM12 line the substrate translocation pathway (Dawson and Locher 2006; Ward et al. 2007; Aller et al. 2009). Indeed, numerous experimental data have indicated that both TM6 and TM12 form part of the chloride permeation pathway in the CFTR channel (see below). Third, CFTR's two NBDs show canonical architectures including Walker A and B sequences and an ABC signature motif conserved among all ABC proteins (Lewis et al.

2004, 2005) and, just like other ABC proteins, ATP binding to CFTR's NBDs induces dimerization of the two NBDs in a head-to-tail configuration (Mense et al. 2006). Finally, a wealth of data in the literature have indicated the presence of an anion-binding site in the CFTR pore that can accommodate large hydrophobic anionic channel blockers as if this apparent imperfect design of the pore represents an evolutionary vestige of a primordial ABC exporter evolving into an ion channel (Chen and Hwang 2008).

Based on the ATP-bound configuration of ABC exporters, several homology models of CFTR's "open state" were composed (Serohijos et al. 2008; Alexander et al. 2009; Mornon et al. 2009; Norimatsu et al. 2012). However, these homology models, like their counterparts of ABC exporters, all lack an uninterrupted pathway for ion permeation. Although these structural models make sense for an active transporter, which prohibits simultaneous exposure of the substrate translocation pathway to both sides of the membrane, a lack of continuous diffusive pathway for a presumed "open" channel conformation is troublesome for an ion channel. Several articles published in recent years have proposed that the gate of CFTR is located at the extracellular end of the pore; whereas the gate at the cytoplasmic end is degraded so that a transporter protein can metamorphose into an ion channel (Gadsby et al. 2006; Chen and Hwang

2008; Jordan et al. 2008). This proposition is supported by a very recent report providing strong evidence that the gate of CFTR is indeed located more to the extracellular part of the pore (Bai et al. 2011; but cf. El Hiani and Linsdell 2010; Wang and Linsdell 2012).

GATING OF CFTR CHANNELS BY ATP BINDING AND HYDROLYSIS

CFTR channels typically open when ATP docks in the ATP-binding site on each NBD. Unlike classical ligand-gated ion channels (e.g., the nicotinic acetylcholine receptor), CFTR's ligand ATP is consumed during the gating cycle to mediate channel closure. This phenomenon, unique to CFTR, likely reflects the evolutionary relationship of CFTR; an ABC transporter turned an ion channel (Gadsby et al. 2006; Chen and Hwang 2008). It is also interesting that, although NBD2, with all the conserved motifs, is capable of catalyzing ATP hydrolysis, NBD1, in contrast, is degenerated in the head subdomain with several substitutions that abolish its ATPase activity (Aleksandrov et al. 2002; Basso et al. 2003; Stratford et al. 2007). Because Mg^{2+} is an essential cofactor for enzyme-catalyzed ATP hydrolysis, this evolutionary relationship also provides an explanation for why it is MgATP, not ATP, which gates CFTR effectively (Dousmanis et al. 2002). Mg^{2+} not only helps in ATP binding by interacting with the Walker B aspartate, but is also essential for ATP hydrolysis that plays a major role in closing the channel (Gunderson and Kopito 1994; Hwang et al. 1994; Ramjeesingh et al.1999; Zeltwanger et al. 1999; Vergani et al. 2005; Stratford et al. 2007).

The strategies Mother Nature employs to stabilize ATP in the ATP-binding site of an NBD monomer were revealed by crystallographic studies of NBDs from ABC proteins, including those of CFTR (Lewis et al. 2004, 2005). First, upstream of the Walker A motif, the side chain of an aromatic amino acid (CFTR: NBD1, tryptophan 401 or W401; NBD2, tyrosine 1219 or Y1219) forms a π-electron stacking interaction with the adenine ring of ATP. Second, the positive charge of the Walker A lysine (CFTR: NBD1, K464; NBD2, K1250) forms an electrostatic in-

teraction with the negative charge of the γ-phosphate. Third, the side chain of the Walker B aspartate (D572 in NBD1 and D1370 in NBD2 of CFTR) coordinates the Mg^{2+} ion. These structural studies of ABC proteins in the past decade (Hung et al. 1998; Diederichs et al. 2000; Gaudet and Wiley 2001; Karpowich et al. 2001; Yuan et al. 2001; Smith et al. 2002; Chen et al. 2003; Schmitt et al. 2003; Verdon et al. 2003; Lewis et al. 2004, 2005) have provided rich molecular details that can be applied to facilitate functional exploration of CFTR. For example, Zhou et al. (2006) used site-directed mutagenesis and high-resolution single-channel recording to investigate the contribution of the aromatic residues W401 in NBD1 and Y1219 in NBD2 that interact with the adenine ring of ATP. Their data suggest that the interaction of ATP with NBD2 is critical for channel opening by ATP, whereas the interaction of ATP with NBD1 is less important for channel opening (Zhou et al. 2006). Similar functional asymmetry of CFTR's two NBDs was observed when the Walker A lysine residues in NBD1 (K464) and 2 (K1250) were mutated (Powe et al. 2002; Vergani et al. 2003). The critical role of this π-electron stacking interaction between ATP and the binding pocket also explains why pyrophosphate (PPi), a ligand without the adenine ring for this stacking interaction, shows a very low apparent affinity as a channel opener (Tsai et al. 2009). Lately, capitalizing on both structural information and bioinformatics, Szollosi et al. (2010) were able to experimentally show an induced fit mechanism following ATP binding to the head subdomain of NBD2 during CFTR gating.

For channel closing, it has long been proposed that hydrolysis of ATP is a prerequisite for fast channel closure as nonhydrolyzable ATP analogs such as AMP–PNP or pyrophosphate (PPi) lock the CFTR Cl^- channel in a stable open state for tens of seconds in the presence of ATP (Gunderson and Kopito 1994; Hwang et al. 1994). Moreover, this locked-open phenotype is also observed with ATP as a ligand when hydrolysis is abolished by mutations at residue K1250 or glutamate 1371 (E1371), both essential for effective ATP hydrolysis (Gunderson and Kopito 1995; Ramjeesingh et al. 1999; Zeltwanger et al. 1999; Powe et al. 2002; Bompadre et al. 2005b; Vergani et al. 2005; Stratford et al. 2007). But is hydrolysis per se closing the gate as suggested by a thermodynamic study of CFTR gating (Csanády et al. 2006)? Or is it the release of the hydrolytic products that is associated with gate closure (Wang et al. 2009)? These questions remain to be answered.

Roles of NBD Dimerization in CFTR Gating

Perhaps the most intriguing mystery about CFTR gating is how the channel uses ATP binding and hydrolysis to drive conformational changes. Using mutant cycle analysis, Vergani and her colleagues provided convincing evidence supporting the idea that ATP binding induces the formation of an NBD dimer, which coincides with gate opening (Vergani et al. 2005). Cross-linking studies also showed that CFTR's two NBDs may form a canonical dimer seen in other members of the ABC proteins (Mense et al. 2006). If dimerized NBDs represent the open channel configuration, it seems natural to contemplate the idea that separation of the

dimer closes the gate (Vergani et al. 2005), based on the analogous structural information from other ABC proteins (Fig. 1B). Two recent studies provide experimental evidence contradicting this popular theory.

Using nonhydrolyzable ATP analogs (e.g., PPi or AMP–PNP) as ligands, Tsai et al. (2009) showed that contrary to ATP-gated channels that show an open time of ~400 ms (Zeltwanger et al. 1999), the open time for PPi-opened channels is ~30 sec (inferred from deactivation rates) when this ligand is applied shortly after removal of ATP, but the open time becomes much shorter (~1.5 sec) if PPi is applied minutes after ATP washout (Tsai et al. 2009). Thus, CFTR channels, once primed by ATP, respond to PPi differently depending on the washout time. The two open states distinguished by their respective lifetime of 1.5 sec and 30 sec demand the presence of two closed states C_1 and C_2, from which PPi exerts its effects (Fig. 2).

As the stability of this C_2 closed state is decreased by mutations at the composite site 1 (formed by the head subdomain of NBD1 and the tail subdomain of NBD2), but increased by

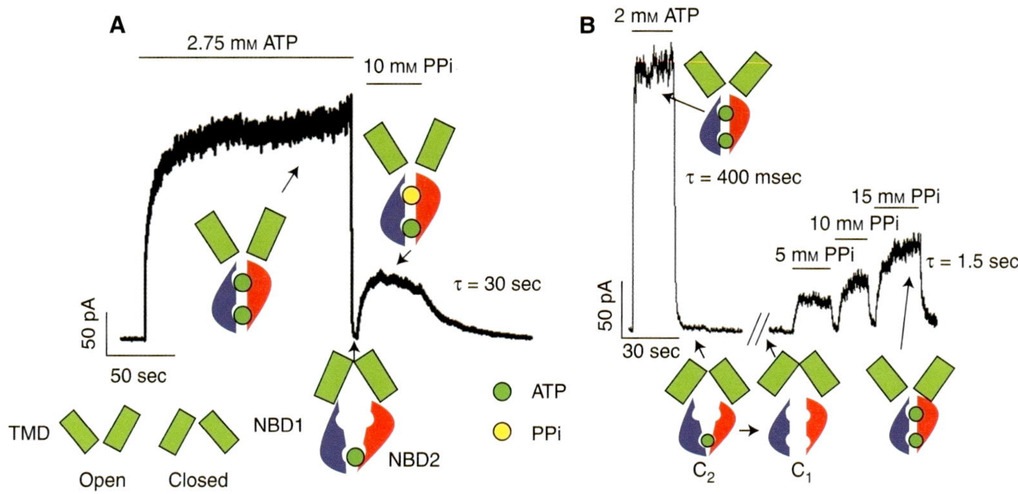

Figure 2. State-dependent modulation of CFTR gating by PPi. (*A*) A continuous macroscopic recording of WT-CFTR in an inside-out patch showing a slow response to PPi 10 sec after removal of ATP. However, once the channels were opened by PPi, they closed very slowly with a relaxation time constant (τ) of ~30 sec. (*B*) Three minutes after ATP washout, CFTR channels responded to PPi in a dose-dependent manner, but with a much shorter open time ($\tau = 1.5$ sec). However, robust response to PPi can be restored by ATP priming again (not shown). Cartoons depict proposed configurations of NBDs. (Modified from Tsai et al. 2009; reprinted, with permission, from the author.)

   Cite this article as *Cold Spring Harb Perspect Med* doi: 10.1101/cshperspect.a009498

priming the channel with a high-affinity ATP analog, N^6-phenylethyl-ATP (or P-ATP), it was proposed that the C_2 closed state contains a partial NBD dimer with the composite site 1 remaining connected by the bound ATP, whereas the composite site 2 is separated after ATP is hydrolyzed. This hypothesis predicts that if one would replace both ATP molecules bound at the NBD dimer interface, two distinct time courses of ligand substitution should occur. Indeed, in a follow-up study, Tsai et al. (2010) showed that on switching the ligand from ATP to P-ATP, a ligand with a higher potency as well as an increased efficacy compared to ATP (Zhou et al. 2005), open probability (P_o) increases in two steps: an immediate increase of the P_o with an increase of the opening rate, and a ~30-sec delayed increase of the P_o with an increased open time (Tsai et al. 2010). Thus, ATP remains sandwiched between the head of NBD1 and the tail of NBD2 for tens of seconds while the channel has undergone several rounds of the opening/closing gating cycle.

These studies lead to a new model for the coupling mechanism between NBD motions and opening/closing of the gate in TMDs (Fig. 3). This model envisages "limited motions" of the NBD dimer in the completion of one catalysis (and gating) cycle. Of note, this model is supported by a report that employed a completely independent approach (Szollosi et al. 2011). The significance of this new model lies in the fact that all members of the ABCC subfamily possess only one catalysis-competent site in their NBDs. Thus, if these limited motions were shared features of all members of this subfamily, one would have to ask if a complete disengagement of the NBDs is required for all other ABC exporters that hold two catalysis-competent sites (Fig. 1B).

Future Perspectives Regarding ATP-Dependent Channel Gating

The discussion above mainly focuses on gating of CFTR by ATP-induced NBD dimerization and hydrolysis-triggered partial or complete separation of the NBD dimer. The model in Figure 3, derived from years of biochemical and biophys-

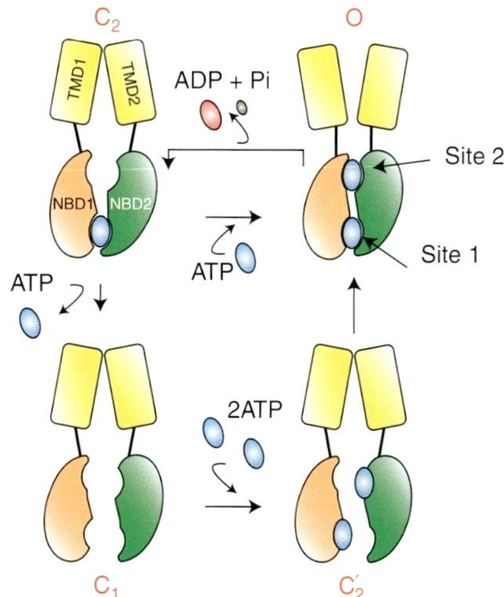

Figure 3. A simplified model for CFTR gating. The diagram illustrates dominant gating transitions ($C_2 \leftrightarrow O$) and a second gating cycle ($C_1 \rightarrow C_2' \rightarrow O \rightarrow C_2 \rightarrow C_1$) involving complete disengagement of the two NBDs. It should be noted that the transition from O to C_2 likely contains many transitional states that await identification. The model also implicates a closed state wherein the two NBDs are separated with only one ATP molecule bound to NBD1. (Adapted from Tsai et al. 2010; reprinted, with permission, from the author.)

ical studies of CFTR gating, is still an oversimplification. For example, there must be several other states existing when the channel is on its way to closure (i.e., $O \rightarrow C_2$) because ATP will be hydrolyzed first followed by opening of the dimer interface for the dissociation of hydrolytic products, ADP and Pi. As the $O \rightarrow C_2$ transition proceeds at a fairly fast rate and hence all those putative states are presumably very unstable, capturing these states with any means will be a challenge. Nevertheless, an early report by Gunderson and Kopito (1995) may shed some light on how to pursue in this direction. By scrutinizing their single-channel data, they were able to discern two conductance levels and an uneven distributions of transitions between these two open channel levels, indicating a violation of microscopic reversibility and hence a necessity of an input of free energy—in this case, from ATP hydrolysis (Gunderson and Kopito 1995). This idea of an infusion of free energy in driving gating transitions of CFTR is further supported by a recent detailed analysis of single-channel open time distributions (Csanády et al. 2010). These investigators showed a paucity of short-lived events that results in an unusual double exponential distributions with a negative component, indicative of a violation of microscopic reversibility. It is interesting to note that in these two pioneering investigations the data imply a strict coupling between the ATP hydrolysis cycle and the gating cycle. A one-to-one stoichiometry between the hydrolysis of one ATP molecule and one opening-closing transition fits nicely with the gating motion proposed for ABC transporters: one TMD-NBD complex moves synchronously on ATP-induced NBD dimerization and hydrolysis-triggered separation of the NBD dimer (Ward et al. 2007; Khare et al. 2009).

However, the concept of an obligatory coupling of NBD and TMD motions is not easily reconciled with other experimental data. In particular, it is known that CFTR channels can open, albeit with a very low frequency, in a complete absence of ATP (Bompadre et al. 2005a). This ATP-independent gating is readily seen with mutants with dysfunctional site 2 (e.g., the G551D mutation) (Bompadre et al. 2007), or interestingly constructs that lack the entire NBD2 (Cui et al. 2007; Wang et al. 2007). Furthermore, gain-of-function (GOF) mutations that enhance ATP-free CFTR gating and increase the otherwise low activities of constructs that are insensitive to ATP (G551D and NBD-deletion mutants) have been produced (Szollosi et al. 2010; Wang et al. 2010). ATP-independent gating also can be strongly enhanced by chemically modifying cysteines placed at specific locations in TM 6 (Bai et al. 2010). These results raise the question of whether NBD/TMD coupling is as tight as proposed.

The aforementioned results also blur the distinction between CFTR channel gating by ATP and the gating of a conventional ligand-gated channel by reversible ligand binding and unbinding (e.g., a nicotinic acetylcholine channel). Ligand binding to the latter increases the probability of channel opening but is not absolutely required for the channel to open (Changeux and Edelstein 2005; Purohit and

Auerbach 2009). Unliganded openings are detectable and GOF mutations that increase this activity are possible. GOF mutations for conventional ligand-gated channels typically locate along the axis that links the ligand-binding site to the pore. The GOF mutations/modifications that have been discovered for CFTR locate near the NBD dimer interface (Szollosi et al. 2010), along the cytosolic loops that connect the NBDs to the TMDs (Wang et al. 2010) and within a TM that has been argued to line the pore (Bai et al. 2010). Thus, the CFTR gating mechanism appears to share features with the gating of conventional ligand-gated channels; notably, unliganded openings and GOF mutations along the axis that links the ligand-binding domains to the pore. Reconciling these properties of CFTR gating with the important role of ATP hydrolysis in channel closing will be essential to developing a complete picture of the CFTR gating process. It is thus safe to say that new approaches are needed to tackle this coupling mechanism that bears ramifications not only on CFTR, but also on ABC proteins at large. In addition, we foresee that understanding CFTR gating at a molecular level will shed light on the mechanism of action for therapeutic reagents such as Vx-770 (Kalydeco or ivacaftor), a recently FDA-approved CFTR potentiator for treating CF patients carrying the G551D mutation (Van Goor et al. 2009; Accurso et al. 2010; Ramsey et al. 2011).

CFTR REGULATION BY PKA PHOSPHORYLATION

The main stimulus of CFTR activity in vivo is its phosphorylation by cyclic nucleotide-dependent protein kinases (e.g., protein kinase A, or PKA) (Sheppard and Welsh 1999). Here we discuss the effects of PKA phosphorylation on channel gating, which is the best characterized mode of CFTR regulation by phosphorylation. Protein kinase C (Chappe et al. 2004) and AMP kinase (King et al. 2009) stimulate and inhibit CFTR activity, respectively, by less well understood mechanisms.

Unphosphorylated CFTR channels open at very low rates, if at all. Channel opening rate in an excised membrane patch is increased by

 Cite this article as *Cold Spring Harb Perspect Med* doi: 10.1101/cshperspect.a009498

adding PKA catalytic subunit in the presence of MgATP (Mathews et al. 1998; Wang et al. 2000). Burst duration also increases under highly phosphorylating conditions which implies that one or more PKA sites modulates channel closing (Csanády et al. 2000). Most of the important PKA sites reside within the large R domain between NBD1 and TM7 (Fig. 4). This approximately 150 residue domain contains eight dibasic PKA sites plus a number of monobasic sites. At least five of the PKA sites in the R domain (S700, S737, S768, S795, and S813) plus another site in the nearby distal portion of NBD1 (S660) are phosphorylated in vivo (Cheng et al. 1991; Hegedus et al. 2009). In general, phosphorylation of the different PKA sites has additive effects on channel activity with no one site being essential. Fifteen sites must be eliminated to create CFTR channels that are completely PKA-insensitive (Seibert et al. 1999; Hegedus et al. 2009). Two PKA sites (S737 and S768) are argued to be inhibitory on the basis of the observation that their disruption increases channel activity (Wilkinson et al. 1997; Csanády et al. 2005). These sites also appear to be phosphorylated by AMP kinase which inhibits CFTR activity possibly under hypoxic conditions (King et al. 2009).

How does the R domain regulate gating? The consensus view is that the unphosphorylated R domain inhibits channel opening, an effect that is relieved by its phosphorylation.

Rich et al. (1991) discovered that channels that lacked large portions of the R domain were more active in the absence of PKA stimulation. Csanády et al. (2000) subsequently showed that split CFTR molecules that lack the R domain and distal NBD1 (missing residues 634–836) formed channels with opening rates in the absence of PKA that approached maximal opening rates for fully phosphorylated wild-type channels. These findings confirm that channel opening is strongly inhibited by the unphosphorylated R domain (perhaps with assistance from distal NBD1). How phosphorylation relieves this inhibition is unclear but the increased negative charge at these sites on phosphorylation appears to be relevant (i.e., replacing serines in PKA sites with acidic residues creates channels that open more frequently in the absence of PKA [Rich et al. 1993]).

Proposed Mechanisms of CFTR Regulation by PKA

We consider three mechanisms: (1) R domain phosphorylation regulates NBD dimerization; (2) R domain phosphorylation affects gating independently of NBD dimerization perhaps via physical interactions among the R domain, the TMDs, and/or their cytosolic extensions; and (3) phosphorylation of PKA sites within NBD1 modulates NBD1–TMD interactions

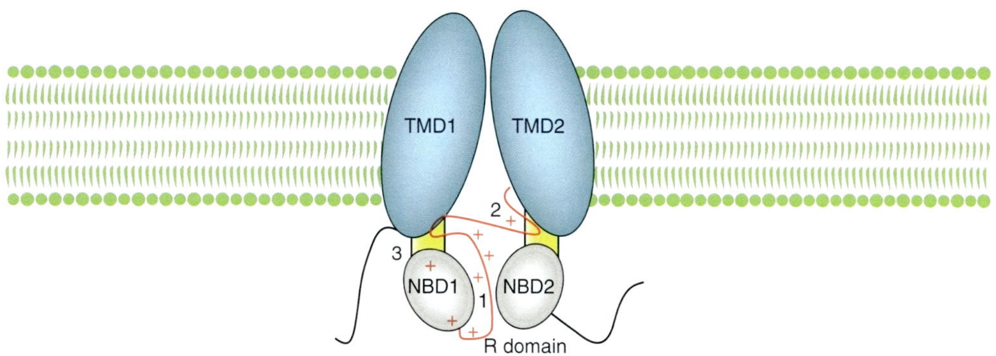

Figure 4. Three proposed mechanisms for CFTR channel regulation by PKA phosphorylation. PKA sites are indicated by the red crosses. (1) Phosphorylation of the R domain (in red) and/or distal NBD1 modulates NBD dimerization. (2) Phosphorylation more directly regulates the flexibility or packing of the TMs by direct physical interactions with the long cytosolic loops (in yellow) or with the TMs themselves. (3) Phosphorylation of sites within NBD1 stabilizes the coupling between the NBDs and the TMDs.

(Fig. 4). These mechanisms are not mutually exclusive given the many PKA sites in CFTR.

Mechanism 1

Mense et al. (2006) showed an effect of PKA on NBD dimerization. They observed that (1) split CFTR molecules could be cross-linked by introducing cysteines at the NBD dimer interface predicted from the crystal structures of other ABC transporters, and (2) PKA activation strongly enhanced cross-linking efficiency (confirmed by He et al. 2008). Baker et al. (2007) provided additional support for this type of mechanism in an NMR structural analysis of an R domain fragment. This fragment bound to recombinant NBD1 in solution with most of the contact points located at or near PKA sites. PKA phosphorylation of this fragment reduced its binding to NBD1. The investigators proposed that the unphosphorylated R domain inhibits CFTR channel activity by binding to NBD1, which could prevent the NBD dimerization that mediates channel opening. Phosphorylation was argued to relieve this inhibition by reducing its affinity for NBD1. This is a plausible model but it does not account for all aspects of how the R domain controls channel gating (see below).

Mechanism 2

The R domain also regulates channel activity independently of ATP binding or NBD dimerization. As noted earlier CFTR truncation mutants that lack NBD2 show detectable channel activity (Cui et al. 2007).Wang et al. (2007, 2010) observed that the channel activity of an NBD2-deletion mutant was strongly dependent on PKA phosphorylation but not ATP binding, an effect that required the R domain. Similarly, G551D–CFTR activity is dependent on PKA phosphorylation but not ATP binding (Bompadre et al. 2007; Wang et al. 2010). These data indicate that the R domain must regulate more than NBD dimerization. Conceivably, portions of the R domain physically interact in a PKA-regulated manner with the cytosolic loops or the TMs with consequent effects on the TM confor-

mational rearrangements that underlie channel gating. The apparent location of the R domain in electrocryomicroscopic images of purified CFTR reported by Zhang et al. (2011) is consistent with this idea. Hegedus et al. (2008) proposed a somewhat related mechanism based on the results of a computational analysis of the R domain. In their model, the R domain regulates the packing or organization of the TM helices by functioning as an "entropic spring" that controls the spacing between NBD1 and TMD2. Phosphorylation, which increases the overall size of the R domain in their simulations, is imagined to change this spacing with concomitant effects on TM packing.

Mechanism 3

The third mechanism is supported by the results of an NMR study by Kanelis et al. (2010). Although most PKA sites reside within the R domain, several sites exist within NBD1 in CFTR-specific regions termed the regulatory extension (e.g., S660) and the regulatory insertion (S422). Kanelis et al. (2010) observed dynamic interactions between the NBD1 core and these CFTR-specific regions that were inhibited by PKA phosphorylation. In addition, they detected a binding interaction between phosphorylated (but not unphosphorylated) NBD1 and a peptide that mimics a region of cytosolic loop 1 that presumably mediates the coupling between NBD1 and TMD1. The investigators concluded that NBD1–TMD interactions are dynamic and regulated by PKA phosphorylation (in this case of sites in NBD1). Such a mechanism could operate in parallel with regulation by the R domain to potentiate CFTR activation by PKA.

In summary, PKA phosphorylation may affect channel gating in multiple ways. Most PKA regulation is attributable to the large R domain with its many PKA sites. Conceivably there is a division of labor among these sites; some sites may regulate NBD dimerization and others may more directly regulate TM flexibility or packing via physical interactions with the TMs or their connecting cytosolic loops. But it is also possible that one predominant mechanism exists with other outcomes secondary to this primary

     Cite this article as *Cold Spring Harb Perspect Med* doi: 10.1101/cshperspect.a009498

mode of regulation. For example, the observed enhancement of NBD1–NBD2 cross-linking by PKA could be a secondary consequence of a more direct effect of PKA phosphorylation on TM packing. Given that the NBD dimer associates primarily with the channel open state, factors that bias the equilibrium toward the open state should also promote NBD dimerization by allosteric coupling (Kirk and Wang 2011).

THE CFTR PORE

We noted earlier that the CFTR permeation pathway likely has some similarities to the translocation pathways of other ABC transporters. But CFTR is an ion channel and not a pump, meaning that its pore must remain open to both sides of the membrane to permit anion diffusion down an electrochemical potential gradient. This property of CFTR, coupled with its preference for small anions, means that its permeation pathway cannot be inferred directly from other ABC transporters. Progress in understanding the CFTR pore has been made, however, by (1) characterizing its ion selectivity; (2) developing biophysical models of the pore based on its selectivity; and (3) using site-directed mutagenesis and chemical modification strategies to identify transmembrane helices (TMs), and residues within these TMs, that influence permeation and selectivity. Such TM residues become candidate pore-lining resides (with some caveats) and, as such, offer a first glimpse into how the pore may be structured.

Anion Selectivity

Chloride channels are less selective for different anions than most cation channels are for different cations (e.g., voltage-gated K channels) (Hille 1973). This difference reflects the necessity of a K channel discriminating between K and the equally abundant Na. The lower selectivity of anion channels means that they can have multiple substrates that are physiologically relevant. Chloride is an important CFTR substrate in tissues that mediate salt and fluid transport such as exocrine glands (Wine and Joo 2004). CFTR-mediated bicarbonate transport also is relevant

in pancreatic ducts and the airways (Ishiguro et al. 2009; Kim and Steward 2009).

We must be precise when we discuss selectivity measurements. The two common measures of ion selectivity are "relative permeability" and "relative conductance." The permeability and conductance sequences of CFTR are quite different, which underscores the importance of understanding each. A CFTR permeability sequence is $SCN > NO_3 > Br > Cl > I >$ acetate; the corresponding conductance sequence is $Cl > NO_3 > Br >$ acetate $> I > SCN$ (Linsdell et al. 2000; McCarty and Zhang 2001). Relative conductance may be the more biologically relevant measure because it is an index of "throughput," i.e., the rate of flow (flux) of an ion through the pore per unit driving force. This flux is measured as a current under conditions when the test ion is the only current-carrying species. Relative permeabilities are measured from shifts in reversal potential (voltage at zero current) when a test ion is added to the side of the membrane opposite Cl. Such reversal potential shifts typically are greatest for those ions that enter the pore most easily, not for ions with the greatest throughput. However, permeability ratios do offer insights into how anions vary with respect to their abilities to enter the pore and how they may interact with the pore walls, as discussed below.

Biophysical Models of CFTR Selectivity: Dielectric Tunnels and Low-Grade Selectivity Filters

The CFTR permeability sequence is ordered principally on the basis of the hydration energies of the permeant anions, termed a lyotropic sequence (Borman et al. 1987; Smith et al. 1999). This indicates that pore entry is dominated by the energy required to remove the ion from water. What about translation through the pore? Dawson and colleagues (Smith et al. 1999; Liu et al. 2003) showed mathematically that, on the basis of its permeability sequence and the known hydration energies of its permeant anions, the CFTR pore can be modeled as a "dielectric or polarizable tunnel" for which no specific interactions between the permeant anions and the pore walls need occur. The only requirement for this conceptual tunnel is that it has a dielectric constant (index of charge screening) intermediate between water ($\sim$80) and lipid ($<$5); namely, about 20 for the CFTR pore. Selectivity arises because those anions that partition more effectively into the pore from water (e.g., SCN) also reside longer on average within the pore (bind more tightly) because of a greater difference between their interaction energies with water (hydration energies) and with the lower dielectric pore interior (solvation energies). To a first approximation this bias-type selectivity mechanism (Smith et al. 1999) describes how permeant anions traverse the CFTR pore with of course the limit that some anions are too big to pass through the pore in the first place (Alexander et al. 2009).

This argument does not exclude the possibility that permeant anions physically interact with pore-lining residues in ways that influence CFTR selectivity. There is good evidence for a narrowing of the CFTR pore that is lined by residues that physically interact with anions as they traverse the channel. This evidence comes from mutational and chemical modification studies of the various TMs with TM6 generating the most enthusiasm. CFTR may have a sort of low-grade selectivity filter—not on a par with the selectivity filters of voltage-gated K channels (Jiang et al. 2003)—but influencing anion selectivity nonetheless.

Pore-Lining TMs

TM mutagenesis experiments were first performed by Anderson et al. (1991b) to determine if the newly cloned CFTR cDNA encoded an anion channel. By mutating two lysines in predicted TMs 1 and 6 (K95 and K335) to acidic residues they altered the relative halide permeability sequence of the cAMP-activated current in CFTR-expressing cells. This was the strongest evidence at the time that CFTR is an anion channel. However, these results do not mean necessarily that K95 and K335 form part of a selectivity filter or even line the pore. A caveat in this sort of analysis is the possibility that an engineered mutation indirectly affects pore architecture. In this regard, K95 substitutions have multiple, complicated effects on permeation. K95 has been argued more recently to attract anions into the pore from the cytosol by residing near the entrance to the pore (inferred from strong voltage-dependent rectification of macroscopic currents for certain K95 mutants) (Linsdell 2006; Zhou et al. 2010). Some K95 mutations (e.g., K95S) also markedly decrease single channel conductance at both depolarizing and hyperpolarizing voltages (Zhou et al. 2010), which implies an effect on pore structure separate from a charge-attracting role for K95.

With the difficulty in interpreting K95/ TM1 data as a cautionary tale, we return to TM6 for which a consensus view has emerged. Linsdell et al. (2000) first reported that mutations at residues F337 and T338 in TM6 strongly affected the permeability sequence. Those findings were extended by testing additional TM6 mutants (McCarty and Zhang 2001; Ge et al. 2004) and by using cysteine scanning to explore the accessibility of TM6 residues (Alexander et al. 2009; Bai et al. 2010; El Hiani and Linsdell 2010). Alexander et al. (2009) compared the effects of two types of reactive compounds on a series of TM6 cysteine mutants: large methanethiosulfonate (MTS) compounds that are not expected to traverse the pore and smaller pseudohalides that both traverse the CFTR pore and are thiol reactive ($Ag(CN)_2$ and $Au(CN)_2$). The impermeant MTS reagents reacted with cysteines at positions 331 to 338 but no deeper into the pore when added to the extracellular side. Conversely, the permeant pseudohalides reacted with cysteines positioned along the entire length of TM6 (to 353). These results confirmed that TM6 residues line the pore and strongly supported the proposed narrowing near position 338 from the extracellular end of the pore (Linsdell et al. 2000; McCarty and Zhang 2001). On the other hand, applying bulky MTS reagents from the cytoplasmic side of the channel identified position 341 as the accessibility limit (Bai et al. 2010; El Hiani and Linsdell 2010). Thus, it appears that the pore is constructed with two fairly accessible internal and external vestibules with a "bottleneck" that traverses only one helical turn (Norimatsu et al. 2012).

 Cite this article as *Cold Spring Harb Perspect Med* doi: 10.1101/cshperspect.a009498

What about the other TMs? We noted above the evidence (with caveats) for a role for TM1. TMs 5, 11, and 12 also have been argued to contribute to the pore based on mutagenesis and cysteine scanning results (Zhang et al. 2000a,b; Fatehi and Linsdell 2009). Recent results from systematic cysteine scanning studies not only suggest that TM12 assumes an α-helical secondary structure, but also support the idea that TM12 is part of the pore-forming domain (Bai et al. 2011; cf. Qian et al. 2011). This latest study also provides new evidence supporting the evolutionary relationship between the CFTR channel and ABC exporters. First, the pattern of state-dependent accessibility for MTS reagents suggests that, during gating transitions, TM12 undergoes a rotational movement, which has long proposed to be part of the molecular motions for ABC transporters (Rosenberg et al. 2001; Gutmann et al. 2010). Second, using MTS reagents with different sizes, Bai et al. (2011) showed that the opening of CFTR's gate is associated with a narrowing of the cytoplasmic entrance of the pore—a result that makes sense if CFTR's gating conformational changes involve movements that are analogous to the flip-flop motion seen in ABC transporters (cf. Wang and Linsdell 2012). As more experimental and structural data emerge (Norimatsu et al. 2012), which TMs contribute to the CFTR pore will become clearer.

CFTR Pore Blockers and Vestibules

CFTR is blocked from the cytosolic side by a diverse group of large organic anions that include a number of commonly used anion channel inhibitors (e.g., glibenclamide) (McDonough et al. 1994; Zhang et al. 2000b; Zhou et al. 2002). The large sizes of some of these blockers and the voltage-dependence of their block indicate the existence of a large intracellular vestibule near the pore entrance (Fig. 5). This vestibule appears to be lined by positively charged residues given that blocker efficacy is reduced by mutating several basic residues in the cytosolic regions of the TMs (e.g., K95, R303, and R352) (Linsdell 2005; St. Aubin and Linsdell 2006; Zhou et al. 2010). As noted

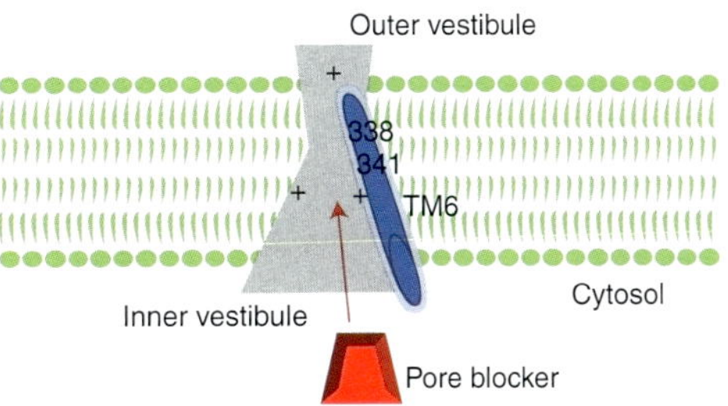

Figure 5. Schematic view of the CFTR permeation pathway. TM6 is indicated as a likely pore-lining helix. The contributions of other TM helices are less clear (see text).

above, the interpretation of such results is complicated by possible effects of mutations on pore structure. In this regard, R352 in TM6 has been proposed to form a salt bridge with D993 in TM9 that stabilizes the open channel conformation (Cui et al. 2008). However, overall, the existing data support the concept of a large, positively charged intracellular vestibule. This charged vestibule should increase the "capture radius" of the pore (Smith et al. 1999) and attract anions to the pore entrance. A similarly charged (but smaller) extracellular vestibule has been proposed to facilitate anion entry from the cell exterior (Smith et al. 2001).

In summary, we have a rudimentary understanding of the pore based on biophysical studies and mutational analyses (Fig. 5). A stringent selectivity filter is not required to explain the CFTR permeability sequence, which is ordered primarily by the hydration energies of the permeant ions. The pore appears to have a narrow region near residues 337–341 in TM6 that may function as a low grade selectivity filter. Positively charged, asymmetric vestibules at the intracellular and extracellular sides attract anions to the pore entrances to facilitate anion permeation.

CONCLUDING REMARKS

The unique duality of CFTR as an ATP-gated channel with an intrinsic enzymatic activity that regulates channel closure has fascinated many investigators. Valuable insights and predictions into how CFTR channel gating is regulated by ATP binding, NBD dimerization, and

ATP hydrolysis have come from recent structural information for its cousins, the ABC exporters. Some features of CFTR channel function are not so easily predicted from the structures of other ABC transporters, however. These more CFTR-specific issues include (1) how phosphorylation of its unique R domain regulates gating; (2) how an anion-selective pore is created from the body plan of an ABC transporter; and (3) the emerging evidence that CFTR gating shares common principles with conventional ligand-gated channels. Resolving the latter issues cannot be achieved solely by borrowing structural information from other ABC transporters. Instead, they await the development of new approaches and tools including high-resolution structures of CFTR per se. As the first drug that targets specifically CFTR gating is in the clinics, mechanistic understanding of CFTR gating will not only shed light on how CFTR potentiators rectify gating defects caused by disease-associated mutations, but also pave the way for the development of complementary strategies for curing cystic fibrosis.

ACKNOWLEDGMENTS

The authors acknowledge the following grant support from the National Institutes of Health: T.-C.H., R01 HL53445 and R01 DK55835; K.L.K., RO1 HL058341 and RO1 DK056796.

REFERENCES

Accurso FJ, Rowe SM, Clancy JP, Boyle MP, Dunitz JM, Durie PR, Sagel SD, Hornick DB, Konstan MW, Donaldson SH, et al. 2010. Effect of VX-770 in persons with cystic fibrosis and the G551D–CFTR mutation. *N Engl J Med* **363:** 1991–2003.

Aleksandrov L, Aleksandrov AA, Chang XB, Riordan JR. 2002. The first nucleotide binding domain of cystic fibrosis transmembrane conductance regulator is a site of stable nucleotide interaction, whereas the second is a site of rapid turnover. *J Biol Chem* **277:** 15419–15425.

Alexander C, Ivetac A, Liu X, Norimatsu Y, Serrano JR, Landstrom A, Sansom M, Dawson DC. 2009. Cystic fibrosis transmembrane conductance regulator: Using differential reactivity toward channel-permeant and channel-impermeant thiol-reactive probes to test a molecular model for the pore. *Biochemistry* **48:** 10078–10088.

Aller SG, Yu J, Ward A, Weng Y, Chittaboina S, Zhuo R, Harrell PM, Trinh YT, Zhang Q, Urbatsch IL, et al. 2009. Structure of P-glycoprotein reveals a molecular basis for poly-specific drug binding. *Science* **323:** 1718–1722.

Anderson MP, Berger HA, Rich DP, Gregory RJ, Smith AE, Welsh MJ. 1991a. Nucleoside triphosphates are required to open the CFTR chloride channel. *Cell* **67:** 775–784.

Anderson MP, Gregory RJ, Thompson S, Souza DW, Paul S, Mulligan RC, Smith AE, Welsh MJ. 1991b. Demonstration that CFTR is a chloride channel by alteration of its anion selectivity. *Science* **253:** 202–205.

Bai Y, Li M, Hwang TC. 2010. Dual roles of the sixth transmembrane segment of the CFTR chloride channel in gating and permeation. *J Gen Physiol* **136:** 293–309.

Bai Y, Li M, Hwang TC. 2011. Structural basis for the channel function of a degraded ABC transporter, CFTR (ABCC7). *J Gen Physiol* **138:** 495–507.

Baker JMR, Hudson RP, Kanelis V, Choy W-Y, Thibodeau PH, Thomas PJ, Forman-Kay JD. 2007. CFTR regulatory region interacts with NBD1 predominately via multiple transient helices. *Nature Struct Mol Biol* **14:** 738–745.

Basso C, Vergani P, Nairn AC, Gadsby DC. 2003. Prolonged nonhydrolytic interaction of nucleotide with CFTR's NH2-terminal nucleotide binding domain and its role in channel gating. *J Gen Physiol* **122:** 333–348.

Bear CE, Li CH, Kartner N, Bridges RJ, Jensen TJ, Ramjeesingh M, Riordan JR. 1992. Purification and functional reconstitution of the cystic fibrosis transmembrane conductance regulator (CFTR). *Cell* **68:** 809–818.

Bompadre SG, Ai T, Cho JH, Wang X, Sohma Y, Li M, Hwang TC. 2005a. CFTR gating I: Characterization of the ATP-dependent gating of a phosphorylation-independent CFTR channel (δR-CFTR). *J Gen Physiol* **125:** 361–375.

Bompadre SG, Cho JH, Wang X, Zou X, Sohma Y, Li M, Hwang TC. 2005b. CFTR gating II: Effects of nucleotide binding on the stability of open states. *J Gen Physiol* **125:** 377–394.

Bompadre SG, Sohma Y, Li M, Hwang TC. 2007. G551D and G1349D, two CF-associated mutations in the signature sequences of CFTR, exhibit distinct gating defects. *J Gen Physiol* **129:** 285–298.

Bormann J, Hamill OP, Sakmann B. 1987. Mechanism of anion permeation through channels gated by glycine and γ-aminobutyric acid in mouse cultured spinal neurons. *J Physiol* **385:** 243–286.

Changeux JP, Edelstein SJ. 2005. Allosteric mechanisms of signal transduction. *Science* **308:** 1424–1428.

Chappe V, Hinkson DA, Howell LD, Evagelidis A, Liao J, Chang X-B, Riordan JR, Hanrahan JW. 2004. Stimulatory and inhibitory protein kinase C consensus sequences regulate the cystic fibrosis transmembrane conductance regulator. *Proc Natl Acad Sci* **101:** 390–395.

Chen TY, Hwang TC. 2008. CLC-0 and CFTR: Chloride channels evolved from transporters. *Physiol Rev* **88:** 351–387.

Chen J, Lu G, Lin J, Davidson AL, Quiocho FA. 2003. A tweezers-like motion of the ATP-binding cassette dimer in an ABC transport cycle. *Mol Cell* **12:** 651–661.

Cheng SH, Rich DP, Marshall J, Gregory RJ, Welsh MJ, Smith AE. 1991. Phosphorylation of the R domain by cAMP-dependent protein kinase regulates the CFTR chloride channel. *Cell* **66:** 1027–1036.

Csanády L, Chan KW, Seto-Young D, Kopsco DC, Nairn AC, Gadsby DC. 2000. Severed channels probe regulation of gating of cystic fibrosis transmembrane conductance regulator by its cytoplasmic domains. *J Gen Physiol* **116:** 477–500.

Csanády L, Seto-Young D, Chan KW, Cenciarelli C, Angel BB, Qin J, McLachlin DT, Krutchinsky AN, Chait BT, Nairn AC, et al. 2005. Preferential phosphorylation of R-domain serine 768 dampens activation of CFTR channels by PKA. *J Gen Physiol* **125:** 171–186.

Csanády L, Nairn AC, Gadsby DC. 2006. Thermodynamics of CFTR channel gating: A spreading conformational change initiates an irreversible gating cycle. *J Gen Physiol* **128:** 523–533.

Csanády L, Vergani P, Gadsby DC. 2010. Strict coupling between CFTR's catalytic cycle and gating of its Cl$^-$ ion pore revealed by distributions of open channel burst durations. *Proc Natl Acad Sci* **107:** 1241–1246.

Cui L, Aleksandrov L, Chang XB, Hou YX, He L, Hegedus T, Gentzsch M, Aleksandrov A, Balch WE, Riordan JR. 2007. Domain interdependence in the biosynthetic assembly of CFTR. *J Mol Biol* **365:** 981–994.

Cui G, Zhang Z-R, O'Brien AR, Song B, McCarty NA. 2008. Mutations at arginine 352 alter the pore architecture of CFTR. *J Membr Biol* **222:** 91–106.

Dawson RJ, Locher KP. 2006. Structure of a bacterial multidrug ABC transporter. *Nature* **443:** 180–185.

Dean M, Annilo T. 2005. Evolution of the ATP-binding cassette (ABC) transporter superfamily in vertebrates. *Annu Rev Genomics Hum Genet* **6:** 123–142.

Diederichs K, Diez J, Greller G, Muller C, Breed J, et al. 2000. Crystal structure of MalK, the ATPase subunit of the trehalose/maltose ABC transporter of the archaeon *Thermococcus litoralis*. *EMBO J* **19:** 5951–5961.

Dousmanis AG, Nairn AC, Gadsby DC. 2002. Distinct Mg^{2+}-dependent steps rate limit opening and closing of a single CFTR Cl$^-$ channel. *J Gen Physiol* **119:** 545–559.

El Hiani Y, Linsdell P. 2010. Changes in accessibility of cytoplasmic substances to the pore associated with activation of the cystic fibrosis transmembrane conductance regulator chloride channel. *J Biol Chem* **285:** 32126–32140.

Fatehi M, Linsdell P. 2009. Novel residues lining the CFTR chloride channel pore identified by functional modification of introduced cysteines. *J Membr Biol* **228:** 151–164.

Gadsby DC, Vergani P, Csanady L. 2006. The ABC protein turned chloride channel whose failure causes cystic fibrosis. *Nature* **440:** 477–483.

Gaudet R, Wiley DC. 2001. Structure of the ABC ATPase domain of human TAP1, the transporter associated with antigen processing. *EMBO J* **20:** 4964–4972.

Ge N, Muise CN, Gong X, Linsdell P. 2004. Direct comparison of the functional roles played by different transmembrane regions in the cystic fibrosis transmembrane conductance regulator channel pore. *J Biol Chem* **279:** 55283–55289.

Gunderson KL, Kopito RR. 1994. Effects of pyrophosphate and nucleotide analogs suggest a role for ATP hydrolysis in cystic fibrosis transmembrane regulator channel gating. *J Biol Chem* **269:** 19349–19353.

Gunderson KL, Kopito RR. 1995. Conformational states of CFTR associated with channel gating: The role TP binding and hydrolysis. *Cell* **82:** 231–239.

Gutmann DA, Ward A, Urbatsch IL, Chang G, van Veen HW. 2010. Understanding polyspecificity of multidrug ABC transporters: Closing in on the gaps in ABCB1. *Trends Biochem Sci* **35:** 36–42.

He L, Aleksandrov AA, Serohijos AWR, Hegedus T, Aleksandrov LA, Cui L, Dokholyan NV, Riordan JR. 2008. Multiple membrane-cytoplasmic domain contacts in the cystic fibrosis transmembrane conductance regulator (CFTR) mediate regulation of channel gating. *J Biol Chem* **283:** 26383–26390.

Hegedus T, Serohijos AWR, Dokholyan NV, He L, Riordan JR. 2008. Computational studies reveal phosphorylation-dependent changes in the unstructured R domain of CFTR. *J Mol Biol* **378:** 1052–1063.

Hegedus T, Aleksandrov A, Mengos A, Cui L, Jensen TJ, Riordan JR. 2009. Role of individual R domain phosphorylation sites in CFTR regulation by protein kinase A. *Biochim Biophys Acta* **1788:** 1341–1349.

Hille B. 1973. Potassium channels in myelinated nerve. Selective permeability to small cations. *J Gen Physiol* **61:** 669–686.

Hung LW, Wang IX, Nikaido K, Liu PQ, Ames GF, Kim SH. 1998. Crystal structure of the ATP-binding subunit of an ABC transporter. *Nature* **396:** 703–707.

Hwang TC, Nagel G, Nairn AC, Gadsby DC. 1994. Regulation of the gating of cystic fibrosis transmembrane conductance regulator C$_1$ channels by phosphorylation and ATP hydrolysis. *Proc Natl Acad Sci* **91:** 4698–4702.

Ishiguro H, Steward MC, Naruse S, Ko SB, Goto H, Case RM, Kondo T, Yamamoto A. 2009. CFTR functions as a bicarbonate channel in pancreatic duct cells. *J Gen Physiol* **133:** 315–326.

Jiang Y, Lee A, Chen J, Ruta V, Cadene M, Chait BT, MacKinnon R. 2003. X-ray structure of a voltage-dependent K$^+$ channel. *Nature* **423:** 33–41.

Jordan IK, Kota KC, Cui G, Thompson CH, McCarty NA. 2008. Evolutionary and functional divergence between the cystic fibrosis transmembrane conductance regulator and related ATP-binding cassette transporters. *Proc Natl Acad Sci* **105:** 18865–18870.

Kanelis V, Hudson RP, Thibodeau PH, Thomas PJ, Forman-Kay JD. 2010. NMR evidence for differential phosphorylation-dependent interactions in WT and ΔF508 CFTR. *EMBO J* **29:** 263–277.

Karpowich N, Martsinkevich O, Millen L, Yuan YR, Dai PL, MacVey K, Thomas PJ, Hunt JF. 2001. Crystal structures of the MJ1267 ATP binding cassette reveal an induced-fit effect at the ATPase active site of an ABC transporter. *Structure* **9:** 571–586.

Khare D, Oldham ML, Orelle C, Davidson AL, Chen J. 2009. Alternating access in maltose transporter mediated by rigid-body rotations. *Mol Cell* **33:** 528–536.

Kim D, Steward MC. 2009. The role of CFTR in bicarbonate secretion by pancreatic duct and airway epithelia. *J Med Invest* **56** (suppl): 336–342.

King JD Jr, Fitch AC, Lee JK, McCane JE, Mak DOD, Foskett JK, Hallows KR. 2009. AMP-activated protein kinase phosphorylation of the R domain inhibits PKA stimulation of CFTR. *Am J Physiol* **297**: C94–C101.

Kirk KL, Wang W. 2011. A unified view of cystic fibrosis transmembrane conductance regulator (CFTR) gating: Combining the allosterism of a ligand-gated channel with the enzymatic activity of an ATP-binding cassette (ABC) transporter. *J Biol Chem* **286**: 12813–12819.

Leier I, Jedlitschky G, Buchholz U, Cole SP, Deeley RG, Keppler D. 1994. The MRP gene encodes an ATP-dependent export pump for leukotriene C4 and structurally related conjugates. *J Biol Chem* **269**: 27807–27810.

Lewis HA, Buchanan SG, Burley SK, Conners K, Dickey M, Dorwart M, Fowler R, Gao X, Guggino WB, Hendrickson WA, et al. 2004. Structure of nucleotide-binding domain 1 of the cystic fibrosis transmembrane conductance regulator. *EMBO J* **23**: 282–293.

Lewis HA, Zhao X, Wang C, Sauder JM, Rooney I, Noland BW, Lorimer D, Kearins MC, Conners K, Condon B, et al. 2005. Impact of the ΔF508 mutation in first nucleotide-binding domain of human cystic fibrosis transmembrane conductance regulator on domain folding and structure. *J Biol Chem* **280**: 1346–1353.

Linsdell P. 2005. Location of a common inhibitor binding site in the cytoplasmic vestibule of the cystic fibrosis transmembrane conductance regulator chloride channel pore. *J Biol Chem* **280**: 8945–8950.

Linsdell P. 2006. Mechanism of chloride permeation in the cystic fibrosis transmembrane conductance regulator chloride channel. *Exp Physiol* **91.1**: 123–129.

Linsdell P, Evagelidis A, Hanrahan JW. 2000. Molecular determinants of anion selectivity in the cystic fibrosis transmembrane conductance regulator chloride channel pore. *Biophys J* **78**: 2973–2982.

Liu X, Smith SS, Dawson DC. 2003. CFTR: What's it like inside the pore? *J Exp Zool* **300A**: 69–75.

Loe DW, Almquist KC, Deeley RG, Cole SP. 1996. Multidrug resistance protein (MRP)-mediated transport of leukotriene C4 and chemotherapeutic agents in membrane vesicles. Demonstration of glutathione-dependent vincristine transport. *J Biol Chem* **271**: 9675–9682.

Mathews CJ, Tabcharani JA, Chang X-B, Jensen TJ, Riordan JR, Hanrahan JW. 1998. Dibasic protein kinase A sites regulate bursting rate and nucleotide sensitivity of the cystic fibrosis transmembrane conductance regulator chloride channel. *J Physiol* **508.2**: 365–377.

McCarty NA, Zhang Z-R. 2001. Identification of a region of strong discrimination in the pore of CFTR. *Am J Lung Cell Mol Physiol* **281**: L852–L867.

McDonough S, Davidson N, Lester HA, McCarty NA. 1994. Novel pore-lining residues in CFTR govern permeation and open-channel block. *Neuron* **13**: 623–634.

Mense M, Vergani P, White DM, Altberg G, Nairn AC, Gadsby DC. 2006. In vivo phosphorylation of CFTR promotes formation of a nucleotide-binding domain heterodimer. *EMBO J* **25**: 4728–4739.

Mornon JP, Lehn P, Callebaut I. 2009. Molecular models of the open and closed states of the whole human CFTR protein. *Cell Mol Life Sci* **66**: 3469–3486.

Nagel G, Hwang TC, Nastiuk KL, Nairn AC, Gadsby DC. 1992. The protein kinase A-regulated cardiac Cl-channel resembles the cystic fibrosis transmembrane conductance regulator. *Nature* **360**: 81–84.

Norimatsu Y, Ivetac AD, Alexander C, Kirkham J, O'Donnell N, Dawson DC, Sansom MS. 2012. CFTR: A molecular model defines the architecture of the anion conduction path and locates a "bottleneck" in the pore. *Biochemistry* **51**: 2199–2212.

Powe AC Jr, Al-Nakkash L, Li M, Hwang TC. 2002. Mutation of Walker-A lysine 464 in cystic fibrosis transmembrane conductance regulator reveals functional interaction between its nucleotide-binding domains. *J Physiol* **539**: 333–346.

Purohit P, Auerbach A. 2009. Unliganded gating of acetylcholine receptor channels. *Proc Natl Acad Sci* **106**: 115–120.

Qian F, El Hiani Y, Linsdell P. 2011. Functional arrangement of the 12th transmembrane region in the CFTR chloride channel pore based on functional investigation of a cysteine-less CFTR variant. *Pflugers Arch* **462**: 559–571.

Ramjeesingh M, Li C, Garami E, Huan LJ, Galley K, Wang Y, Bear CE. 1999. Walker mutations reveal loose relationship between catalytic and channel-gating activities of purified CFTR (cystic fibrosis transmembrane conductance regulator). *Biochemistry* **38**: 1463–1468.

Ramsey BW, Davies J, McElvaney NG, Tullis E, Bell SC, Drevinek P, Griese M, McKone EF, Wainwright CE, Konstan MW, et al. 2011. A CFTR potentiator in patients with cystic fibrosis and the G551D mutation. *N Engl J Med* **365**: 1663–1672.

Rees DC, Johnson E, Lewinson O. 2009. ABC transporters: The power to change. *Nature Rev* **10**: 218–227.

Rich DP, Gregory RJ, Anderson MP, Manavalan P, Smith AE, Welsh MJ. 1991. Effect of deleting the R domain on CFTR-generated chloride channels. *Science* **253**: 205–207.

Rich DP, Berger HA, Cheng SH, Travis SM, Saxena M, Smith AE, Welsh MJ. 1993. Regulation of the cystic fibrosis transmembrane conductance regulator Cl⁻ channel by negative charge in the R domain. *J Biol Chem* **268**: 20259–20267.

Riordan JR, Rommens JM, Kerem B, Alon N, Rozmahel R, Grzelczak Z, Zielenski J, Lok S, Plavsic N, Chou JL, et al. 1989. Identification of the cystic fibrosis gene: Cloning and characterization of complementary DNA. *Science* **245**: 1066–1073.

Rosenberg MF, Mao Q, Holzenburg A, Ford RC, Deeley RG, Cole SP. 2001. The structure of the multidrug resistance protein 1 (MRP1/ABCC1). Crystallization and single-particle analysis. *J Biol Chem* **276**: 16076–16082.

Schmitt L, Benabdelhak H, Blight MA, Holland IB, Stubbs MT. 2003. Crystal structure of the nucleotide-binding domain of the ABC-transporter haemolysin B: Identification of a variable region within ABC helical domains. *J Mol Biol* **330**: 333–342.

Seibert FS, Chang X-B, Aleksandrov AA, Clarke DM, Hanrahan JW, Riordan JR. 1999. Influence of phosphorylation by protein kinase A on CFTR at the cell surface and endoplasmic reticulum. *Biochim Biophys Acta* **1461**: 275–283.

Serohijos AW, Hegedus T, Aleksandrov AA, He L, Cui L, Dokholyan NV, Riordan JR. 2008. Phenylalanine-508 mediates a cytoplasmic-membrane domain contact in the CFTR 3D structure crucial to assembly and channel function. *Proc Natl Acad Sci* **105**: 3256–3261.

Sheppard DN, Welsh MJ. 1999. Structure and function of the CFTR chloride channel. *Physiol Rev* **79** (1 Suppl): S23–S45.

Smith SS, Steinle ED, Meyerhoff ME, Dawson DC. 1999. Cystic fibrosis transmembrane conductance regulator: Physical basis for lyotropic anion selectivity patterns. *J Gen Physiol* **114**: 799–818.

Smith SS, Liu X, Zhang Z-R, Sun F, Kriewall TE, McCarty NA, Dawson DC. 2001. CFTR: Covalent and noncovalent modification suggests a role for fixed charges in anion conduction. *J Gen Physiol* **118**: 407–431.

Smith PC, Karpowich N, Millen L, Moody JE, Rosen J, et al. 2002. ATP binding to the motor domain from an ABC transporter drives formation of a nucleotide sandwich dimer. *Mol Cell* **10**: 139–149.

St. Aubin CN, Linsdell P. 2006. Positive charges at the intracellular mouth of the pore regulate anion conduction in the CFTR chloride channel. *J Gen Physiol* **128**: 535–545.

Stratford FL, Ramjeesingh M, Cheung JC, Huan LJ, Bear CE. 2007. The Walker B motif of the second nucleotide binding domain (NBD2) of CFTR plays a key role in ATPase activity of the NBD1, NBD2 heterodimer. *Biochem J* **401**: 581–586.

Szollosi A, Vergani P, Csanady L. 2010. Involvement of F1296 and N1303 of CFTR in induced-fit conformational change in response to ATP binding at NBD2. *J Gen Physiol* **136**: 407–423.

Szollosi A, Muallem DR, Csanady L, Vergani P. 2011. Mutant cycles at CFTR's non-canonical ATP-binding site support little interface separation during gating. *J Gen Physiol* **137**: 549–562.

Tsai MF, Shimizu H, Sohma Y, Li M, Hwang TC. 2009. State-dependent modulation of CFTR gating by pyrophosphate. *J Gen Physiol* **133**: 405–419.

Tsai MF, Li M, Hwang TC. 2010. Stable ATP binding mediated by a partial NBD dimer of the CFTR chloride channel. *J Gen Physiol* **135**: 399–414.

Van Goor F, Hadida S, Grootenhuis PD, Burton B, Cao D, Neuberger T, Turnbull A, Singh A, Joubran J, Hazlewood A, et al. 2009. Rescue of CF airway epithelial cell function in vitro by a CFTR potentiator, VX-770. *Proc Natl Acad Sci* **106**: 18825–18830.

Verdon G, Albers SV, Dijkstra BW, Driessen AJ, Thunnissen AM. 2003. Crystal structures of the ATPase subunit of the glucose ABC transporter from Sulfolobus solfataricus: Nucleotide-free and nucleotide-bound conformations. *J Mol Biol* **330**: 343–358.

Vergani P, Nairn AC, Gadsby DC. 2003. On the mechanism of MgATP-dependent gating of CFTR Cl- channels. *J Gen Physiol* **121**: 17–36.

Vergani P, Lockless SW, Nairn AC, Gadsby DC. 2005. CFTR channel opening by ATP-driven tight dimerization of its nucleotide-binding domains. *Nature* **433**: 876–880.

Wang W, Linsdell P. 2012. Alternating access to the transmembrane domain of the ATP-binding cassette protein cystic fibrosis transmembrane conductance regulator (ABCC7). *J Biol Chem* **287**: 10156–10165.

Wang F, Zeltwanger S, Hu S, Hwang T-C. 2000. Deletion of phenylalanine 508 causes attenuated phosphorylation-dependent activation of CFTR chloride channels. *J Physiol* **524**: 637–648.

Wang W, Bernard K, Li G, Kirk KL. 2007. Curcumin opens cystic fibrosis transmembrane conductance regulator channels by a novel mechanism that requires neither ATP binding nor dimerization of the nucleotide-binding domains. *J Biol Chem* **282**: 4533–4544.

Wang X, Bompadre SG, Li M, Hwang TC. 2009. Mutations at the signature sequence of CFTR create a Cd^{2+}-gated chloride channel. *J Gen Physiol* **133**: 69–77.

Wang W, Wu J, Bernard K, Li G, Wang G, Bevensee MO, Kirk KL. 2010. ATP-independent CFTR channel gating and allosteric modulation by phosphorylation. *Proc Natl Acad Sci* **107**: 3888–3893.

Ward A, Reyes CL, Yu J, Roth CB, Chang G. 2007. Flexibility in the ABC transporter MsbA: Alternating access with a twist. *Proc Natl Acad Sci* **104**: 19005–19010.

Wilkinson DJ, Strong TV, Mansoura MK, Wood DL, Smith SS, Collins FS, Dawson DC. 1997. CFTR activation: Additive effects of stimulatory and inhibitory phosphorylation sites in the R domain. *Am J Physiol* **273**: L127–L133.

Wine JJ, Joo NS. 2004. Submucosal glands and airway defense. *Proc Am Thorac Soc* **1**: 47–53.

Yuan YR, Blecker S, Martsinkevich O, Millen L, Thomas PJ, Hunt JF. 2001. The crystal structure of the MJ0796 ATP-binding cassette. Implications for the structural consequences of ATP hydrolysis in the active site of an ABC transporter. *J Biol Chem* **276**: 32313–32321.

Zeltwanger S, Wang F, Wang GT, Gillis KD, Hwang TC. 1999. Gating of cystic fibrosis transmembrane conductance regulator chloride channels by adenosine triphosphate hydrolysis. Quantitative analysis of a cyclic gating scheme. *J Gen Physiol* **113**: 541–554.

Zhang Z-R, McDonough SI, McCarty NA. 2000a. Interaction between permeation and gating in a putative pore domain mutant in the cystic fibrosis transmembrane conductance regulator. *Biophys J* **79**: 298–313.

Zhang Z-R, Zeltwanger S, McCarty NA. 2000b. Direct comparison of NPPB and DPC as probes of CFTR expressed in *Xenopus* oocytes. *J Membrane Biol* **175**: 35–52.

Zhang L, Aleksandrov LA, Riordan JR, Ford RC. 2011. Domain location within the cystic fibrosis transmembrane conductance regulator protein investigated by electron microscopy and gold labeling. *Biochim Biophys Acta* **808**: 399–404.

Zhou Z, Hu S, Hwang T-C. 2002. Probing an open CFTR pore with organic anion blockers. *J Gen Physiol* **120**: 647–662.

Zhou Z, Wang X, Li M, Sohma Y, Zou X, Hwang TC. 2005. High affinity ATP/ADP analogues as new tools for studying CFTR gating. *J Gen Physiol* **569**: 447–457.

Zhou Z, Wang X, Liu HY, Zou X, Li M, Hwang TC. 2006. The two ATP binding sites of cystic fibrosis transmembrane conductance regulator (CFTR) play distinct roles in gating kinetics and energetics. *J Gen Physiol* **128**: 413–422.

Zhou J-J, Li M-S, Qi J, Linsdell P. 2010. Regulation of conductance by the number of fixed charges in the intracellular vestibule of the CFTR chloride channel pore. *J Gen Physiol* **135**: 229–245.

Cystic Fibrosis Transmembrane Conductance Regulator (ABCC7) Structure

John F. Hunt[1], Chi Wang[1], and Robert C. Ford[2]

[1]Department of Biological Sciences, Columbia University, New York, New York 10027

[2]Faculty of Life Sciences, The University of Manchester, Manchester M13 9PT, United Kingdom

Correspondence: jfhunt@biology.columbia.edu; robert.ford@manchester.ac.uk

Structural studies of the cystic fibrosis transmembrane conductance regulator (CFTR) are reviewed. Like many membrane proteins, full-length CFTR has proven to be difficult to express and purify; hence, much of the structural data available is for the more tractable, independently expressed soluble domains. Therefore, this chapter covers structural data for individual CFTR domains in addition to the sparser data available for the full-length protein. To set the context for these studies, we will start by reviewing structural information on model proteins from the ATP-binding cassette (ABC) transporter superfamily, to which CFTR belongs.

INSIGHT FROM STRUCTURES OF MODEL ABC TRANSPORTERS

Domain Organization of ABC Transporters

The ABC transporter superfamily is characterized by a stereotyped ATP-binding cassette (ABC), alternatively called a nucleotide-binding domain (NBD), that is readily identified via simple sequence-homology searches (Higgins 1992; Linton and Higgins 1998; Dassa and Bouige 2001; Dean et al. 2001; Kerr 2002; Davidson and Chen 2004; Jones and George 2004; Dean 2005; Dean and Annilo 2005). This domain contains a series of signature sequences involved in binding ATP (described below) that enable an expert to identify most representatives based on visual inspection of the amino acid sequence (Hung et al. 1998; Karpowich et al. 2001; Smith et al. 2002). The first reported sequence of a mammalian ABC transporter was that of P-glycoprotein (Pgp, also called ABCB1) (Dean et al. 2001; Dean and Annilo 2005), a multidrug resistance protein overexpressed in a variety of advanced cancers. The publication of the Pgp sequence in 1986 represented a breakthrough in the application of molecular biology methods to understanding human disease (Fojo et al. 1985; Gerlach et al. 1986; Chen et al. 1986; Roninson et al. 1986). The identification of the human cystic fibrosis transmembrane conductance regulator (CFTR, also called ABCC7) (Dean et al. 2001; Dean and Annilo 2005), reported in 1989 (Riordan et al. 1989), represented another breakthrough marking the dawn of the era of genomic medicine. Although protein sequencing had led to the characterization of the molecular basis of other human diseases, the discovery of CFTR represented the first report of the DNA sequence of a human disease-causing gene. Most remarkably, the sequence of CFTR showed strong homology to that of Pgp as well as to a series of bacterial proteins involved in transmembrane active transport (Riordan et al. 1989; Higgins 1992; Dean et al. 2001). As sequence databases expanded, it became clear that these proteins belong to a ubiquitously distributed protein superfamily that came to be known as the ABC transporters (Higgins 1992; Saurin et al. 1999; Dean et al. 2001). Although some superfamily members are involved in DNA repair and protein translation (Kerr 2004), the vast majority are transmembrane proteins that function as ATP-dependent active transporters (Higgins 1992; Saurin et al. 1999; Dean et al. 2001; Davidson and Chen 2004; Jones and George 2004). Although the NBDs from ABC transporters share strongly conserved structures, the transmembrane domains (TMDs) are more diverse and come from a variety of sequence/structure families.

Although CFTR functions as an ATP-gated chloride channel (Rich et al 1990; Anderson et al. 1991a,b; Drumm et al. 1991) and is not believed to mediate active transport (Gadsby et al. 2006; Riordan 2008; Kim Chiaw et al. 2011), it shares the conserved domain architecture (Fig. 1A–D) characteristic of the ATP-dependent transporters that represent the vast majority of the ABC superfamily (Higgins 1992; Davidson and Chen 2004; Jones and George 2004; Hollenstein et al. 2007a; Cui and Davidson 2011; Zolnerciks et al. 2011; George and Jones 2012). This architecture involves a dimeric organization of TMDs (Fig. 1B,D) and tightly associated NBDs (Fig. 1B–D) that are encoded in anything from one to four polypeptide chains (Higgins 1992; Dean et al. 2001). In all ABC transporters, the TMDs and NBDs both interact with one another with approximate twofold symmetry (Fig. 1B,D), but the interacting domains can be either identical in sequence (i.e., making a homodimeric interaction) or homologous but divergent in sequence (i.e., making a

heterodimeric interaction). Although bacterial ABC importers exhibit more diverse schemes of covalent organization (Higgins 1992; Linton and Higgins 1998; Saurin et al. 1999), human ABC transporters like bacterial ABC exporters have their TMDs and NBDs fused in a single polypeptide chain (Dean et al. 2001; Dean and Annilo 2005). In some human ABC transporters, including CFTR (Fig. 1A), both TMDs (TMD1 and TMD2) and both NBDs (NBD1 and NBD2) are fused in a single polypeptide chain at least 1100 amino acids in length, producing an effectively heterodimeric functional organization. The binding and hydrolysis of two ATP molecules at the interface between the homodimeric or heterodimeric NBDs drives functional conformational changes in ABC proteins (Fig. 1D) (Jones and George 1999; Smith et al. 2002; Davidson and Chen 2004; Vergani et al. 2005; Riordan 2008), as discussed in greater detail below. In the case of CFTR, the ATP-gated chloride channel resides in the interface between the heterodimeric TMDs (Anderson et al. 1991b; Mansoura et al. 1998; Linsdell et al. 2000, 2006; Liu et al. 2003, 2004; Alexander et al. 2009; Liu and Dawson 2011). The transported substrate binds at the equivalent location in the ATP-dependent transporters in the ABC superfamily (Davidson and Chen 2004; Hollenstein et al. 2007a; Cui and Davidson 2011; Kim Chiaw et al. 2011; Zolnerciks et al. 2011; George and Jones 2012).

Some ABC transporters contain additional domains, most frequently at their amino terminus or immediately after their first NBD. CFTR has such a domain called the regulatory or R domain, which is ∼240 residues in length and inserted between NBD1 and TMD2. Its presence produces the following overall amino-terminal to carboxy-terminal domain organization in CFTR: TMD1, NBD1, R, TMD2, and finally NBD2 (Riordan et al. 1989; Gadsby et al. 2006; Riordan 2008) (Fig. 1A). CFTR additionally has an amino-terminal extension of ∼80 residues and a carboxy-terminal extension of ∼30 residues. These terminal sequences are likely disordered except when interacting with specific protein-binding partners. Similarly, the R domain is predominantly disordered (Ostedgaard et al. 1997; Baker et al. 2007; Kanelis et al. 2010; Lewis et al. 2010), at least in isolation from the other domains of CFTR, which has led to the proposal that it should be renamed the R region rather than the R domain. Low-resolution electron microscopy data on CFTR molecules with the R domain labeled with 1.8-nm immunogold spheres has identified a relatively well-defined location for some of its residues relative to the other domains in full-length CFTR (Zhang et al. 2009, 2010), as discussed further below. Furthermore, NMR data show some interaction of the isolated R domain with isolated CFTR nucleotide-binding domains when mixed together in vitro (Baker et al. 2007).

The R domain plays a central role in mediating activation of CFTR by protein kinase A (PKA) (Gregory et al. 1990; Zhu et al. 2002; Chappe et al. 2005; Csanády et al. 2005) and may function in other regulatory processes as well (Chappe et al. 2004). It has a series of phosphorylation sites and is believed to act as an integrator of PKA signaling and perhaps also signaling from some other physiological systems. Although full-length CFTR has minimal basal activity prior to phosphorylation by PKA, deletion of the R domain leads to 50% of maximal activity in the absence of phosphorylation (Chappe et al. 2005; Csanády et al. 2000). Therefore, the R domain acts primarily although perhaps not exclusively by inhibiting channel activation when it is in the unphosphorylated state. These inhibitory interactions probably involve binding to interdomain interfaces in CFTR and thereby blocking the changes in interdomain interactions required to open the chloride channel (Lewis et al. 2004; Baker et al. 2007). One specific site on CFTR believed to participate in such an inhibitory interaction is schematized in Figure 1D and described below.

Structural Organization of the NBDs from ABC Transporters

Several key features of the structural organization of the NBDs from ABC transporters were inferred from sequence analysis prior to the first experimental structure determination for any superfamily member (Hobson et al. 1984;

Thomas et al. 1991; Mourez et al. 1997; Hung et al. 1998). Their sequences contain the canonical Walker A and Walker B motifs (Fig. 1A,C, and D) present in a variety of different ATPase superfamilies (Walker et al. 1982), including the superfamily I and II helicases (Hunt et al. 2002) and the AAA ATPases (Snider and Houry 2008). The residues from these motifs, which were first identified in the F1 ATPase (Walker et al. 1982), ligate the Mg-ATP substrate in a conserved manner in all F1-like ATPase superfamilies including the ABC transporters. The Walker A motif (GxxGxGKT, with x being any residue) ligates the α- and β-phosphates of ATP, while the aspartate residue at the end of the Walker B motif (ϕ_4D, with ϕ being a hydrophobic residue) ligates its tightly bound Mg^{2+} cofactor. In ABC domains, the Walker B is followed immediately by a glutamate residue (i.e., ϕ_4DE) that was recognized as being invariant in sequence alignments long before crystallographic studies showed that it is the catalytic base that activates a water molecule for hydrolysis of ATP (Hung et al. 1998; Smith et al. 2002; Moody et al. 2002). The conservation of the Walker A and B motifs in the NBDs from ABC transporters combined with secondary structure analyses led to the inference that the ATP-binding core of the domain would have the same fold as F1 ATPase and other F1-like ATPase superfamilies, an inference later verified by the crystal structures of isolated NBDs from bacterial ATP transporters (e.g., as shown for the NBDs from CFTR in Fig. 1B,C) (Hung et al. 1998; Karpowich et al. 2001).

These structures also verified the sequence-based inference that an α-helical subdomain unique to the ABC superfamily is inserted into topology of the F1-like ATP-binding core in the NBDs from ABC transporters (Fig. 1B,C) (Mourez et al. 1997). This "ABCα subdomain" (Karpowich et al. 2001) contains the so-called "signature sequence" of ABC transporters, a nearly invariant pentapeptide motif with sequence LSGGQ that precedes the Walker B motif in ABC domains by $\sim$20 amino acids (Higgins 1992; Jones and George 1999). Enzymological studies of model ABC transporters showed that this sequence, sometimes called the C motif, is

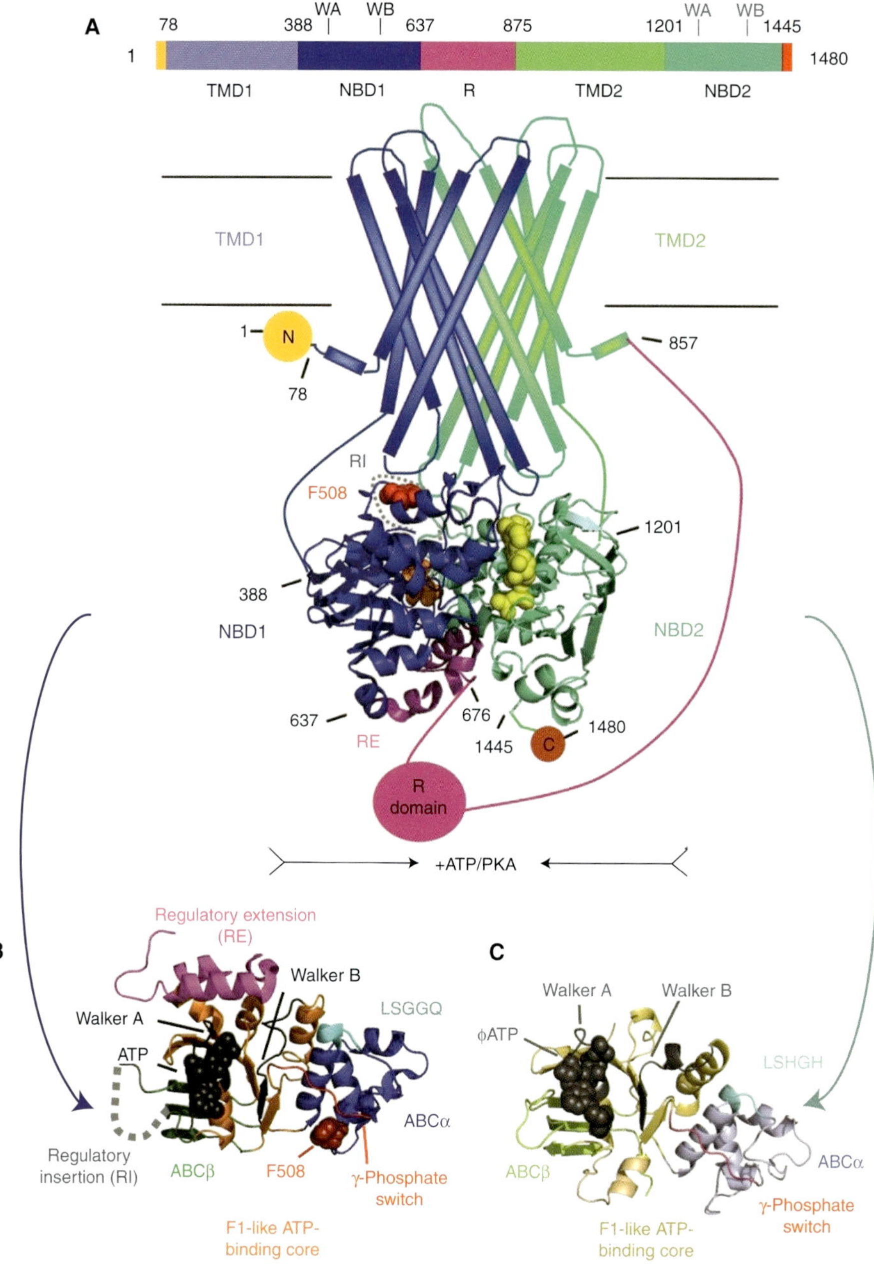

Figure 1. Domain organization of CFTR and simplified consensus gating model. (*A*) The three-dimensional domain organization of CFTR is schematized below a bar showing its linear sequence organization. NBD stands for nucleotide-binding domain, TMD for transmembrane domain, R for regulatory domain, RI for the regulatory insertion in NBD1, RE for the regulatory extension at the junction between NBD1 and the R domain, and N and C for the amino- and carboxy-terminal cytoplasmic tails, respectively. ATP is shown in orange (bound to the Walker motifs in NBD1) or yellow (bound to the Walker motifs in NBD2) space-filling representation. The NBDs are represented by ribbon diagrams of crystal structures of the isolated human domains (PDB IDs 2BBO for hNBD1 and 3GD7 for hNBD2), which are shown interacting in an ATP-sandwich heterodimer conformation modeled based on least-squares alignment to the crystal structure of the low-affinity homodimer of hNBD1-Δ(RI,RE) (PDB ID 2PZE). (*Legend continues on following page.*)

Cite this article as *Cold Spring Harb Perspect Med* doi: 10.1101/cshperspect.a009514

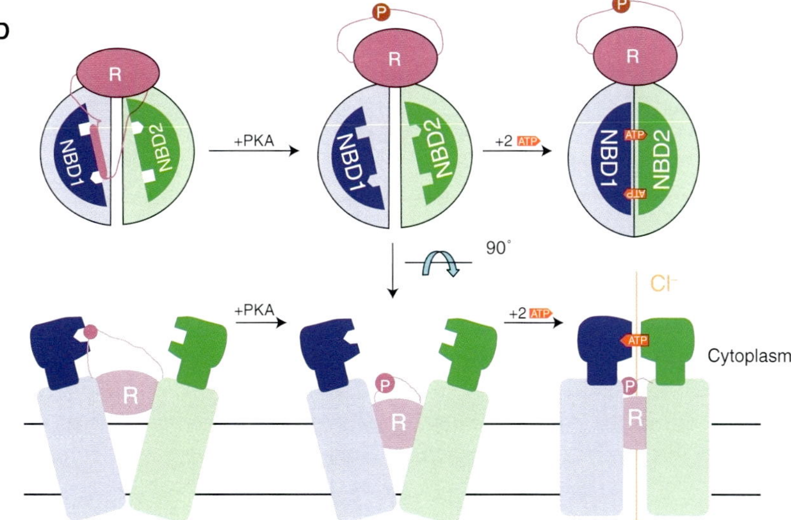

Figure 1. (*Continued*) PKA stands for protein kinase A, which uses ATP to phosphorylate and thereby activate the CFTR channel; phosphorylation presumably occurs primarily on residues in the R domain. (*B*) Ribbon diagram showing the subdomain organization of hNBD1 (PDB ID 2BBO), with the F1-like ATP-binding core subdomain shown in orange, the ABCβ subdomain in dark green, the ABCα subdomain in blue, the γ-phosphate switch in light red, the LSGGQ signature sequence in light cyan, the RE in magenta, the RI in gray (depicted as a dotted line), and Mg-ATP in black space-filling representation. Residue F508 (dark red) and its Walker A and B motifs are labeled. (*C*) Ribbon diagram showing the subdomain organization of hNBD2 (PDB ID 3GD7), with the F1-like ATP-binding core subdomain shown in orange, the ABCβ subdomain in light green, the ABCα subdomain in gray-blue, the γ-phosphate switch in magenta, the LSHGH signature sequence in dark cyan, and Mg-φATP (phenyl-ATP) in black space-filling representation. Its Walker A and B motifs are labeled. (*D*) Schematic diagrams illustrating a simplified version of the accepted model for the mechanochemistry of CFTR channel gating. The *upper* row shows a top view looking down on the NBDs from the cytoplasm, whereas the *lower* row shows a side view of the transmembrane structure (with the phospholipid bilayer schematized as two parallel black lines). The coloring is equivalently to panel *A* except for ATP, which is shown here in red. The opening of the transmembrane chloride channel in CFTR is gated by mechanical reorientation of the NBDs upon ATP binding at their interface, which drives formation of an ATP-sandwich heterodimer structure. Phosphorylation of the R domain by PKA is believed to facilitate channel opening at least in part by disrupting binding interactions between polypeptide segments in the R domain and surfaces in the NBDs mediating tight interdomain structural interactions in the open conformation of the channel.

essential for catalysis of ATP hydrolysis as well as substrate transport (as reviewed in Davidson and Chen 2004), leading to expectations that it would form part of the ATPase active site in ABC transporters (Jones and George 1999). In transporters with heterodimeric NBDs, mutation of one of the two LSGGQ motifs was shown to be sufficient to block activity (Davidson and Sharma 1997). This observation reinforced the inference, from Michaelis-Menten analyses showing twofold kinetic cooperativity (Davidson et al. 1996), that the functional organization of ABC transporters involves participation of the signature sequence in some kind of dimeric structural interaction (Jones and George 1999). Insightful analyses of bacterial ABC transporters suggested that the ABCα subdomain also makes important structural contacts to the TMDs (Mourez et al. 1997), another inference later verified by structural studies (Locher et al. 2002; Dawson and Locher 2006). These contacts are proximal to the site of the predominant disease-causing F508del mutation in CFTR (Fig. 1B), as discussed in detail below.

The first X-ray crystal structures of isolated NBDs from model bacterial ABC transporters verified most of these prior inferences concerning domain architecture but also provided some surprises (Hung et al. 1998; Karpowich et al. 2001). The core of the NBD indeed has an identical topology to that of the catalytic subunit of F1 ATPase, and it ligates the α/β-phosphates and Mg^{2+}-cofactor of ATP with equivalent stereochemistry (Fig. 1B,C). As predicted, this F1-like ATP-binding core of the NBD is interrupted by the insertion of a bundle of three α-helices, the ABCα subdomain, between two of the six conserved β-strands in the parallel β-sheet forming the core of the domain (Karpowich et al. 2001).

Unexpectedly, the amino terminus of the NBD contains an antiparallel β-sheet, the ABCβ subdomain, formed in large part by residues preceding the F1-like ATP-binding core (Karpowich et al. 2001). Like the ABCα subdomain, the ABCβ subdomain is unique to the NBDs from ABC superfamily members and is not found in other F1-like ATPase superfamilies. The ABCβ subdomain contributes to forming the ATP-binding site in ABC transporters. A conserved aromatic residue at the carboxyl terminus of the first β-strand in the ABCβ subdomain makes a base-stacking interaction with the adenine of the ATP, and this interaction positions the entirety of the nucleotide in a solvent-exposed position on the surface of the domain, in a substantially different geometry from that observed in other F1-like ATPases (Hung et al. 1998; Karpowich et al. 2001).

The most surprising feature in the crystal structures of isolated NBDs from a series of model bacterial ABC transporters was their failure to form a consistent dimeric structure (Hung et al. 1998; Diederichs et al. 2000; Karpowich et al. 2001). In the protomer structures, the LSGGQ signature sequence is at least 15 Å away from the ATP-binding site (Fig. 1B,C). Given the well-established importance of this motif for catalytic activity (Davidson and Chen 2004), this structural observation suggested that there are likely to be conserved oligomeric interactions in functional ABC transporters that were not recapitulated in the early structures of isolated NBDs. One specific prediction that proved to be highly insightful was that the LSSGQ signature sequence in one protomer would interact with the ATP molecule bound to the Walker motifs in a second protomer in the functional dimeric conformation of ABC transporters (Jones and George 1999).

Another noteworthy feature in the crystal structures of isolated NBDs from ABC transporters was a wide variation in the alignment of the ABCα subdomain relative to the F1-like ATP-binding core of the domain (Diederichs et al. 2000; Karpowich et al. 2001; Yuan et al. 2001). Different crystallographic snapshots, even of the same NBD, showed up to 20° relative rotations of the ABCα subdomain. These rotations are clearly correlated with a change in the conformation of the protein loop connecting the F1-like core to the amino terminus of the ABCα subdomain. In ATP-bound structures, an invariant glutamine at the amino terminus of this loop contacts the Mg^{2+} counterion bridging the β- and γ- phosphates of the ATP molecule (Hung et al. 1998; Smith et al. 2002), and this interaction holds the ABCα subdomain in a consistent orientation in close proximity to the F1-like ATP-binding core of the domain. In contrast, in ADP-bound structures, in which the position of the Mg^{2+} cofactor shifts because of the absence of a γ-phosphate group, the ABCα subdomain is observed in a variety of different orientations, usually further away from the ATP-binding core (Diederichs et al. 2000; Karpowich et al. 2001; Yuan et al. 2001). Given the associated conformational change in the protein loop connecting the F1-like core subdomain to the ABCα subdomain, combined with the critical role played by the contact between the γ-phosphate group of ATP and the invariant glutamine at the amino terminus of this loop in controlling its conformation, this loop is alternatively called the γ-phosphate switch (Karpowich et al. 2001; Smith et al. 2002) or the Q-loop (Hopfner et al. 2000). The function of the γ-phosphate-dependent ABCα subdomain rotation is not understood in detail, but it has been proposed to function in facilitating release of the ADP product from the active site or in allosteric activation of

ATPase activity in response to the binding of transport substrate (i.e., by modulating ATP-binding affinity in response to allosteric communication from the TMDs) (Karpowich et al. 2001; Yuan et al. 2001).

NBD Mechanochemistry

The foundation of the mechanochemistry of ABC transporters is the formation of an "ATP-sandwich dimer" by NBDs upon binding Mg-ATP (Jones and George 1999; Smith et al. 2002), as schematized in Figure 1D. This structure, which is sometimes called a "head-to-tail" NBD dimer, has two Mg-ATP molecules bound in the inter-NBD interface, with each molecule sandwiched between the Walker A/B motifs in one protomer and the LSGGQ signature sequence in the other protomer. The NBDs forming this structure can be either homodimeric (e.g., in most bacterial ABC transporters) or heterodimeric (as shown for CFTR in Fig. 1D). The binding of Mg-ATP to nucleotide-free NBDs drives their reorientation to form the ATP-sandwich dimer, and this ATP-driven reorientation provides a mechanical "power stroke" that is propagated to the attached TMDs to drive solute translocation in transporters (Smith et al. 2002; Oldham et al. 2011). A similar ATP-driven conformational rearrangement of the TMDs coupled to formation of the ATP-sandwich dimer by NBD1 and NBD2 is believed to mediate channel gating in CFTR (Fig. 1D) (Aleksandrov et al. 2002a,b, 2009; Vergani et al. 2005; Mense et al. 2006; Riordan 2008).

This widely accepted mechanochemical model was established primarily via studies of glutamate-to-glutamine (E-to-Q) mutations in the catalytic base that activates water for hydrolytic attack on ATP (Moody et al. 2002; Smith et al. 2002). As described above, in the NBDs from ABC transporters, this catalytic glutamate immediately follows the aspartate residue at the end of the Walker B motif, yielding a conserved hexapeptide sequence ϕ_4DE (with ϕ being a hydrophobic residue). In a wide variety of ABC transporters, the E-to-Q mutation of the catalytic glutamate blocks the hydrolysis of ATP while preserving its high-affinity binding (Smith et al. 2002; Vergani et al. 2005; Oldham et al. 2007). The mechanochemistry of ABC transporters relies on the energy of ATP binding to drive functional conformational changes during formation of the ATP-sandwich dimer by the NBDs (as schematized in Fig. 1D). The energy of ATP binding to wild-type NBDs is very difficult to assess because of the transient lifetime of the prehydrolysis complex, preventing quantitative evaluation of the effect of the E-to-Q mutation on ATP-binding affinity. However, this mutation has enabled kinetic trapping and characterization of the functional ATP-bound conformation of a wide variety of ABC transporters (Smith et al. 2002; Janas et al. 2003; Vergani et al. 2005; Oldham et al. 2007; Zutz et al. 2010) and other ATPases in the ABC superfamily (Barthelme et al. 2011) (G Boel, PCM Smith, and JF Hunt, unpubl.). In some NBDs, similar thermodynamic, kinetic, and structural results are produced by introducing a histidine-to-alanine (H-to-A) mutation in an invariant active-site residue that hydrogen bonds (H-bonds) directly to the γ-phosphate of ATP (Zaitseva et al. 2005). In contrast, nonhydrolyzable analogs of ATP, all of which change the identity of one of the five atoms in its γ-phosphate group (i.e., the atoms distinguishing it from its hydrolysis product ADP), show variable and often poor efficacy in recapitulating the functional consequences of ATP binding (Hung et al. 1998; Moody et al. 2002; Smith et al. 2002; Horn et al. 2003). In this context, the binding of unmodified ATP to NBDs harboring the E-to-Q or H-to-A mutation in their active site is assumed to preserve affinity better than the binding of nonhydrolyzable analogs of ATP to wild-type NBDs.

A dramatic example of the difference between the two approaches to stabilizing the functional ATP-bound conformation of ABC proteins is provided by the early studies of isolated NBDs that led to the now standard model for ABC mechanochemistry. When expressed in the absence of their cognate TMDs, the NBDs from ABC transporters consistently purify as monomers with a minimal tendency to interact with one another. They remain monomeric in the presence of ADP or nonhydrolyzable analogs of ATP (Hung et al. 1998; Moody et al. 2002). In contrast, several NBDs harboring the E-to-Q mutation in their catalytic base, but not the corresponding wild-type NBDs, form hyperstable dimers on binding unmodified ATP (Moody et al. 2002; Smith et al. 2002). This phenomenon was used to determine the first crystal structure of an ATP-sandwich dimer of the NBDs from an ABC transporter, that of the MJ0796 protein from *Methanococcus jannaschii* (Smith et al. 2002). This model NBD comes from a homodimeric transporter likely involved in lipoprotein transport. Its crystal structure confirmed the earlier prediction that the LSGGQ signature sequence in one NBD directly ligates the ATP molecule bound to the Walker A/B motifs in another NBD in the functional dimeric state of ABC transporters (Jones and George 1999).

Crystallographic and thermodynamic studies on the E-to-Q mutant of MJ0796 suggest that electrostatic charge balance in the active site plays a critical role in modulating ABC mechanochemistry (Smith et al. 2002). The mutant protein, which removes a negative charge from the active site, binds sodium ATP (Na-ATP) rather than Mg-ATP in the ATP-sandwich dimer. Therefore, by compensating for the loss of a negative charge on the protein by the loss of a positive charge on the cofactor of the bound nucleotide, this hyperstable E-to-Q dimer complex maintains the same net electrostatic charge balance in the active site as the wild-type NBD bound to Mg-ATP. Electrostatic analyses of the interface of the ATP-sandwich dimer of MJ0796 showed that there should be substantial electrostatic repulsion between the two NBDs forming the complex (Smith et al. 2002). This observation suggests that a second electrostatically driven power stroke could push apart the NBDs in ABC transporters following ATP hydrolysis. However, the occurrence of such an electrostatic power stroke has yet to be clearly documented in functional studies of any superfamily member.

Extensive functional data on CFTR, summarized elsewhere in this book, suggests that, in the open state of the CFTR channel, NBD1 and NBD2 of CFTR form an ATP-sandwich complex stereochemically equivalent to that formed by the ATP-bound E-to-Q mutant of MJ0796 (as schematized in Fig. 1D) (Gadsby 2004; Gadsby et al. 2006; Mense et al. 2006; Riordan 2008; Aleksandrov et al. 2008, 2009; Hwang et al. 2009). One key observation supporting this model is that electrophysiological studies show that the E-to-Q mutation in NBD2 of CFTR (E1371Q) produces a very long-lived open state of the channel in the presence of ATP (Vergani et al. 2005; Gadsby et al. 2006; Hwang et al. 2009). However, the isolated domains have yet to be observed to form such a complex in vitro, as discussed below.

Structures of Homologous Transmembrane Proteins

The structure of CFTR probably resembles the structures of the bacterial efflux transporters Sav1866 (Dawson and Locher 2006, 2007) and MsbA (Ward et al. 2007) and the mammalian multidrug resistance transporter Pgp (ABCB1) (Aller et al. 2009), which are illustrated in Figure 2. This class of ABC transporter has two TMDs, each containing six transmembrane α-helices. Four of these six α-helices have long cytoplasmic extensions that form two so-called intracellular loops (ICLs). Two of these loops, each comprising two of the cytoplasmic helical extensions and the short polypeptide segment connecting them, bind to one of the cytoplasmic NBDs, with one loop from each TMD crossing over to contact one loop from the other TMD. The resulting pair of loops coming from different TMDs binds in the groove on the surface of a single NBD (Dawson and Locher 2006, 2007).

Examination of the structures of Sav1866, MsbA, and Pgp (Fig. 2) shows that the overall folds of the TMDs and NBDs are broadly similar. However, the existing structures of Pgp and some MsbA orthologs show a wide separation of the NBDs and an inward-facing (i.e., cytoplasm-facing) orientation of the two TMDs (Fig. 2) (Dawson and Locher 2007; Aller et al. 2009). In contrast, Sav1866 and other MsbA orthologs display closely associated NBDs and a more outward-facing configuration of the TMDs (Dawson and Locher 2006, 2007; Ward et al. 2007). Therefore, these structural studies

 Cite this article as *Cold Spring Harb Perspect Med* doi: 10.1101/cshperspect.a009514

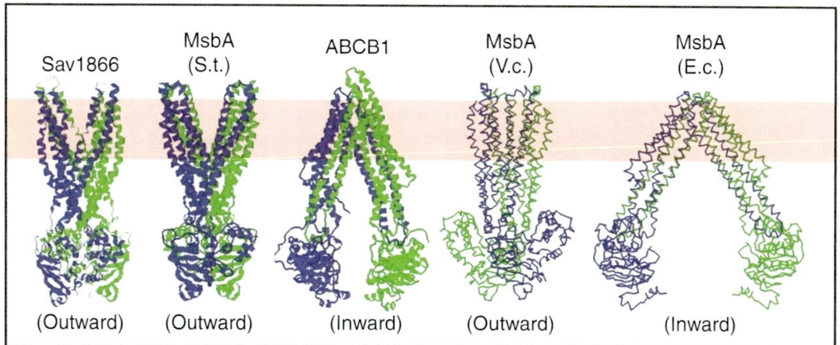

Figure 2. Published crystal structures of ABC transporters homologous to CFTR (Dawson and Locher 2006, 2007; Ward et al. 2007; Aller et al. 2009). Ribbon diagrams are shown for three structures in which side chains were modeled (*left*), whereas backbone traces are shown for the two structures in which the limited resolution of the diffraction data prevented side chain assignment (*right*) (Ward et al. 2007). Structures have been determined for the MsbA orthologs from *Salmonella typhimurium* (S.t.), *Vibrio cholera* (V.c.), and *Escherichia coli* (E.c). All structures shown are homodimers, with the exception of ABCB1 (also called P-glycoprotein or Pgp), which has two NBDs and two transmembrane domains fused in a single polypeptide chain. The two subunits in each homodimer are colored differently (i.e., blue vs. green), as are the amino-terminal and carboxy-terminal halves of ABCB1. The approximate location of the lipid bilayer is illustrated by a pink band. The SAV1866, MsbA (S.t.), and MsbA (V.c.) structures are in outward-facing conformations, whereas the ABCB1 and MsbA (E.c.) structures are in inward-facing conformations.

suggest that major conformational changes can occur in the relative orientation of the four domains in this type of ABC protein. Similar, but smaller, conformational changes have also been observed in a separate group of bacterial ABC proteins, which carry out the import of substances into bacterial cells (Locher et al. 2002; Davidson and Chen 2004; Hollenstein et al. 2007b; Oldham et al. 2007; Pinkett et al. 2007; Kadaba et al. 2008; Khare et al. 2009). Large conformational changes were previously proposed to be part of the substrate translocation mechanism for ABC transporters (Rosenberg et al. 2001).

Inward versus Outward Facing Conformations of CFTR

If CFTR's molecular architecture can be understood in terms of the available structures for intact ABC transporters shown in Figure 2, then relatively coarse structural data ($\sim 1/35$ Å^{-1}) should be reliable to detect the wide separation of the NBDs displayed by the Pgp conformation. The structural data published so far for full-length CFTR (see Fig. 5), which exceeds

this resolution, does not show a discrete separation of two globular domains (Awayn et al. 2005; Zhang et al. 2009, 2010), implying that CFTR more likely resembles the closed conformation observed in Sav1866. However, it is possible that CFTR undergoes major conformational rearrangements but that conditions have not yet been explored that favor the inward-facing conformation with separated NBDs.

CFTR Models Based on ABC Protein Structures

Several structural homology models for CFTR based on Sav1866 and Pgp have been published (Callebaut et al. 2004; Mornon et al. 2008, 2009; Serohijos et al. 2008). Although the two models based on Sav1866 are superficially very similar, especially in the NBD regions of the models (Callebaut et al. 2004; Serohijos et al. 2008), significant differences exist at the level of the individual residues, including alternative rotamers for side chains and differing relative positions (phasing) of amino acids in the transmembrane regions. These discrepancies are not surprising, because the sequence homology between CFTR and Sav1866 or Pgp is low in the transmembrane regions, which limits the stereochemical accuracy of homology models. Inevitably, the beginning and end of transmembrane helices is rather uncertain, and a discrepancy of only two residues can result in a $\sim$180Å change in the orientation of a side chain versus the long axis of the transmembrane α-helix.

Evolution of CFTR Ion Channel Function

Hypotheses for the evolution of the channel function of CFTR suggest that the ATP-bound, outward-facing transmembrane conformation observed in the model ABC transporters is likely to be similar to that adopted by CFTR when its channel is open, whereas the inward-facing transmembrane conformation is likely to be similar to that adopted by CFTR when its channel is closed (Vergani et al. 2003, 2005; Gadsby et al. 2006; Kos and Ford 2009). These hypotheses fit well with the observations that ATP is required to initiate CFTR channel opening both in patch-clamp measurements in vivo and in single-channel measurements from black lipid membranes in vitro (i.e., from such membranes fused with CFTR-containing vesicles) (Sheppard et al. 1993; Sheppard and Welsh 1999; Sheppard 2004; Vergani et al. 2003, 2005; Gadsby et al. 2006; Aleksandrov et al. 2007). Furthermore, CFTR channels show long quiescent periods approximating the expected rate of ADP release and ATP rebinding by NBD2 in CFTR (Sheppard et al. 1993; Sheppard and Welsh 1999; Vergani et al. 2003, 2005; Sheppard 2004; Gadsby et al. 2006). Similarly, the likely timescale of the large structural changes between the observed inward- and outward-facing conformations of the model ABC transporters seems to be consistent with the moderate ATPase velocity of these proteins, supporting the hypothesis that their large-scale conformational changes are coupled to ATP binding and hydrolysis (Davidson and Chen 2004; Dawson et al. 2007; Davidson and Maloney 2007; Oldham et al. 2007; McDevitt et al. 2008; Khare et al. 2009). On the other hand, electrophysiological recordings from CFTR can also show very rapid transitions from open to closed states and between subconductance states, on a timescale much faster than its rate of ATP hydrolysis (Cai et al. 2003; Da Paula et al. 2010). These observations suggest that small conformational shifts could produce transiently or partially closed states of the channel while the NBDs of CFTR remain in the ATP-bound state.

STRUCTURAL STUDIES OF THE NBDS FROM CFTR

Variations in ABC Domain Structure in NBD1 from CFTR

For many years, efforts to express and purify NBD1 from CFTR were impeded by ambiguities in its domain boundaries (Thomas et al. 1991; Qu et al. 1997). A breakthrough was made by the company Structural GenomiX Inc., in a project that was initiated and funded by the U.S. Cystic Fibrosis Foundation. This long-term project applied high-throughput robotic cloning and protein expression methods to evaluate many possible domain boundaries using CFTR orthologs from a variety of vertebrate species, leading eventually to the isolation and crystallization of soluble protein constructs containing NBD1 (Lewis et al. 2004). Insights developed in the course of determining crystal structures of NBD1 from murine CFTR (mNBD1) enabled subsequent determination of crystal structures of NBD1 from human CFTR (hNBD1—Figs. 1A–C, 3A,C,E, and 4) (Lewis et al. 2005, 2010). To date, nine crystal structures of mNBD1 and eight of hNBD1 have been deposited in the Protein Data Bank (PDB). (The mNBD1 structures spanning residues 389–673 or 389–655 are available under accession codes (IDs) 1Q3H, 1R0W, 1R0X, 1R0Y, 1R0Z, 1R10, 1XF9, 1XFA, and 3SI7, while the hNBD1 structures spanning residues 389–678 or 389–646 are available under IDs 1XMI, 1XMJ, 2BBO, 2BBS, 2BBT, 2PZE, 2PZF, and 2PZG.)

Even using domain boundaries equivalent to those in the crystal structures of mNBD1, the orthologous human domain could only be purified in sufficient yield and in monodisperse form after introducing a series of "solubilizing" point mutations (Lewis et al. 2005, 2010). Many sets of such mutations were evaluated using robotic mutagenesis and expression methods, and several sets yielded significant improvements in the apparent solubility of the domain during purification and concentration for crystallization. One such set, called the Teem set, comprised three point mutations (G550E/R553Q/R555K) that previously were shown to promote improved biogenesis of F508del CFTR in tissue cultures cells, presumably by improving the stability of the protein (Teem et al. 1993; DeCarvalho et al. 2002). (This mutation set is found in combination with the F409L, F429S, F433L, and H667R mutations in PDB IDs 1XMJ and 2BBO.) The other mutation sets that improved the yield of soluble hNBD1 involved substitution of surface-exposed residues in hNBD1 with more polar residues occurring at the same position in CFTR orthologs from other species (F429S/F494N/Q637R found in PDB ID 2BBS, F494N/Q637R found in PDB ID 2BBT, and F429S/H667R found in PDB ID 1XMI). Eventually, it was shown that the mutations that significantly improve the purification of hNBD1 all thermodynamically stabilize the native conformation of the domain compared to an aggregation-prone "molten globule" conformation (described further below) (Protasevich et al. 2010; Wang et al. 2010). Therefore, what had been perceived as a solubility problem, a technical problem related to requirements for purification in vitro, turned out to be a stability problem directly related to the molecular pathology caused by the F508del mutation (as explained below). Thus, the research program focused on crystallographic characterization of CFTR domains eventually provided vital insight not only into the atomic structure of CFTR but also into the molecular pathogenesis of CF.

There are no significant differences in the crystal structures of mNBD1 and hNBD1, irrespective of the presence or absence of stabilizing/solubilizing mutations in the human domain, even though their structures were determined in solvent environments with dramatic differences in salt concentration and in

some cases pH (Lewis et al. 2010). Both structures show the canonical fold for the NBD from an ABC transporter, as described above. There are local conformational variations in NBD1 at the carboxyl terminus of the ABCβ subdomain and immediately preceding the LSGGQ signature sequence, in regions that show significant structural variations among different members of the ABC transporter superfamily (Zaitseva et al. 2005). However, in addition to these well-precedented structural variations, the crystal structures revealed two noteworthy structural variations specific to NBD1 from CFTR. The first of these, which was anticipated from sequence alignments with NBDs from ABC transporters, involves the substitution of two highly conserved residues in the ATPase active site with noncanonical amino acids (Lewis et al. 2004). The catalytic glutamate residue is substituted with a serine (S573), rendering the Walker B sequence ϕ_4DS instead of ϕ_4DE, and a catalytically vital histidine residue that contacts the γ-phosphate of ATP is also substituted with serine (S605). Although these changes do not produce a significant change in active site conformation relative to NBDs from model ABC transporters, they prevent ATP hydrolysis by NBD1 (Lewis et al. 2004; Thibodeau et al. 2005). This property allowed determination of both the murine (Lewis et al. 2004) and human (Lewis et al. 2005) structures bound to unmodified ATP with a physiological Mg^{2+} cofactor, which has not been possible with any catalytically active NBD. Enzymological studies of the domain in solution verified the lack of significant ATPase activity in purified preparations of NBD1 (Lewis et al. 2004; Thibodeau et al. 2005).

The second noteworthy structural variation revealed by the crystal structures of NBD1 from CFTR is the addition of two protein segments not found in NBDs from model ABC transporters (Lewis et al. 2004, 2010). One of these is an ~35-residue insertion in the loop between the first two β-strands in domain (i.e., at the amino terminus of the ABCβ subdomain). NBD1 preserves the aromatic residue at the end of the first β-strand in the ABCβ subdomain (Lewis et al. 2005), which makes stacking interactions

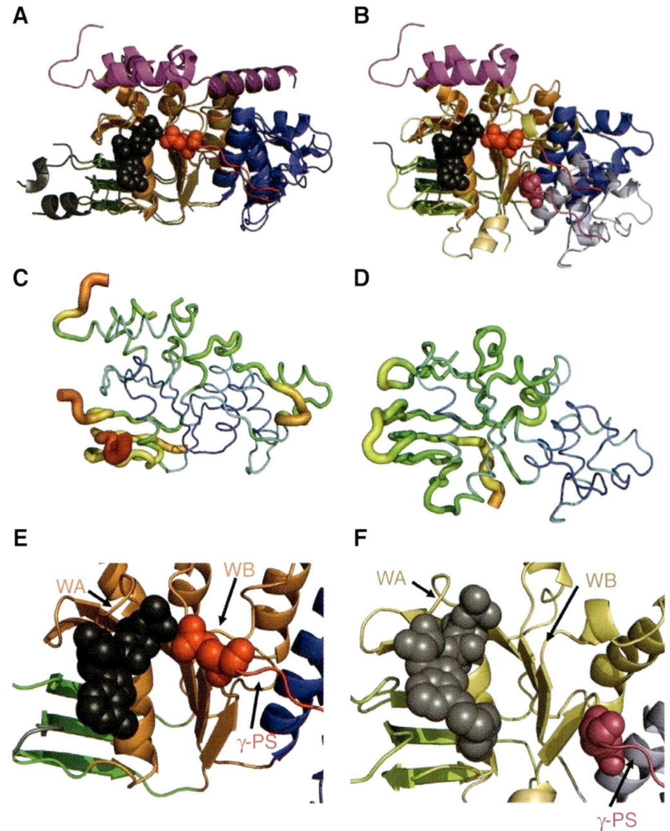

Figure 3. Subdomain reorientation and backbone B-factors in hNBD1 and hNBD2 crystal structures. All panels show the hNBD1 (panels *A*, *C*, and *E* on the *left* and panel *B* on the *right*) and hNBD2 (panels *B*, *D*, and *F* on the *right*) domains in the same orientation after least-squares alignment of the β-strands in their ATP-binding cores (i.e., the ABCβ subdomain plus the F1-like core subdomain). The ribbon diagrams in panels *A,B* and *E,F* are colored equivalently to Fig. 1*B,C* (i.e., orange or yellow for the F1-like ATP-binding core subdomain, shades of blue for the ABCα subdomain, shades of green for the ABCβ subdomain, and red or magenta for the γ-phosphate switch or Q-loop). The bound ATP molecules (black or gray) and the invariant glutamines in the γ-phosphate switch (Q493 in hNBD1 in red and Q1291 in hNBD2 in magenta) are shown in space-filling representation. (*A,B*) Least-squares superpositions of two full-length F508-hNBD1 structures (panel *A*—PDB IDs 2BBO and 1 XMI) and of one of these with hNBD2 (panel *B*—PDB IDs 2BBO and 3GD7, respectively). The 7° rotation of the ABCα subdomain between the two hNBD1 structures shown here represents the largest displacement observed when comparing any two hNBD1 crystal structure, all of which have ATP bound in the active site. As discussed in the text, the 17° rotation and ∼10 Å displacement of the ABCα subdomain in hNBD2 relative to the ATP-bound conformation in hNBD1 likely prevents formation of a hydrolytically active ATP-sandwich heterodimer with hNBD1 in vitro, even though extensive evidence supports formation of such a complex coupled to opening of the CFTR channel in vivo (as schematized in Fig. 1D) (Gadsby 2004; Mense et al. 2006; Aleksandrov et al. 2008, 2009; Hwang et al. 2009). (*C,D*) Equivalent views of the hNBD1 (panel *C*) and hNBD2 (panel *D*) structures from panel *B* with the thickness and color of the backbone trace representing their mean backbone B-factors. B-factors quantify the variation or disorder in atomic position in a crystal structure, with larger B-factors indicating more disorder. The default encoding scheme in PyMOL was used to generate these images (i.e., 0.6–4.0 Å backbone thickness and a linear blue-to-green-to-red color ramp for B-factors running from 32–80 Å^2 for hNBD1 and from 19–76 Å^2 for hNBD2). (*E,F*) Close-up views of the ATP-binding sites in the crystal structures of F508-hNBD1-Δ(RI,RE) (panel *E*—PDB ID 2PZE) and hNBD2 (panel *F*—PDB ID 3GD7). (*Legend continues on following page.*)

with the adenine of ATP exactly as observed in NBDs from model ABC transporters (Karpowich et al. 2001). However, the ∼5-residue tight turn that follows this conserved aromatic in most NBDs is replaced by a 35-residue loop that is largely disordered based on crystallographic analyses as well as hydrogen-deuterium exchange mass spectrometry (HDX-MS) studies conducted on the domain in solution (Lewis et al. 2010). This loop spanning approximately residues 402–436 in human CFTR has been called the regulatory insertion or RI (Figs. 1A, B and 3A,C). Its unexpected presence following the first ∼12 residues in NBD1 is largely responsible for the inability to accurately identify the amino terminus of the domain based on sequence analysis (Lewis et al. 2010).

A second protein segment that is not shared with NBDs from model ABC transporters is an ∼35-residue extension appended at the carboxyl terminus of NBD1 (Figs. 1A,B and 3C,E) (Lewis et al. 2004). This regulatory extension or RE segment spanning approximately residues 638–672 in human CFTR could be considered to comprise the amino terminus of the R domain. However, its inclusion improves the solubility of NBD1 protein constructs (Lewis et al. 2004, 2005), and it adopts a consistent folded conformation (Figs. 1B, and 3A,C) in most crystal structures of murine and human NBD1 constructs, including structures determined under widely varying conditions of ionic strength and pH (Lewis et al. 2010). HDX-MS studies show that the RE rapidly explores unfolded conformations, suggesting that it is likely to be predominantly unfolded in the isolated domain in solution (Lewis et al. 2010). However, elegant NMR dynamics studies support the inference from crystallographic studies that it

also populates a folded state in which it interacts at specific sites on the surface of NBD1 (Baker et al. 2007; Kanelis et al. 2010).

Eventually, a construct of hNBD1 was designed with the RI (residues 405–436) and the RE (residues after Gly646) both deleted (i.e., yielding a construct spanning residues 389–646 in human CFTR with a 32-residue internal deletion) (Atwell et al. 2010). Although removal of the RE has little influence on the stability of the domain, removal of the RI strongly stabilizes the domain in vitro (Protasevich et al. 2010) and significantly improves biogenesis of F508del-CFTR in tissue culture cells (Aleksandrov et al. 2010). This effect is presumably attributable to favorable energetic interaction of the RI with the molten globule conformation of NBD1, which reduces the free energy difference between the native state and the unfolded state (Protasevich et al. 2010; Wang et al. 2010). The hNBD1 construct without either the RI or RE, called NBD1-Δ(RI,RE), has been an important tool for biochemical and biophysical studies of CFTR because of its improved stability compare to the full-length domain (Aleksandrov et al. 2010; Protasevich et al. 2010; Wang et al. 2010). A crystal structure has been determined for this NBD1-Δ(RI,RE) construct in the presence of Mg-ATP (Atwell et al. 2010), which shows it forming a homodimeric ATP-sandwich complex with two ATP molecules bound between the Walker motif in one protomer and the LSGGQ signature sequence in the other protomer. This structure is similar in conformation and oligomeric organization to the physiological ATP-sandwich complexes formed by the NBDs from model homodimeric bacterial ABC transporters. However, the NBD1-Δ(RI,RE) construct, like all other NBD1

Figure 3. (*Continued*) Labels indicate the locations of the Walker A (WA) and Walker B (WB) motifs in the F1-like core subdomain and the γ-phosphate switch (γ-PS) linking the core subdomain to the amino terminus of the ABCα subdomain. Although hNBD1 is catalytically inactive, due at least in part to the substitution of a serine for the catalytic glutamate residue adjacent to the Walker B motif, its γ-PS adopts a catalytically active conformation in which its amino-terminal glutamine (Q493, red spheres) contacts the Mg^{2+} cofactor of ATP, as observed in ATP-bound structures of catalytically active NBDs from a wide variety of ABC transporters (Hung et al. 1998; Smith et al. 2002). In contrast, the γ-PS in hNBD2 adopts a different conformation resulting in an ∼17 Å displacement of the amide group of its amino-terminal glutamine (Q1291, magenta spheres) from the ATP molecule bound in the active site.

constructs characterized to date, does not homodimerize in solution in the presence of ATP (Atwell et al. 2010). Thus, the crystal structure of NBD1-Δ(RI,RE) has trapped a low affinity and likely nonphysiological homodimer that nonetheless provides a potentially useful stereochemical model for the physiological ATP-sandwich heterodimer formed with NBD2 in CFTR.

A Structural Paradigm for CFTR Regulation by the R Domain

The crystal structures of NBD1 constructs including the RI and RE segments have provided a model for the structural basis of CFTR activation by covalent modification of the R domain (Lewis et al. 2004, 2010). In different crystal structures, both segments show variable conformations (Lewis et al. 2010), consistent with biophysical assays showing that these segments are dynamic in solution (Baker et al. 2007; Lewis et al. 2010). In some structures, each segment is observed to adopt a conformation interacting with the NBD2-binding surface of NBD1, suggesting that the R domain could inhibit channel activation by physically blocking formation of the ATP-sandwich complex by NBD1 and NBD2 (Fig. 1D). Although it is unclear whether the RI functions in this manner physiologically, substantial evidence exists supporting participation of the RE in such a regulatory mechanism. Most crystal structures of NBD1 show a long segment of the RE forming an α-helix docked at the same site on the NBD2-binding surface of NBD1, which would clearly block formation of any ATP-sandwich complex by NBD1 (Lewis et al. 2004). This conformation of the RE is observed in all crystal structures of both the murine and human domains determined at neutral pH, including structures from crystals grown at drastically different ionic strengths ranging from low millimolar to several molar (Lewis et al. 2010).

NMR studies show that the RE is largely disordered in solution but populates some partially folded and likely α-helical conformational states (Baker et al. 2007). Chemical-shift perturbation studies show that several regions of NBD1, including some observed to bind to the RE in crystal structures, interact with unphosphorylated RE, and these interactions are weakened by phosphorylation of the RE by PKA. These studies suggest that multiple segments of the RE interact with multiple regions of NBD1 to enable integration of diverse signaling pathways mediating covalent modifications of different segments of the R domain (Csanády et al. 2006; Baker et al. 2007; Kanelis et al. 2010). One such interaction is likely to involve the crystallographically observed docking of the RE on the NBD2-binding surface of NBD1 (Lewis et al. 2004, 2010), which will block formation of the physiological ATP-sandwich heterodimer with NBD2 until this interaction is somehow disrupted by phosphorylation of the RE or an allosterically coupled site elsewhere in the R domain (Fig. 1D).

Structural Consequences of the F508del Mutation in NBD1

The deletion of the phenylalanine residue at position 508 in NBD1 of CFTR is the predominant CF-causing mutation in the human population (Cutting 1993, 2005). This F508del mutation accounts for at least one mutant allele in ∼90% of CF patients, making it a critical priority to understand the molecular defects that it causes and to develop methods to remediate them. Extensive efforts to stabilize NBD1 for structural studies, as described above, eventually led to successful purification and crystallization of the domain harboring the destabilizing F508del mutation (Lewis et al. 2005). The crystal structure of the mutant protein shows conformational perturbations only in the immediate vicinity of the mutation (Fig. 4A) (Lewis et al. 2005, 2010), and hydrogen–deuterium exchange studies show that even perturbations in backbone dynamics are limited to modest enhancements at sites immediately adjacent to F508 in the primary sequence (Lewis et al. 2010). Extensive in vitro biophysical studies of NBD1 (Protasevich et al. 2010; Wang et al. 2010), correlated with a wide variety of in vivo studies (DeCarvalho et al 2002; Pissarra et al. 2008; Aleksandrov et al. 2010, 2012), have established that

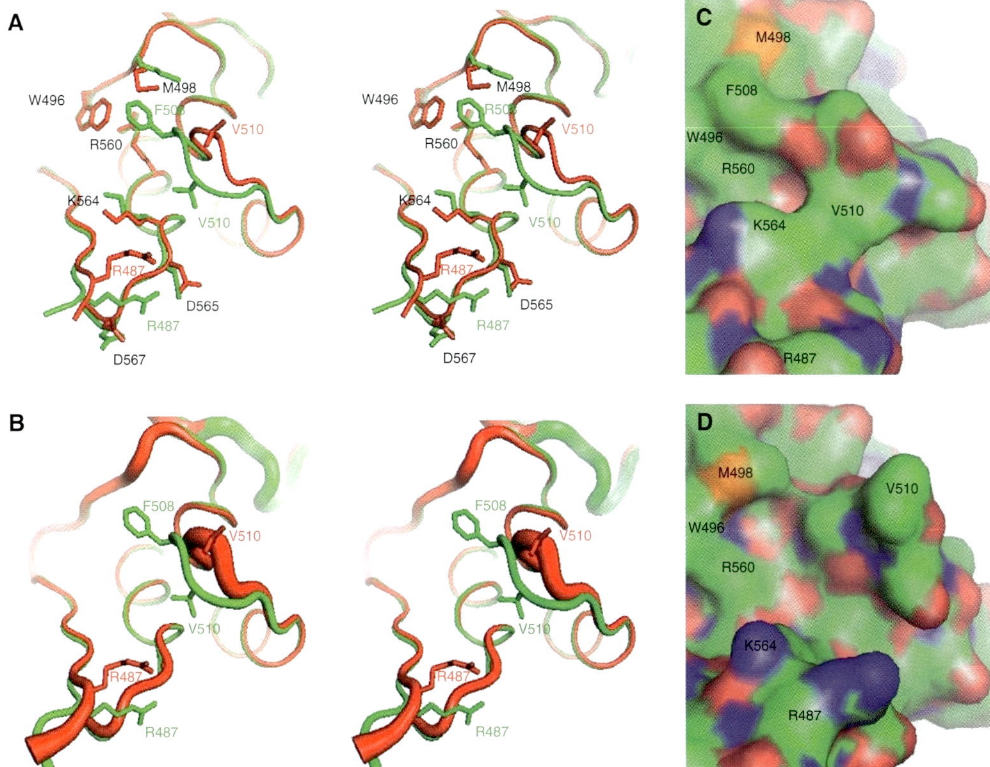

Figure 4. Structural consequences of the F508del mutation in hNBD1. Structures of hNBD1 with and without the F508del mutation were aligned based on least-squares superposition of the three α-helices comprising the core of the ABCα subdomain. (*A*) Stereopair showing the interface between the F1-like core subdomain and the ABCα subdomain in F508 (green, PDB ID 2BBO) and F508del (red, PDB ID 1XMJ) hNBD1 structures. Selected side chains are shown in ball-and-stick representation. The site of attachment of hNBD1 to the transmembrane domains of CFTR is proximal to the viewer in this orientation, and the γ-PS is at the *upper left* (i.e., the backbone segment including residues W496 and M498). (*B*) Stereopair showing the same view of the same hNBD1 structures using the same color scheme, but with the thickness of the backbone traces encoding their mean backbone B-factors (using the default scheme in PyMOL, with 0.6–4.0 Å radius representing 33–77 Å^2 and 12–69 Å^2 for the F508 and F508del structures, respectively). (*C,D*) Surface representations of the same structures showing the alterations in local surface topography caused by the F508del mutation. These two panels show the same view of the molecular surfaces of F508 (panel *C*) and F508del (panel *D*) hNBD1, with carbon atoms colored green, oxygen atoms colored red, nitrogen atoms colored blue, and sulfur atoms colored orange. Note that the structures shown here contain seven point mutations included in hNBD1 constructs because of their beneficial influence on yield during purification—F409L, F429S, F433L, G550E, R553Q, R555K, and H667R. Extensive analyses led to the conclusion that these mutations do not significantly perturb the native ground-state conformation of hNBD1 (Lewis et al. 2010) but that they do stabilize it thermodynamically compared to an aggregation-prone molten globule intermediate that accounts for both the instability of F508del-CFTR in vivo and the progressive loss of hNBD1 constructs in vitro during purification (Protasevich et al. 2010; Wang et al. 2010). (The images shown here are from Lewis et al. 2010; adapted, with permission, from the authors.)

thermodynamic destabilization of NBD1 by the F508del mutation plays a central role in the molecular pathogenesis of CF, as outlined in the next section. However, the local structural perturbations caused by the F508del mutation in the folded conformation of NBD1 are likely to contribute at least indirectly to pathogenesis by disrupting structural interactions between NBD1 and the TMDs of CFTR (Lewis et al. 2005, 2010; Serohijos et al. 2008; He et al. 2010; Loo et al. 2010; Thibodeau et al. 2010). Recently published work suggests that developing effective drugs to correct the molecular defects caused by the F508del mutation in CFTR may require correction of these disrupted interdomain interactions (Mendoza et al. 2012; Rabeh et al. 2012).

The F508 residue packs onto the surface of the ABCα subdomain in NBD1 in a broad but shallow groove at the interface between this subdomain and the F1-like ATP-binding core (Fig. 4A) (Lewis et al. 2004, 2010), as described briefly above. Crystal structures of full-length ABC transporters show that this region of the NBD is the principal site of packing interactions with the cytoplasmic loops of the TMDs (Fig. 1A) (Locher et al. 2002; Dawson and Locher 2006; Serohijos et al. 2008; Aller et al. 2009). The conservation of equivalent interdomain interactions in CFTR is supported by the crosslinking of pairs of cysteines replacing residues on either side of the predicted interface in an otherwise cysteine-less CFTR construct (Serohijos et al. 2008). Moreover, the functional significance of this interface is supported by electrophysiological experiments showing reversible arrest of single-channel gating on addition of a bifunctional sulfhydryl crosslinker to such a CFTR construct with cysteines replacing residues F508 in NBD1 and F1068 in TMD2 (Serohijos et al. 2008).

These important interdomain structural interactions between NBD1 and the TMDs of CFTR are likely to be significantly perturbed by the deletion of F508, even though the structural consequences of this mutation are relatively modest and localized within NBD1 (Lewis et al. 2004, 2010). The F508del mutation changes the conformation of the adjacent loop that connects the first and second α-helices in the ABCα sub-

domain, effectively shortening it (Fig. 4A,B) (Lewis et al. 2010). The glycine residue at position 509, which is invariant in CFTR orthologs and conserved at this site in many other ABC transporters, participates in a "capping motif" that H-bonds to the carboxyl terminus of the α-helix containing F508. These H-bonding interactions are altered by the deletion of F508, which causes a significant change in the backbone conformation of the loop spanning residues 509–511 in NBD1. This change flips the side chain of the valine residue at position 510 from being partially buried on the surface of the ABCα domain to projecting outward (Fig. 4) into the likely interface between NBD1 and the TMDs of CFTR (Fig. 1A). The removal of the side chain of residue F508 combined with the conformational change in the adjacent loop produces a substantial change in the topography and qualitative chemical characteristics of the surface of NBD1 at its likely site of interaction with the TMDs of CFTR (Fig. 4C,D) (Serohijos et al. 2008; Lewis et al. 2010).

These structural variations suggest that the F508del mutation is likely to perturb the interaction between NBD1 and the TMDs of CFTR (Lewis et al. 2004, 2010; Serohijos et al. 2008). Although the inability to date to purify the TMDs of CFTR in a native conformational state has prevented direct biophysical evaluation of this hypothesis, its validity is supported by the results from mutagenesis experiments conducted by several different laboratories (He et al. 2010; Loo et al. 2010; Thibodeau et al. 2010; Mendoza et al. 2012; Rabeh et al. 2012). These experiments provide evidence that the defective trafficking of F508del-CFTR in tissue culture cells can be suppressed by strengthening the interaction between the perturbed surface on NBD1 and the cognate region of the TMD. A second-site R1070W mutation in the proximal region of the TMDs produces significant suppression of the defect (Thibodeau et al. 2010), presumably by increasing hydrophobic interactions with the altered conformation of residue V510 in F508del-NBD1. Moreover, restoration of the trafficking of F508del-NBD1 by the V510D suppressor mutation, which introduces a negative charge into a generally apolar region of the interdomain interface (Fig. 4C,D), is strongly attenuated by introducing the R1070A or R1070D mutations that remove a complementary positive charge from the proximal surface of the TMD (Loo et al. 2010). In addition to providing insight into the molecular pathologies caused by the F508del mutation, these observations provide additional evidence that homology models based on model ABC transporters accurately predict the structural interactions between NBD1 and the TMDs of CFTR (Serohijos et al. 2008; Alexander et al. 2009; Mornon et al. 2009; Dalton et al. 2012).

Two recent papers both concluded that strengthening the interface between NBD1 and the TMDs in F508del-CFTR would be required for effective pharmacological correction of its trafficking defect (Mendoza et al. 2012; Rabeh et al. 2012). However, another recent paper directly challenges this critical pharmacological inference (Aleksandrov et al. 2012). The two papers supporting this inference both present systematic evaluations of different combinations of suppressor mutations in correcting the trafficking defect in F508del-CFTR in tissue culture cells. In these studies, none of the suppressor mutation sets confined to NBD1 restored maturation to wild-type levels, but full restoration or correction of the F508del defect was observed when combining suppressor mutations in NBD1 with mutations likely to strengthen the interdomain interface between NBD1 and the TMDs in F508del-CFTR (Mendoza et al. 2012; Rabeh et al. 2012). In contrast, the third recent paper showed essentially wild-type levels of maturation and stability in F508del-CFTR containing second-site suppressor mutations exclusively in NBD1 (i.e., the I539T mutation plus four proline substitutions found in chicken CFTR, which is naturally more thermostable than human CFTR) (Aleksandrov et al. 2012). Based on this latter finding, small molecules that bind tightly to properly folded NBD1, and thereby significantly stabilize its native conformation, could be sufficient to correct the molecular pathologies caused by the F508del mutation and cure the disease in patients having this mutation. However, the conflicting interpretation offered in the other two recent papers

implies that effective pharmacological correction would additionally require stabilization of the interface between NBD1 and the TMDs in F508del-CFTR. Given the critical implications of this controversy for CF drug development, future cell biological and pharmacological investigations should focus on establishing the operative constraints in a system that accurately reflects the dynamics of CFTR maturation and degradation in human lungs.

Perturbation of the NBD1 Folding Pathway by F508del

The folding pathway of NBD1 plays an important role in the molecular pathology caused by the F508del mutation. Recent publications describe this pathway in detail (Protasevich et al. 2010; Wang et al. 2010), so only its central features will be summarized here. The native conformation of NBD1 (Figs. 1 and 3) exists in equilibrium with a relatively low energy "molten globule" conformation. The molten globule, first described by Privalov and coworkers (Griko and Privalov 1994; Griko et al. 1994), is a dynamic species that retains a substantial amount of native-like 2Å structure but does not form stable tertiary structure (Arai and Kuwajima 2000; Englander 2000; Redfield 2004; Sampson et al. 2011). Its thermodynamic hallmark, which has been clearly observed in hNBD1, is a very low enthalpy of unfolding for a species with a substantial amount of regular secondary structure. Increasing temperature decreases the stability of the native conformation of NBD1 compared to the molten globule, while the binding of ATP has the opposite effect and substantially increases the stability of the native conformation compared to the molten globule (Protasevich et al. 2010; Wang et al. 2010).

Extensive thermodynamic analyses show that the F508del mutation destabilizes the native conformation of NBD1 compared to the molten globule, making the latter species the dominant conformation of the isolated domain at $37°C$ even in the presence of physiological concentrations of ATP (Protasevich et al. 2010). In contrast, only a small proportion of the population of the wild-type domain adopts the molten globule conformation under these conditions. The molten globule of NBD1 aggregates aggressively and irreversibly in vitro (Wang et al. 2010). A series of second-site mutations in NBD1 have parallel effects in rescuing the trafficking defect in CFTR in vivo (DeCarvalho et al. 2002; Pissarra et al. 2008; Aleksandrov et al. 2010) and inhibiting molten globule formation by isolated NBD1 in vitro (G550E/R553Q/R555K, F494N/Q637R, or V510D) (Protasevich et al. 2010; Wang et al. 2010). These results suggest that the influence of F508del in promoting formation of the aggregation-prone molten globule conformation is likely to be the principal factor underlying the molecular pathologies caused by the F508del mutation in vivo.

In this context, small molecules that bind to NBD1 and stabilize it in the native conformational state should be able to remediate the trafficking defect caused by the F508del mutation in CFTR. The efficacy of this approach is supported by observations on two related compounds that display significant corrector activity in tissue culture cells and that also appear to stabilize purified NBD1 against thermal denaturation in vitro (Sampson et al. 2011). Broader and deeper investigations are merited of this pharmacological approach focusing on thermodynamic correction of the folding defect in F508del-NBD1. It will be important to identify the location of compound binding sites in natively folded F508del-NBD1 and also to evaluate whether stabilization of NBD1 must be accompanied by stabilization of the NBD1-TMD interface to produce effective correction of the molecular defect caused by the F508del mutation, a claim supported by some studies cited in the last section (Mendoza et al. 2012; Rabeh et al. 2012) but questioned by others (He et al. 2010; Aleksandrov et al. 2012).

The Existing Crystal Structure of NBD2 Shows an Inactive Conformation

A crystal structure of NBD2 from human CFTR (hNBD2) bound to unhydrolyzed phenyl-ATP (ϕATP) has been solved and deposited into the Protein Data Bank by Structural GenomiX Inc., the same company that produced the crystal structures of NBD1 described above. They used equivalent high-throughput robotic cloning and expression methods in combination with more aggressive and ambitious protein-engineering methods to generate a crystal structure of an hNBD2 construct. This construct spans residues 1193–1427 in human CFTR and has the regulatory domain from the *E. coli* MalK protein fused at its carboxyl terminus (PDB ID 3GD7; X Zhao, S Atwell, and S Emtage, unpubl.). This domain, which is found at the carboxyl terminus of the NBD from the *E. coli* maltose transporter (Diederichs et al. 2000; Oldham et al. 2007), was observed to mediate extensive crystal-packing contacts in earlier crystal structures containing the domain (Chen et al. 2003). These observations suggested that fusion of this domain at an equivalent position in hNBD2 might help promote its crystallization. Therefore, a homology model was constructed for hNBD2 and used to engineer a fusion joint likely to preserve the rigid orientation of the regulatory domain of MalK relative to the F1-like ATP-binding subdomain of the NBD. The design strategy involved superimposing the carboxy-terminal α-helices from the homology model of hNBD2 and the experimentally determined crystal structure of MalK to create a precise in-frame fusion construct. This fusion construct, like all other constructs of hNBD2 characterized to date, suffers from poor stability and solubility during purification in vitro. Improvements in the yield of soluble protein were obtained by introducing the hydrolytically inactivating H1402A mutation in ATPase active site of hNBD2 plus a series of "solubilizing" mutations on its surface (Q1280E/Y1307N/W1310H/Q1411D) (X Zhao, S Atwell, JF Hunt, et al., unpubl.). Two additional "surface entropy reduction" mutations (E1308A/Q1309A) were introduced into the construct to promote crystallization. The H1402A mutation as well as at least some of the "solubilizing" mutations seem likely to stabilize the native conformation of the domain thermodynamically, as observed for the solubilizing surface mutations that improved the stability and yield of preparations of hNBD1 (as described above). However, because of the limited stability of the domain in the absence of these mutations, it has not been possible to verify this inference experimentally.

Nonetheless, indirect evidence supports the hypothesis that the H1402A mutation stabilizes hNBD2. Based on the ATP-bound crystal structures of model NBDs from bacterial transporters, histidine 1402 in hNBD2 contacts the γ-phosphate of ATP in the prehydrolysis complex, while glutamate 1371 is the catalytic base that actives water for hydrolytic attack on the γ-phosphate of ATP (Hung et al. 1998; Smith et al. 2002; Zaitseva et al. 2005). Electrophysiological studies of the H1402A and E1371Q mutations in intact human CFTR support these inferences concerning catalytic geometry by showing that the H1402A (Kloch et al. 2010) and E1371Q (Vergani et al. 2005) mutations both greatly increase the lifetime of the open state of the CFTR chloride channel, presumably because they block ATP hydrolysis and stabilize ATP-sandwich heterodimer formed by hNBD1 and hNBD2. This interpretation is consistent with the effects of equivalent mutations in a wide variety of model NBDs and ABC transporters (Smith et al. 2002; Janas et al. 2003; Vergani et al. 2005; Oldham et al. 2007; Zutz et al. 2010), as explained above. However, despite their similar influence on the gating properties of intact CFTR, the H1402A and E1371Q mutations have dramatically different effects on the stability and yield of hNBD2 during purification, with the first greatly improving yield compared to the second (X Zhao, S Atwell, JF Hunt, et al., unpubl.). In silico energetic calculations suggest that the H1402A mutation stabilizes isolated hNBD2 bound to Mg-ATP, while the E1371Q mutation destabilizes it (P Kumar, C Wang, JF Hunt, et al., unpubl.). Future experimental studies will be required to evaluate this computational inference that the influence of these mutations on the yield of purified hNBD2 reflects their differential effects on the thermodynamic stability of the native conformation of the domain.

Although the crystal structure of the extensively engineered hNBD2 construct shows a canonical fold, without any elaborations like the RI in hNBD1, it adopts a surprising and likely inactive conformation. This structure (PDB id 3GD7) shows the ABCα subdomain pulled away from the F1-like ATP-binding core in hNBD2 by $\sim$10 Å compared to its position in hNBD1 (Fig. 3B), even though unhydrolyzed Mg-ϕATP is observed in the active site. (In this crystal structure, the phenyl group covalently attached to the adenine ring in ϕATP mediates intermolecular packing contacts that seem likely to stabilize the lattice, presumably explaining why the structure could be obtained only with ϕATP; X Zhao, S Atwell, and S Emtage, unpubl.) Furthermore, the γ-phosphate switch (or Q-loop) in hNBD2 adopts a different conformation that moves the terminal atoms in the side chain of the invariant glutamine at its amino terminus out of the active site, to a position 17 Å away from the position of the equivalent residue in hNBD1 (i.e., comparing Gln1291 in hNBD2 in Fig. 3F to Gln493 in hNBD1 in Fig. 3E). Although qualitatively similar displacements of the ABCα subdomain and alternative conformations of the γ-phosphate switch have been observed in nucleotide-free and Mg-ADP-bound structures of a variety of NBDs from model ABC transporters (Diederichs et al. 2000; Karpowich et al. 2001; Yuan et al. 2001), as described above, a conformation of this kind has not previously been observed in any structure with unhydrolyzed ATP or a non-hydrolyzable analog of ATP bound in the active site. The observed conformation of the ABCα subdomain and γ-phosphate switch are likely to be incompatible with binding to hNBD1 and formation of the functional ATP-sandwich heterodimer complex believed to gate opening of the CFTR channel (as described above and schematized in Fig. 1D).

All previously determined NBD structures in which a γ-phosphate group or an analog of this group are present in the active site show the ABCα domain aligned in close apposition to the F1-like core subdomain (Hung et al. 1998; Smith et al. 2002; Chen et al. 2003; Oldham et al. 2007; Atwell et al. 2010), with the conserved glutamine residue at the amino terminus of the γ-phosphate switch contacting the metal cofactor bound to the nucleotide (e.g., as shown for Gln493 in hNBD1 in Fig. 3E). Among the 17 different crystal structure of NBD1 currently available in the PDB, there is at most a 7° rotation of the ABCα subdomain producing minor displacements of its putative NBD2-interacting surface (Fig. 3A), even though these structures come from two different species and were determined in some cases in solutions with drastically different pHs and ionic strengths (Lewis et al. 2010). In contrast, the ~10 Å displacement of the ABCα subdomain observed in all four copies of hNBD2 present in the asymmetric unit of its crystal structure moves the domain diagonally away from the likely location of its interface with hNBD1 (i.e., based on the geometry of the ATP-sandwich dimer complexes observed in structures of NBDs from model ABC transporters as well as the structure of the low-affinity hNBD1 homodimer described above) (Atwell et al. 2010). The ABCα subdomain in hNBD2 appears to move primarily as a rigid body, based on both comparison of hNBD2 to hNBD1 and analysis of the smaller 1–2 Å variations in its location in the four different molecules of hNBD2 present in its crystal structure (data not shown). However, the α-helix following the Walker B motif, which makes extensive packing interactions with the ABCα subdomain, shows more complex variations in conformation involving smaller displacements in the same direction as the ABCα subdomain, which are coupled to differential bending of its helical axis (Fig. 3F and additional data not shown). These conformational changes in the α-helix following the Walker B and in the ABCα subdomain seem very likely to block functional interaction of hNBD2 with hNBD1.

Indeed, a wide variety of experiments conducted on the crystallized hNBD2 construct have failed to detect interaction with purified hNBD1 in solution (X Zhao, C Wang, S Atwell, et al., unpubl.). These experiments suggest that, even in solution, this extensively engineered hNBD2 construct adopts an inactive conformation. Therefore, the unusual conformation observed in its crystal structure might accurately reflect its inherent conformational tendencies. It is unclear whether wild-type hNBD2 would adopt a similarly inactive conformation in isolation from the other domains with which it interacts in CFTR. However, none of the mutations present in the crystallized construct provide an obvious explanation for its adoption of an inactive conformation with an ATP analog bound in the active site. Electrophysiological (Vergani et al. 2005; Kloch et al. 2010) and biochemical (Mense et al. 2006; Aleksandrov et al. 2009) experiments on CFTR strongly support formation of a canonical ATP-sandwich heterodimer by hNBD1 and hNBD2 in the open state of the CFTR channel (as discussed above and schematized in Fig. 1A) (Gadsby et al. 2006; Riordan 2008; Hwang et al. 2009), and the electron crystallography studies reviewed below provide additional circumstantial support for this model (Rosenberg et al. 2004, 2011; Ford et al. 2011). In this context, future studies will be required to determine whether hNBD2 has evolved to adopt an inactive conformation in the absence of proper interdomain packing interactions in intact CFTR or whether the sequence variations present in the crystallized hNBD2 construct are responsible for its adoption of an inactive conformation.

STRUCTURAL DATA FOR FULL-LENGTH CFTR

Crystallization of Full-Length CFTR

A method for two-dimensional (2D) crystallization of full-length CFTR has been published by Rosenberg and coworkers (2004). The method uses detergent (dodecylmaltoside) solubilized CFTR that was produced in stably-transfected baby hamster kidney (BHK) cells. CFTR obtained in this way has been relatively well characterized, and it appears to be fully glycosylated but mostly dephosphorylated (Zhang et al. 2009). This lack of phosphorylation could imply that CFTR is kept inactive in the BHK cells (which may be needed for cell growth when they express significant levels of this ion channel). However, it also seems possible that CFTR is rapidly dephosphorylated by cellular phosphatases at an early stage in the purification procedure. Dephosphorylation of CFTR may be important for 2D crystallization, as it probably removes some heterogeneity in the properties of the purified protein (i.e., in surface charge characteristics). There is some evidence that PKA-catalyzed phosphorylation of purified CFTR yields protein that crystallizes poorly (Awayn 2008). Glycosylation of CFTR will also yield heterogeneity in the physicochemical properties of the protein, but interestingly, this form of heterogeneity appears to be compatible with crystallization (Rosenberg et al. 2004; Zhang et al. 2009), perhaps because 2D crystals can accommodate the extracellular sugar polymers outside the planar lattice. The only method published so far for the crystallization of CFTR involves epitaxial (surface) crystallization in which the protein is slowly brought out of solution by concentration of solutes within the crystallization droplet. This methodology is essentially an extension of the sitting-drop vapor-diffusion method for generating three-dimensional (3D) crystals of biomolecules (Scarborough 1994; Auer et al. 1998, 1999; Rosenberg et al. 2001), but it is conducted at a 10- to 100-fold lower protein concentration. The method can also yield thin, multilayered 3D crystals at the carbon–water interface as well as 2D crystals (Scarborough 1994; Auer et al. 1998, 1999; Rosenberg et al. 2001), but the low concentration of protein ensures that most of the generated crystals are thin enough for electron crystallography (i.e., <50 nm) (Ford and Holzenburg 2008).

Structural data derived from 2D crystals of full-length CFTR using electron crystallography methods were published in 2004 (Rosenberg et al. 2004). The crystals were grown in the presence of the nonhydrolyzable ATP analog AMP-PNP (Rosenberg et al. 2004), and two separate crystal forms were detected. The crystals were stained with uranyl acetate to provide contrast and studied in the high vacuum of a conventional transmission electron microscope. The resolution of the data generated by these studies was ∼20 Å, which is typical for such specimens because visualization of molecular details is limited by the granular nature of the uranyl acetate stain and by dehydration of the specimen (Harris 1999; Ford and Holzenburg 2008). Furthermore, stain penetration tends to be uneven, which causes some parts of the specimen to be better preserved or contrasted than others (Harris 1999; Ford and Holzenburg 2008). Despite these drawbacks, the structural data from these electron crystallographic analyses showed that CFTR is in a monomeric state in the 2D crystals, and that its conformation bears strong similarity to that of Pgp (ABCB1), as previously predicted (Rosenberg et al. 2003).

Crystallization conditions have recently been identified that yield larger, better ordered, and more reproducible 2D arrays of full-length CFTR, and higher resolution electron crystallography data have been obtained for these 2D crystals in an unstained and frozen-hydrated state. Such crystals, which only grow in the absence of nucleotide in the crystallization reaction, have the protein packed in two crystalline layers stacked on top of each other.

Single-Particle Structural Studies of Full-Length CFTR

Structural data for purified CFTR has also been obtained for noncrystalline specimens using various microscopy techniques. An early atomic force microscopy (AFM) study of CFTR-expressing *Xenopus* oocytes observed donut-shaped objects within the plasma membrane that were associated with gold-labeled secondary antibodies bound to primary anti-CFTR antibodies (Schillers et al. 2004). However, the large dimensions (∼60 nm diameter) of these objects in the plasma membrane were difficult to reconcile with the expected molecular structure of CFTR. For example, the central pore of the donut-shaped objects was larger (∼20 nm diameter) than the dimensions of any ABC transporter characterized to date at atomic resolution. AFM of antibody-labeled CFTR has also been applied to try to assess its expression level in erythrocyte membranes (Schillers 2008). Despite the above-mentioned reservations concerning the AFM studies, they potentially represent structural data for CFTR resident in its native environment in the plasma membrane. If the large objects observed by AFM are indeed antibody-labeled CFTR, then our understanding of the behavior of CFTR in the plasma membrane will need to be revised. The objects detected in these studies would imply that several CFTR molecules encircle a large central cavity, and that CFTR expression in plasma membranes is at much higher levels than previously thought (Haggie et al. 2006). Further structural studies on CFTR localized to the plasma membrane will needed to evaluate the significance of these AFM observations.

Electron microscopy (EM) has also been employed to study the structure of purified full-length CFTR solubilized in the nonionic detergent β-dodecylmaltoside (β-DDM) or in the charged detergent *lyso*-phosphatidyl glycerol (*l*-PG). Dimeric complexes in β-DDM were reported in an early study conducted using single-particle averaging of purified CFTR embedded in a heavy-atom stain (Awayn et al. 2005). A subsequent study of purified CFTR particles in β-DDM also reported dimeric complexes, although these complexes were unexpectedly hollow in the center (Mio et al. 2008), perhaps because of lack of stain penetration or other problems associated with negative staining (Harris 1999; Ford and Holzenburg 2008). Later, EM structural data were obtained for single particles of purified CFTR in the unstained and frozen-hydrated state solubilized in either β-DDM or *l*-PG (Fig. 5), which provide similar images (Zhang et al. 2009). These data, in which stain artifacts are eliminated, confirmed the dimeric nature of the purified CFTR preparation but provided no evidence for a hollow cavity in the center of the dimer. Instead, they showed that the 3D structure was consistent with two side-by-side ABC transporters with a similar structure to Sav1866. Labeling of these CFTR particles using 1.8-nm diameter immuno-gold particles verified many conclusions from the earlier studies (Zhang et al. 2010), including the potential location of the R domain in the complex.

Regions Unique to CFTR

The interpretation of the available structural data for full-length CFTR has relied on the substantially higher resolution structures available

for the homologous ABC exporters Sav1866, MsbA, and Pgp (Dawson and Locher 2006, 2007; Ward et al. 2007; Aller et al. 2009). This set of proteins has been assumed to represent a structural archetype for most eukaryotic ABC transporters. The Sav1866 structure has been used most frequently for homology modeling of such structures, including that of CFTR. Recently, these homology models have begun to be tested experimentally in mutagenesis studies (Stenham et al. 2003; Zolnerciks et al. 2007; Serohijos et al. 2008; Alexander et al. 2009; Mornon et al. 2009; He et al. 2010). However, Sav1866 is a homodimeric protein, with each protomer consisting of a single TMD fused to a single NBD. Hence, Sav1866 lacks the R domain that covalently connects TMD1/NBD1 at the amino terminus of CFTR to TMD2/NBD2 at its carboxyl terminus. Despite the extensive evidence discussed above that the R domain is substantially disordered (Ostedgaard et al. 1997; Baker et al. 2007; Kanelis et al. 2010; Lewis et al. 2010), EM imaging of CFTR labeled with immuno-gold suggests that part of the R region may be spatially localized in the full-length protein, at a site spanning the interface between NBD1 and NBD2 and proximal to the cytoplasmic surface of the plasma membrane (Fig. 5) (Zhang et al. 2009, 2010).

When activated, CFTR must contain a hydrophilic channel that is selective for negatively charged ions (anions), especially chloride. This channel must run continuously through its TMDs and be open to both sides of the membrane for significant periods of time when CFTR is activated by PKA and ATP. Comparison of the available high-resolution structures for ABC exporters shows that they are all sealed to either the inside or the outside surface of the membrane. A partially sealed conformation of this kind is presumably a requirement of the translocation mechanism, because the transported substrate should be exposed alternatively to one side of the membrane or the other, but not simultaneously to both of them. Hence, at least when the anion channel in CFTR is open, its detailed transmembrane structure must be significantly different from that of Sav1866 and other ABC exporters. Hypotheses have been

     Cite this article as *Cold Spring Harb Perspect Med* doi: 10.1101/cshperspect.a009514

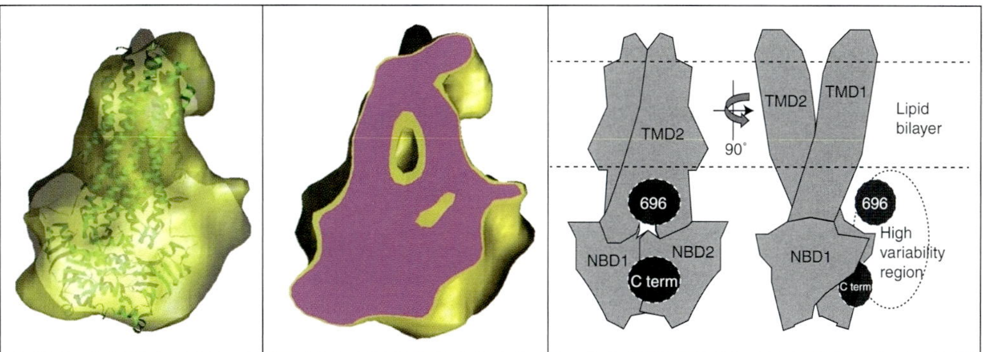

Figure 5. Cryo electron microscopy (EM) analysis of human CFTR. The yellow surface in the *left* panel shows structural data (Coulomb density map) derived from single-particle EM analysis of CFTR purified in the nonionic detergent β-dodecylmaltoside (β-DDM) but eluted from the final chromatography column in the charged detergent *lyso*-phosphatidylglycerol (*l*-PG) just above its critical micelle concentration (cmc) (Zhang et al. 2009). Similar images are obtained from samples eluted from the final column in β-DDM (Zhang et al. 2009). The crystal structure of Sav1866 (green ribbon trace) has been docked into the map. The *center* panel shows a slice through the same map (purple surface), revealing its internal features. The *right* panel shows two orthogonal views of a model for the domain organization of CFTR derived from the cryo-EM map shown in the other panels. The locations of the carboxyl terminus (C term) and the region around residue 696 in the R region were inferred from immuno-gold-labeling of poly-Histidine tags inserted into CFTR at these positions (Zhang et al. 2010). A high variability region revealed by cryo-EM (Zhang et al. 2009) is hypothesised to represent the R region.

advanced to explain how the anion-selective channel in CFTR could have evolved and how it may be gated so that anion flux can be controlled by the cell (Jordan et al. 2008; Alexander et al. 2009).

CFTR's Channel and Gate

Examination of the outward-facing configuration of ABC exporters, as exemplified by the structure of Sav1866, suggests an appealing model to explain the existence of a continuous transmembrane channel in CFTR. The 12 transmembrane α-helices in the functional Sav1866 dimer form a cone-shape molecular surface surrounding a central water-filled cavity that is connected via a wide (∼10 Å diameter) mouth to the extracellular milieu (as shown on the left side of Fig. 2) (Dawson and Locher 2006). This cavity is sealed on the cytoplasmic side of the membrane by a group of amino acid residues in the ICLs of Sav1866. A significant conformational change, presumably coupled to reorientation or separation of its NBDs, would be

needed to open this interface in Sav1866 to the cytoplasm. Notably, such a configuration of the cytoplasmic ICLs is found in the structure of the Pgp dimer (Aller et al. 2009), which is in an inward-facing conformation. However, the central cavity in this structure is blocked on the extracellular side of the membrane (center of Fig. 2). Intriguingly, the sealed region on the extracellular side of the transmembrane domains in this structure is relatively short. Hence, conceptually, it is easier to envision an inward-facing, Pgp-like conformation as being the active form of CFTR and residues on the extracellular side of its transmembrane domains as forming the gate of its channel. However, the existing mutagenesis and electrophysiological studies have been interpreted to imply that an outward-facing, Sav1866-like conformation is more likely to represent the active form of CFTR and that residues on the cytoplasmic side of the central portal are likely to form the gate of its channel. Nonetheless, experimental evidence shows that some ion selectivity is conferred by residues near the extracellular surface of CFTR (Jordan et al. 2008; Alexander et al. 2009). (The electrophysiological properties of the CFTR channel and predictions about its structure from homology models are covered in greater depth in a separate article in this work.) Further structural and functional studies will be required to reconcile these various inferences about the structure of the CFTR channel and understand the mechanism by which it is gated.

The molecular graphics images in Figs. 1, 3, and 4 in this article were prepared using PyMOL (www.pymol.org). The atomic models and 3D maps in Figs. 2 and 5 were displayed using Chimera (Pettersen et al. 2004) (www.cgl.ucsf.edu/chimera). Semi-automated fitting of the Sav1866 structure (PDB ID 2HYD) was performed by placing the models within the map by hand and then refining using the *Fit-in-Map* routine with optimization for correlation between the experimental CFTR maps and a simulated map based on the fitted models, which was arbitrarily restricted to a resolution corresponding to the reported resolution of the experimental map (Zhang et al. 2009). Both positive and negative regions of the map were included in the correlation calculation.

ACKNOWLEDGMENTS

We thank the Cystic Fibrosis Foundation (U.S.) for financial support via grants FORD08XX0 and HUNT09XX0.

REFERENCES

Aleksandrov AA, Aleksandrov L, Riordan JR. 2002a. Nucleoside triphosphate pentose ring impact on CFTR gating and hydrolysis. *FEBS Lett* **518**: 183–188.

Aleksandrov L, Aleksandrov AA, Chang XB, Riordan JR. 2002b. The first nucleotide binding domain of cystic fibrosis transmembrane conductance regulator is a site of stable nucleotide interaction, whereas the second is a site of rapid turnover. *J Biol Chem* **277**: 15419–15425.

Aleksandrov AA, Aleksandrov LA, Riordan JR. 2007. CFTR (ABCC7) is a hydrolyzable-ligand-gated channel. *Pflugers Arch* **453**: 693–702.

Aleksandrov L, Aleksandrov A, Riordan JR. 2008. Mg^{2+}-dependent ATP occlusion at the first nucleotide-binding domain (NBD1) of CFTR does not require the second (NBD2). *Biochem J* **416**: 129–136.

Aleksandrov AA, Cui L, Riordan JR. 2009. Relationship between nucleotide binding and ion channel gating in cystic fibrosis transmembrane conductance regulator. *J Physiol* **587**: 2875–2886.

Aleksandrov AA, Kota P, Aleksandrov LA, He L, Jensen T, Cui L, Gentzsch M, Dokholyan NV, Riordan JR. 2010. Regulatory insertion removal restores maturation, stability and function of δF508 CFTR. *J Mol Biol* **401**: 194–210.

Aleksandrov AA, Kota P, Cui L, Jensen T, Alekseev AE, Reyes S, He L, Gentzsch M, Aleksandrov LA, Dokholyan NV, et al. 2012. Allosteric modulation balances thermodynamic stability and restores function of δF508 CFTR. *J Mol Biol* **419**: 41–60.

Alexander C, Ivetac A, Liu X, Norimatsu Y, Serrano JR, Landstrom A, Sansom M, Dawson DC. 2009. Cystic fibrosis transmembrane conductance regulator: Using differential reactivity toward channel-permeant and channel-impermeant thiol-reactive probes to test a molecular model for the pore. *Biochemistry* **48**: 10078–10088.

Aller SG, Yu J, Ward A, Weng Y, Chittaboina S, Zhuo R, Harrell PM, Trinh YT, Zhang Q, Urbatsch IL, et al. 2009. Structure of P-glycoprotein reveals a molecular basis for poly-specific drug binding. *Science* **323**: 1718–1722.

Anderson MP, Berger HA, Rich DP, Gregory RJ, Smith AE, Welsh MJ. 1991a. Nucleoside triphosphates are required to open the CFTR chloride channel. *Cell* **67**: 775–784.

Anderson MP, Gregory RJ, Thompson S, Souza DW, Paul S, Mulligan RC, Smith AE, Welsh MJ. 1991b. Demonstration that CFTR is a chloride channel by alteration of its anion selectivity. *Science* **253**: 202–205.

Arai M, Kuwajima K. 2000. Role of the molten globule state in protein folding. *Adv Protein Chem* **53**: 209–282.

Atwell S, Brouillette CG, Conners K, Emtage S, Gheyi T, Guggino WB, Hendle J, Hunt JF, Lewis HA, Lu F, et al. 2010. Structures of a minimal human CFTR first nucleotide-binding domain as a monomer, head-to-tail homodimer, and pathogenic mutant. *Protein Eng Des Sel* **23**: 375–384.

Auer M, Scarborough GA, Kuhlbrandt W. 1998. Three-dimensional map of the plasma membrane H^+-ATPase in the open conformation. *Nature* **392**: 840–843.

Auer M, Scarborough GA, Kuhlbrandt W. 1999. Surface crystallisation of the plasma membrane H^+-ATPase on a carbon support film for electron crystallography. *J Mol Biol* **287**: 961–968.

Awayn NH. 2008. Structural analyses of the human cystic fibrosis transmembrane conductance regulator (CFTR) protein using electron microscopy, PhD thesis, University of Manchester, Manchester, UK.

Awayn NH, Rosenberg MF, Kamis AB, Aleksandrov LA, Riordan JR, Ford RC. 2005. Crystallographic and single-particle analyses of native- and nucleotide-bound forms of the cystic fibrosis transmembrane conductance regulator (CFTR) protein. *Biochem Soc Trans* **33**: 996–999.

Baker JM, Hudson RP, Kanelis V, Choy WY, Thibodeau PH, Thomas PJ, Forman-Kay JD. 2007. CFTR regulatory region interacts with NBD1 predominantly via multiple transient helices. *Nat Struct Mol Biol* **14**: 738–745.

Barthelme D, Dinkelaker S, Albers SV, Londei P, Ermler U, Tampé R. 2011. Ribosome recycling depends on a mechanistic link between the FeS cluster domain and a conformational switch of the twin-ATPase ABCE1. *Proc Natl Acad Sci* **108:** 3228–3233.

Cai Z, Scott-Ward TS, Sheppard DN. 2003. Voltage-dependent gating of the cystic fibrosis transmembrane conductance regulator Cl- channel. *J Gen Physiol* **122:** 605–620.

Callebaut I, Eudes R, Mornon JP, Lehn P. 2004. Nucleotide-binding domains of human cystic fibrosis transmembrane conductance regulator: Detailed sequence analysis and three-dimensional modeling of the heterodimer. *Cell Molec Life Sci* **61:** 230–242.

Chappe V, Hinkson DA, Howell LD, Evagelidis A, Liao J, Chang XB, Riordan JR, Hanrahan JW. 2004. Stimulatory and inhibitory protein kinase C consensus sequences regulate the cystic fibrosis transmembrane conductance regulator. *Proc Natl Acad Sci* **101:** 390–395.

Chappe V, Irvine T, Liao J, Evagelidis A, Hanrahan JW. 2005. Phosphorylation of CFTR by PKA promotes binding of the regulatory domain. *EMBO J* **24:** 2730–2740.

Chen CJ, Chin JE, Ueda K, Clark DP, Pastan I, Gottesman MM, Roninson IB. 1986. Internal duplication and homology with bacterial transport proteins in the mdr1 (P-glycoprotein) gene from multidrug-resistant human cells. *Cell* **47:** 381–389.

Chen J, Lu G, Lin J, Davidson AL, Quiocho FA. 2003. A tweezers-like motion of the ATP-binding cassette dimer in an ABC transport cycle. *Mol Cell* **12:** 651–661.

Csanády L, Chan KW, Seto-Young D, Kopsco DC, Nairn AC, Gadsby DC. 2000. Severed channels probe regulation of gating of cystic fibrosis transmembrane conductance regulator by its cytoplasmic domains. *J Gen Physiol* **116:** 477–500.

Csanády L, Seto-Young D, Chan KW, Cenciarelli C, Angel BB, Qin J, McLachlin DT, Krutchinsky AN, Chait BT, Nairn AC, et al. 2005. Preferential phosphorylation of R-domain Serine 768 dampens activation of CFTR channels by PKA. *J Gen Physiol* **125:** 171–186.

Csanády L, Nairn AC, Gadsby DC. 2006. Thermodynamics of CFTR channel gating: A spreading conformational change initiates an irreversible gating cycle. *J Gen Physiol* **128:** 523–533.

Cui J, Davidson AL. 2011. ABC solute importers in bacteria. *Essays Biochem* **50:** 85–99.

Cutting GR. 1993. Spectrum of mutations in cystic fibrosis. *J Bioenerg Biomembr* **25:** 7–10.

Cutting GR. 2005. Modifier genetics: Cystic fibrosis. *Annu Rev Genomics Hum Genet* **6:** 237–260.

Dalton J, Kalid O, Schushan M, Ben-Tal N, Villa-Freixa J. 2012. New model of CFTR proposes active channel-like conformation. *J Chem Inf Model* doi: 10.1021/ci2005884.

Da Paula AC, Sousa M, Xu Z, Dawson ES, Boyd AC, Sheppard DN, Amaral MD. 2010. Folding and rescue of a cystic fibrosis transmembrane conductance regulator trafficking mutant identified using human-murine chimeric proteins. *J Biol Chem* **285:** 27033–27044.

Dassa E, Bouige P. 2001. The ABC of ABCS: A phylogenetic and functional classification of ABC systems in living organisms. *Res Microbiol* **152:** 211–229.

Davidson AL, Chen J. 2004. ATP-binding cassette transporters in bacteria. *Annu Rev Biochem* **73:** 241–268.

Davidson AL, Maloney PC. 2007. ABC transporters: How small machines do a big job. *Trends Microbiol* **15:** 448–455.

Davidson AL, Sharma S. 1997. Mutation of a single MalK subunit severely impairs maltose transport activity in *Escherichia coli*. *J Bacteriol* **179:** 5458–5464.

Davidson AL, Laghaeian SS, Mannering DE. 1996. The maltose transport system of *Escherichia coli* displays positive cooperativity in ATP hydrolysis. *J Biol Chem* **271:** 4858–4863.

Dawson RJ, Locher KP. 2006. Structure of a bacterial multidrug ABC transporter. *Nature* **443:** 180–185.

Dawson RJ, Locher KP. 2007. Structure of the multidrug ABC transporter Sav1866 from *Staphylococcus aureus* in complex with AMP-PNP. *FEBS Lett* **581:** 935–938.

Dawson RJ, Hollenstein K, Locher KP. 2007. Uptake or extrusion: Crystal structures of full ABC transporters suggest a common mechanism. *Mol Microbiol* **65:** 250–257.

Dean M. 2005. The genetics of ATP-binding cassette transporters. *Methods Enzymol* **400:** 409–429.

Dean M, Annilo T. 2005. Evolution of the ATP-binding cassette (ABC) transporter superfamily in vertebrates. *Annu Rev Genomics Hum Genet* **6:** 123–142.

Dean M, Rzhetsky A, Allikmets R. 2001. The human ATP-binding cassette (ABC) transporter superfamily. *Genome Res* **11:** 1156–1166.

DeCarvalho AC, Gansheroff LJ, Teem JL. 2002. Mutations in the nucleotide binding domain 1 signature motif region rescue processing and functional defects of cystic fibrosis transmembrane conductance regulator δ f508. *J Biol Chem* **277:** 35896–35905.

Diederichs K, Diez J, Greller G, Müller C, Breed J, Schnell C, Vonrhein C, Boos W, Welte W. 2000. Crystal structure of MalK, the ATPase subunit of the trehalose/maltose ABC transporter of the archaeon *Thermococcus litoralis*. *EMBO J* **19:** 5951–5961.

Drumm ML, Wilkinson DJ, Smit LS, Worrell RT, Strong TV, Frizzell RA, Dawson DC, Collins FS. 1991. Chloride conductance expressed by δ F508 and other mutant CFTRs in *Xenopus* oocytes. *Science* **254:** 1797–1799.

Englander SW. 2000. Protein folding intermediates and pathways studied by hydrogen exchange. *Annu Rev Biophys Biomol Struct* **29:** 213–238.

Fojo AT, Whang-Peng J, Gottesman MM, Pastan I. 1985. Amplification of DNA sequences in human multidrug-resistant KB carcinoma cells. *Proc Natl Acad Sci* **82:** 7661–7665.

Ford RC, Holzenburg A. 2008. Electron crystallography of biomolecules: Mysterious membranes and missing cones. *Trends Biochem Sci* **33:** 38–43.

Ford RC, Birtley J, Rosenberg MF, Zhang L. 2011. CFTR three-dimensional structure. *Methods Mol Biol* **741:** 329–346.

Gadsby DC. 2004. Ion transport: Spot the difference. *Nature* **427:** 795–797.

Gadsby DC, Vergani P, Csanady L. 2006. The ABC protein turned chloride channel whose failure causes cystic fibrosis. *Nature* **440:** 477–483.

George AM, Jones PM. 2012. Perspectives on the structure-function of ABC transporters: The switch and constant contact models. *Prog Biophys Mol Biol* **109:** 95–107.

Gerlach JH, Endicott JA, Juranka PF, Henderson G, Sarangi F, Deuchars KL, Ling V. 1986. Homology between P-glycoprotein and a bacterial haemolysin transport protein suggests a model for multidrug resistance. *Nature* **324:** 485–489.

Gregory RJ, Cheng SH, Rich DP, Marshall J, Paul S, Hehir K, Ostedgaard L, Klinger KW, Welsh MJ, Smith AE. 1990. Expression and characterization of the cystic fibrosis transmembrane conductance regulator. *Nature* **347:** 382–386.

Griko YV, Privalov PL. 1994. Thermodynamic puzzle of apomyoglobin unfolding. *J Mol Biol* **235:** 1318–1325.

Griko YV, Gittis A, Lattman EE, Privalov PL. 1994. Residual structure in a staphylococcal nuclease fragment. Is it a molten globule and is its unfolding a first-order phase transition? *J Mol Biol* **243:** 93–99.

Haggie PM, Kim JK, Lukacs GL, Verkman AS. 2006. Tracking of quantum dot-labeled CFTR shows near immobilization by C-terminal PDZ interactions. *Mol Biol Cell* **17:** 4937–4945.

Harris JR. 1999. Negative staining of thinly spread biological particulates. *Methods Mol Biol* **117:** 13–30.

He L, Aleksandrov LA, Cui L, Jensen TJ, Nesbitt KL, Riordan JR. 2010. Restoration of domain folding and interdomain assembly by second-site suppressors of the δF508 mutation in CFTR. *FASEB J* **24:** 3103–3112.

Higgins CF. 1992. ABC transporters: From microorganisms to man. *Annu Rev Cell Bio* **8:** 67–113.

Hobson AC, Weatherwax R, Ames GF. 1984. ATP-binding sites in the membrane components of histidine permease, a periplasmic transport system. *Proc Natl Acad Sci* **81:** 7333–7337.

Hollenstein K, Dawson RJ, Locher KP. 2007a. Structure and mechanism of ABC transporter proteins. *Curr Opin Struct Biol* **17:** 412–418.

Hollenstein K, Frei DC, Locher KP. 2007b. Structure of an ABC transporter in complex with its binding protein. *Nature* **446:** 213–216.

Hopfner KP, Karcher A, Shin DS, Craig L, Arthur LM, Carney JP, Tainer JA. 2000. Structural biology of Rad50 ATPase: ATP-driven conformational control in DNA double-strand break repair and the ABC-ATPase superfamily. *Cell* **101:** 789–800.

Horn C, Bremer E, Schmitt L. 2003. Nucleotide dependent monomer/dimer equilibrium of OpuAA, the nucleotide-binding protein of the osmotically regulated ABC transporter OpuA from *Bacillus subtilis. J Mol Biol* **334:** 403–419.

Hung LW, Wang IX, Nikaido K, Liu PQ, Ames GF, Kim SH. 1998. Crystal structure of the ATP-binding subunit of an ABC transporter. *Nature* **396:** 703–707.

Hunt JF, Weinkauf S, Henry L, Fak JJ, McNicholas P, Oliver DB, Deisenhofer J. 2002. Nucleotide control of interdomain interactions in the conformational reaction cycle of SecA. *Science* **297:** 2018–2026.

Hwang TC, Sheppard DN. 2009. Gating of the CFTR Cl-channel by ATP-driven nucleotide-binding domain dimerisation. *J Physiol* **587:** 2151–2161.

Janas E, Hofacker M, Chen M, Gompf S, van der Does C, Tampé R. 2003. The ATP hydrolysis cycle of the nucleotide-binding domain of the mitochondrial atp-binding cassette transporter Mdl1p. *J Biol Chem* **278:** 26862–26869.

Jones PM, George AM. 1999. Subunit interactions in ABC transporters: Towards a functional architecture. *FEMS Microbiol Lett* **179:** 187–202.

Jones PM, George AM. 2004. The ABC transporter structure and mechanism: Perspectives on recent research. *Cell Mol Life Sci* **61:** 682–699.

Jordan IK, Kota KC, Cui G, Thompson CH, McCarty NA. 2008. Evolutionary and functional divergence between the cystic fibrosis transmembrane conductance regulator and related ATP-binding cassette transporters. *Proc Natl Acad Sci* **105:** 18865–18870.

Kadaba NS, Kaiser JT, Johnson E, Lee A, Rees DC. 2008. The high-affinity *E. coli* methionine ABC transporter: Structure and allosteric regulation. *Science* **321:** 250–253.

Kanelis V, Hudson RP, Thibodeau PH, Thomas PJ, Forman-Kay JD. 2010. NMR evidence for differential phosphorylation-dependent interactions in WT and δF508 CFTR. *EMBO J* **29:** 263–277.

Karpowich N, Martsinkevich O, Millen L, Yuan YR, Dai PL, MacVey K, Thomas PJ, Hunt JF. 2001. Crystal structures of the MJ1267 ATP binding cassette reveal an induced-fit effect at the ATPase active site of an ABC transporter. *Structure* **9:** 571–586.

Kerr ID. 2002. Structure and association of ATP-binding cassette transporter nucleotide-binding domains. *Biochim Biophys Acta* **1561:** 47–64.

Kerr ID. 2004. Sequence analysis of twin ATP binding cassette proteins involved in translational control, antibiotic resistance, and ribonuclease L inhibition. *Biochem Biophys Res Commun* **315:** 166–173.

Khare D, Oldham ML, Orelle C, Davidson AL, Chen J. 2009. Alternating access in maltose transporter mediated by rigid-body rotations. *Mol Cell* **33:** 528–536.

Kim Chiaw P, Eckford PD, Bear CE. 2011. Insights into the mechanisms underlying CFTR channel activity, the molecular basis for cystic fibrosis and strategies for therapy. *Essays Biochem* **50:** 233–248.

Kloch M, Milewski M, Nurowska E, Dworakowska B, Cutting GR, Dolowy K. 2010. The H-loop in the second nucleotide-binding domain of the cystic fibrosis transmembrane conductance regulator is required for efficient chloride channel closing. *Cell Physiol Biochem* **25:** 169–180.

Kos V, Ford RC. 2009. The ATP-binding cassette family: A structural perspective. *Cell Mol Life Sci* **66:** 3111–3126.

Lewis HA, Buchanan SG, Burley SK, Conners K, Dickey M, Dorwart M, Fowler R, Gao X, Guggino WB, Hendrickson WA, et al. 2004. Structure of nucleotide-binding domain 1 of the cystic fibrosis transmembrane conductance regulator. *EMBO J* **23:** 282–293.

Lewis HA, Zhao X, Wang C, Sauder JM, Rooney I, Noland BW, Lorimer D, Kearins MC, Conners K, Condon B, et al. 2005. Impact of the δF508 mutation in first nucleotide-binding domain of human cystic fibrosis transmembrane conductance regulator on domain folding and structure. *J Biol Chem* **280:** 1346–1353.

Lewis HA, Wang C, Zhao X, Hamuro Y, Conners K, Kearins MC, Lu F, Sauder JM, Molnar KS, Coales SJ, et al. 2010. Structure and dynamics of NBD1 from CFTR characterized using crystallography and hydrogen/deuterium exchange mass spectrometry. *J Mol Biol* **396:** 406–430.

Linsdell P. 2006. Mechanism of chloride permeation in the cystic fibrosis transmembrane conductance regulator chloride channel. *Exp Physiol* **91:** 123–129.

Linsdell P, Evagelidis A, Hanrahan JW. 2000. Molecular determinants of anion selectivity in the cystic fibrosis transmembrane conductance regulator chloride channel pore. *Biophys J* **78:** 2973–2982.

Linton KJ, Higgins CF. 1998. The *Escherichia coli* ATP-binding cassette (ABC) proteins. *Mol Microbiol* **28:** 5–13.

Liu X, Dawson DC. 2011. Cystic fibrosis transmembrane conductance regulator: Temperature-dependent cysteine reactivity suggests different stable conformers of the conduction pathway. *Biochemistry* **50:** 10311–10317.

Liu X, Smith SS, Dawson DC. 2003. CFTR: What's it like inside the pore? *J Exp Zoolog A Comp Exp Biol* **300:** 69–75.

Liu X, Zhang ZR, Fuller MD, Billingsley J, McCarty NA, Dawson DC. 2004. CFTR: A cysteine at position 338 in TM6 senses a positive electrostatic potential in the pore. *Biophys J* **87:** 3826–3841.

Locher KP, Lee AT, Rees DC. 2002. The E. coli BtuCD structure: A framework for ABC transporter architecture and mechanism. *Science* **296:** 1091–1098.

Loo TW, Bartlett MC, Clarke DM. 2010. The V510D suppressor mutation stabilizes δF508-CFTR at the cell surface. *Biochemistry* doi: 10.1021/bi100807h.

Mansoura MK, Smith SS, Choi AD, Richards NW, Strong TV, Drumm ML, Collins FS, Dawson DC. 1998. Cystic fibrosis transmembrane conductance regulator (CFTR) anion binding as a probe of the pore. *Biophys J* **74:** 1320–1332.

McDevitt CA, Shintre CA, Grossmann JG, Pollock NL, Prince SM, Callaghan R, Ford RC. 2008. Structural insights into P-glycoprotein (ABCB1) by small angle X-ray scattering and electron crystallography. *FEBS Lett* **582:** 2950–2956.

Mendoza JL, Schmidt A, Li Q, Nuvaga E, Barrett T, Bridges RJ, Feranchak AP, Brautigam CA, Thomas PJ. 2012. Requirements for efficient correction of δF508 CFTR revealed by analyses of evolved sequences. *Cell* **148:** 164–174.

Mense M, Vergani P, White DM, Altberg G, Nairn AC, Gadsby DC. 2006. In vivo phosphorylation of CFTR promotes formation of a nucleotide-binding domain heterodimer. *EMBO J* **25:** 4728–4739.

Mio K, Ogura T, Mio M, Shimizu H, Hwang TC, Sato C, Sohma Y. 2008. Three-dimensional reconstruction of human cystic fibrosis transmembrane conductance regulator chloride channel revealed an ellipsoidal structure with orifices beneath the putative transmembrane domain. *J Biol Chem* **283:** 30300–30310.

Moody JE, Millen L, Binns D, Hunt JF, Thomas PJ. 2002. Cooperative, ATP-dependent association of the nucleotide binding cassettes during the catalytic cycle of ATP-binding cassette transporters. *J Biol Chem* **277:** 21111–21114.

Mornon JP, Lehn P, Callebaut I. 2008. Atomic model of human cystic fibrosis transmembrane conductance regulator: Membrane-spanning domains and coupling interfaces. *Cell Mol Life Sci* **65:** 2594–2612.

Mornon JP, Lehn P, Callebaut I. 2009. Molecular models of the open and closed states of the whole human CFTR protein. *Cell Mol Life Sci* **66:** 3469–3486.

Mourez M, Hofnung M, Dassa E. 1997. Subunit interactions in ABC transporters: A conserved sequence in hydrophobic membrane proteins of periplasmic permeases defines an important site of interaction with the ATPase subunits. *EMBO J* **16:** 3066–3077.

Oldham ML, Chen J. 2011. Crystal structure of the maltose transporter in a pretranslocation intermediate state. *Science* **332:** 1202–1205.

Oldham ML, Khare D, Quiocho FA, Davidson AL, Chen J. 2007. Crystal structure of a catalytic intermediate of the maltose transporter. *Nature* **450:** 515–521.

Ostedgaard LS, Rich DP, DeBerg LG, Welsh MJ. 1997. Association of domains within the cystic fibrosis transmembrane conductance regulator. *Biochemistry* **36:** 1287–1294.

Pettersen EF, Goddard TD, Huang CC, Couch GS, Greenblatt DM, Meng EC, Ferrin TE. 2004. UCSF Chimera—a visualization system for exploratory research and analysis. *J Comput Chem* **25:** 1605–1612.

Pinkett HW, Lee AT, Lum P, Locher KP, Rees DC. 2007. An inward-facing conformation of a putative metal-chelate-type ABC transporter. *Science* **315:** 373–377.

Pissarra LS, Farinha CM, Xu Z, Schmidt A, Thibodeau PH, Cai Z, Thomas PJ, Sheppard DN, Amaral MD. 2008. Solubilizing mutations used to crystallize one CFTR domain attenuate the trafficking and channel defects caused by the major cystic fibrosis mutation. *Chem Biol* **15:** 62–69.

Protasevich I, Yang Z, Wang C, Atwell S, Zhao X, Emtage S, Wetmore D, Hunt JF, Brouillette CG. 2010. Thermal unfolding studies show the disease causing F508 deletion mutation in cystic fibrosis transmembrane conductance regulator (CFTR) thermodynamically destabilizes nucleotide-binding domain 1. *Protein Sci* **19:** 1917–1931.

Qu BH, Strickland EH, Thomas PJ. 1997. Localization and suppression of a kinetic defect in cystic fibrosis transmembrane conductance regulator folding. *J Biol Chem* **272:** 15739–15744.

Rabeh WM, Bossard F, Xu H, Okiyoneda T, Bagdany M, Mulvihill CM, Du K, di Bernardo S, Liu Y, Konermann L, et al. 2012. Correction of both NBD1 energetics and domain interface is required to restore δF508 CFTR folding and function. *Cell* **148:** 150–163.

Redfield C. 2004. Using nuclear magnetic resonance spectroscopy to study molten globule states of proteins. *Methods* **34:** 121–132.

Rich DP, Anderson MP, Gregory RJ, Cheng SH, Paul S, Jefferson DM, McCann JD, Klinger KW, Smith AE, Welsh MJ. 1990. Expression of cystic fibrosis transmembrane conductance regulator corrects defective chloride channel regulation in cystic fibrosis airway epithelial cells. *Nature* **347:** 358–363.

Riordan JR. 2008. CFTR function and prospects for therapy. *Annu Rev Biochem* **77:** 701–726.

Riordan JR, Rommens JM, Kerem B, Alon N, Rozmahel R, Grzelczak Z, Zielenski J, Lok S, Plavsic N, Chou JL. 1989. Identification of the cystic fibrosis gene: Cloning and characterization of complementary DNA. *Science* **245:** 1066–1073.

Roninson IB, Chin JE, Choi KG, Gros P, Housman DE, Fojo A, Shen DW, Gottesman MM, Pastan I. 1986. Isolation of human mdr DNA sequences amplified in multidrug-resistant KB carcinoma cells. *Proc Natl Acad Sci* **83:** 4538–4542.

Rosenberg MF, Velarde G, Ford RC, Martin C, Berridge G, Kerr ID, Callaghan R, Schmidlin A, Wooding C, Linton KJ, et al. 2001. Repacking of the transmembrane domains of P-glycoprotein during the transport ATPase cycle. *EMBO J* **20:** 5615–5625.

Rosenberg MF, Kamis AB, Callaghan R, Higgins CF, Ford RC. 2003. Three-dimensional structures of the mammalian multidrug resistance P-glycoprotein demonstrate major conformational changes in the transmembrane domains upon nucleotide binding. *J Biol Chem* **278:** 8294–8299.

Rosenberg MF, Kamis AB, Aleksandrov LA, Ford RC, Riordan JR. 2004. Purification and crystallization of the cystic fibrosis transmembrane conductance regulator (CFTR). *J Biol Chem* **279:** 39051–39057.

Rosenberg MF, O'Ryan LP, Hughes G, Zhao Z, Aleksandrov LA, Riordan JR, Ford RC. 2011. The cystic fibrosis transmembrane conductance regulator (CFTR): 3D structure and localisation of a channel gate. *J Biol Chem* **286:** 42647–42654.

Sampson HM, Robert R, Liao J, Matthes E, Carlile GW, Hanrahan JW, Thomas DY. 2011. Identification of a NBD1-binding pharmacological chaperone that corrects the trafficking defect of F508del-CFTR. *Chem Biol* **18:** 231–242.

Saurin W, Hofnung M, Dassa E. 1999. Getting in or out: Early segregation between importers and exporters in the evolution of ATP-binding cassette (ABC) transporters. *J Mol Evol* **48:** 22–41.

Scarborough GA. 1994. Large single-crystals of the neurospora crassa plasma membrane H⁺-Atpase: An approach to the crystallization of integral membrane-proteins. *Acta Crystallogr D Biol Crystallogr* **50:** 643–649.

Schillers H. 2008. Imaging CFTR in its native environment. *Pflugers Arch* **456:** 163–177.

Schillers H, Shahin V, Albermann L, Schafer C, Oberleithner H. 2004. Imaging CFTR: A tail to tail dimer with a central pore. *Cell Physiol Biochem* **14:** 1–10.

Serohijos AW, Hegedus T, Aleksandrov AA, He L, Cui L, Dokholyan NV, Riordan JR. 2008. Phenylalanine-508 mediates a cytoplasmic-membrane domain contact in the CFTR 3D structure crucial to assembly and channel function. *Proc Natl Acad Sci* **105:** 3256–3261.

Sheppard DN. 2004. CFTR channel pharmacology: Novel pore blockers identified by high-throughput screening. *J Gen Physiol* **124:** 109–113.

Sheppard DN, Welsh MJ. 1999. Structure and function of the CFTR chloride channel. *Physiol Rev* **79:** S23–S45.

Sheppard DN, Rich DP, Ostedgaard LS, Gregory RJ, Smith AE, Welsh MJ. 1993. Mutations in CFTR associated with mild-disease-form Cl- channels with altered pore properties. *Nature* **362:** 160–164.

Smith PC, Karpowich N, Millen L, Moody JE, Rosen J, Thomas PJ, Hunt JF. 2002. ATP binding to the motor domain from an ABC transporter drives formation of a nucleotide sandwich dimer. *Mol Cell* **10:** 139–149.

Snider J, Houry WA. 2008. AAA⁺ proteins: Diversity in function, similarity in structure. *Biochem Soc Trans* **36:** 72–77.

Stenham DR, Campbell JD, Sansom MS, Higgins CF, Kerr ID, Linton KJ. 2003. An atomic detail model for the human ATP binding cassette transporter P-glycoprotein derived from disulfide cross-linking and homology modeling. *FASEB J* **17:** 2287–2289.

Teem JL, Berger HA, Ostedgaard LS, Rich DP, Tsui LC, Welsh MJ. 1993. Identification of revertants for the cystic fibrosis δ F508 mutation using STE6-CFTR chimeras in yeast. *Cell* **73:** 335–346.

Thibodeau PH, Brautigam CA, Machius M, Thomas PJ. 2005. Side chain and backbone contributions of Phe508 to CFTR folding. *Nat Struct Mol Biol* **12:** 10–16.

Thibodeau PH, Richardson JM, Wang W, Millen L, Watson J, Mendoza JL, Du K, Fischman S, Senderowitz H, Lukacs GL, et al. 2010. The cystic fibrosis-causing mutation δF508 affects multiple steps in cystic fibrosis transmembrane conductance regulator biogenesis. *J Biol Chem* **285:** 35825–35835.

Thomas PJ, Shenbagamurthi P, Ysern X, Pedersen PL. 1991. Cystic fibrosis transmembrane conductance regulator: Nucleotide binding to a synthetic peptide. *Science* **251:** 555–557.

Vergani P, Nairn AC, Gadsby DC. 2003. On the mechanism of MgATP-dependent gating of CFTR Cl- channels. *J Gen Physiol* **121:** 17–36.

Vergani P, Lockless SW, Nairn AC, Gadsby DC. 2005. CFTR channel opening by ATP-driven tight dimerization of its nucleotide-binding domains. *Nature* **433:** 876–880.

Walker JE, Saraste M, Runswick MJ, Gay NJ. 1982. Distantly related sequences in the α- and β-subunits of ATP synthase, myosin, kinases and other ATP-requiring enzymes and a common nucleotide binding fold. *EMBO J* **1:** 945–951.

Wang C, Protasevich I, Yang Z, Seehausen D, Skalak T, Zhao X, Atwell S, Spencer Emtage J, Wetmore DR, Brouillette CG, et al. 2010. Integrated biophysical studies implicate partial unfolding of NBD1 of CFTR in the molecular pathogenesis of F508del cystic fibrosis. *Protein Sci* **19:** 1932–1947.

Ward A, Reyes CL, Yu J, Roth CB, Chang G. 2007. Flexibility in the ABC transporter MsbA: Alternating access with a twist. *Proc Natl Acad Sci* **104:** 19005–19010.

Yuan YR, Blecker S, Martsinkevich O, Millen L, Thomas PJ, Hunt JF. 2001. The crystal structure of the MJ0796 ATP-binding cassette. Implications for the structural consequences of ATP hydrolysis in the active site of an ABC transporter. *J Biol Chem* **276:** 32313–32321.

Zaitseva J, Jenewein S, Jumpertz T, Holland IB, Schmitt L. 2005. H662 is the linchpin of ATP hydrolysis in the nucleotide-binding domain of the ABC transporter HlyB. *EMBO J* **24:** 1901–1910.

Zhang L, Aleksandrov LA, Zhao Z, Birtley JR, Riordan JR, Ford RC. 2009. Architecture of the cystic fibrosis transmembrane conductance regulator protein and structural

Cite this article as *Cold Spring Harb Perspect Med* doi: 10.1101/cshperspect.a009514

changes associated with phosphorylation and nucleotide binding. *J Struct Biol* **167:** 242–251.

Zhang L, Aleksandrov LA, Riordan JR, Ford RC. 2010. Domain location within the cystic fibrosis transmembrane conductance regulator protein investigated by electron microscopy and gold labelling. *Biochim Biophys Acta* **1808:** 399–404.

Zhu T, Hinkson DA, Dahan D, Evagelidis A, Hanrahan JW. 2002. CFTR regulation by phosphorylation. *Methods Mol Med* **70:** 99–109.

Zolnerciks JK, Wooding C, Linton KJ. 2007. Evidence for a Sav1866-like architecture for the human multidrug transporter P-glycoprotein. *FASEB J* **21:** 3937–3948.

Zolnerciks JK, Andress EJ, Nicolaou M, Linton KJ. 2011. Structure of ABC transporters. *Essays Biochem* **50:** 43–61.

Zutz A, Hoffmann J, Hellmich UA, Glaubitz C, Ludwig B, Brutschy B, Tampé R. 2010. Asymmetric ATP hydrolysis cycle of the heterodimeric multidrug ABC transport complex TmrAB from *Thermus thermophilus*. *J Biol Chem* **286:** 7104–7115.

Dynamics Intrinsic to Cystic Fibrosis Transmembrane Conductance Regulator Function and Stability

P. Andrew Chong[1], Pradeep Kota[2], Nikolay V. Dokholyan[2], and Julie D. Forman-Kay[1,3]

[1]Program in Molecular Structure and Function, Hospital for Sick Children, Toronto, Ontario M5G 1X8, Canada

[2]Department of Biochemistry and Biophysics, University of North Carolina at Chapel Hill, Chapel Hill, North Carolina 27599

[3]Department of Biochemistry, University of Toronto, Toronto, Ontario M5S 1A8, Canada

Correspondence: forman@sickkids.ca

The cystic fibrosis transmembrane conductance regulator (CFTR) requires dynamic fluctuations between states in its gating cycle for proper channel function, including changes in the interactions between the nucleotide-binding domains (NBDs) and between the intracellular domain (ICD) coupling helices and NBDs. Such motions are also linked with fluctuating phosphorylation-dependent binding of CFTR's disordered regulatory (R) region to the NBDs and partners. Folding of CFTR is highly inefficient, with the marginally stable NBD1 sampling excited states or folding intermediates that are aggregation-prone. The severe CF– causing F508del mutation exacerbates the folding inefficiency of CFTR and leads to impaired channel regulation and function, partly as a result of perturbed NBD1–ICD interactions and enhanced sampling of these NBD1 excited states. Increased knowledge of the dynamics within CFTR will expand our understanding of the regulated channel gating of the protein as well as of the F508del defects in folding and function.

Understanding how proteins function often requires knowledge of structure *and* conformational stability and dynamics. Unlike mechanical parts, proteins are self-assembling dynamic components that require flexibility for function. Conformational changes are required during the folding process and for progression through functional cycles. Aberrations in stability and/or dynamics can lead to protein misfolding or dysfunction, primary causes of many diseases including cystic fibrosis (CF). Changes in stability can be independent of significant structural changes to a protein's ground state, as is the case for the cystic fibrosis transmembrane conductance regulator (CFTR). As discussed in previous articles, deletion of F508 in CFTR (F508del), widely observed in CF patients, results in a significant decrease in folding and processing efficiency as well as in gating defects (Riordan et al. 1989; Cheng et al. 1990; Dalemans et al. 1991; Drumm et al. 1991). An early indication that the mutation does not sterically inhibit proper folding or remove a critical structural component was the observation that the processing efficiency could be significantly improved by expressing the protein at 26°C instead of 37°C (Denning et al. 1992). The mutation decreases CFTR's thermostability, but a small amount of properly processed mutant protein does arrive at the cell membrane and has residual activity. Although the mutation does not prohibit the conformations required for proper gating, the activity of F508del is rapidly lost with increase in temperature, indicating that the mutation affects CFTR stability and dynamics. The pathology is characterized by the inefficiency of folding and processing and a reduction in the lifetime of the open conformation during the gating cycle.

CFTR comprises two repeated units, each containing a membrane-spanning domain (MSD1 and MSD2) and a cytoplasmic nucleotide-binding domain (NBD1 and NBD2) (Riordan et al. 1989). The carboxyl terminus of NBD1 is connected to the amino terminus of MSD2 via a 200-residue intrinsically disordered regulatory (R) region (Fig. 1) (Riordan et al. 1989; Dulhanty and Riordan 1994; Ostedgaard et al. 2000). Folding of wild-type (WT) CFTR is generally found to be inefficient, although efficiency varies with cell type (Lukacs et al. 1994; Ward and Kopito 1994; Varga et al. 2004). The dominant F508del mutation exacerbates the inefficiency. Improperly folded CFTR is retained by endoplasmic reticulum (ER) quality control mechanisms, a process that is readily measured by monitoring the degree of glycosylation by SDS polyacrylamide gel electrophoresis (Cheng et al. 1990). Although some reports indicate that CFTR domains initiate their folding cotranslationally in an independent fashion (Kleizen et al. 2005; Hoelen et al. 2010), interdomain interactions are required for final processing and escape from the ER quality control machinery (Cui et al. 2007). Notably, acquisition of a protease-resistant NBD2 fold has been shown to require appropriate interactions with the preceding domains (Du et al. 2005). Deletion of F508 in NBD1 negatively affects proper folding of NBD2, possibly by introducing a nonphysiological tight interaction between NBD1 and NBD2

(Du et al. 2005). Interestingly, NBD2 is not required for transportation out of the ER (Cui et al. 2007) or CFTR gating (Wang et al. 2010b). Misfolding of NBD2 in F508del may result in CFTR being retained in the ER by the quality control machinery, unable to proceed to the cell surface.

At the cell surface, MSD1 and MSD2 of CFTR form a regulated anion channel that cycles through a closed state, an open-ready state, and an open state. The cytoplasmic extensions of the transmembrane helices form helical bundle structural regions, which collectively constitute the intracellular domains (ICDs) believed to transmit information from the NBDs to the transmembrane pore. Gating of the channel is known to require phosphorylation of the R region (Cheng et al. 1991; Berger et al. 1993), binding of ATP at the two sites at the NBD heterodimer interface, and hydrolysis of ATP at the site formed by the ATP-binding core of NBD2 (Cheung et al. 2008). Phosphorylation is believed to relieve R-region inhibition of channel opening, but the conformational change effected by the phosphorylation is poorly understood. NBD1 and -2 form a functional dimer (reviewed by Cheung et al. 2008), but the changes in the dimer conformation that open and close the channel are also poorly understood.

Here we will discuss gating cycle dynamics, the dynamics of the R region and NBD1, and changes in CFTR's dynamic behavior resulting from deletion of F508.

FULL-LENGTH CFTR

Unlike the other ATP-binding cassette (ABC) transporters to which it is related, CFTR functions as a channel rather than as an active transporter (Anderson et al. 1991). Despite this difference, CFTR shows significant sequence similarity with ABC transporters (Riordan et al. 1989) and likely undergoes similar conformational rearrangements. As a channel, CFTR exists in a closed conformation, which does not permit chloride transport, and an open conformation, which permits rapid chloride flow through the pore formed by the MSDs. Single-channel recordings of CFTR indicate that the

 Cite this article as *Cold Spring Harb Perspect Med* doi: 10.1101/cshperspect.a009522

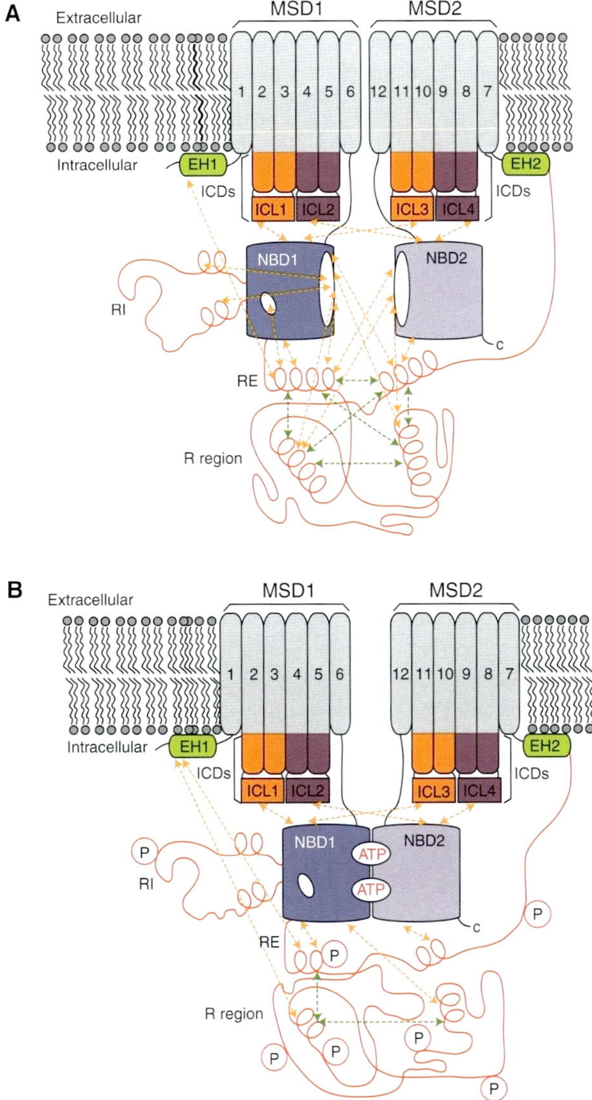

Figure 1. Schematic representation of nonphosphorylated (*A*) and phosphorylated (*B*) CFTR. The membrane-spanning domains (MSDs) and nucleotide-binding domains (NBDs) are shown in gray and blue, respectively. The R region, regulatory insertion (RI), and regulatory extension (RE) are shown in red. The intracellular domains (ICDs) are shown in orange or purple with the coupling helices or intracellular loops (ICLs) identified. The elbow helices (EHs) are shown in green. Intramolecular interactions are indicated by dashed arrows.

channel has two phases, a burst interval, during which the channel is open but flickers rapidly through a closed state, and an interburst interval, during which the channel is closed (Winter et al. 1994), indicating that CFTR has at least three conformations: a closed state, an open state, and an open-ready state. The open-ready state does not allow chloride flow but, unlike the closed state, is poised to rapidly transition to the open state. As a large polytopic membrane protein containing multiple domains and disordered segments, full-length CFTR has resisted efforts to directly resolve the structure of any of these conformations by X-ray crystallography.

Insight into the structure and dynamics of full-length CFTR has come from crystal structures of related ABC transporters, molecular modeling, cross-linking studies, electrophysiology experiments, and electron microscopy. In this section we review what is known about the dynamics of full-length CFTR as it passes through the gating cycle.

CFTR-related ABC transporters representing different stages of the transport cycle have been crystallized (Dawson and Locher 2006; Hollenstein et al. 2007; Pinkett et al. 2007; Ward et al. 2007). The structures suggest that the MSDs collectively form two discrete helical bundles that move with respect to each other, forming alternately outward- and inward-facing conformations. In the outward-facing conformation (as exemplified by the Sav1866 structure [Dawson and Locher 2006]), the helical bundles point away from each other on the extracellular side of the protein, diverging near the midpoint of the membrane, and the NBDs form tight dimers with both composite ATP-binding sites tightly juxtaposed and nucleotide-filled (Fig. 2). In the inward-facing conformation (as exemplified by the MsbA nucleotide-free structures [Ward et al. 2007]), the helical bundles point away from each other on the intracellular side of the membrane and the NBDs are at least partially separated and nucleotide-free (Hollenstein et al. 2007; Ward et al. 2007). A range of separation is observed for the inward-facing conformation, suggesting a great deal of flexibility in the inward-facing conformation (Ward et al. 2007). In the MsbA "open apo" structure, the NBDs are completely separated. The outward-facing conformation appears to be promoted by nucleotide-induced NBD tight dimer formation transmitted through an interaction between the NBDs and the ICDs. The inward-facing conformation may be triggered by nucleotide hydrolysis and release.

The relationship between the inward- and outward-facing conformations observed in these crystal structures of related transporters and the range of channel motions in CFTR has not been

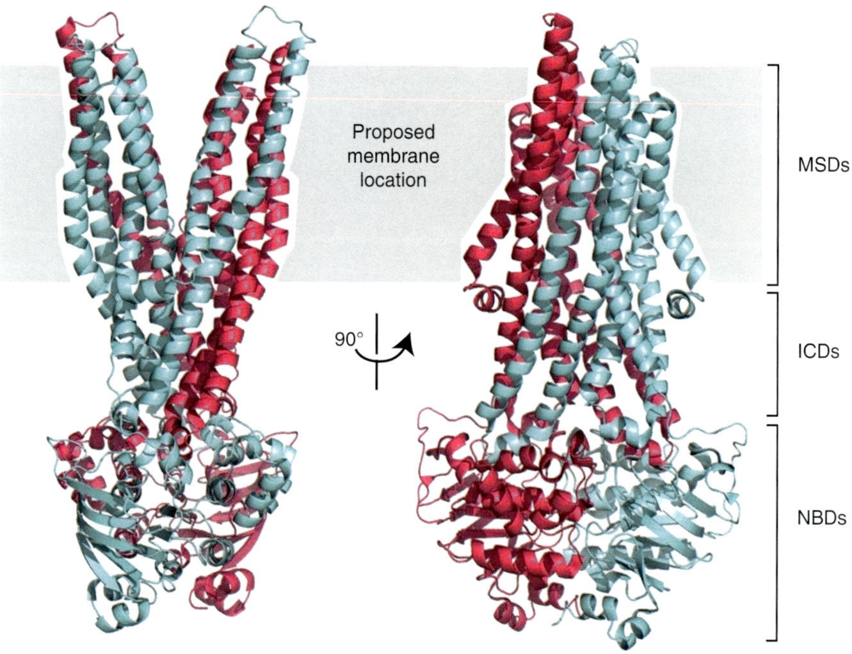

Figure 2. Ribbon diagram representation of the outward-facing conformation of Sav1866 in two orthogonal views. The subunits of the dimer are shown in blue and purple. Note that the two transmembrane helical bundles move apart as they near the extracellular side of the membrane (PDB 2HYD [Dawson and Locher 2006]).

Cite this article as *Cold Spring Harb Perspect Med* doi: 10.1101/cshperspect.a009522

conclusively established. The outward-facing conformation has been assumed to be representative of the open conformation of the CFTR channel in part because it is the nucleotide-bound state in the observed structures (Mornon et al. 2009). By analogy, the closed conformation may be represented by the inward-facing conformation. ABC transporters operate by an "alternating access" mechanism (Mitchell 1957; Jardetzky 1966). They bind substrate in one conformation and then undergo a conformational change that allows release of the substrate on the opposite side of the membrane. The substrate is never able to freely pass through the transporter. Because CFTR is closely related to these ABC transporters, it is envisioned that it undergoes a similar conformational change. However, as it is a channel and not an active transporter, it is not clear that the analogy holds and that CFTR maintains the same dynamic modes. In particular, CFTR is believed to have three different conformations rather than two, although one or more of these conformations may be very similar. As with other ABC transporters, channel gating has been shown to be tightly coupled to ATP hydrolysis (Csanady et al. 2009), but the effect of nucleotide binding on the position of the NBDs with respect to each other has not been resolved. Channel opening may require induced dimerization or a more moderate conformational change in preexisting dimers (Smith et al. 2002; Dawson and Locher 2006; Jones and George 2007, 2009).

Electron microscopy images obtained from single-particle analysis of noncrystalline as well as two-dimensional crystals of CFTR in a nucleotide-free state and a nucleotide-bound phosphorylated state lend some support to CFTR having inward- and outward-facing conformations (Rosenberg et al. 2004; Awayn et al. 2005; Zhang et al. 2009). The images suggest that in the absence of nucleotide the helical bundles have a greater separation on the cytoplasmic side, whereas in the nucleotide-bound form the helical bundles are closer to the outward-facing model. The role of nucleotide binding and conformational changes at the NBDs is not discernible from these low-resolution images. However, the images do suggest that the NBDs do not completely separate as they do in one of the MsbA structures (open apo) (Ward et al. 2007).

The crystal structures of Sav1866 and MsbA, bacterial ABC transporters, were used to construct structural models of the opened and closed states of full-length CFTR that have been partially validated experimentally (Mornon et al. 2008, 2009; Serohijos et al. 2008a; Alexander et al. 2009; Huang et al. 2009; Kanelis et al. 2010; Moran 2010). The channel pore architecture has been experimentally probed using cysteine mutants and thiol-reactive reagents to determine the accessibility of particular sites in the pore. Taken together, these studies are strongly supportive of the Sav1866-based homology models (Zhang et al. 2005; Fatehi and Linsdell 2008; Alexander et al. 2009). They show a pore that is wide on the extracellular side and becomes more inaccessible to molecular probes, especially anionic probes, moving toward the cytoplasmic surface. There is also evidence that the pore architecture undergoes significant conformational change during the gating cycle (Zhang et al. 2005; Fatehi and Linsdell 2008). Activation of the channel (by forskolin and 3-isobutyl-1-methylxantine [IBMX]) appears to significantly enhance access of anions to the pore (Fatehi and Linsdell 2008). Accessibility for cations is less stringent, and even large organic cations can reach cysteines introduced into the putative inner vestibule of the pore in the inactive state. Fatehi and Linsdell further propose that there are two functionally distinct closed states: the inactivated channel and the "activated but still closed" channel conformation. They suggest that the latter conformation is activated by protein kinase A (PKA) but still requires ATP-dependent conformational changes at the NBDs to open the channel to chloride flow. The activated but still closed conformation is likely related to the open-ready conformation observed in single-channel gating studies.

The homology models also identify specific interactions between the ICDs and the NBDs, which have been tested experimentally by chemical cross-linking studies. At a gross level, these studies confirm close interactions between the NBDs and the ICD "coupling helices" that lie at the NBD interface (called ICLs or CLs), which supports the notion that the positioning of the MSDs is controlled by interactions between the NBDs and ICLs (Serohijos et al. 2008a). At a specific level, cross-linking studies confirm interactions between ICL1, ICL3, and ICL4 and NBD1 (He et al. 2008, 2010). These studies also show that cross-linking NBD1 to ICL3 or -4, but not ICL1, strongly inhibits channel gating (He et al. 2008), a result that suggests that the NBD–ICL interface must change during the gating cycle. Similar results were noted for NBD2–ICL interfaces (He et al. 2008). Significant flexibility in NBD–ICL interactions is also indicated by the large distribution of linker lengths that can cross-link particular NBD–ICL cysteine pairs introduced into a Cys-less CFTR (Serohijos et al. 2008a). Although cross-linking can inhibit channel gating, suggesting a requirement for flexibility, NBD1–ICL4 and NBD2–ICL2 cross-linking are not strongly influenced by PKA phosphorylation or adenosine $5'$-(β, γ-imido)triphosphate (AMPPNP) binding. These findings imply that the interfaces between NBD1 and ICL4 and between NBD2 and ICL2, respectively, remain relatively tight during the gating cycle (He et al. 2008). On the other hand, nuclear magnetic resonance (NMR) evidence indicates that ICL1 only binds to the PKA-phosphorylated state of mouse NBD1 (mNBD1) (phosphorylated on the RI and RE; see below) (Kanelis et al. 2010), suggesting that activation of CFTR by PKA does alter the NBD1–ICL1 interaction.

A critical point made by these models is that F508 forms part of the interface between NBD1 and ICL4. As discussed below, deletion of F508 has a limited effect on the ground state structure of NBD1 itself. The structure of the mutant NBD1 is the same as that of the WT, with the exception of altered surface properties in the vicinity of F508 and structural rearrangement of the F494–W496 loop. Deletion of F508 and the resulting change in NBD1 surface properties were hypothesized to disrupt the interaction with the ICL4 coupling helix that contains R1070. In F508del CFTR, mutation of V510, which is near the position of F508 in the WT, to aspartic acid partially corrects the CF defect and promotes CFTR maturation (Loo et al.

2010). The V510D mutation likely introduces a salt bridge with R1070, strengthening this interface. These data support the hypothesis that F508 disrupts this critical ICL4 interaction. Similar results are seen with an R1070W mutation, which might be expected to enhance the hydrophobic contacts at this interface or fill the void introduced by deletion of F508, or with a combined R1070D/V510R mutation, which would reverse the proposed salt bridge (Farinha et al. 2010; Loo et al. 2010). These additional mutations support the idea that the F508del phenotype can be attributed at least partially to a disruption of this interface, although as we discuss below, the V510D mutation also thermodynamically stabilizes NBD1 itself. The data on the V510D mutation indicate that interdomain interactions between NBD1 and ICL4 are required for proper folding and maturation of CFTR. Thus, restoring this interface via therapeutic compounds designed to enhance this interaction might be expected to partially correct the F508del defect. Examining the effects of stabilizing and destabilizing mutations at this interface may be a useful approach for exploring the dynamics of this interface through the gating cycle.

Molecular modeling of CFTR based on related ABC transporters, electron microscopy, cross-linking, and electrophysiology experiments have yielded significant insight into the overall organization and dynamics of full-length CFTR. Disruption of the NBD1–ICL4 interaction by F508del provides a partial answer for how the mutation disrupts processing and gating. Further explanation of how F508del disrupts CFTR processing arose from a closer look at the dynamics and thermodynamics of the isolated NBDs. CFTR's processing efficiency has been shown to be related to the instability and aggregation tendencies of the NBDs. Deletion of F508, which is found in NBD1, further decreases the stability of NBD1. The R region, which is an important site of phosphodependent regulation, is also highly dynamic. The highly dynamic natures of the R region and NBD1 are evidence of a shallow energetic landscape, with an equilibrium between multiple accessible conformations that is sensitive to subtle

Cite this article as *Cold Spring Harb Perspect Med* doi: 10.1101/cshperspect.a009522

sequence changes. As a result of this thermodynamic sensitivity, NBD1 dynamics and interactions are highly responsive to destabilizing and stabilizing mutations. The next section focuses on the complex dynamics of the R region and NBD1 and how these impact CFTR processing and function.

CFTR REGIONS AND DOMAINS

R Region

The R region is the principal physiological regulator of CFTR function, and its mechanism of action has been the focus of intense study. A dominant view is that the nonphosphorylated R region inhibits channel opening. This view is supported by evidence indicating that removal of portions of the R region encompassing residues 760–783 or 817–838 produces channels that open without activation by phosphorylation (Baldursson et al. 2001; Xie et al. 2002). Activation of the channel requires PKA phosphorylation of multiple serines (there are >10 sites), most of which are in the R region (Picciotto et al. 1992; Chappe et al. 2004), although there are also protein kinase C (PKC) (Picciotto et al. 1992; Chappe et al. 2004) and inhibitory adenosine monophosphate-stimulated kinase (AMPK) sites (King et al. 2009; Kongsuphol et al. 2009). Critically, there is no requirement for phosphorylation at any one specific site (Cheng et al. 1991; Rich et al. 1993). Phosphorylation at an increasing number of sites enhances the level of channel activation in a roughly additive manner (Chang et al. 1993; Rich et al. 1993; Wilkinson et al. 1997). The R region is disordered, as shown by NMR and circular dichroism experiments (Dulhanty and Riordan 1994; Ostedgaard et al. 2000; Baker et al. 2007), providing a key component in understanding the plasticity in its behavior.

As a disordered protein segment, the R region is highly dynamic and does not assume a single ground state conformation whether or not it is phosphorylated. Rather, NMR experiments indicate that the isolated R region rapidly exchanges between many different conformations (Baker et al. 2007). Individual segments of the R region can transiently populate secondary structural elements such as α-helices (Baker et al. 2007). Phosphorylation reduces the global α-helical content of the R region, which likely alters R-region binding preferences. R-region structural ensembles generated using discrete molecular dynamics (MD) simulations suggest that phosphorylation increases the R-region radius of gyration (Hegedus et al. 2008); however, NMR-based diffusion measurements indicate that phosphorylation actually decreases the hydrodynamic radius of the R region (Bozoky et al. 2010). Electron microscopy images provide evidence of a high-variability region of density adjacent to the ICDs that is not accounted for by the Sav1866 protein structure and suggests one possible location for a dynamic R region (Zhang et al. 2009). Electron microscopy images of nanogold-labeled CFTR also support a position of the R region near the ICDs.

R-region flexibility makes phosphorylation sites highly accessible and is likely to play an important role in its regulatory function. Disordered proteins often function as protein interaction hubs, which integrate protein inputs (Dunker et al. 2005). Besides interactions with PKA, PKC, and AMPK, the R region has been reported to interact with NBD1, NBD2, the disordered carboxy-terminal portion of CFTR, and the STAS domain of SLC26A3 in a dynamic fashion (Bozoky et al. 2010). NMR evidence for the interactions indicates that they are transient (lifetimes of less than milliseconds) and that multiple sections of the R region are involved. The R region may serve to integrate various cellular control parameters to enable generation of an appropriate level of channel response. A particularly important set of R interactions may be with the NBDs. The R region interacts with the NBDs and causes changes on multiple NBD1 surfaces, including some that overlap with the proposed NBD2 interaction interface. R-region phosphorylation was shown to decrease binding to NBD1 (Baker et al. 2007), possibly by reducing the population of NBD1-interacting α-helices. It is important to note that the data suggest that multiple portions of the R region interact with the NBDs in a dynamic fashion. No specific segments of the R region have been identified as critical for NBD1 interaction. This points to a model in which various segments of the nonphosphorylated R region bind to the NBD dimer interface in a transient manner and sterically inhibit formation of a productive heterodimer, thus preventing channel opening. Phosphorylation reduces R region–NBD interactions, facilitating dimer formation.

Another hypothesis is that the R region directly controls the position of the MSDs in the gating cycle (Hegedus et al. 2008), rather than exerting its regulatory effect through the NBDs. In this distance constraint model, the size of the R-region ensemble directly controls the physical separation of the MSDs through its amino-terminal tether to NBD1 and its carboxy-terminal tether to MSD2. Consistent with this, channel gating may not require NBD dimerization in and of itself (Cui et al. 2007; Wang et al. 2007, 2010b). Channels with NBD2 removed have been found to have a low open probability (Cui et al. 2007; Wang et al. 2007; Zhang et al. 2011). Electron microscopy positioning of the R region near to the ICDs is also consistent with this idea. Computer models of phosphorylated and nonphosphorylated R regions and their interactions with various binding partners based on NMR and small-angle X-ray scattering data are in progress to more clearly define the R-region conformational ensemble within the context of CFTR.

NBD1

As NBD1 is the site of the dominant F508del mutation, this domain has been the subject of intensive structural, dynamic, and thermodynamic analyses. Crystal structures of NBD1 show that it is divided into the three basic regions seen in nucleotide-binding domains of other ABC transporters (Lewis et al. 2004, 2005; Thibodeau et al. 2005; Atwell et al. 2010): the β-sheet subdomain, the ATP-binding core, and the α-helical subdomain (Fig. 3). The ATP-binding core contains the Walker A and Walker B motifs and forms the primary contacts with ATP in the monomeric form. Two additional elements containing phosphorylation sites could be resolved in some structures: the regulatory extension (RE), which is formed

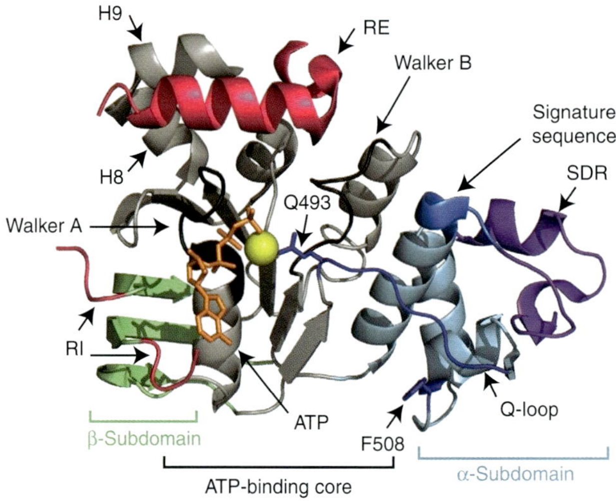

Figure 3. Ribbon diagram representation of CFTR NBD1. The ATP-binding core, α-subdomain, and β-subdomain are shown in gray, blue, and green, respectively. The regulatory extension (RE) and the amino and carboxyl termini of the regulatory insertion (RI) are shown in pink. The central portion of the RI is disordered and is not shown. ATP and magnesium are shown in orange and yellow, respectively. Other features, including the structurally diverse region (SDR), the Q-loop within the α-subdomain, and helices 8 and 9 (H8 and H9), are indicated by the labeled arrows. (PDB 2BBO [Lewis et al. 2010].)

          Cite this article as *Cold Spring Harb Perspect Med* doi: 10.1101/cshperspect.a009522

by the first 30 residues of the R region and is alternatively classified as the carboxyl terminus of NBD1, and the regulatory insertion (RI), a 30-residue, largely disordered insertion between the first and second β-strands unique to CFTR's NBD1. Analysis of several published NBD1 crystal structures reveals that F508del does not grossly perturb the structure of the ground state of the core NBD1 domain. It only causes significant structural changes near the location of the mutation (Lewis et al. 2004, 2005, 2010; Thibodeau et al. 2005; Atwell et al. 2010). Nevertheless, several lines of evidence suggest that the mutation does alter NBD1 dynamic modes or excited states.

The RI and RE show the most pronounced conformational variability in comparisons of the different crystal structures (Lewis et al. 2010), strongly suggesting that they are mobile. Residues 654–669 in the RE consistently form an α-helix, but the helix packs in different positions in the various crystal structures. One dominant interaction site is near the Walker A motif, an interaction that would inhibit NBD heterodimerization. Hydrogen/deuterium (H/D) exchange experiments show that the amide protons in this α-helix readily exchange in solution, indicating that the helix is unstable (Lewis et al. 2010). NMR-derived chemical shifts provide evidence that the preceding two helices (H8 and H9) also exchange with a coil conformation (Hudson et al. 2012). Crystal structures show short helices at both the amino and carboxyl termini of the RI; however, electron density is absent for the central portion, implying that this region of the RI is disordered. The two short helices at either end of the RI are also found in different orientations in the crystal structures, again implying mobility. The RI affects the geometry of ATP binding in some structures (Lewis et al. 2004). H/D exchange experiments confirm that amide protons in the RI are highly solvent-exposed, indicating that the terminal helices and interaction with ATP are not stable.

The RI and RE contain phosphoregulatory sites at residues 422, 660, and 670 (Townsend et al. 1996; Csanady et al. 2005). NMR spectral changes caused by phosphorylation of mNBD1 were shown to be similar to changes caused by

removal of the RE, indicating that phosphorylation disrupts interactions between the NBD1 core and the RE (Kanelis et al. 2010). Phosphorylation also seems to disrupt the RI interaction with the NBD1 core, so that the RI appears to sample more disordered conformations. Kanelis et al. proposed a model in which the RE and RI form transient inhibitory interactions with the NBD1 core, much like the R region does. These inhibitory interactions reduce dimerization probability and are in turn inhibited by phosphorylation. Phosphorylation of the RI may also enhance ATP affinity by disrupting the RI's interaction with the ATP-binding core, a possible regulatory mechanism. Deletion of F508 reduces the magnitude of the changes resulting from phosphorylation, suggesting that the phosphorylated RI and RE interact more strongly with the NBD1 core in the mutant (Kanelis et al. 2010). This strengthened interaction was hypothesized to reduce NBD heterodimerization propensity and channel open probability. It was also observed that F508del inhibits the interaction between ICL1 and the phosphorylated mNBD1. Collectively, these data suggest that F508del alters interactions between the regulatory regions and NBD1 in the absence of large structural changes to the ground state of NBD1, implying that properties other than the altered surface near F508 have been changed. These additional altered properties must result from changes in dynamic modes or excited states that are not observed in the static crystal structures.

NMR evidence also indicates that NBD1 is a highly dynamic protein with extensive motion in the millisecond-to-picosecond timescales. Even the ATP-binding core of the protein experiences millisecond-to-microsecond timescale dynamics, as shown by loss of NMR signal intensities (Chong and Forman-Kay 2010; Kanelis et al. 2010). The exchanging population in the ATP-binding core is likely to be small, as H/D exchange reveals that exchange in this region is relatively slow (Lewis et al. 2010). Analysis of crystallographic data indicates that the α-subdomain has variable orientations with respect to the ATP-binding core (up to 7° rotation), strongly suggesting a flexible linkage (Lewis et al. 2010). This was not unexpected, because the orientation of the α-subdomain with respect to the ATP-binding core has been shown to be nucleotide-dependent in other ABC transporters (Smith et al. 2002; Jones and George 2007, 2009). This rotation is postulated to play a critical role in the power stroke of ABC transporters (Hopfner et al. 2000; Smith et al. 2002), but its role in CFTR is unknown. Whether the CFTR α-subdomain reorients in response to nucleotide changes is also unknown. Conformational variability in the Q-loop, which links the ATP-binding core subdomain with the amino terminus of the α-subdomain, is not observed in the crystal structures, possibly because most of the structures have bound ATP or a nonhydrolyzable analog. One MD study indicates that deletion of F508 destabilizes the Q-loop, causing it to dissociate from the body of the domain, and leads to α-subdomain orientational freedom (Wieczorek and Zielenkiewicz 2008). Crystallographic conformational variability and/or high B-factors are observed for the 509–511 loop, a portion of the structurally diverse region (SDR residues 541–547) that likely interacts with NBD2, the Walker B loop (572–579), the α-helix following the H-loop, and the loop preceding the Walker B β-strand. High B-factors are observed for the final β-strand of the β-subdomain, which likely binds to the ICDs. Comparison of H/D exchange for WT and F508del indicates that the mutation results in a significant increase in H/D exchange rate in the sequence vicinity of the 508 mutation, extending from before residue 504 to beyond residue 517 (Lewis et al. 2010). Changes in the H/D exchange rate elsewhere in the domain are minimal, at least when Mg-ATP is bound to the domain. This is consistent with the loose packing of the F508 residue and the variable conformations for this region observed in the crystal structures. It is also consistent with the minimal structural changes observed in the ground state structure on deletion of F508.

Thus, two mechanisms explain how F508del negatively affects processing and gating. First, the mutation directly changes the ICL-interacting surface of NBD1, negatively affecting these interactions that are critical for processing and likely for gating as well. Second, F508del changes the dynamic and thermodynamic properties of NBD1, including enhanced interactions with the RI, the RE, and the R region. This second mechanism likely involves a change in the overall energetic landscape of NBD1 on deletion of F508, resulting in perturbed excited states and differential sampling of these excited states. Because of the changes in the energetic landscape, this second mechanism also relates to the folding and unfolding properties of NBD1. Several lines of evidence indicate that the change in the NBD1 energetic landscape is a key part of the explanation for the negative effects of F508del.

The nature of a subset of the suppressor mutations for the F508del defect, in particular, suggests that NBD1 stability is central to the defect. F508del causes a substantial, $\sim 6°C$–$7°C$ drop in thermal melting temperature of purified NBD1 (Protasevich et al. 2010). Many of the suppressor mutations, like G550E, R553Q, and R555K (Teem et al. 1993, 1996), which are relatively distant from the F508 position, increase NBD1 thermal stability and reverse some of the processing and functional defects, presumably without reverting the surface changes caused by deletion of F508. V510D, which may rescue F508del by facilitating the NBD1–ICL4 interaction, also stabilizes NBD1 (Protasevich et al. 2010), suggesting a dual mechanism of action. Recently, substitution of proline residues at key positions in the Q-loop, the SDR, and the RI in context of the I539T suppressor mutation has been shown to restore channel function and thermostability to full-length F508del CFTR. MD studies support a role for these proline mutations in reducing flexibility and increasing the temperatures of unfolding transitions for F508del NBD1 (Aleksandrov et al. 2012). Most convincingly, for a subset of suppressor mutants, a good correlation is observed between the improvement in NBD1 stability and CFTR processing efficiency (Thibodeau et al. 2010; Rabeh et al. 2011).

In addition to stabilizing point mutations, complete removal of the RI, which is relatively distant from the α-subdomain, also dramatically enhances the solution stability of NBD1 (Atwell et al. 2010). Furthermore, removal of

   Cite this article as *Cold Spring Harb Perspect Med* doi: 10.1101/cshperspect.a009522

the RI has been shown to suppress trafficking and gating defects associated with F508del (Aleksandrov et al. 2010). Not only does removal of RI improve solution stability of NBD1, it also restores function to F508del CFTR at physiologically relevant temperatures, as shown by iodide efflux and electrophysiological measurements (Aleksandrov et al. 2010). Partial removal of RI did not recapitulate this effect on NBD1 stability and restoration of function, indicating that the dynamics of RI (residues 401–436) plays an inhibitory role in CFTR function. As mentioned earlier, NBD1 crystal structures present only portions of the RI, and these in multiple conformations. The complete RI could not be resolved because of its intrinsic flexibility. Discrete MD simulations of NBD1 with a structural model for the RI revealed that the RI dynamically couples different regions of NBD1 (Fig. 4A,B) (Aleksandrov et al. 2010). Residuewise root mean square fluctuation (RMSF) during the simulation reiterates that the RI is intrinsically flexible and dynamic (Fig. 4A). Most importantly, dynamic coupling exists between the F508-containing loop and the ATP-binding core subdomain of NBD1, which is lost on deletion of F508 (Fig. 4B). Such coupling is partially restored on deletion of the RI, indicating that the RI influences the dynamics of the F508-containing loop (Fig. 4B). Therefore, deletion of F508 could cause dynamic effects that influence the NBD1–ICL3–ICL4 interfaces and hence perturb the overall gating characteristics of the channel (Fig. 4C,D). Given that the structure of NBD1 with and without F508 is not significantly different, NBD1 dynamics likely play a crucial role in trafficking, maturation, and function of CFTR. Based on these results, one might hypothesize that conformational restriction of the RI region in NBD1 would improve maturation, trafficking, and function of F508del CFTR.

As we have suggested, changes to the NBD1 energetic landscape not only alter its interaction with regulatory regions, but they also affect its self-association. Extensive chemical and thermal denaturation studies monitored by a variety of biophysical techniques have identified an aggregation-prone, partially unfolded intermediate

state on the unfolding pathway that is enhanced by F508del (Richardson et al. 2007; Protasevich et al. 2010; Wang et al. 2010a). F508del does not seem to affect unfolding of this partially unfolded intermediate, suggesting that the F508 region is already unfolded in the intermediate. F508 and many suppressor mutations are located in the α-helical subdomain, implying that the partially unfolded state involves this subdomain (Wang et al. 2010a). The RI is not located in or near the α-helical subdomain; however, it still may affect the conformation of this subdomain. One possible mechanism is that ATP affinity is enhanced on removal of the RI. ATP binding in turn may affect the α-helical subdomain orientation and stability via its interaction with the Q-loop. A second possibility is dynamic coupling between the RI and the SDR region as observed in MD simulations (Aleksandrov et al. 2010). Deletion of the RI stabilizes the SDR region in these MD studies and may impart an effect on NBD1 stability similar to α-helical subdomain suppressor mutations (Aleksandrov et al. 2010). The partially unfolded intermediate identified by the Hunt and Brouillette groups appears to be aggregation-prone and thus likely reduces both processing and gating efficiency. A partially unfolded configuration for the α-subdomain, which may be similar to that identified by Wang et al. (2010a), can be observed in MD simulations, suggesting that the dynamics of NBD1 can be accurately captured using simulations. In addition, MD folding simulations led to the conclusion that F508 deletion alters the kinetics of NBD1 folding (Serohijos et al. 2008b). These studies provide atomistic explanations for the experimentally observed decrease in NBD1 folding yields on deletion of F508 (Thibodeau et al. 2005). Different folding intermediates are observed and different transient contact patterns are formed for WT and F508del NBD1 in these simulations (Serohijos et al. 2008b). In this study, the H5–S6 loop and the S7–H5 loop near the α-subdomain–ATP-binding core interface were found to represent significant structural differences between WT and F508del NBD1. NMR dispersion experiments have also identified a weakly-populated excited state for WT NBD1 protein that is enhanced by F508del (Chong and

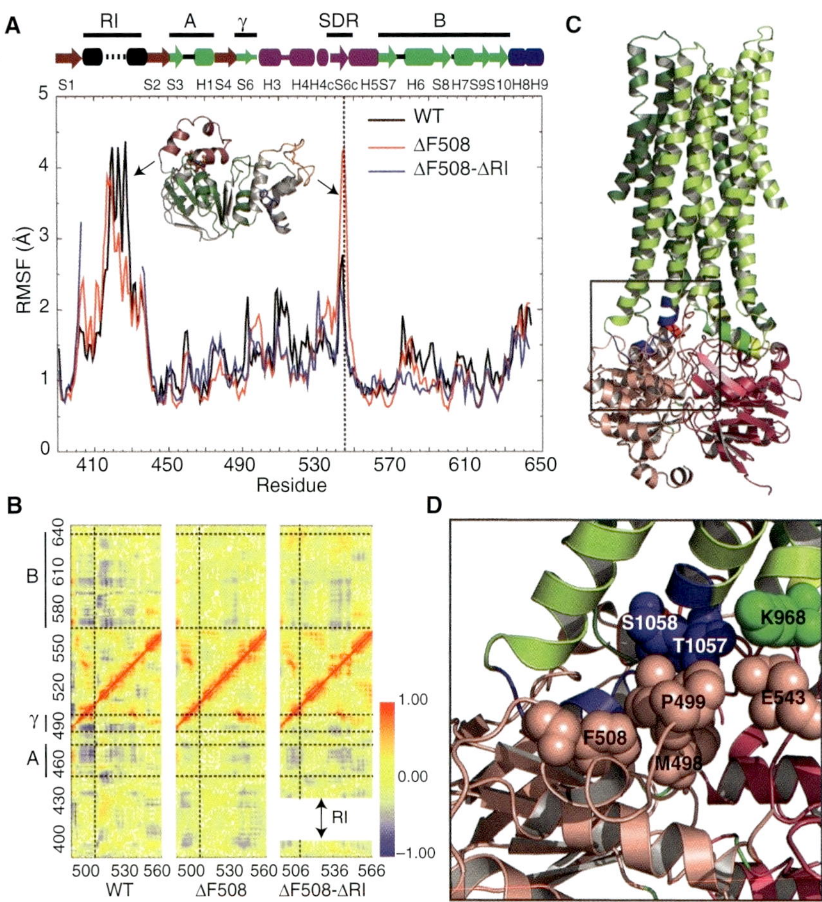

Figure 4. CFTR NBD1 dynamics. (*A*) Root mean square fluctuations (RMSFs) of every residue in NBD1 as a function of time, obtained from MD simulations. The vertical dotted line indicates fluctuations of the SDR loop. Black arrows represent RMSF peaks for the corresponding structural regions in the plot. The ATP-binding core subdomain (A, G451–L475; B, D565–Q637) and γ-switch (Q493–P499) are colored green and labeled by the bold lines at the top for reference. (*B*) Pairwise Cα correlation map for residues 490–560 in comparison with those of all other regions in NBD1. Horizontal dotted lines separate the ATP-binding core subdomain (A and B) and γ-switch, as indicated by the bold lines on the left of the *y*-axis. The vertical dotted line corresponds to position F508. The shift in the *y*-axis of the F508del–ΔRI correlation map is a consequence of RI deletion. Red dots (correlation coefficient = 1) indicate residue pairs that move in concert in the same direction, and blue dots (correlation coefficient = −1) indicate residue pairs that move with opposite velocities all the time. (*C*) Structural model of full-length CFTR (http://dokhlab.unc.edu/research/CFTR/home.html). The area of the NBD1–ICL3–ICL4 interface is shown in the *inset*. (*D*) A detailed view of the location and orientation of the residues in the NBD1–ICL3–ICL4 interface from *C*. Residues F508, P499, M498, and E543 from NBD1 (salmon), S1058 and T1057 from ICL4 (blue), and K968 from ICL3 (cyan) are shown as spheres. (From Aleksandrov et al. 2010; reproduced, with permission.)

Forman-Kay 2010). The relationship between the excited states observed by these different methods and groups is unclear, but collectively they support the notion that F508del profoundly perturbs the NBD1 energetic landscape.

The nature of some of the NBD1 excited states might be investigated by analysis of NBD1 fragments that have been found to be competent to bind ATP. An early report identified a 67-residue fragment comprising residues

450 through 516 of NBD1 that could bind to ATP with submillimolar affinity (Thomas et al. 1991). Subsequently, a fragment comprising residues 404–589 was shown to be capable of folding and binding to ATP with an affinity of ~90 μM (Qu et al. 1997). This fragment is missing the first β-strand of NBD1 and the carboxy-terminal portion of the ATP-binding core. More recently, in vitro translation experiments monitored by fluorescence resonance energy transfer confirm that NBD1 residues 389 through 500, which comprise the β-subdomain and parts of the ATP-binding core, can fold autonomously into a subdomain that can bind to ATP (Khushoo et al. 2011). This study further showed that ATP stimulates proper folding of this fragment. The ATP-binding capacity of these NBD1 fragments suggests that these fragments can form folded structures. However, they are unlikely to assume the conformation observed in the ground state structures, as critical portions are missing in each instance. Thus, these fragments may form nonnative conformations that may be in equilibrium with the ground state conformation in the context of the full-length NBD1, providing insight into the accessible energy landscape.

In summary, NBD1 stability and changes in the energetic landscape of NBD1 on deletion of F508 are likely a critical part of the F508 defect. Suppression of the processing defect by stabilizing mutations and low-temperature rescue support the key role these changes in NBD1 dynamics play in the F508 defect. CFTR channel opening and closing has been shown to be tightly coupled to ATP binding and hydrolysis at the NBDs (Csanady et al. 2009). The challenge for the future will be to understand at a molecular level how the nucleotide-binding status of the NBDs, dimer conformation, and dynamics are allosterically linked to MSD conformation and channel gating. It will also be interesting to see how the absence of NBD2 will affect these linkages.

by PKA phosphorylation and in response to nucleotide binding by the NBDs, forming a closed state, an open-ready state, and an open state. The precise nature of these conformational changes awaits higher-resolution electron microscopy images and more channel-gating experiments building on already extensive studies. MSD interactions with NBD1 via the ICDs are required for proper folding and gating. Disruption of ICD and NBD1 interactions is one mechanism by which F508del disrupts folding and gating. F508del also appears to alter the transient and complex dynamic interactions between the NBD1 and the highly flexible, disordered phosphoregulatory elements of CFTR that include the R region, RI, and RE, further impairing channel gating. NMR data-based computational models of these regulatory interactions will be useful for understanding exactly how the phosphoregulatory elements and their regulatory interactions are changed by phosphorylation and how they are affected by deletion of F508. NBD1 itself is highly dynamic. Although F508del causes only minor changes to the NBD1 ground state structure, it alters the stability, folding pathway, and binding properties of this marginally stable domain. A variety of data indicate that F508del changes the energy landscape for this domain and alters the frequency with which higher energy states are accessed. The dynamics and thermodynamics of NBD1 remain exciting areas of study, both because changes in NBD1 dynamics have been directly linked to misprocessing, instability, and channel-gating defects and because some of these higher energy states may have functional relevance. As such, information about dynamics will be key for building a model to describe how the R region and NBDs allosterically control the MSDs and channel activity. In addition, knowledge of the dynamics involved in domain and full-length CFTR folding and interactions with cellular processing machinery will be critical for mechanistic understanding of the F508del defects.

CONCLUDING REMARKS

CFTR shows a wide range of dynamics that enable proper folding and gating. The MSDs and ICDs likely rearrange in response to activation

ACKNOWLEDGMENTS

The authors thank Drs. John R. Riordan, Adrian W.R. Serohijos, Tamas Hegedus, Lihua He,

Andrei Aleksandrov, Rhea Hudson, Jennifer Dawson, Robert Vernon, and Zoltán Bozóky for valuable discussions. In addition, funding from the Cystic Fibrosis Foundation Therapeutics and Cystic Fibrosis Canada to J.D.F.-K. and from the National Institutes of Health (R01GM080742) to N.V.D. is gratefully acknowledged.

REFERENCES

Aleksandrov AA, Kota P, Aleksandrov LA, He L, Jensen T, Cui L, Gentzsch M, Dokholyan NV, Riordan JR. 2010. Regulatory insertion removal restores maturation, stability and function of ΔF508 CFTR. *J Mol Biol* **401:** 194–210.

Aleksandrov AA, Kota P, Cui L, Jensen T, Alekseev AE, Reyes S, He L, Gentzsch M, Aleksandrov LA, Dokholyan NV, et al. 2012. Allosteric modulation balances stability and restores function of ΔF508 CFTR. *J Mol Biol* **419:** 41–60.

Alexander C, Ivetac A, Liu X, Norimatsu Y, Serrano JR, Landstrom A, Sansom M, Dawson DC. 2009. Cystic fibrosis transmembrane conductance regulator: Using differential reactivity toward channel-permeant and channel-impermeant thiol-reactive probes to test a molecular model for the pore. *Biochemistry* **48:** 10078–10088.

Anderson MP, Gregory RJ, Thompson S, Souza DW, Paul S, Mulligan RC, Smith AE, Welsh MJ. 1991. Demonstration that CFTR is a chloride channel by alteration of its anion selectivity. *Science* **253:** 202–205.

Atwell S, Brouillette CG, Conners K, Emtage S, Gheyi T, Guggino WB, Hendle J, Hunt JF, Lewis HA, Lu F, et al. 2010. Structures of a minimal human CFTR first nucleotide-binding domain as a monomer, head-to-tail homodimer, and pathogenic mutant. *Protein Eng Des Sel* **23:** 375–384.

Awayn NH, Rosenberg MF, Kamis AB, Aleksandrov LA, Riordan JR, Ford RC. 2005. Crystallographic and single-particle analyses of native- and nucleotide-bound forms of the cystic fibrosis transmembrane conductance regulator (CFTR) protein. *Biochem Soc Trans* **33:** 996–999.

Baker JM, Hudson RP, Kanelis V, Choy WY, Thibodeau PH, Thomas PJ, Forman-Kay JD. 2007. CFTR regulatory region interacts with NBD1 predominantly via multiple transient helices. *Nat Struct Mol Biol* **14:** 738–745.

Baldursson O, Ostedgaard LS, Rokhlina T, Cotten JF, Welsh MJ. 2001. Cystic fibrosis transmembrane conductance regulator Cl⁻ channels with R domain deletions and translocations show phosphorylation-dependent and -independent activity. *J Biol Chem* **276:** 1904–1910.

Berger HA, Travis SM, Welsh MJ. 1993. Regulation of the cystic fibrosis transmembrane conductance regulator Cl⁻ channel by specific protein kinases and protein phosphatases. *J Biol Chem* **268:** 2037–2047.

Bozoky Z, Baker JM, Thomas PJ, Bear CE, Forman-Kay JD. 2010. The regulatory R region of CFTR serves as a dynamic integrator. *Pediatr Pulmonol* **45** (Suppl): 220.

Chang XB, Tabcharani JA, Hou YX, Jensen TJ, Kartner N, Alon N, Hanrahan JW, Riordan JR. 1993. Protein kinase A (PKA) still activates CFTR chloride channel after mutagenesis of all 10 PKA consensus phosphorylation sites. *J Biol Chem* **268:** 11304–11311.

Chappe V, Hinkson DA, Howell LD, Evagelidis A, Liao J, Chang XB, Riordan JR, Hanrahan JW. 2004. Stimulatory and inhibitory protein kinase C consensus sequences regulate the cystic fibrosis transmembrane conductance regulator. *Proc Natl Acad Sci* **101:** 390–395.

Cheng SH, Gregory RJ, Marshall J, Paul S, Souza DW, White GA, O'Riordan CR, Smith AE. 1990. Defective intracellular transport and processing of CFTR is the molecular basis of most cystic fibrosis. *Cell* **63:** 827–834.

Cheng SH, Rich DP, Marshall J, Gregory RJ, Welsh MJ, Smith AE. 1991. Phosphorylation of the R domain by cAMP-dependent protein kinase regulates the CFTR chloride channel. *Cell* **66:** 1027–1036.

Cheung JC, Kim Chiaw P, Pasyk S, Bear CE. 2008. Molecular basis for the ATPase activity of CFTR. *Arch Biochem Biophys* **476:** 95–100.

Chong PA, Forman-Kay JD. 2010. Alteration of CFTR NBD1 dynamics upon deletion of F508 underlying the basic defect. *Pediatr Pulmonol* **45** (Suppl): 240.

Csanady L, Chan KW, Nairn AC, Gadsby DC. 2005. Functional roles of nonconserved structural segments in CFTR's NH₂-terminal nucleotide binding domain. *J Gen Physiol* **125:** 43–55.

Csanady L, Vergani P, Gadsby DC. 2009. Strict coupling between CFTR's catalytic cycle and gating of its Cl⁻ ion pore revealed by distributions of open channel burst durations. *Proc Natl Acad Sci* **107:** 1241–1246.

Cui L, Aleksandrov L, Chang XB, Hou YX, He L, Hegedus T, Gentzsch M, Aleksandrov A, Balch WE, Riordan JR. 2007. Domain interdependence in the biosynthetic assembly of CFTR. *J Mol Biol* **365:** 981–994.

Dalemans W, Barbry P, Champigny G, Jallat S, Dott K, Dreyer D, Crystal RG, Pavirani A, Lecocq JP, Lazdunski M. 1991. Altered chloride ion channel kinetics associated with the ΔF508 cystic fibrosis mutation. *Nature* **354:** 526–528.

Dawson RJ, Locher KP. 2006. Structure of a bacterial multidrug ABC transporter. *Nature* **443:** 180–185.

Denning GM, Anderson MP, Amara JF, Marshall J, Smith AE, Welsh MJ. 1992. Processing of mutant cystic fibrosis transmembrane conductance regulator is temperature-sensitive. *Nature* **358:** 761–764.

Drumm ML, Wilkinson DJ, Smit LS, Worrell RT, Strong TV, Frizzell RA, Dawson DC, Collins FS. 1991. Chloride conductance expressed by ΔF508 and other mutant CFTRs in *Xenopus* oocytes. *Science* **254:** 1797–1799.

Du K, Sharma M, Lukacs GL. 2005. The ΔF508 cystic fibrosis mutation impairs domain-domain interactions and arrests post-translational folding of CFTR. *Nat Struct Mol Biol* **12:** 17–25.

Dulhanty AM, Riordan JR. 1994. Phosphorylation by cAMP-dependent protein kinase causes a conformational change in the R domain of the cystic fibrosis transmembrane conductance regulator. *Biochemistry* **33:** 4072–4079.

Dunker AK, Cortese MS, Romero P, Iakoucheva LM, Uversky VN. 2005. Flexible nets. The roles of intrinsic disorder in protein interaction networks. *FEBS J* **272**: 5129–5148.

Farinha CM, Da Paula A, Amaral MD. 2010. Rescuing F508del-CFTR by genetic revertants, low temperature and small molecules. *Pediatr Pulmonol* **45** (Suppl): 230.

Fatehi M, Linsdell P. 2008. State-dependent access of anions to the cystic fibrosis transmembrane conductance regulator chloride channel pore. *J Biol Chem* **283**: 6102–6109.

He L, Aleksandrov AA, Serohijos AW, Hegedus T, Aleksandrov LA, Cui L, Dokholyan NV, Riordan JR. 2008. Multiple membrane-cytoplasmic domain contacts in the cystic fibrosis transmembrane conductance regulator (CFTR) mediate regulation of channel gating. *J Biol Chem* **283**: 26383–26390.

He L, Aleksandrov LA, Cui L, Jensen TJ, Nesbitt KL, Riordan JR. 2010. Restoration of domain folding and interdomain assembly by second-site suppressors of the ΔF508 mutation in CFTR. *FASEB J* **24**: 3103–3112.

Hegedus T, Serohijos AW, Dokholyan NV, He L, Riordan JR. 2008. Computational studies reveal phosphorylation-dependent changes in the unstructured R domain of CFTR. *J Mol Biol* **378**: 1052–1063.

Hoelen H, Kleizen B, Schmidt A, Richardson J, Charitou P, Thomas PJ, Braakman I. 2010. The primary folding defect and rescue of ΔF508 CFTR emerge during translation of the mutant domain. *PLoS ONE* **5**: e15458.

Hollenstein K, Frei DC, Locher KP. 2007. Structure of an ABC transporter in complex with its binding protein. *Nature* **446**: 213–216.

Hopfner KP, Karcher A, Shin DS, Craig L, Arthur LM, Carney JP, Tainer JA. 2000. Structural biology of Rad50 ATPase: ATP-driven conformational control in DNA double-strand break repair and the ABC-ATPase superfamily. *Cell* **101**: 789–800.

Huang SY, Bolser D, Liu HY, Hwang TC, Zou X. 2009. Molecular modeling of the heterodimer of human CFTR's nucleotide-binding domains using a protein–protein docking approach. *J Mol Graph Model* **27**: 822–828.

Hudson RP, Chong PA, Protasevich II, Vernon R, Noy E, Bihler H, Li AJ, Kalid O, Sela-Culang I, Mense M, et al. 2012. Conformational changes relevant to channel activity and folding within the first nucleotide binding domain of the cystic fibrosis transmembrane conductance regulator. *J Biol Chem* **287**: 28480–28494.

Jardetzky O. 1966. Simple allosteric model for membrane pumps. *Nature* **211**: 969–970.

Jones PM, George AM. 2007. Nucleotide-dependent allostery within the ABC transporter ATP-binding cassette: A computational study of the MJ0796 dimer. *J Biol Chem* **282**: 22793–22803.

Jones PM, George AM. 2009. Opening of the ADP-bound active site in the ABC transporter ATPase dimer: Evidence for a constant contact, alternating sites model for the catalytic cycle. *Proteins* **75**: 387–396.

Kanelis V, Hudson RP, Thibodeau PH, Thomas PJ, Forman-Kay JD. 2010. NMR evidence for differential phosphorylation-dependent interactions in WT and ΔF508 CFTR. *EMBO J* **29**: 263–277.

Khushoo A, Yang Z, Johnson AE, Skach WR. 2011. Ligand-driven vectorial folding of ribosome-bound human CFTR NBD1. *Mol Cell* **41**: 682–692.

King JD Jr, Fitch AC, Lee JK, McCane JE, Mak DO, Foskett JK, Hallows KR. 2009. AMP-activated protein kinase phosphorylation of the R domain inhibits PKA stimulation of CFTR. *Am J Physiol Cell Physiol* **297**: C94–C101.

Kleizen B, van Vlijmen T, de Jonge HR, Braakman I. 2005. Folding of CFTR is predominantly cotranslational. *Mol Cell* **20**: 277–287.

Kongsuphol P, Cassidy D, Hieke B, Treharne KJ, Schreiber R, Mehta A, Kunzelmann K. 2009. Mechanistic insight into control of CFTR by AMPK. *J Biol Chem* **284**: 5645–5653.

Lewis HA, Buchanan SG, Burley SK, Conners K, Dickey M, Dorwart M, Fowler R, Gao X, Guggino WB, Hendrickson WA, et al. 2004. Structure of nucleotide-binding domain 1 of the cystic fibrosis transmembrane conductance regulator. *EMBO J* **23**: 282–293.

Lewis HA, Zhao X, Wang C, Sauder JM, Rooney I, Noland BW, Lorimer D, Kearins MC, Conners K, Condon B, et al. 2005. Impact of the ΔF508 mutation in first nucleotide-binding domain of human cystic fibrosis transmembrane conductance regulator on domain folding and structure. *J Biol Chem* **280**: 1346–1353.

Lewis HA, Wang C, Zhao X, Hamuro Y, Conners K, Kearins MC, Lu F, Sauder JM, Molnar KS, Coales SJ, et al. 2010. Structure and dynamics of NBD1 from CFTR characterized using crystallography and hydrogen/deuterium exchange mass spectrometry. *J Mol Biol* **396**: 406–430.

Loo TW, Bartlett MC, Clarke DM. 2010. The V510D suppressor mutation stabilizes ΔF508-CFTR at the cell surface. *Biochemistry* **49**: 6352–6357.

Lukacs GL, Mohamed A, Kartner N, Chang XB, Riordan JR, Grinstein S. 1994. Conformational maturation of CFTR but not its mutant counterpart (ΔF508) occurs in the endoplasmic reticulum and requires ATP. *EMBO J* **13**: 6076–6086.

Mitchell P. 1957. A general theory of membrane transport from studies of bacteria. *Nature* **180**: 134–136.

Moran O. 2010. Model of the cAMP activation of chloride transport by CFTR channel and the mechanism of potentiators. *J Theor Biol* **262**: 73–79.

Mornon JP, Lehn P, Callebaut I. 2008. Atomic model of human cystic fibrosis transmembrane conductance regulator: Membrane-spanning domains and coupling interfaces. *Cell Mol Life Sci* **65**: 2594–2612.

Mornon JP, Lehn P, Callebaut I. 2009. Molecular models of the open and closed states of the whole human CFTR protein. *Cell Mol Life Sci* **66**: 3469–3486.

Ostedgaard LS, Baldursson O, Vermeer DW, Welsh MJ, Robertson AD. 2000. A functional R domain from cystic fibrosis transmembrane conductance regulator is predominantly unstructured in solution. *Proc Natl Acad Sci* **97**: 5657–5662.

Picciotto MR, Cohn JA, Bertuzzi G, Greengard P, Nairn AC. 1992. Phosphorylation of the cystic fibrosis transmembrane conductance regulator. *J Biol Chem* **267**: 12742–12752.

Pinkett HW, Lee AT, Lum P, Locher KP, Rees DC. 2007. An inward-facing conformation of a putative metal-chelate-type ABC transporter. *Science* **315:** 373–377.

Protasevich I, Yang Z, Wang C, Atwell S, Zhao X, Emtage S, Wetmore D, Hunt JF, Brouillette CG. 2010. Thermal unfolding studies show the disease causing F508del mutation in CFTR thermodynamically destabilizes nucleotide-binding domain 1. *Protein Sci* **19:** 1917–1931.

Qu BH, Strickland EH, Thomas PJ. 1997. Localization and suppression of a kinetic defect in cystic fibrosis transmembrane conductance regulator folding. *J Biol Chem* **272:** 15739–15744.

Rabeh WM, Di Bernardo S, Okiyoneda T, Liu Y, Mulvihill CM, Du K, Xu H, Konermann L, Lukacs G, Bossard F. 2011. Conformational stabilization of the NBD1 is necessary, but not sufficient for CFTR biogenesis. *Pediatr Pulmonol* **45** (Suppl)**:** 220.

Rich DP, Berger HA, Cheng SH, Travis SM, Saxena M, Smith AE, Welsh MJ. 1993. Regulation of the cystic fibrosis transmembrane conductance regulator Cl⁻ channel by negative charge in the R domain. *J Biol Chem* **268:** 20259–20267.

Richardson JM, Thibodeau PH, Watson J, Thomas PJ. 2007. Identification of a non-native state of NBD1 that is affected by ΔF508. *Pediatr Pulmonol* **45** (Suppl)**:** 203.

Riordan JR, Rommens JM, Kerem B, Alon N, Rozmahel R, Grzelczak Z, Zielenski J, Lok S, Plavsic N, Chou J-L, et al. 1989. Identification of the cystic fibrosis gene: Cloning and characterization of complementary DNA. *Science* **245:** 1066–1073.

Rosenberg MF, Kamis AB, Aleksandrov LA, Ford RC, Riordan JR. 2004. Purification and crystallization of the cystic fibrosis transmembrane conductance regulator (CFTR). *J Biol Chem* **279:** 39051–39057.

Serohijos AW, Hegedus T, Aleksandrov AA, He L, Cui L, Dokholyan NV, Riordan JR. 2008a. Phenylalanine-508 mediates a cytoplasmic-membrane domain contact in the CFTR 3D structure crucial to assembly and channel function. *Proc Natl Acad Sci* **105:** 3256–3261.

Serohijos AW, Hegedus T, Riordan JR, Dokholyan NV. 2008b. Diminished self-chaperoning activity of the ΔF508 mutant of CFTR results in protein misfolding. *PLoS Comput Biol* **4:** e1000008.

Smith PC, Karpowich N, Millen L, Moody JE, Rosen J, Thomas PJ, Hunt JF. 2002. ATP binding to the motor domain from an ABC transporter drives formation of a nucleotide sandwich dimer. *Mol Cell* **10:** 139–149.

Teem JL, Berger HA, Ostedgaard LS, Rich DP, Tsui LC, Welsh MJ. 1993. Identification of revertants for the cystic fibrosis ΔF508 mutation using STE6-CFTR chimeras in yeast. *Cell* **73:** 335–346.

Teem JL, Carson MR, Welsh MJ. 1996. Mutation of R555 in CFTR-ΔF508 enhances function and partially corrects defective processing. *Receptors Channels* **4:** 63–72.

Thibodeau PH, Brautigam CA, Machius M, Thomas PJ. 2005. Side chain and backbone contributions of Phe508 to CFTR folding. *Nat Struct Mol Biol* **12:** 10–16.

Thibodeau PH, Richardson JM III, Wang W, Millen L, Watson J, Mendoza JL, Du K, Fischman S, Senderowitz H, Lukacs GL, et al. 2010. The cystic fibrosis-causing mutation ΔF508 affects multiple steps in cystic fibrosis transmembrane conductance regulator biogenesis. *J Biol Chem* **285:** 35825–35835.

Thomas PJ, Shenbagamurthi P, Ysern X, Pedersen PL. 1991. Cystic fibrosis transmembrane conductance regulator: Nucleotide binding to a synthetic peptide. *Science* **251:** 555–557.

Townsend RR, Lipniunas PH, Tulk BM, Verkman AS. 1996. Identification of protein kinase A phosphorylation sites on NBD1 and R domains of CFTR using electrospray mass spectrometry with selective phosphate ion monitoring. *Protein Sci* **5:** 1865–1873.

Varga K, Jurkuvenaite A, Wakefield J, Hong JS, Guimbellot JS, Venglarik CJ, Niraj A, Mazur M, Sorscher EJ, Collawn JF, et al. 2004. Efficient intracellular processing of the endogenous cystic fibrosis transmembrane conductance regulator in epithelial cell lines. *J Biol Chem* **279:** 22578–22584.

Wang W, Bernard K, Li G, Kirk KL. 2007. Curcumin opens cystic fibrosis transmembrane conductance regulator channels by a novel mechanism that requires neither ATP binding nor dimerization of the nucleotide-binding domains. *J Biol Chem* **282:** 4533–4544.

Wang C, Protasevich I, Yang Z, Seehausen D, Skalak T, Zhao X, Atwell S, Spencer Emtage J, Wetmore DR, Brouillette CG, et al. 2010a. Integrated biophysical studies implicate partial unfolding of NBD1 of CFTR in the molecular pathogenesis of F508del cystic fibrosis. *Protein Sci* **19:** 1932–1947.

Wang W, Wu J, Bernard K, Li G, Wang G, Bevensee MO, Kirk KL. 2010b. ATP-independent CFTR channel gating and allosteric modulation by phosphorylation. *Proc Natl Acad Sci* **107:** 3888–3893.

Ward CL, Kopito RR. 1994. Intracellular turnover of cystic fibrosis transmembrane conductance regulator. Inefficient processing and rapid degradation of wild-type and mutant proteins. *J Biol Chem* **269:** 25710–25718.

Ward A, Reyes CL, Yu J, Roth CB, Chang G. 2007. Flexibility in the ABC transporter MsbA: Alternating access with a twist. *Proc Natl Acad Sci* **104:** 19005–19010.

Wieczorek G, Zielenkiewicz P. 2008. ΔF508 mutation increases conformational flexibility of CFTR protein. *J Cyst Fibros* **7:** 295–300.

Wilkinson DJ, Strong TV, Mansoura MK, Wood DL, Smith SS, Collins FS, Dawson DC. 1997. CFTR activation: Additive effects of stimulatory and inhibitory phosphorylation sites in the R domain. *Am J Physiol* **273:** L127–L133.

Winter MC, Sheppard DN, Carson MR, Welsh MJ. 1994. Effect of ATP concentration on CFTR Cl⁻ channels: A kinetic analysis of channel regulation. *Biophys J* **66:** 1398–1403.

Xie J, Adams LM, Zhao J, Gerken TA, Davis PB, Ma J. 2002. A short segment of the R domain of cystic fibrosis transmembrane conductance regulator contains channel stimulatory and inhibitory activities that are separable by sequence modification. *J Biol Chem* **277:** 23019–23027.

Zhang ZR, Song B, McCarty NA. 2005. State-dependent chemical reactivity of an engineered cysteine reveals conformational changes in the outer vestibule of the cystic fibrosis transmembrane conductance regulator. *J Biol Chem* **280:** 41997–42003.

Zhang L, Aleksandrov LA, Zhao Z, Birtley JR, Riordan JR, Ford RC. 2009. Architecture of the cystic fibrosis transmembrane conductance regulator protein and structural changes associated with phosphorylation and nucleotide binding. *J Struct Biol* **167:** 242–251.

Zhang L, Aleksandrov LA, Riordan JR, Ford RC. 2011. Domain location within the cystic fibrosis transmembrane conductance regulator protein investigated by electron microscopy and gold labelling. *Biochim Biophys Acta* **1808:** 399–404.

The Influence of Genetics on Cystic Fibrosis Phenotypes

Michael R. Knowles[1] and Mitchell Drumm[2]

[1]Cystic Fibrosis-Pulmonary Research and Treatment Center, University of North Carolina at Chapel Hill, Chapel Hill, North Carolina 27514

[2]Department of Pediatrics and Genetics and Genome Sciences, Case Western Reserve University, Cleveland, Ohio 44106-4948

Correspondence: michael_knowles@med.unc.edu

Technological advances in genetics have made feasible and affordable large studies to identify genetic variants that cause or modify a trait. Genetic studies have been carried out to assess variants in candidate genes, as well as polymorphisms throughout the genome, for their associations with heritable clinical outcomes of cystic fibrosis (CF), such as lung disease, meconium ileus, and CF-related diabetes. The candidate gene approach has identified some predicted relationships, whereas genome-wide surveys have identified several genes that would not have been obvious disease-modifying candidates, such as a methionine sulfoxide transferase gene that influences intestinal obstruction, or a region on chromosome 11 proximate to genes encoding a transcription factor and an apoptosis controller that associates with lung function. These unforeseen associations thus provide novel insight into disease pathophysiology, as well as suggesting new therapeutic strategies for CF.

Cystic fibrosis (CF) is a Mendelian "monogenic" recessive genetic disorder caused by mutations in the cystic fibrosis transmembrane conductance regulator (*CFTR*) gene (Welsh et al. 2001). There is a broad range of age-of-onset and disease activity for different organ systems in CF, including lung disease, meconium ileus, diabetes, and liver disease, even for CF patients who are homozygous for the most common mutation, *F508del* (Cutting 2010). Therefore, non-*CFTR* genetic variation and/or environmental influences must contribute to the variability of clinical phenotypes.

To enhance understanding of the pathogenesis of organ-system disease in CF and identify novel therapies, it is key to determine the magnitude of environmental and genetic effects of non-*CFTR* gene modifiers on clinical phenotypes. If variability in disease largely reflects environmental influences, there should be intensified focus on these factors, which might include the intensity of medical care, treatment of different types of lung infections, socioeconomic status, and effect of geographical temperature (Schechter 2004; Collaco et al. 2010, 2011). If the variability in clinical phenotypes largely reflects genetic variation in the genome, it is key to determine which genes are involved, and the mechanism of biological effects. Recent evidence indicates that there is a strong

---

influence of non-*CFTR* genetic variants on clinical phenotype in CF, but most of the genetic variation has not yet been defined (Mekus et al. 2000; Vanscoy et al. 2007).

Recent advances in genetic technology and the formation of The International CF Gene Modifier Consortium put CF at the cutting edge of gene-modifier research in "monogenic" disorders. This combination of events provides a critical opportunity for CF, because identification of the most important gene modifiers will provide insight into the mechanisms of disease pathogenesis, and offer the potential for novel prognostic approaches and therapies. This article provides an overview and update on the status of gene modifier research in CF.

HETEROGENEITY OF CF CLINICAL PHENOTYPES RELATED TO *CFTR* MUTATIONS

Overview

CF is the most common autosomal recessive disorder among Caucasians, and the median age of survival is only ~39 years, despite improved treatments. The cystic fibrosis transmembrane regulator (*CFTR*) gene encodes a 1480-amino acid protein that functions as a cAMP-mediated Cl^- channel, which plays a key role in hydrating airway secretions and regulating other cellular functions, including Na^+ transport in airway epithelia (Welsh et al. 2001; Cutting 2010). Because clinical disease in CF results primarily from mutations in *CFTR*, CF is termed a "monogenic" disorder. There are >1500 mutations in *CFTR* (see Cystic Fibrosis Gene Analysis Consortium, www.genet.sickkids.on.ca/cftr/) and the most prevalent mutation results in deletion of phenylalanine at position 508 (F508del), which occurs on 70% of CF chromosomes in the United States.

Heterogeneity of Disease Related to Mutations in *CFTR*

There is strong correlation between the general type of *CFTR* mutation and disease phenotype. Specifically, those mutations without residual function, such as F508del, are associated with pancreatic exocrine insufficiency, whereas ~10% of *CFTR* mutations retain some residual function and are associated with pancreatic exocrine sufficiency. Indeed, patients with at least one copy of a mutant *CFTR* allele with residual function have better nutritional status and milder lung disease, even though there is broad heterogeneity of disease severity (Mickle and Cutting 1993, 2000; The Cystic Fibrosis Genotype-Phenotype Consortium 1993). The development of meconium ileus, CF related diabetes (CFRD), and severe CF liver disease with portal hypertension (CFLD) is largely confined to patients with *CFTR* mutations with no residual function (Blackman et al. 2006, 2009b; Sontag et al. 2006; Bartlett et al. 2009; Moran et al. 2009). However, among patients carrying two mutations with no residual function, there is also a very broad range of lung disease severity, and there are strikingly different prevalences of meconium ileus, diabetes, and liver disease.

HERITABILITY OF CLINICAL PHENOTYPES

Overview

Before extensive genetic data was available, the prevailing concept was that the heterogeneity of clinical phenotypes reflected differing magnitudes of genetic versus environmental influences (Fig. 1) (Castaldo et al. 2001; Drumm 2001; Davies et al. 2005). For example, obstructive azoospermia (congenital bilateral absence of the vas deferens, CBAVD), occurs in nearly all CF males, regardless of whether the *CFTR* mutation has some residual function, or not; therefore, CBAVD is determined by genetics underlying mutations in *CFTR* (Cutting 2010). CF-related diabetes (CFRD) and severe CF liver disease with portal hypertension (CFLD) is limited to patients with two *CFTR* mutations without residual function (Bartlett et al. 2009; Blackman et al. 2009b; Moran et al. 2009). CFLD occurs before adulthood in the vast majority of CF patients who develop CFLD. The prevalence of CFRD is age-dependent, and occurs in ~25%–35% of CF adults. Lung disease occurs in patients regardless of whether *CFTR* mutations have residual function, or not, but the

 Cite this article as *Cold Spring Harb Perspect Med* doi: 10.1101/cshperspect.a009548

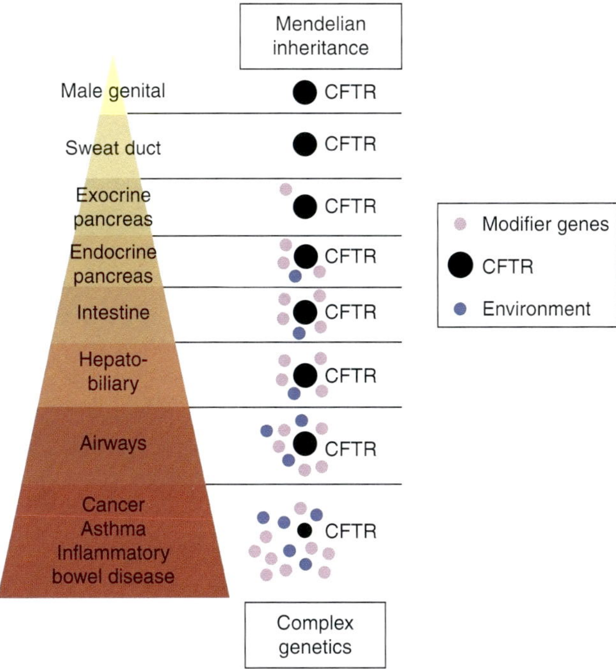

Figure 1. Schematic display of the magnitude of the effect of mutant *CFTR* versus environmental influences and modifier genes. (Image courtesy of Peter Durie, Toronto.)

range of disease severity is very great; thus, there are likely major contributions from both non-*CFTR* genetic modifiers and environmental influences (Mekus et al. 2000; Vanscoy et al. 2007).

Formal studies of genetic heritability of different clinical phenotypes have largely confirmed previous speculations (see above), and clearly show major roles for non-*CFTR* genetic variation in the different clinical phenotypes (Table 1). Genes that cause many "monogenic" disorders have been described, but there is little knowledge of the other specific factors that contribute to variability of the clinical phenotype. Investigations of inbred mice provide evidence that overall genetic background has a major effect on clinical phenotype in monogenic

Table 1. Heritability estimates for CF-related phenotypes

Phenotype	Lung function	Persistent infection	Diabetes	Meconium ileus	BMI	Weight for height	References
Heritability, all subjects	0.54−1.0	0.76−0.85	≤1	≤1	∼0.6	∼0.8	Blackman et al. 2006, 2008, 2009b; Vanscoy et al. 2007; Stanke et al. 2011; Green et al. 2012
Heritability, F508del only	0.56−0.86	No estimate	No estimate	No estimate	No estimate	∼0.6	Vanscoy et al. 2007; Stanke et al. 2011

disorders, which is a concept that relates to non-*CFTR* genetic influences on disease phenotype in CF (Nadeau 2001). There are also *CFTR*-related "mono-organ" disorders, such as male infertility, pancreatitis and sinusitis, which implicate complex genetic variation as important in mono-organ disease. Therefore, the study of modifier genes in CF is well-grounded from a conceptual perspective.

Lung

There is a broad range of lung disease severity, even for patients who are *F508del* homozygotes. This is illustrated in a representative population of Canadian CF patients by great variability in forced expired volume in 1 sec (FEV$_1$) (Fig. 2), which is a measure of airflow obstruction that correlates with lung disease severity and survival in CF. Two twin/sibling studies have concluded that there is a strong non-*CFTR* genetic contribution (0.54–1.0) to the CF lung phenotype (Table 1—"Heritability"). A key feature for studying non-*CFTR* genotype-phenotype relationships is the ability to define a meaningful clinical phenotype in quantitative, or clearcut binary, terms. Successful studies involve substantial effort to define the pertinent phenotype (see section below, "Challenges: Phenotypes").

Meconium Ileus

About 15%–20% of CF neonates are afflicted by an intestinal blockage that develops in utero, a concretion composed of meconium and mucus, and characteristically forms in the distal ileum. Evidence of this condition, known as meconium ileus (MI), is often detected before birth by ultrasound as a dense region on the sonogram. Histologically, there is an increase in the number and size of goblet cells, and elongated villi and distended crypts are seen as well. MI appears to be under nearly complete control of genetic factors, as heritability of this trait approaches 100% (Table 1).

CF-Related Diabetes

CF-related diabetes (CFRD) is a common comorbidity in CF, with about a fifth of adolescents affected and nearly half of adults (Moran et al. 2009). As a clinical entity, CFRD is distinct, although sharing properties of type 1 and type 2 diabetes (T1DM and T2DM, respectively). Type 1 features include reduced β-cell mass and diminished insulin secretion, but the etiology is not one of autoimmunity. Similar to type 2 diabetes, the pancreases of CFRD patients display amyloid deposition (Couce et al. 1996). Also, insulin resistance is common and ketosis

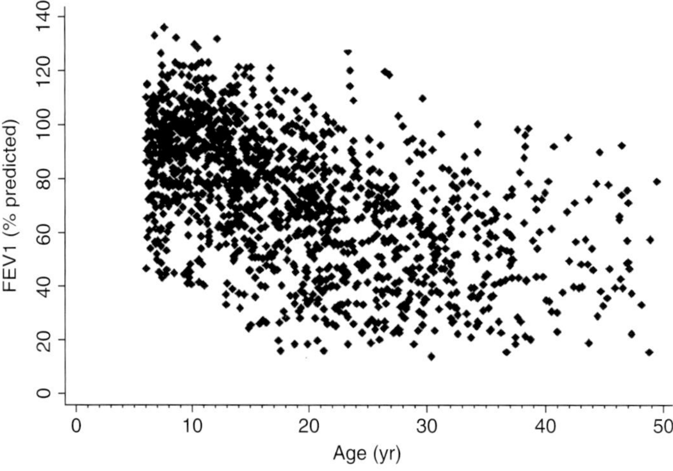

Figure 2. Lung disease severity assessed by FEV$_1$ versus age in 1357 Canadian CF patients with biallelic loss-of-function (pancreatic insufficient) mutations. (Image courtesy of Peter Durie, Toronto.)

is rare, consistent with insulin secretion being reduced but not absent. Unlike type 2 diabetes, obesity is not a risk factor, nor are the dyslipidemia profiles characteristic of type 2 diabetes.

Whereas the morphometric and biochemical risk factors for T2DM do not appear to be risk factors for CFRD, the genetics point to a relationship between T2DM and CFRD. For example, having a relative with T2DM increases the risk of a CF patient for developing CFRD (Blackman et al. 2009a). In fact, the heritability of CFRD is high, approaching 1 (Table 1), substantially higher than for T2DM, suggesting that CFRD is much less influenced by environmental factors, such as diet, than T2DM with which it appears to share genetic etiology.

CF Liver Disease

CF liver disease begins with loss of *CFTR* (Cl$^-$ channel) function on the apical membrane of cholangiocytes. This dysfunction leads to defective (sluggish) bile flow and an inflammatory response with activation and proliferation of hepatic stellate cells, which results in cholangitis and fibrosis in focal portal tracts (Rowland and Bourke 2011). However, only a small subset (~5%) of patients with CF develops severe liver disease with portal hypertension (CFLD). Moreover, CFLD occurs at a median age of only 10 years, and > 90% of CFLD occurs by age 20 years; thus, non-*CFTR* genetic variations likely contribute to risk for CFLD (Bartlett et al. 2009), but there are no formal estimates of heritability. Finally, there is also an increased prevalence of CFLD in males (2:1) over females, suggesting a role for gender and/or male/female hormones in this disorder.

MODIFIER GENETICS AND RECESSIVE MENDELIAN DISORDERS

Approaches to Study Genetic Modifiers in CF

The concept of gene modifiers in a Mendelian disorder such as cystic fibrosis shares some features, but is fundamentally different from the concept of genetic variants in non-Mendelian, complex genetic disorders, such as asthma. In complex genetic disorders, genetic variants interact with each other and the environment to cause the disorder (Van Heyningen and Yeyati 2004). In contrast, cystic fibrosis results from mutations in *CFTR*, and genetic variants that affect pathogenesis of mutant *CFTR* modify the severity of the phenotype (Drumm 2001; Cutting 2010). Genetic variation that has little effect in normal subjects may modify disease in cystic fibrosis.

Two different types of genetic studies have been used to discover gene modifiers in cystic fibrosis, i.e., case-control association studies and family-based (linkage) studies (Cutting 2010). The case-control (association) study is usually more robust for detecting common genetic variants, in part because of the larger number of subjects that can be enrolled in association studies, as compared with family-based studies. Enrollment into family-based twin/sibling linkage studies is more challenging, but it allows better definition of environmental versus genetic effects on the phenotype. Thus, complementary studies employing both the association and family-based methodologies would be optimal, and this strategy has been undertaken in cystic fibrosis by the International CF Gene Modifier Consortium.

Challenges: Phenotypes

Studies of genetic effects on phenotype require a meaningful and accurate definition of phenotype. For lung disease, FEV_1 relates to clinical disease severity and outcome; therefore, the development of a quantitative lung phenotype has been developed for CF, using multiple measures of FEV_1 and adjusting for mortality attrition (Taylor et al. 2011). This rigorous, quantitative lung phenotype has been used in a GWAS study of ~3500 CF patients (Wright et al. 2011). For meconium ileus, a binary (yes/no) approach is used to define phenotype, but requires careful review of source documents to be accurate (Wood et al. 2007; Sun et al. 2012). A recent publication has also determined that the heritability of chronic respiratory infection with *Pseudomonas aeruginosa* is very strong (0.76–0.85), which provides another lung phenotype that can be studied for genetic modifiers (Green et al. 2012). For CF-related diabetes there is an increasing prevalence with age; therefore, genetic studies use age-of-onset ("survival") type analyses (Blackman et al. 2009b). For liver disease, a binary (yes/no) approach is used, but this requires extensive review of source documents to accurately determine which patients have severe liver disease with portal hypertension (CFLD) (Bartlett et al. 2009). Many genetic studies have been flawed by nonmeaningful or imprecise definition of phenotype, but the CF community has focused on establishing meaningful and accurate phenotypes.

IDENTIFIED MODIFIERS OF DISEASE PHENOTYPES BY CANDIDATE GENE APPROACH

Concept

Initial searches for non-*CFTR* genetic variants that influence pulmonary phenotype used the candidate gene approach, which is based on biological plausibility. This approach is attractive because testing a small number of carefully selected candidates avoids severe penalties from multiple comparisons during statistical analysis. However, the candidate gene approach has substantial bias, because of incomplete understanding of the pathophysiology of CF lung disease. Because an association between a genetic variant and disease phenotype can reflect relatively small differences in allele frequency, it is critical to replicate any association in a separate population, preferably several times. In CF, many candidate genes have been tested, but only a few candidates have been tested in >1000 CF patients, and only a few have been replicated (Table 2). A comprehensive review of all gene modifier studies in CF is beyond the scope of this article, so we report a summary of candidate genes that have been replicated in at least one organ in ≥2 populations entailing at least 500 individuals total (Table 2; see text below). Several recent publications review all candidate genes that have been tested in CF (Cutting 2005; Davies et al. 2005; Knowles 2006; Collaco and Cutting 2008).

Lung Disease

Lung disease in CF is linked to defective ion transport, which leads to inadequate hydration of airway mucus and results in defective mucociliary and cough clearance. Acute and chronic airway infection occurs secondary to defective mucociliary clearance, which contributes to an inflammatory response and adversely affects the phenotype. These defects of ion transport, mucus clearance, and infection and inflammation leads to the development of bronchiectasis. Therefore, genetic variation in many genes in these pathways could modify the clinical phenotype, and be associated with more severe (or milder) lung disease, and affect survival.

Multiple studies have tested whether functional variants in mannose binding lectin (MBL2) associate with lung disease severity in CF. MBL2 is biologically plausible, because of its role in innate immunity and the predilection of MBL2 deficient patients to bacterial and viral infection. Six studies involving >2000 CF patients have shown that MBL-deficient genotypes (OO or AO) are associated with worse lung function (Table 2) (Gabolde et al. 1999; Garred et al. 1999; Davies et al. 2004; Yarden et al. 2004; Trevisiol et al. 2005; Dorfman et al. 2008), and three studies reported earlier age of infection with *Pseudomonas aeruginosa* (Ps. a.) (Table 2) (Trevisiol et al. 2005; Dorfman et al. 2008; McDougal et al. 2010). Because severity of lung disease, Ps. a. infection, and age are correlated, a modeling framework addressed these confounding variables, and concluded that deficiency in MBL predisposes to earlier infection with Ps. a. and thereby leads to worse lung disease (McDougal et al. 2010). Several studies did not report an association of MBL deficient genotypes with CF lung disease, which might reflect large numbers of older patients and/or other factors related to bacterial infection (Cutting 2010).

Genetic variation in the endothelial receptor type A (EDNRA) has been associated with CF lung disease severity (Table 2), and replicated in three additional populations of CF patients (1478 total) (Darrah et al. 2010). The variant EDNRA (rs5335) is in the 3′ UTR, and the risk allele for worse lung disease is associated

   Cite this article as *Cold Spring Harb Perspect Med* doi: 10.1101/cshperspect.a009548

Table 2. Candidate gene modifiers of clinical phenotypes in cystic fibrosis

Gene	Lung function (FEV$_1$)	Ps. a. acquisition	Meconium ileus	CF-related diabetes	CF liver disease	References
MBL2	REPLICATED[a]	REPLICATED				Gabolde et al. 1999; Garred et al. 1999; Davies et al. 2004; Yarden et al. 2004; Trevisiol et al. 2005; Dorfman et al. 2008; McDougal et al. 2010
EDNRA	REPLICATED					Darrah et al. 2010
TGF-β1	REPLICATED	Negative[c]			Probably negative[d]	Arkwright et al. 2000, 2003; Drumm et al. 2005; Brazova et al., 2006; Bremer et al. 2008; Corvol et al. 2008; Dorfman et al. 2008; Bartlett et al. 2009; Faria et al. 2009
IFRD1	Replicated[b]					Gu et al. 2009
IL8	Replicated					Hillian et al. 2008
MSRA			REPLICATED			Henderson et al. 2012
ADIPOR2			Replicated			Dorfman et al. 2009
TCF7L2				REPLICATED		Blackman et al. 2009a
SERPINA1	Negative	Probably negative			Replicated	Arkwright et al. 2000, 2003; Drumm et al. 2005; Brazova et al. 2006; Bremer et al. 2008; Corvol et al. 2008 Dorfman et al. 2008; Bartlett et al. 2009; Faria et al. 2009

[a]REPLICATED: seen in ≥3 populations with total ≥1000 individuals.
[b]Replicated: seen in ≥2 populations with total ≥500 individuals.
[c]Negative: negative in ≥3 populations with total ≥1000 individuals.
[d]Probably negative: negative in ≥2 populations with total ≥500 individuals.

with increased expression of EDNRA in cultured human tracheal smooth muscle cells, as well as increased proliferation of smooth muscle cells that are homozygous for the risk variant. Thus, the mechanism for adversely modifying CF lung disease likely reflects increased smooth muscle activity in airways and/or vasculature.

Transforming growth factor β1 (TGF-β1) modifies risk for airways disease in asthma and COPD, and has a key role in regulating airway inflammation and remodeling; hence, TGF-β1 has been repeatedly studied as a modifier of CF lung disease (Table 2). Alleles in the promotor (-509) and first exon (codon 10) of TGF-β1 are associated with worse lung function, and was replicated in the Genetic Modifier Study (GMS) of >1300 CF patients (Drumm et al. 2005). This discovery was independently replicated when haplotypes of alleles at -509 and codon 10 were analyzed in a separate population (Bremer et al. 2008). Further, other gene and environmental interactions are relevant to

TGF-β1, because interactions with variants in MBL2 and second-hand smoke exposure modify TGF-β1 association with lung disease. A few small studies did not report the same association of TGF-β1 and CF lung disease (Cutting 2010).

Genetic variation in the interferon-related developmental regulator 1 (IFRD1) gene was identified as one of six candidate regions, based on study of 320 CF patients in the GMS study with "mild" ($n = 160$) versus "severe" ($n = 160$) lung disease (Table 2). Polymorphisms in IFRD1 replicated in the larger GMS sample, and then showed association in a family-based (twin and siblings) population (Gu et al. 2009). In vitro and studies in mice indicated that IFRD1 acts to modify neutrophil function during bacterial infection, via transcriptional mechanisms, but this gene has not been further replicated in additional CF populations.

IL-8 is a cytokine that is well-recognized to be associated with airways inflammation in CF, and known to mediate neutrophil chemotaxis. The association of IL-8 variants and CF lung disease was first noted in 737 CF patients and was most strongly shown in males (Table 2). The replication was verified for males in an additional 385 patients. The mechanism appears to reflect differential gene expression through IL-8 promoter variants.

Meconium Ileus

The heritability of MI (Table 1) clearly indicates that genes other than CFTR are involved in its etiology, but the candidate gene approach has not been applied to this CF trait. A novel candidate *region* approach was used however, in which a segment of mouse chromosome 7 that was found to influence intestinal obstruction in CF mice (Rozmahel et al. 1996) was examined for the corresponding region of the human genome for its effect on MI. Initial reports indicated that variants in this region did, in fact, associate with MI (Zielenski et al. 1999), but a causative gene was never identified and subsequent studies failed to replicate this finding (Blackman et al. 2006). Although the genetic contribution to MI seems clear, the genes responsible remained elusive.

CF-Related Diabetes

The observation that family history of T2DM increases risk for CFRD, as described above, made obvious the need to test specific genes associated with T2DM for similar associations with CFRD. Thus, a candidate gene approach was initiated and showed that single nucleotide polymorphisms near the transcription factor 7-like 2 (*TCF7L2*) gene showed an association with type 2 diabetes with an estimated odds ratio of ~1.5 per allele (Blackman et al. 2009b).

In addition to showing a specific genetic relationship between T2DM and CFRD, this study showed an important concept regarding the confounding effects that nongenetic, or environmental factors may have on genetic studies, sometimes obscuring them. In this case, the nongenetic influences appear to be pharmacology, specifically steroidal anti-inflammatory drugs. Glucocorticoids are commonly used as an anti-inflammatory treatment for CF, and it is well established that a side effect of glucocorticoids is hyperglycemia, albeit transiently. Thus, this important pharmacotherapy for the lungs has the potential to mask differences in glycemic levels that would be caused by genetic variation. When controlling for the effect of this steroidal treatment by examining patients not recently using glucocorticoids, the risk associated with TCF7L2 alleles increased to 2.9 per allele (Blackman et al. 2009b).

Other evidence of similarities to T2DM is emerging as well. A screen of six genes associated with T2DM found that a polymorphism in Calpain-10 associated nominally with response to a glucose tolerance test (Derbel et al. 2006). Other T2DM genes are sure to be tested in the future, but as the effect of many of them is quite small it remains to be seen whether the cohort sizes available for CF studies will have the power to detect them.

CF Liver Disease

CF liver disease with portal hypertension (CFLD) occurs early in life in CF patients with biallelic loss-of-function (pancreatic insufficient) mutations in *CFTR*. The largest two-stage study of CFLD focused on five candidate genes that had previously been studied in CF "liver disease," but not all patients in previous studies had portal hypertension (Bartlett et al. 2009). The two-stage study showed strong association (and replication) of the Z-allele of the α_1-antiprotease (*SERPINA1*) gene with CFLD in a total of 260 patients with CFLD versus 1931 CF patients without CFLD (Table 2). The mechanism is unclear, but probably does not reflect cumulative efforts of the folding mutations in *CFTR* (F508del) and *SERPINA1* (Z-allele), because these two genes are predominately expressed in two different cell types (cholangiocytes and hepatocytes, respectively) in the liver. Variants in TGF-β1 were associated with CFLD in stage 1, but did not replicate in stage 2 of the CFLD study (Bartlett et al. 2009).

IDENTIFIED MODIFIERS OF DISEASE PHENOTYPE BY GWAS ASSOCIATION AND LINKAGE

The ability to perform unbiased study of relatively common genetic polymorphisms across the entire human genome in a cost-effective fashion has recently come-of-age. To maximize power to detect novel modifiers of CF phenotypes using genome-wide approaches, three different study groups have joined forces and created a consortium with >3500 well-phenotyped CF patients to study modifiers in CF. This consortium includes three different study designs: (1) Gene Modifier Study (GMS), which enrolled F508del homozygotes with extremes-of-lung phenotype (Drumm et al. 2005); (2) Canadian Genetic Study (CGS), which enrolled population-based CF patients; and (3) family-based, twins and siblings (TSS), which enrolled patients from families with more than one CF child. The GMS and CGS are best-suited for association analysis of unrelated patients, whereas the TSS (family-based) study can be used for both linkage and association analyses. Using this complementary study design allows for discovery by association (or by linkage) and then testing for replication in the alternate population(s).

Lung Disease

The combined genome-wide association and linkage study (GWALS) genotyped 3467 CF patients (Illumina 610 Quad platform) to identify genomic regions associated with CF lung disease severity. The strongest association signal ($p = 3 \times 10^{-8}$) was located at chromosome 11p13, intergenic to *EHF* and *APIP* and replicated in the TSS (family-based) patients (Fig. 3) (Wright et al. 2011). *EHF* is an epithelia-specific Ets transcription factor, and may be an important regulator of differentiation under stress conditions and inflammation. *APIP* is a known inhibitor of apoptosis, and eQTLs for *APIP* in lymphocytes suggests the risk SNP in this region may be associated with increased expression of *APIP*, implying that inhibition of apoptosis worsens CF lung disease. This is congruent with the notion that inhibition of apoptosis in airways delays clearance of neutrophils and leads to a hyper-inflammatory state (Dibbert et al. 1999; Harris et al. 2005). Inhibition of apoptosis may also contribute to goblet cell metaplasia, which is a feature of CF airways disease (Dibbert et al. 1999; Harris et al. 2005). Polymorphisms in the intergenic region of chr11p13 may also affect regulatory domains, which could alter expression of *EHF* and *APIP* and other genes in the region, including *ELF5*, *PDHX*, and *CD44*. Studies are ongoing to elucidate the mechanism of disease pathogenesis from genomic polymorphisms at chromosome 11p13.

There was also a near-significant genome-wide association at the chromosome 6p21.3 HLA class II DR locus. Genetic variation at this locus has been widely recognized to play a role in multiple infections and inflammatory disorders, which would be relevant to the infections and inflammatory nature of CF airway disease (Handunnetthi et al. 2010). This locus is very genetically heterogeneous, and studies are underway to define the specific DR allele that is associated with severity of CF lung disease.

There were five other regions of suggestive association (Wright et al. 2011), which will be further analyzed by an ongoing GWAS in an additional ~3500 CF patients. Two previously

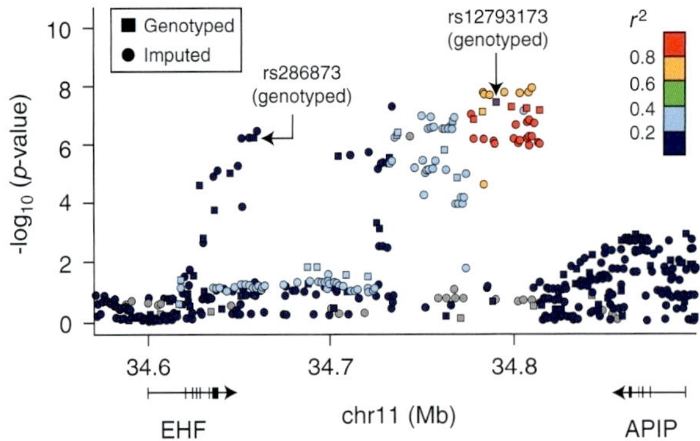

Figure 3. A locus-zoom plot of association in the GMS and CGS F508del homozygotes at chr11p13 EHF/APIP region, showing both genotyped and HapMap-imputed SNPs. Colors represent HapMap CEU linkage disequilibrium r^2 with the most significant genotyped SNP (rs12793173; $p = 3 \times 10^{-8}$). The secondary genotyped peak (rs286873) has low r^2 with the primary peak. Addition of the imputed SNPs to genotyped SNPs illustrates the coverage and LD blocks across the region.

reported modifiers of CF lung disease, *TGF-β1* and *IFRD1*, did not achieve genome-wide significance. However, TGF-β1 achieved *p*-values of 10^{-3} to 10^{-4} in GMS patients, depending on the covariates used in the analyses.

Linkage analysis of 486 sibling pairs from the family-based study identified a robust linkage signal at chr20q13.2 (LOD = 5.03). A region below the linkage peak was analyzed for association in the GMS and CGS patients, and replicated under a region-wide analytical approach ($p = 1 \times 10^{-4}$). The chr20p13.2 linkage region contains five genes (*CBLN4; MC3R; CASS4; CSTF1; AURKA*) that are expressed in fetal or adult lung, or bronchial epithelia. *MC3R* is of particular interest, because it has been implicated in regulation of energy balance and modulating neutrophil accumulation in lung inflammation, which are both relevant to CF lung disease (Savastano et al. 2009). Studies are underway to define the mechanism of lung disease in CF that reflects the linkage signal at chr20p13.2.

Meconium Ileus

The high heritability of MI certainly suggested that genes other than CFTR must be involved, and the availability of high density markers throughout the genome provided the opportu-

nity to carry out linkage studies at high resolution. Using this approach, Blackman and colleagues found by genome-wide analyses several regions of suggestive linkage (logarithm of the odds of linkage >2.0) for contributing to, or protecting from development of, meconium ileus. These regions were found on chromosomes 4q35.1, 8p23.1, and 11q25 (MI contributing loci) and chromosomes 20p11.22 and 21q22.3 (protective loci) (Blackman et al. 2006).

The locus on chromosome 8 is a gene-dense region and was followed up by haplotype analysis and identified a three polymorphism haplotype that most strongly associated and corresponded to a gene, *MSRA*, encoding an enzyme, methionine sulfoxide reductase A, that maintains the amino acid methonine in a reduced state. This candidate was then examined functionally by assessing its effect on obstruction in a CF mouse model lacking functional MsrA, and found to dramatically decrease the incidence of obstruction compared to CF mice with functional MsrA (Henderson et al. 2012).

Another linkage study provided evidence for MI-influencing loci on chromosome 12p13.3. Using densely spaced markers in the 12p13.3 region in a cohort of pancreatic insufficient Canadian CF patients with and without MI, one gene, ADIPOR2 in MI ($p = 0.002$), was implicated. The frequency of alleles in this gene was distorted from Hardy–Weinberg proportions, but only in the MI group. The deviation from expected proportions suggests a causative effect and the association was replicated in an independent sample of CF families. A second locus, but this one protective, was found on chromosome 4q13.3 and one candidate gene in the region, SLC4A4, a member of the solute transport family of proteins, showed evidence of association ($p = 0.002$) (Dorfman et al. 2009).

Association studies can also be used to identify genes involved in meconium ileus, and a novel modification of the association approach has identified several candidate genes. This approach, termed "informed GWAS" incorporates functional annotation into the analyses, giving higher import to those genes with a particular function attributed to them. Using a model in which genes influencing epithelial function were the focus, polymorphisms near SLC6A14 on the X chromosome and SLC26A9 on chromosome 1 accounted for ∼5% of variance in MI and these associations were replicated in an independent sample of affected individuals (Sun et al. 2012).

Other

Genome-wide association data have not yet been reported for CFRD or CFLD, although the Consortium is analyzing data from GWAS1, and awaiting data from GWAS2.

CONCLUDING REMARKS

The ability to identify genes, or genomic regions, that influence disease processes hold enormous promise for better understanding those processes and for manipulating them for therapeutic gain. However, an important caveat of genetic screens is that they may identify genes involved in a process, but as such they are only markers; they do not provide information about the mechanisms by which the variants exert their phenotypic effects. Thus, for the results of any genetic study to have more than prognostic value, the genes must be followed up by functional studies to have clinical utility.

The potential for clinical application is not only great, it is near. There are already approved drugs that target the protein product encoded by several of the genes identified (Table 2), including antagonists to the endothelin and angiotensin receptors, drugs that alter TGF-β1 signaling and small molecule screening provides the potential to rapidly identify new drugs that will target these gene products or the pathways in which they act.

ACKNOWLEDGMENTS

This work is supported in part by the U.S. National Institutes of Health grants R01HL068890 and P30DK27651, and Cystic Fibrosis Foundation grants KNOWLE00A0 and DRUMM0A00. The authors thank Fred A. Wright, Lisa J. Strug, Scott M. Blackman, Rhonda G. Pace, Johanna M. Rommens, Jaclyn R. Stonebraker, Wanda K. O'Neal, Peter R. Durie, Garry R. Cutting, and Tony T. Dang for contributions; plus Syanne D. Olson for editorial contributions and Elizabeth A. Godwin for administrative support.

REFERENCES

Arkwright PD, Laurie S, Super M, Pravica V, Schwarz MJ, Webb AK, Hutchinson IV. 2000. TGF-β1 genotype and accelerated decline in lung function of patients with cystic fibrosis. *Thorax* **55:** 459–462.

Arkwright PD, Pravica V, Geraghty PJ, Super M, Webb AK, Schwarz M, Hutchinson IV. 2003. End-organ dysfunction in cystic fibrosis: Association with angiotensin I converting enzyme and cytokine gene polymorphisms. *Am J Respir Crit Care Med* **167:** 384–389.

Bartlett JR, Friedman KJ, Ling SC, Pace RG, Bell SC, Bourke B, Castaldo G, Castellani C, Cipolli M, Colombo C, et al. 2009. Genetic modifiers of liver disease in cystic fibrosis. *JAMA* **302:** 1076–1083.

Blackman SM, Deering-Brose R, McWilliams R, Naughton K, Coleman B, Lai T, Algire M, Beck S, Hoover-Fong J, Hamosh A, et al. 2006. Relative contribution of genetic and nongenetic modifiers to intestinal obstruction in cystic fibrosis. *Gastroenterology* **131:** 1030–1039.

Blackman SM, Vanscoy LL, Collaco JM, Naughton KM, Cutting GR. 2008. Variability in body mass index in cystic fibrosis is determined partly by a genetic locus on chromosome 5. *Pediatr Pulmonol Suppl* **31:** 271.

Blackman SM, Hsu S, Ritter SE, Naughton KM, Wright FA, Drumm ML, Knowles MR, Cutting GR. 2009a. A susceptibility gene for type 2 diabetes confers substantial risk for diabetes complicating cystic fibrosis. *Diabetologia* **52:** 1858–1865.

Blackman SM, Hsu S, Vanscoy LL, Collaco JM, Ritter SE, Naughton K, Cutting GR. 2009b. Genetic modifiers play a substantial role in diabetes complicating cystic fibrosis. *J Clin Endocrinol Metab* **94:** 1302–1309.

Brazova J, Sismova K, Vavrova V, Bartosova J, Macek M Jr, Lauschman H, Sediva A. 2006. Polymorphisms of TGF-β1 in cystic fibrosis patients. *Clin Immunol* **121:** 350–357.

Bremer LA, Blackman SM, Vanscoy LL, McDougal KE, Bowers A, Naughton KM, Cutler DJ, Cutting GR. 2008. Interaction between a novel TGF-β1 haplotype and CFTR genotype is associated with improved lung function in cystic fibrosis. *Hum Mol Genet* **17:** 2228–2237.

Castaldo G, Fuccio A, Salvatore D, Raia V, Santostasi T, Leonardi S, Lizzi N, La Rosa M, Rigillo N, Salvatore F. 2001. Liver expression in cystic fibrosis could be modulated by genetic factors different from the cystic fibrosis transmembrane regulator genotype. *Am J Med Genet* **98:** 294–297.

Collaco JM, Cutting GR. 2008. Update on gene modifiers in cystic fibrosis. *Curr Opin Pulm Med* **14:** 559–566.

Collaco JM, Blackman SM, McGready J, Naughton KM, Cutting GR. 2010. Quantification of the relative contribution of environmental and genetic factors to variation in cystic fibrosis lung function. *J Pediatr* **157:** 802–807.

Collaco JM, McGready J, Green DM, Naughton KM, Watson CP, Shields T, Bell SC, Wainwright CE, Cutting GR. 2011. Effect of temperature on cystic fibrosis lung disease and infections: A replicated cohort study. *PLoS ONE* **6:** e27784.

Corvol H, Boelle PY, Brouard J, Knauer N, Chadelat K, Henrion-Caude A, Flamant C, Muselet-Charlier C, Boule M, Fauroux B, et al. 2008. Genetic variations in inflammatory mediators influence lung disease progression in cystic fibrosis. *Pediatr Pulmonol* **43:** 1224–1232.

Couce M, O'Brien TD, Moran A, Roche PC, Butler PC. 1996. Diabetes mellitus in cystic fibrosis is characterized by islet amyloidosis. *J Clin Endocrinol Metab* **81:** 1267–1272.

Cutting GR. 2005. Modifier genetics: Cystic fibrosis. *Annu Rev Genomics Hum Genet* **6:** 237–260.

Cutting GR. 2010. Modifier genes in Mendelian disorders: The example of cystic fibrosis. *Ann NY Acad Sci* **1214:** 57–69.

Darrah R, McKone E, O'Connor C, Rodgers C, Genatossio A, McNamara S, Gibson R, Stuart EJ, Ennis M, Gallagher CG, et al. 2010. EDNRA variants associate with smooth muscle mRNA levels, cell proliferation rates, and cystic fibrosis pulmonary disease severity. *Physiol Genomics* **41:** 71–77.

Davies JC, Turner MW, Klein N, London MBL CF Study Group. 2004. Impaired pulmonary status in cystic fibrosis adults with two mutated *MBL-2* alleles. *Eur Respir J* **24:** 298–804.

Davies JC, Griesenbach U, Alton E. 2005. Modifier genes in cystic fibrosis. *Pediatr Pulmonol* **39:** 383–391.

Derbel S, Doumaguet C, Hubert D, Mosnier-Pudar H, Grabar S, Chelly J, Bienvenu T. 2006. Calpain 10 and development of diabetes mellitus in cystic fibrosis. *J Cyst Fibros* **5:** 47–51.

Dibbert B, Weber M, Nikolaizik WH, Vogt P, Schoni MH, Blaser K, Simon HU. 1999. Cytokine-mediated Bax deficiency and consequent delayed neutrophil apoptosis: A general mechanism to accumulate effector cells in inflammation. *Proc Natl Acad Sci* **96:** 13330–13335.

Dorfman R, Sandford A, Taylor C, Huang B, Frangolias D, Wang Y, Sang R, Pereira L, Sun L, Berthiaume Y, et al. 2008. Complex two-gene modulation of lung disease severity in children with cystic fibrosis. *J Clin Invest* **118:** 1040–1049.

Dorfman R, Li W, Sun L, Lin F, Wang Y, Sandford A, Pare PD, McKay K, Kayserova H, Piskackova T, et al. 2009. Modifier gene study of meconium ileus in cystic fibrosis: Statistical considerations and gene mapping results. *Hum Genet* **126:** 763–778.

Drumm ML. 2001. Modifier genes and variation in cystic fibrosis. *Respir Res* **2:** 125–128.

Drumm ML, Konstan MW, Schluchter MD, Handler A, Pace R, Zou F, Zariwala M, Fargo D, Xu A, Dunn JM, et al. 2005. Gene modifiers of lung disease in cystic fibrosis. *N Engl J Med* **353:** 1443–1453.

Faria EJ, Faria IC, Ribeiro JD, Ribeiro AF, Hessel G, Bertuzzo CS. 2009. Association of MBL2, TGF-β1 and CD14 gene polymorphisms with lung disease severity in cystic fibrosis. *J Bras Pneumol* **35:** 334–342.

Gabolde M, Guilloud-Bataille M, Feingold J, Besmond C. 1999. Association of variant alleles of mannose binding lectin with severity of pulmonary disease in cystic fibrosis: Cohort study. *Br Med J* **319:** 1166–1167.

Garred P, Pressler T, Madsen HO, Frederiksen B, Svejgaard A, Hoiby N, Schwartz M, Koch C. 1999. Association of mannose-binding lectin gene heterogeneity with severity of lung disease and survival in cystic fibrosis [see comments]. *J Clin Invest* **104:** 431–437.

Green DM, Collaco JM, McDougal KE, Naughton KM, Blackman SM, Cutting GR. 2012. Heritability of respiratory infection with Pseudomonas aeruginosa in cystic fibrosis. *J Pediatr* doi: 10.1016/j.jpeds.2012.01.042.

Gu Y, Harley IT, Henderson LB, Aronow BJ, Vietor I, Huber LA, Harley JB, Kilpatrick JR, Langefeld CD, Williams AH, et al. 2009. Identification of IFRD1 as a modifier gene for cystic fibrosis lung disease. *Nature* **458:** 1039–1042.

Handunnetthi L, Ramagopalan SV, Ebers GC, Knight JC. 2010. Regulation of major histocompatibility complex class II gene expression, genetic variation and disease. *Genes Immun* **11:** 99–112.

Harris JF, Fischer MJ, Hotchkiss JR, Monia BP, Randell SH, Harkema JR, Tesfaigzi Y. 2005. Bcl-2 sustains increased mucous and epithelial cell numbers in metaplastic airway epithelium. *Am J Respir Crit Care Med* **171:** 764–772.

Henderson LB, Doshi VK, Blackman SM, Naughton KM, Pace RG, Moskovitz J, Knowles MR, Durie PR, Drumm ML, Cutting GR. 2012. Variation in MSRA modifies risk of neonatal intestinal obstruction in cystic fibrosis. *PLoS Genet* **8:** e1002580.

Hillian AD, Londono D, Dunn JM, Goddard KAB, Pace RG, Knowles MR, Drumm ML, CF Gene Modifier Study Group. 2008. Modulation of cystic fibrosis lung disease by variants in interleukin-8. *Genes Immun* **9:** 501–508.

Knowles MR. 2006. Gene modifiers of lung disease. *Curr Opin Pulm Med* **12:** 416–421.

McDougal KE, Green DM, Vanscoy LL, Fallin MD, Grow M, Cheng S, Blackman SM, Collaco JM, Henderson LB,

Naughton K, et al. 2010. Use of a modeling framework to evaluate the effect of a modifier gene (MBL2) on variation in cystic fibrosis. *Eur J Hum Genet* **18:** 680–684.

Mekus F, Ballmann M, Bronsveld I, Bijman J, Veeze H, Tummler B. 2000. Categories of deltaF508 homozygous cystic fibrosis twin and sibling pairs with distinct phenotypic characteristics. *Twin Res* **3:** 277–293.

Mickle JE, Cutting GR. 2000. Genotype-phenotype relationships in cystic fibrosis. *Med Clin North Am* **84:** 197–202.

Moran A, Dunitz J, Nathan B, Saeed A, Holme B, Thomas W. 2009. Cystic fibrosis-related diabetes: Current trends in prevalence, incidence, and mortality. *Diabetes Care* **32:** 1626–1631.

Nadeau JH. 2001. Modifier genes in mice and humans. *Nat Rev Genet* **2:** 165–174.

Rowland M, Bourke B. 2011. Liver disease in cystic fibrosis. *Curr Opin Pulm Med* **17:** 461–466.

Rozmahel R, Wilschanski M, Matin A, Plyte S, Oliver M, Auerbach W, Moore A, Forstner J, Durie P, Nadeau J, et al. 1996. Modulation of disease severity in cystic fibrosis transmembrane conductance regulator deficient mice by a secondary genetic factor. *Nat Genet* **12:** 280–287.

Savastano DM, Tanofsky-Kraff M, Han JC, Ning C, Sorg RA, Roza CA, Wolkoff LE, Anandalingam K, Jefferson-George KS, Figueroa RE, et al. 2009. Energy intake and energy expenditure among children with polymorphisms of the melanocortin-3 receptor. *Am J Clin Nutr* **90:** 912–920.

Schechter MS. 2004. Non-genetic influences on CF lung disease: The role of sociodemographic characteristics, environmental exposures and healthcare interventions. *Pediatr Pulmonol Suppl* **26:** 82–85.

Sontag MK, Corey M, Hokanson JE, Marshall JA, Sommer SS, Zerbe GO, Accurso FJ. 2006. Genetic and physiologic correlates of longitudinal immunoreactive trypsinogen decline in infants with cystic fibrosis identified through newborn screening. *J Pediatr* **149:** 650–657.

Stanke F, Becker T, Kumar V, Hedtfeld S, Becker C, Cuppens H, Tamm S, Yarden J, Laabs U, Siebert B, et al. 2011. Genes that determine immunology and inflammation modify the basic defect of impaired ion conductance in cystic fibrosis epithelia. *J Med Genet* **48:** 24–31.

Sun L, Rommens JM, Corvol H, Li W, Li X, Chiang TA, Lin F, Dorfman R, Busson PF, Parekh RV, et al. 2012. Multiple apical plasma membrane constituents are associated with susceptibility to meconium ileus in individuals with cystic fibrosis. *Nat Genet* **44:** 562–569.

Taylor C, Commander CW, Collaco JM, Strug LJ, Li W, Wright FA, Webel AD, Pace RG, Stonebraker JR, Naughton K, et al. 2011. A novel lung disease phenotype adjusted for mortality attrition for cystic fibrosis genetic modifier studies. *Pediatr Pulmonol* **46:** 857–869.

The Cystic Fibrosis Genotype-Phenotype Consortium. 1993. Correlation between genotype and phenotype in patients with cystic fibrosis. *N Engl J Med* **329:** 1308–1313.

Trevisiol C, Boniotto M, Giglio L, Poli F, Morgutti M, Crovella S. 2005. MBL2 polymorphisms screening in a regional Italian CF Center. *J Cyst Fibros* **4:** 189–191.

Van Heyningen V, Yeyati PL. 2004. Mechanisms of non-Mendelian inheritance in genetic disease. *Hum Mol Genet* **13:** R225–R233.

Vanscoy LL, Blackman SM, Collaco JM, Bowers A, Lai T, Naughton K, Algire M, McWilliams R, Beck S, Hoover-Fong J, et al. 2007. Heritability of lung disease severity in cystic fibrosis. *Am J Respir Crit Care Med* **175:** 1036–1043.

Welsh MJ, Ramsey BW, Accurso FJ, Cutting GR. 2001. Cystic fibrosis. In *The metabolic and molecular bases of inherited disease* (ed. Scriver CR, et al.), pp. 5121–5188. McGraw-Hill, New York.

Wood SD, Adams S, Dunn J, Baler R, Xu A, Pace RG, Kohl A, Yeatts J, Stonebraker JR, Konstan MW, et al. 2007. Inaccuracy of reporting of meconium ileus on case report forms of the gene modifier study. *Pediatr Pulmonol Suppl* **30:** 272.

Wright FA, Strug LJ, Doshi VK, Commander CW, Blackman SM, Sun L, Berthiaume Y, Cutler D, Cojocaru A, Collaco JM, et al. 2011. Genome-wide association and linkage identify modifier loci of lung disease severity in cystic fibrosis at 11p13 and 20q13.2. *Nat Genet* **43:** 539–546.

Yarden J, Radojkovic D, De Boeck K, Macek M Jr, Zemkova D, Vavrova V, Vlietinck R, Cassiman JJ, Cuppens H. 2004. Polymorphisms in the mannose binding lectin gene affect the cystic fibrosis pulmonary phenotype. *J Med Genet* **41:** 629–633.

Zielenski J, Corey M, Rozmahel R, Markiewicz D, Aznarez I, Casals T, Larriba S, Mercier B, Cutting GR, Krebsova A, et al. 1999. Detection of a cystic fibrosis modifier locus for meconium ileus on human chromosome 19q13. *Nat Genet* **22:** 128–129.

Status of Fluid and Electrolyte Absorption in Cystic Fibrosis

M.M. Reddy[1] and M. Jackson Stutts[2]

[1]Department of Pediatrics, UCSD School of Medicine, University of California, San Diego, La Jolla, California 92093

[2]Department of Physiology, University of North Carolina, Chapel Hill, North Carolina 27599

Correspondence: hoopster@med.unc.edu

Salt and fluid absorption is a shared function of many of the body's epithelia, but its use is highly adapted to the varied physiological roles of epithelia-lined organs. These functions vary from control of hydration of outward-facing epithelial surfaces to conservation and regulation of total body volume. In the most general context, salt and fluid absorption is driven by active Na^+ absorption. Cl^- is absorbed passively through various available paths in response to the electrical driving force that results from active Na^+ absorption. Absorption of salt creates a concentration gradient that causes water to be absorbed passively, provided the epithelium is water permeable. Key differences notwithstanding, the transport elements used for salt and fluid absorption are broadly similar in diverse epithelia, but the regulation of these elements enables salt absorption to be tailored to very different physiological needs. Here we focus on salt absorption by exocrine glands and airway epithelia. In cystic fibrosis, salt and fluid absorption by gland duct epithelia is effectively prevented by the loss of cystic fibrosis transmembrane conductance regulator (CFTR). In airway epithelia, salt and fluid absorption persists, in the absence of CFTR-mediated Cl^- secretion. The contrast of these tissue-specific changes in CF tissues is illustrative of how salt and fluid absorption is differentially regulated to accomplish tissue-specific physiological objectives.

ABSORPTION IN GLANDS AND INTESTINES

Exocrine glands are generally involved in two major physiological functions: secretion of macromolecules such as enzymes and electrolyte fluid into the lumen, and postsecretory modification of primary secretions by a process involving absorption of fluid and electrolytes as they pass through the lumen of the ductal system (Fig. 1). For the sake of simplicity, we can take the sweat gland as a representative example of exocrine structure and function. The sweat gland is a morphologically and physiologically simple exocrine model system. The sweat gland has two morphologically distinct regions: a secretory coil that secretes isotonic fluid into its lumen and a reabsorptive duct that reabsorbs most of the secreted Na^+, Cl^-, and HCO_3^- from the lumen as the primary sweat passes on to the skin surface to evaporate for cooling (Quinton 1987). Secretion of salt into the lumen of the secretory coil serves the primary function of

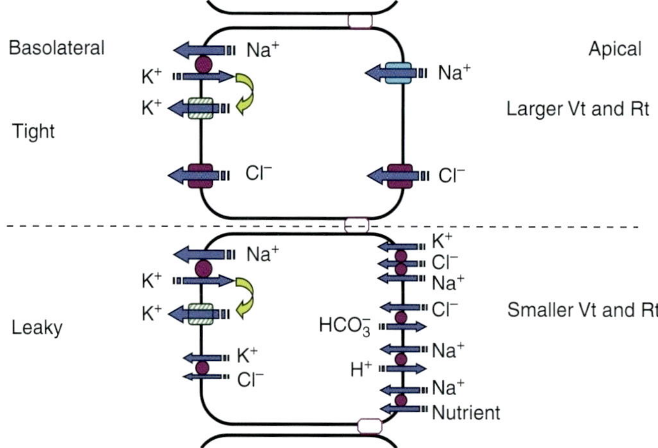

Figure 1. A simplified model depicting significant transport processes in tight and leaky epithelia. Notice that the tight epithelia predominantly depend on the amiloride-sensitive epithelial Na^+ channel and CFTR anion channel in the apical membranes for transporting Na^+ and Cl^- ions into the cell down the electrochemical gradient. These epithelia are characterized by relatively high electrical resistance (Rt) and transepithelial potentials (Vt). In contrast, the leaky epithelia predominantly depend on carrier-mediated transporters such as NaK_2Cl and Na nutrient (e.g., Na^+-glucose) cotransporters, Na^+/H^+, and Cl^-/HCO_3^- exchangers across the apical membranes. Leaky epithelia are also characterized by relatively low values of Rt and Vt. Although most epithelia generate a lumen-negative Vt of varying magnitude, proximal convoluted tubules can generate a lumen-positive Vt (about +3 mV, attributed to paracellular Cl^- diffusion potential) in the absence of nutrient transport and decreased luminal $NaHHCO_3^-$ concentration (Kokko 1973). Although the Na^+/K^+ pump plays a major role in the extrusion of Na^+ from the cell in the salt absorption process, the mechanism of Cl^- exit is either mediated by CFTR or by a carrier-mediated transporter such as the KCl cotransporter in the basolateral membrane.

creating a serosa-to-lumen osmotic gradient to facilitate secretion of water. Because the primary function of the sweat gland is to wet the skin surface for evaporative cooling, losing electrolytes along with water from the body is highly undesirable because it can lead to severe dehydration, electrolyte imbalance, and associated multi-organ disturbances. This physiological threat has been effectively solved in primates as the primary sweat passes through the reabsorptive duct, where most of the secreted salt is returned by active absorption to the extracellular fluid compartment before the sweat deposits onto the skin surface. Most exocrine glands and organs use some variant of this basic principle; that is, first, a primary secretion of an isotonic fluid, followed by a subsequent modification by reabsorption to create a tissue-specific final secretory product so that different exocrine glands exploit this process to create

an optimal extracellular aqueous medium to assist in specific physiological functions. For example, although conservation of salt and/ or maintenance of electrolyte balance are the primary goals of the absorptive process in the sweat gland and the kidney tubules (Quinton 1990; Bhalla and Hallows 2008), airway epithelial cells may exploit this absorptive process to maintain proper volume and composition of the airway surface liquid (ASL), which is critical for keeping the airways clear of infections and blockage (Widdicombe and Widdicombe 1995; Welsh 1996; Boucher 2001). The physiological significance of this process is emphasized by the fact that abnormal electrolyte absorptive function can lead to severe pathological conditions such as cystic fibrosis (CF), hypertension, pseudohypoaldosteronism (PHA), and Liddle syndrome (Quinton 1990; Bhalla and Hallows 2008).

Absorptive Mechanisms Are Tissue Specific

Epithelial cells welded together by tight junctions can accomplish vectorial transport of NaCl and water. Epithelial cells are polarized by an apical membrane (APM) and a basolateral membrane (BLM). Each membrane incorporates specific ion channels and transport proteins such that in an absorptive epithelium, NaCl can enter the cell at the APM and exits the cell across the BLM. Epithelial cells of different exocrine organs use diverse physiological strategies involving coupled ion cotransporters, counter-ion exchangers, and ion channels to absorb salt from the lumen to blood. Significant differences exist in the mechanisms of salt absorption across the apical membranes of different epithelial cells so as to be in compliance with their respective physiological roles in a particular tissue. Because epithelial water permeability varies among tissues, fluid and electrolyte absorption can be either hypo-osmotic (luminal fluid has lower osmolality compared with plasma) (Bijman and Quinton 1984; Quinton 1984) or isotonic (luminal fluid has the same osmolality as plasma). So-called leaky epithelia depend on carrier-mediated Na^+-absorptive mechanisms, and they are generally involved in isotonic fluid absorption. In contrast, epithelia using ENaC-dependent active salt transport (e.g., sweat glands, cortical collecting duct, and distal colon) generally tend to be involved in hypo-osmotic absorption (Quinton 1984; Devuyst and Guggino 2002), as described in greater detail in the following discussion.

Isotonic Absorption

Most "leaky" absorptive epithelia (so-called because of their electrically conductive tight junctions) such as proximal tubules, small intestine, and gallbladder are characterized by the presence of small trans-epithelial electrical potentials and relatively high water permeability. These epithelial cells generally use combinations of carrier-mediated transport components, such as $Na^+/K^+/2Cl^-$ and Na^+/Cl^- cotransporters, as well as Na^+/H^+ and Cl^-/HCO_3^- exchangers and Na^+-coupled nutrient transporters such as Na^+-glucose and Na^+-amino acid cotransporters, in the apical membrane (Berschneider et al. 1988; Reuss 1985; Reuss et al. 1991; Kunzelmann and Mall 2002). For example, fluid absorption in the intestine and proximal tubules is always coupled to the movement of solutes from the lumen. There are significant regional differences with respect to the relative role of the aforementioned carrier-mediated transport processes even within the gastrointestinal tract and kidney tubules. For example, Na^+-dependent glucose and amino acid transporters play a significant role in the process of absorption in the jejunum and ileal regions. In contrast, varying degrees of contribution from Na^+/H^+ (NHE) and Cl^-/HCO_3^- exchangers to fluid absorption is observed from the duodenum to the distal colon (Berschneider et al. 1987; Quinton 1999). Similarly, along the nephron, Na^+-dependent nutrient transporters (e.g., Na^+-glucose cotransporters) in the proximal tubule, NHE in the early proximal convoluted tubules, NaCl cotransporters in the distal convoluted tubules, and $Na^+/K^+/2Cl^-$ cotransporters and NHE in the thick ascending limb (Quinton 2007) play tissue-specific roles in absorptive processes. Even though defective functions of these transporters are associated with several disease conditions in various leaky epithelia, most of the abnormalities in carrier-mediated transport processes in cystic fibrosis appear to be secondary to abnormal CFTR (cystic fibrosis transmembrane conductance regulator)–mediated Cl^- permeability, as discussed below.

Hypertonic Absorption

In contrast to leaky epithelia, the so-called tight epithelia are generally characterized by the presence of relatively large lumen negative trans-epithelial potentials (< -10 mV, lumen negative) and ion channels involved in electrogenic absorption of NaCl (Boucher et al. 1991; Quinton 1999; Reddy et al. 1999; Kunzelmann and Mall 2002; Reddy and Quinton 2002). These epithelia are mostly, if not exclusively, involved in hypertonic (with reference to plasma) salt absorption and therefore tend to be relatively water impermeable. The human sweat duct, cortical collecting duct, colon, and salivary glandular ducts are examples that share these properties. The absorptive process in tight epithelia is predominantly mediated by ENaC channels and CFTR anion channels in the APM (Reuss et al. 1991; Stutts et al. 1995; Reddy et al. 1999; Kunzelmann and Mall 2002), which involves the following steps: Na^+ enters the cells via ENaC channels down a favorable electrochemical gradient, making the cytosolic side of the apical membrane more positive and thereby creating a more favorable electrochemical gradient for negatively charged Cl^- ions to enter the cells passively via CFTR Cl^- channels. Intracellular Na^+ is pumped out of the cell by the Na^+/K^+ pump in the BLM, which pumps three Na^+ ions out of the cell in exchange for two K^+ ions. Hence, the Na^+/K^+ pump exports one net positive charge out of the cell, thereby contributing slightly to the negative electrical potential of the cell. Cl^- entering the cell is driven out across the BLM through Cl^- channels down a favorable electrical gradient created by a more negative basolateral membrane potential (Welsh 1987; Reddy and Quinton 1991, 1994; Smith et al. 1996; Kunzelmann and Mall 2002). The BLM of epithelial cells expresses a predominant K^+ ion selectivity. The basolateral K^+ channels not only provide a leakage pathway for extrusion of excess K^+ accumulated during pump activity but also play a significant role in maintaining trans-epithelial electrical potential due to significantly larger hyperpolarization of BLM relative to the apical membrane potential, thereby providing a significant driving force for vectorial salt transport by epithelial cells (Reddy and Quinton 1987, 1991; Kunzelmann and Mall 2002). Although the salt absorption creates an osmotic gradient from lumen to serosa (blood side), water movement across the tight epithelia down the osmotic gradient is limited by the epithelial barrier to water.

CFTR in Salt Absorption

CFTR plays a crucial role in diverse physiological functions (regarding absorption and secretion of salt) by numerous epithelial organs (Quinton 1999). CFTR is also unique among ion channels with respect to (1) complex molecular structure (Riordan et al. 1989; Welsh et al. 1992); (2) multiple regulatory controls involving PKC (Berger et al. 1993; Jia et al. 1997), PKA (Cheng et al. 1991; Berger et al. 1993; Dulhanty and Riordan 1994), PKG (Picciotto et al. 1992; French et al. 1995; Vaandrager et al. 1996, 1998), Ca^{2+} calmodulin-dependent protein kinases (Picciotto et al. 1992), protein tyrosine kinases (Gadsby and Nairn 1999), and different protein phosphatases (Berger et al. 1993; Fischer et al. 1998; Luo et al. 1998; Gadsby and Nairn 1999); and (3) multifunctional behavior involving several transport functions (Schwiebert et al. 1999). Structurally, it is distinguished from most ion channels by the presence of a regulatory domain (R-domain) with multiple consensus phosphorylation sites and two nucleotide-binding domains (NBDs) (Riordan et al. 1989; Welsh et al. 1992). Even though these studies implicated several kinases and phosphatases regulating this channel, we do not know whether all of these regulatory processes are tissue specific or coexist within most epithelial cells. Even if these control mechanisms do exist in vivo, the purpose of such complex regulation remains puzzling. CFTR regulation involves both hydrolytic and nonhydrolytic ATP binding to NBDs (Anderson et al. 1991; Quinton and Reddy 1991; Ko and Pedersen 1995; Reddy and Quinton 1996; Randak et al. 1997; Sheppard and Welsh 1999). Nonhydrolytic ATP binding may be required for coupling trans-epithelial electrolyte transport to cellular energy charge (Quinton and Reddy 1991). However, it is intriguing that, unlike most ion channels that allow passive diffusion of ions (as opposed to energy-dependent active transport), CFTR uses ATP energy during the transport process. Recent studies have indicated that these multiple controls may play a role in determining the anion selectivity of the CFTR channel to HCO_3^- and Cl^- (Reddy and Quinton 2003b). Thus, understanding the regulatory mechanisms is essential not only to better understand how different mutations selectively affect epithelial Cl^- and HCO_3^- conductances in CF (Reddy and Quinton 2003b), but also to develop effective therapeutic strategies to treat diseases like CF.

     Cite this article as *Cold Spring Harb Perspect Med* doi: 10.1101/cshperspect.a009555

ENaC in Salt Absorption

The central role that ENaC plays in salt absorption in several diseases including pseudohypoaldosteronism (PHA), Liddle syndrome, and cystic fibrosis (Schild et al. 1995; Rossier 1996; Hummler and Horisberger 1999; Pilewski and Frizzell 1999; Schreiber et al. 1999; Schwiebert et al. 1999; de La Rosa et al. 2000) emphasizes the clinical and physiological significance of this cation channel in epithelial tissues (Quinton 2007). ENaC is expressed in the APM of cells in the salivary gland ducts, sweat gland ducts, cortical collecting tubules of the kidney, the distal colon, and the airways (Duc et al. 1994; Fuller et al. 1995; Stutts et al. 1995; Rossier 1996). ENaC channels in the kidney are directly involved in the maintenance of extracellular volume and blood pressure and indirectly involved in the homeostasis of K^+ and H^+ ions. The reabsorption of Na^+ in the cortical collecting tubule generates a lumen-negative potential that facilitates the secretion of K^+ and H^+ by the distal nephron. Blocking ENaC with amiloride or inactivation by mutation leads to Na^+ wasting, hyperkalemia, and metabolic acidosis. Many regulatory mechanisms such as the renin–angiotensin–aldosterone axis, antidiuretic hormone, and catecholamines modulate the activity of the ENaC channels (de La Rosa et al. 2000). ENaC channels are expressed in the respiratory tract from the nose to the alveoli (de La Rosa et al. 2000). In the lung, ENaC activity is necessary for fluid handling, in particular at birth when the transition from a liquid-filled lung occurs (de La Rosa et al. 2000). The composition of the fluid produced by secretory organs such as salivary glands and sweat glands is also modified by ENaC channels located along the ducts that reabsorb luminal Na^+.

CFTR–ENaC Coordination in the Apical Membrane

Studies on the sweat duct indicated that ENaC and CFTR activation are closely coupled but possibly independent of contemporaneous phosphorylation by PKA, which may suggest molecular interactions between the two channel proteins. The carboxy-terminal motif interacting with PDZ1 or PDZ2 of EBP50 (ERM-binding phosphoprotein 50) or other related proteins may be important for some aspects of CFTR function (Hall et al. 1998; Short et al. 1998), which might include CFTR–ENaC interactions. However, we do not know how closely the two channels function in vivo. Recent studies also indicated that activating G-proteins and A and G-kinases stimulate both CFTR and ENaC simultaneously in the human sweat duct. However, we do not know whether such activation is due to direct action of the agonists on the individual channels or to indirect effects through CFTR interacting with ENaC. Secondary changes in intracellular ion composition seem to control ENaC as well as CFTR channel functions in a complex fashion. Several studies have shown that increasing intracellular Cl^- concentration inhibits ENaC activity (Cook et al. 1998; Kunzelmann and Mall 2002; Reddy and Quinton 2003a). It is important to note that activation of CFTR was shown to inhibit ENaC activity in some experimental model systems indicating that changes in intracellular Cl^- activity could be one of the mechanisms by which ENaC is inhibited (Dinudom et al. 1993; Reddy and Quinton 2002). Other studies have also indicated that ENaC activity and resultant increases in intracellular Na^+ cause feedback inhibition of ENaC (Dinudom et al. 1993). Changing intracellular Na^+ could have indirect effects on ENaC activity through changing intracellular pH mediated by basolateral NHE activity (Fig. 2) (Reddy et al. 2008).

Cell Specificity of ENaC–CFTR Interaction

Studies on the human airway epithelium found increased Na^+ absorption in CF airways, and some studies in heterologous systems (Boucher 1994b; Knowles et al. 1997; Guggino 1999; Pilewski and Frizzell 1999; Schwiebert et al. 1999) detected a negative influence on ENaC activity. No evidence of reciprocal coupling between ENaC exists in other native tissues where it has been studied, such as the sweat duct. It is possible that physiological tissue-specific functions such as absorption versus secretion determine

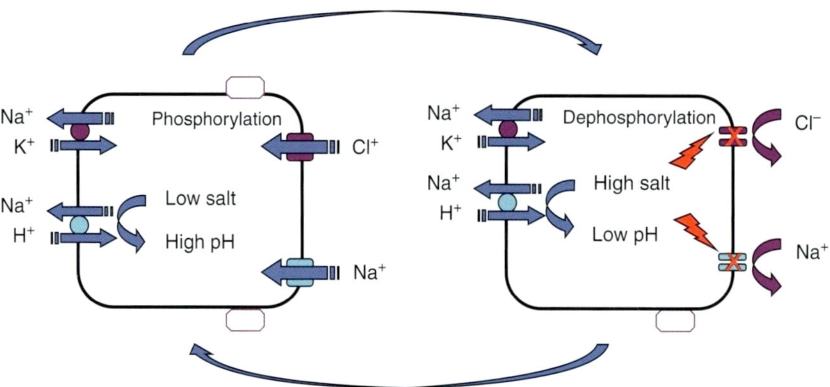

Figure 2. A simplified model depicting feedback regulation of salt absorption. The model illustrates that as intracellular Na^+ concentration rises as a function of ENaC activity, the chemical driving force for proton extrusion by the Na^+/H^+ exchanger (NHE) in the basolateral membrane decreases and allows cytosolic acidification that would negatively feedback and inhibit both ENaC and CFTR activity and limit salt influx across the apical membrane. The feedback inhibition of salt entry would relax as Na^+ is extruded by the Na^+/K^+ pump in the basolateral membrane. Prevention of salt influx in excess of pump capacity would protect the absorptive cell from disruptive changes in cell volume during trans-epithelial salt absorption.

the type of CFTR–ENaC interaction or complicate the interpretation of results. However, in heterologous systems, overexpressed proteins may very well skew or adversely affect normal physiological function. For example, in CFPAC-1 cells, hyperexpressing CFTR actually depressed outward-rectifying Cl^- currents and inward-rectifying K^+ currents (Mohammad-Panah et al. 1998). Native airway tissues and primary cultures of airway epithelia consist of multiple cell types, some of which may secrete while others absorb NaCl (Welsh 1987; Pilewski and Frizzell 1999). This possibility raises the question of whether the proportional activities of ENaC and CFTR seen in the sweat duct might be present in some cells but overshadowed by functions of others, in more complicated tissues. That is, cAMP-mediated secretion is decreased or absent in CF epithelia (Boucher 1994b; Sato and Sato 1984; Quinton 1999) so that a relatively greater loss of secretory transport in the presence of simultaneous absorptive transport could result in an apparent net increase in Na^+ absorption relative to normal tissue (Boucher et al. 1988). In fact, some studies on human airways suggested an increase in salt concentration in CF airway surface fluid (Joris et al. 1993; Zabner et al. 1998; Quinton 1999).

These observations seem consistent with recent reports suggesting that increased salt concentration in airway surface fluid may compromise airway defense mechanisms in CF (Smith et al. 1996; Goldman et al. 1997). Furthermore, recent studies involving human airway epithelia (Itani et al. 2011), the $CFTR^{-/-}$, and $CFTR^{\Delta F508/\Delta F508}$ (Chen et al. 2010; Ostedgaard et al. 2011) have shown reduced Cl^- but not Na^+ conductance in CF airways. Furthermore, a recent study on the relationship between airway epithelial Cl^- secretion and Na^+ absorption balance has led to the conclusion that hCFTR overexpression increases basal secretion but does not regulate Na^+ transport in wild-type mice (Grubb et al. 2012). A consensus on this topic is still much needed.

EFFECT OF CF ON THE ABSORPTIVE PROCESSES

Sweat Duct

Absorption of NaCl follows the previously described classic model of salt absorption by epithelia involved in hypertonic absorption. Na^+ and Cl^- ions enter the cell via ENaC and CFTR channels in the APM. Intracellular Na^+ is pumped out by the Na^+/K^+ pump in the BLM. Cl^- diffuses out of the cells through basolateral CFTR Cl^- channels. It was shown that the CFTR anion channels are absent in both the BLM and APM of sweat ducts isolated from CF patients. Lack of functional expression of CFTR prevents Cl^- exit across the BLM of CF cells (Reddy and Quinton 1992). In addition, absence of CFTR in the APM of CF cells prevents lumen-to-cell Cl^- diffusion and causes significant depolarization (makes the membrane more positive) of the apical membrane potential (Reddy and Quinton 1989). APM depolarization abolishes the electrical driving force and prevents passive diffusion of Na^+. As a consequence, both Na^+ and Cl^- ions are retained in the lumen, causing significant loss of electrolytes during sweating, particularly in patients afflicted by overheating (Quinton 1999). Thus, elevated sweat NaCl concentration is the basis of the classic pilocarpine-induced sweat test as a diagnostic feature of CF disease (Quinton 1999).

Salivary Glands

The physiology of salivary glands seems to follow the classic pattern of exocrine function in which the acinar cells of the salivary glands secrete an isotonic fluid, and the ductal system reabsorbs the salt from the primary secretion. The functional properties of salivary gland ducts appear to mimic closely those of the duct cells that are involved in hypertonic absorption of salt from the lumen using ENaC and CFTR channels. Salt absorption by the salivary ducts were significantly inhibited by amiloride and $CFTR_{inh}$-172 in mouse salivary gland ducts (Catalan et al. 2010). Unlike in the airways, where CFTR seems to inhibit ENaC activity, there is a coordinated activation of CFTR and ENaC in the salivary duct cells so that inhibition of CFTR in the CF mice resulted in the increase in both Na^+ and Cl^- concentrations (Reddy et al. 1999; Catalan et al. 2010). There are inconsistent and conflicting reports on the effect of CF on salivary gland functions. Abnormal Cl^- absorption and HCO_3^- secretions by the submandibular gland ducts were ob-

served from CF patients (Blomfield et al. 1973; Davies et al. 1989, 1991; McPherson et al. 1992). Several studies have indicated functional and pathological abnormalities in salivary gland functions. It is well known that CFTR and ENaC channels play a predominant role in hypertonic salt absorption in human salivary gland ducts. Early studies indicate elevated NaCl concentration in the parotid gland secretion from CF patients (Prader and Gautier 1955; di Sant'Agnese 1956; Johnson 1956; Barbero and Sibinga 1962). Increased levels of NaCl (and defective β-adrenergic secretions) (Bergler et al. 1994a,b; Catalan et al. 2010) were also observed in the submandibular secretions from mice with the $\Delta F508$ CFTR mutation (Catalan et al. 2010).

Small Intestine

There are conflicting reports on the effect of CF on absorptive process in the intestine. Na^+-dependent nutrient transport is the predominant mechanism of absorption in the small intestine. Early reports indicated enhanced Na^+-dependent glucose cotransport (Frase et al. 1985). These observations were supported by the subsequent studies showing increased nutrient-dependent short circuit currents (Baxter et al. 1988, 1989, 1990a,b; Hardcastle et al. 1990; Teune et al. 1996). It is well known that secretion of electrolyte fluid is severely compromised throughout the intestine including the jejunum and ileum (Berschneider et al. 1987, 1988; Taylor et al. 1987, 1988). Because the nutrient absorption is driven by the driving force for Na^+, absence of a Cl^- shunt due to lack of CFTR anion channels in the plasma membranes of CF intestinal epithelium seems to increase the driving force for Na^+, causing a significant surge in glucose and other nutrient absorption. To have an increased driving force for Na^+ glucose cotransport, it seems necessary that both CFTR and Na^+-dependent cotransporters must be present in the plasma membranes of the same cell. In fact, a Cl^- conductance resembling CFTR is colocalized with Na^+/glucose cotransport in rat and human small intestine, supporting the possibility that abnormalities in glucose absorption in cystic fibrosis may be a secondary effect of defects in Cl^- channel function (Baxter et al. 1990a,b; Hardcastle et al. 1990). Russo et al. (2003) have reported that fluid absorption processes are, in fact, reduced in CF jejunum because of a marked depression of passive chloride absorption, but that Na^+-glucose cotransport in the CF jejunum is normal. In contrast to human intestine, intestinal glucose absorption seems reduced in transgenic mouse models of CF with the ΔF508 mutation. Therefore, it was suggested that the transgenic mouse models of CF might not accurately reflect all aspects of intestinal dysfunction in the human disease (Hardcastle et al. 2004). It was further reported that the *Cftr*-null mice lacking one or both copies of the NHE3 (Na^+/H^+ exchanger) gene exhibited increased fluidity of their intestinal contents, which prevented the formation of obstructions and increased survival (Bradford et al. 2009), indicating that salt absorption in the face of reduced secretions caused by CFTR mutations plays a significant role in the development of intestinal pathology in CF mice. However, increased Na^+-dependent nutrient and fluid absorption in the face of reduced CFTR-dependent Cl^- secretion in CF intestinal epithelium appears to be a primary cause of dehydrated luminal content leading to the pathogenesis of meconium ileus in CF patients (Berschneider et al. 1988).

Colon

Electrolyte and fluid absorption is the predominant function of the colonic epithelium. The human colon absorbs large quantities of fluid and electrolytes per day ($\sim$1.5 L/d) (Debongnie and Phillips 1978). Absorptive processes in the colon are complex and exhibit significant regional differences. In the ascending colon, electroneutral transport involving Na^+/H^+ and Cl^-/HCO_3^- exchangers seems to be the predominant mechanism of absorption. In contrast, electrogenic absorption involving ENaC and CFTR channels appears to play a significant role in electrolyte fluid absorption in the descending colon (Clauss et al. 1985; Kunzelmann et al. 2001; Kunzelmann and Mall 2002). Defects in the absorptive processes were observed in both electroneutral as well as electrogenic components of colonic absorptive processes (Kunzelmann and Mall 2001, 2002). Early studies on CF mouse colons show increased electrogenic sodium absorption compared with wild-type tissues. Elevated plasma aldosterone in CF mice was suspected to be responsible for part or all of the increased sodium absorption in CF mouse colons (Cuthbert et al. 1994, 1999). It was also reported that in both CFTR knockout mice and in CF patients, the absence of CFTR-mediated inhibition of ENaC seems responsible for increased amiloride-sensitive short-circuit currents and Na^+ absorption in the colonic epithelium compared with non-CF controls (Grubb and Boucher 1997; Briel et al. 1998; Mall et al. 1999; Kunzelmann and Mall 2001; Kunzelmann et al. 2001).

Isotonic Salt Absorption by Airway Epithelia

The essential physiological purpose of human airways is to conduct inspired air to and from the alveolar surfaces where gas exchange occurs. Structurally, the lower airways dichotomously branch from the trachea ($\sim$2 cm diameter) to thousands of terminal airways (0.5–1 mm diameter) (Kilburn 1974). The large aggregate surface of these conducting airways is lined by a continuous epithelium that is protected from desiccation by a thin ($\sim$10 μm) liquid layer called airway surface liquid (ASL). Additionally, ASL supports the innate immune defense mechanism of mucociliary clearance of particles deposited from inhaled air, including infectious agents (Knowles and Boucher 2002). The airway mucosal surface is kept hydrated and mucociliary clearance is maintained by regulation of ASL volume and composition through the fluid transport activities and passive barrier properties of airway epithelia (Boucher 1994a). Airway epithelia are equipped both to absorb Na^+ and secrete Cl^-, indicative of the importance of regulating airway surface hydration. Electroneutrality requires that Na^+ cannot be absorbed without a co-ion, predominantly Cl^-, and neither can Cl^- be secreted without an co-ion, predominantly Na^+. Thus, ASL

Cite this article as *Cold Spring Harb Perspect Med* doi: 10.1101/cshperspect.a009555

homeostasis is achieved by balanced isotonic salt absorption and secretion.

Path for Na^+ Absorption

Airway epithelia are composed of ciliated and nonciliated cells (Kreda et al. 2005). Ciliated cells not only elaborate the cilia that move ASL cephalad, but are also the cells that accomplish salt transport. Nonciliated cells secrete macromolecules, used in innate defense, including mucins, whose physicochemical properties are dependent on hydration (Kesimer et al. 2010). The path for Na^+ absorption through ciliated cells consists of entry at the apical membrane and extrusion from the basolateral membrane. Entry of Na^+ from ASL is rate limiting for Na^+ absorption and is primarily mediated by the epithelial Na^+ channel (ENaC). Other means of Na^+ entry have been considered, such as Na^+/proton exchange or Na^+/glucose cotransport, but have not been found to play a major role in airway epithelia. Na^+ extrusion across the basolateral membrane is accomplished by Na^+–K^+-ATPase (Knowles et al. 1984). This transport element is obviously essential for Na^+ absorption by all epithelia, but it does not appear to be the point of regulation in airway epithelia. Thus, Na^+ absorption in airways is regulated principally at the ENaC-mediated entry step, as discussed below.

Path for Cl^- Absorption

Isotonic absorption of fluid and salt in response to electrochemical driving forces generated by active Na^+ transport requires a Cl^- conductance. Cl^- is predicted to diffuse across the epithelium through available cellular and paracellular routes. The paracellular path of airway epithelia has the properties of free solution (Cotton et al. 1983). The further observation that the ratio of Cl^- to mannitol trans-epithelial permeability coefficients exceeds the ratio of their free solution diffusivities suggests that a trans-cellular, in addition to a paracellular, route is available for passive Cl^- diffusion (Stutts et al. 1988). Trans-cellular cellular Cl^- absorption requires Cl^- entry across the apical

membrane. Although most work on the elements of Cl^- movement across airway epithelia has focused on Cl^- secretion, ion channels in the apical cell membrane including CFTR and other Cl^- channels, such as ano1, provide an apical Cl^- conductance for Cl^- to enter or leave the cell, depending on the existing Cl^- electrochemical potential (Willumsen et al. 1989; Tarran et al. 2002; Ferrera et al. 2010). Moreover, airway epithelia express SLC family members that exchange intracellular HCO_3^- for extracellular Cl^- (Mount and Romero 2004). Cl^- movement from cell to interstitium is thought to require a Cl^- basolateral cell membrane conductance, for which electrophysiological data exist, but for which no molecular identity is known (Reddy and Quinton 1992; Uyekubo et al. 1998).

Relationship between Na^+ Absorption and Cl^- Secretion in Airways

Studies in thin film preparations confirm that both Na^+ absorption and Cl^- secretion are used to control the depth of liquid on the airway surface (Tarran et al. 2002, 2006). This duality of salt transport mechanisms is extensively documented in studies of freshly excised airway epithelia of many species (Olver et al. 1975), including human (Boucher et al. 1991). Ion flux data from decades of studies show an absorption flux of Na^+ that exceeds passive backflux under both short circuit and open circuit conditions (Boucher 1994a,b; Mall 2009), consistent with a classic passive Na^+ entry from the luminal solution via ENaC with active extrusion of Na^+ by the basolateral Na^+/K^+ ATPase (Ussing et al. 1974). This net flux is eliminated by amiloride-like ENaC blockers and by ouabain, an inhibitor of the Na^+/K+ pump (Olver et al. 1975). The same studies have characterized Cl^- secretion by airway epithelia as a secondary active ion transport process (Olver et al. 1975; Boucher et al. 1980). That is, Cl^- is not actively transported per se but is accumulated in airway cells above its predicted electrochemical potential by a basolateral cotransporter that use Na^+, K^+, and Cl^- gradients supported by the Na^+/K^+ pump (Frizzell et al. 1979). Passive Cl^- conductance in the apical membrane allows

Cl^- accumulated above its electrochemical potential to exit into the luminal solution, creating net Cl^- secretory transport (cf. Frizzell and Hanrahan 2012). Consistent with the passive nature of this process, Cl^- secretion by airway epithelia responds to maneuvers that hyperpolarize the apical membrane potential. Such maneuvers prominently include Na^+ channel blockers and clamping trans-epithelial potential difference to zero (so-called short circuit conditions). Even so, under open circuit conditions and in the absence of amiloride, net absorption of both Na^+ and Cl^- has been documented in native human (Knowles et al. 1984) and canine bronchial epithelium (Boucher et al. 1981).

Water Permeability of Airway Epithelia

A striking characteristic of airway epithelia that sets it apart from sweat ductal epithelia is its significant water permeability (Matsui et al. 2000; Verkman et al. 2000). Decreased water permeability is detected in alveoli and airways mice with knocked-down expression of aquaporins, but ASL hydration was unaffected (Verkman 2007). This result was attributed to the relatively small flux of water across airway epithelia, compared with the overall non-aquaporin routes available for water diffusion across airway epithelia (Verkman 2007). Accordingly, water permeability is not expected to be limiting for fluid absorption by airway epithelia.

Regulation of ENaC

Net salt and fluid absorption, that is, the net of active Na^+ absorption (with passive Cl^-) and Cl^- secretion (with passive Na^+), is determined by the driving forces established in response to Na^+ conductance of the apical membrane. When Cl^- conductance is predominant owing to the activation of CFTR and Na^+ conductance is blocked or very small, the apical membrane should hyperpolarize sufficiently to support Cl^- secretion. As Na^+ conductance increases, the apical cell membrane must depolarize and diminish the driving force on Cl^- to passing from cell to lumen (secretion). At some point

as Na^+ conductance increases, the net driving force on Cl^- reverses and Cl^- is no longer secreted but moves in the absorptive direction to support net Na^+ and Cl^- (and fluid) absorption. Because apical Na^+ conductance is principally due to ENaC, with a small component of nonselective cation conductance of uncertain molecular origin, regulation of ENaC in airway epithelia is effectively the regulator of fluid absorption.

ENaC Structure and Stoichiometry

ENaC is a heteromeric channel composed of α-, β-, and γ-subunits (Canessa et al. 1994b). Each subunit has two membrane-spanning regions with short amino and carboxyl cytoplasmic tails. The membrane-spanning regions of each subunit are joined by a large extracellular domain (Canessa et al. 1994b). ENaC belongs to a family of channels with similar topology that include acid-sensing channels (ASICs) (Kellenberger and Schild 2002). ASIC1 has been crystallized, revealing a homotrimeric structure (Jasti et al. 2007). This was taken as a strong indication that ENaC has a trimeric structure of α-, β-, and γ-subunits (Stockand et al. 2008), but at this time, ENaC stoichiometry remains a controversial question (Anantharam and Palmer 2007).

ENaC trafficking to and from the apical membrane involves regulated processes that determine the density of ENaC in the apical membrane (Palmer et al. 2012). Liddle's disease of inherited amiloride-sensitive hypertension is caused by mutations in the carboxy-terminal cytosolic tails of β- and γ-ENaC, which contain highly conserved PY protein interaction motifs that mediate the association of ENaC with the E3 ubiquitin ligase, Nedd4-L (Schild et al. 1996; Staub et al. 2000). Nedd4-L catalyzes ubiquitinylation of ENaC cytosolic amino-terminal lysines, and this posttranslational modification causes ENaC internalization (Rotin 2000). Numerous signaling pathways, including SGK1, PKA, and MAP kinases, adjust the strength of the ENaC–Nedd4-L association through the phosphorylation status of specific residues, giving rise to wide tissue variability

 Cite this article as *Cold Spring Harb Perspect Med* doi: 10.1101/cshperspect.a009555

in the impact of this regulatory axis (Debonnevilles et al. 2001). For example, the hypertension of patients with Liddle's disease is attributed to excess ENaC at the apical membrane of renal tubules (Palmer et al. 2012). However, Liddle's disease patients show no evidence of increased Na^+ absorption in respiratory epithelia (Knowles and Boucher 2002), suggesting that other mechanisms exist for controlling Na^+ conductance in certain tissues. Nonetheless, Nedd4-L must provide the underlying mechanism of retrieval of ENaC from the surface of airway epithelial cells, because mice with lung-specific Nedd4-L knockdown developed Na^+ hyperabsorption and fatal lung disease (Kimura et al. 2011).

ENaC activity at the apical membrane is also subject to regulation by signaling that influences channel open probability (P_O). ENaC is stimulated by proteases that cleave sites within the extracellular domains of ENaC (Vuagniaux et al. 2000; Hughey et al. 2004). This action acutely increases P_O (Caldwell et al. 2004, 2005). Although still not completely understood, it now appears that cleavage at defined sites within the extracellular domains of α- and γ-ENaC results in conformational changes that allow greater stability of the open state of the channel (Kashlan et al. 2011). It is also evident that numerous proteases can cleave and activate ENaC and that their catalytic activity is opposed by cognate protease inhibitors (Myerburg et al. 2008; Passero et al. 2008; Svenningsen et al. 2009). Thus, the stimulation of ENaC through extracellular proteolysis is likely to be intricately balanced and will depend on the coincidence of proteases and inhibitor expression with ENaC.

ENaC P_O is strongly stimulated by binding to membrane phosphoinositides through basic residues in the cytosolic amino termini, especially β- and γ-ENaC (Ma et al. 2002; Yue et al. 2002). Numerous reports have established that exogenous PIP2 or PIP3 is a strong stimulate of ENaC, such that signaling mechanisms that affect the abundance of these phosphoinositides clearly influence ENaC-mediated Na^+ absorption (Kunzelmann et al. 2004; Wang et al. 2008).

Regulation of ASL Depth by Embedded Signals in Normal and CF Airways

ENaC-mediated Na^+ conductance strongly affects the electrical potential difference across the apical membrane of airway epithelia (Willumsen et al. 1989; Willumsen and Boucher 1991), which determines the balance between isotonic fluid absorption and fluid secretion. Tarran and others, using "thin film" or "air–liquid interface" culture techniques, showed that airway cells cultured under conditions that mimic the ~ 10-μm depth of ASL maintained in vivo rely on soluble signals in the small volume on the culture surface to regulate ENaC (Tarran et al. 2001, 2005; Song et al. 2009). In fact, this work identified both protease inhibitors as well as nucleotides as the major inhibitory regulators of ENaC. As ASL depth is reduced, the concentration of these ENaC inhibitory signals increases, slowing NaCl absorption (Myerburg et al. 2006, 2010; Tarran et al. 2006). Moreover, ATP can stimulate Ca^{2+}-dependent Cl^- conductance in the apical membrane, and adenosine arising from the catabolism of ATP can stimulate CFTR activity. With ENaC inhibited, driving forces for Cl^- secretion can be achieved. In this manner, airway epithelia can temper, or even reverse, the absorption of NaCl and H_2O to maintain optimal airway surface hydration. These approaches revealed a striking difference between ASL homeostasis in differentiated cultures derived from normal and CF airways. CF cultures were unable to switch from absorption to secretion, leading to collapse of ASL depth (Tarran et al. 2006). This result was not unpredicted, given the absence of CFTR to mediate Cl^- secretion in the CF cultures. However, it was also clear that ENaC remained hyperactive, even as ASL height decreased. A potential basis for ENaC hyperactivity in CF was detected in two laboratories as a much greater extent of ENaC cleavage in differentiated CF cultures compared with normal cultures (Myerburg et al. 2006; Gentzsch et al. 2010). Coexpression of CFTR in *Xenopus* oocytes with ENaC strongly depressed ENaC proteolysis and amiloride-sensitive currents. Taken together, these results confirm the role of CFTR-mediated Cl^- secretion in homeostatic control of ASL volume and further implicate loss of CFTR-mediated inhibition of ENaC proteolysis/stimulation as a potential cause of the dysregulated Na^+ absorption that has been reported in CF airways.

REFERENCES

*Reference is also in this collection.

Anantharam A, Palmer LG. 2007. Determination of epithelial Na^+ channel subunit stoichiometry from single-channel conductances. *J Gen Physiol* **130:** 55–70.

Anderson MP, Berger HA, Rich DP, Gregory RJ, Smith AE, Welsh MJ. 1991. Nucleoside triphosphates are required to open the CFTR chloride channel. *Cell* **67:** 775–784.

Barbero GJ, Sibinga MS. 1962. Enlargement of the submaxillary salivary glands in cystic fibrosis. *Pediatrics* **29:** 788–793.

Baxter PS, Dickson JAS, Variend S, Taylor CJ. 1988. Intestinal disease in cystic fibrosis. *Arch of Dis Child* **63:** 1496–1497.

Baxter PS, Read NW, Hardcastle PT, Wilson AJ, Hardcastle J, Taylor CJ. 1989. Abnormal jejunal potential difference in cystic fibrosis. *Lancet* 464–466.

Baxter P, Goldhill J, Hardcastle J, Hardcastle PT, Taylor CJ. 1990a. Enhanced intestinal glucose and alanine transport in cystic fibrosis. *Gut* **31:** 817–820.

Baxter PS, Goldhill J, Hardcastle J, Hardcastle PT, Taylor CJ. 1990b. Enhanced intestinal glucose and alanine transport in cystic fibrosis. *Gut* **31:** 817–820.

Berger HA, Travis SM, Welsh MJ. 1993. Regulation of the cystic fibrosis transmembrane conductance regulator Cl^- channel by specific protein kinases and protein phosphatases. *J Biol Chem* **268:** 2037–2047.

Bergler MK, Dorin JR, Porteous DJ, Quinton PM. 1994a. Defective salivary gland function in the CF mouse reveals three distinct phenotypes. *Ped Pulmon* **10:** 211.

Bergler MK, Dorin JR, Porteous DJ, Quinton PM. 1994b. Defective salivary gland function in the CF mouse. *European Cystic Fibrosis Conference* Paris, France, June.

Berschneider HM, Azizhan RG, Knowles M, Boucher RC, Powell DW. 1987. Intestinal electrolyte in cystic fibrosis. *Gastroenterology* **92:** 1315.

Berschneider HM, Knowles MR, Azizkhan RJ, Boucher RC, Tobey NA, Orlando RC, Powell DW. 1988. Altered intestinal chloride transport in cystic fibrosis. *FASEB J* **2:** 2625–2629.

Bhalla V, Hallows KR. 2008. Mechanisms of ENaC regulation and clinical implications. *J Am Soc Nephrol* **19:** 1845–1854.

Bijman J, Quinton PM. 1984. Influence of abnormal Cl^- impermeability on sweating in cystic fibrosis. *Am J Physiol* **247:** C3–C9.

Blomfield J, Warton KL, Brown JM. 1973. Flow rate and inorganic components of submandibular saliva in cystic fibrosis. *Arch Dis Child* **48:** 267–274.

Boucher RC. 1994a. Human airway ion transport. Part one. *Am J Respir Crit Care Med* **150:** 271–281.

Boucher RC. 1994b. Human airway ion transport. Part two. *Am J Respir Crit Care Med* **150:** 581–593.

Boucher RC. 2001. Pathogenesis of cystic fibrosis airways disease. *Trans Am Clin Climatol Assoc* **112:** 99–107.

Boucher RC Jr, Bromberg PA, Gatzy JT. 1980. Airway transepithelial electric potential in vivo: Species and regional differences. *J Appl Physiol* **48:** 169–176.

Boucher RC, Stutts MJ, Gatzy JT. 1981. Regional differences in bioelectric properties and ion flow in excised canine airways. *J Appl Physiol* **51:** 706–714.

Boucher RC, Cotton CU, Gatzy JT, Knowles MR, Yankaskas JR. 1988. Evidence for reduced Cl^- and increased Na^+ permeability in cystic fibrosis human primary cell cultures. *J Physiol (Paris)* **405:** 77–103.

Boucher RC, Chinet T, Willumsen N, Knowles MR, Stutts MJ. 1991. Ion transport in normal and CF airway epithelia. *Adv Exp Med Biol* **290:** 105–115.

Bradford EM, Sartor MA, Gawenis LR, Clarke LL, Shull GE. 2009. Reduced NHE3-mediated Na^+ absorption increases survival and decreases the incidence of intestinal obstructions in cystic fibrosis mice. *Am J Physiol Gastrointest Liver Physiol* **296:** G886–G898.

Briel M, Greger R, Kunzelmann K. 1998. Cl^- transport by cystic fibrosis transmembrane conductance regulator (CFTR) contributes to the inhibition of epithelial Na^+ channels (ENaCs) in *Xenopus* oocytes co-expressing CFTR and ENaC. *J Physiol* **508:** 825–836.

Caldwell RA, Boucher RC, Stutts MJ. 2004. Serine protease activation of near-silent epithelial Na^+ channels. *Am J Physiol Cell Physiol* **286:** C190–C194.

Caldwell RA, Boucher RC, Stutts MJ. 2005. Neutrophil elastase activates near-silent epithelial Na^+ channels and increases airway epithelial Na^+ transport. *Am J Physiol Lung Cell Mol Physiol* **288:** L813–L819.

Canessa CM, Merillat AM, Rossier BC. 1994a. Membrane topology of the epithelial sodium channel in intact cells. *Am J Physiol* **267:** C1682–C1690.

Canessa CM, Schild L, Buell G, Thorens B, Gautschi I, Horisberger JD, Rossier BC. 1994b. Amiloride-sensitive epithelial Na^+ channel is made of three homologous subunits. *Nature* **367:** 463–467.

Catalan MA, Nakamoto T, Gonzalez-Begne M, Camden JM, Wall AM, Clarke LL, Melvin JE. 2010. Cftr and ENaC ion channels mediate NaCl absorption in the mouse submandibular gland. *J Physiol* **588:** 713–724.

Chen JH, Stoltz DA, Karp PH, Ernst SE, Pezzulo AA, Moninger TO, Rector MV, Reznikov LR, Launspach JL, Chaloner K, et al. 2010. Loss of anion transport without increased sodium absorption characterizes newborn porcine cystic fibrosis airway epithelia. *Cell* **143:** 911–923.

Cheng SH, Rich DP, Marshall J, Gregory RJ, Welsh MJ, Smith AE. 1991. Phosphorylation of the R domain by cAMP-dependent protein kinase regulates the CFTR chloride channel. *Cell* **66:** 1027–1036.

Clauss W, Dürr J, Skadhauge E, Hörnicke H. 1985. Effects of aldosterone and dexamethasone on apical membrane properties and Na-transport of rabbit distal colon in vitro. *Pflugers Arch* **403:** 186–192.

Cook DI, Dinudom A, Komwatana P, Young JA. 1998. Control of Na^+ transport in salivary duct epithelial cells by cytosolic Cl^- and Na^+. *Eur J Morphol* **36:** 67–73.

Cotton CU, Lawson EE, Boucher RC, Gatzy JT. 1983. Bioelectric properties and ion transport of airways excised from adult and fetal sheep. *J Appl Physiol* **55**: 1542–1549.

Cuthbert AW, MacVinish LJ, Hickman ME, Ratcliff R, Colledge WH, Evans NJ. 1994. Ion-transporting activity in the murine colonic epithelium of normal animals and animals with cystic fibrosis. *Pflugers Arch* **428**: 508–515.

Cuthbert AW, Hickman ME, MacVinish LJ. 1999. Formal analysis of electrogenic sodium, potassium, chloride and bicarbonate transport in mouse colon epithelium. *Br J Pharmacol* **126**: 358–364.

Davies H, Bagg J, Muxworthy S, Goodchild MC, McPherson MA. 1989. Electrolyte concentrations in control and cystic fibrosis submandibular saliva. *Biochem Soc Trans* **18**: 447–448.

Davies H, Bagg J, Goodchild MC, McPherson MA. 1991. Defective regulation of electrolyte and protein secretion in submandibular saliva of cystic fibrosis patients. *Acta Paediatr Scand* **80**: 1094–1095.

Debongnie JC, Phillips SF. 1978. Capacity of the human colon to absorb fluid. *Gastroenterology* **74**: 698–703.

Debonneville C, Flores SY, Kamynina E, Plant PJ, Tauxe C, Thomas MA, Münster C, Chraïbi A, Pratt JH, Horisberger JD, et al. 2001. Phosphorylation of Nedd4-2 by Sgk1 regulates epithelial Na$^+$ channel cell surface expression. *EMBO J* **20**: 7052–7059.

de La Rosa DA, Canessa CM, Fyfe GK, Zhang P. 2000. Structure and regulation of amiloride sensitive sodium channels. *Annu Rev Physiol* **62**: 573–594.

Devuyst O, Guggino WB. 2002. Chloride channels in the kidney: Lessons learned from knockout animals. *Am J Physiol Renal Physiol* **283**: F1176–F1191.

Dinudom A, Young JA, Cook DI. 1993. Na$^+$ and Cl$^-$ conductances are controlled by cytosolic Cl$^-$ concentration in the intralobular duct cells of mouse mandibular glands. *J Membr Biol* **135**: 289–295.

di Sant'Agnese PA. 1956. Cystic fibrosis of the pancreas. *Am J Med* **21**: 406–422.

Duc C, Farman N, Canessa MC, Bonvalet JP, Rossier BC. 1994. Cell-specific expression of epithelial sodium channel α, β, and γ subunits in aldosterone-responsive epithelia from the rat: Localization by in situ hybridization and immunocytochemistry. *J Cell Biol* **127**: 1907–1921.

Dulhanty AM, Riordan JR. 1994. Phosphorylation by cAMP-dependent protein kinase causes a conformational change in the R domain of the cystic fibrosis transmembrane conductance regulator. *Biochemistry* **33**: 4072–4079.

Ferrera L, Caputo A, Galletta LJ. 2010. TMEM16A protein: A new identity for Ca^{2+}-dependent Cl$^-$ channels. *Physiology* **25**: 357–363.

Fischer H, Illek B, Machen TE. 1998. Regulation of CFTR by protein phosphatase 2B and protein kinase C. *Pflugers Arch* **436**: 175–181.

Frase LL, Strickland AD, Kachel GW, Krejs GJ. 1985. Enhanced glucose absorption in the jejunum of patients with cystic fibrosis. *Gastroenterology* **88**: 478–484.

French PJ, Bijman J, Edixhoven M, Vaandrager AB, Scholte BJ, Lohmann SM, Nairn AC, De Jonge HR. 1995. Isotype-specific activation of cystic fibrosis transmembrane conductance regulator-chloride channels by cGMP-dependent protein kinase II. *J Biol Chem* **270**: 26626–26631.

* Frizzell RA, Hanrahan JW. 2012. Physiology of epithelial chloride and fluid secretion. *Cold Spring Harb Perspect Biol* **2**: a009563.

Frizzell RA, Field M, Schultz SG. 1979. Sodium-coupled chloride transport by epithelial tissues. *Am J Physiol* **263**: F1–F8.

Fuller CM, Awayda MS, Arrate MP, Bradford AL, Morris RG, Canessa CM, Rossier BC, Benos DJ. 1995. Cloning of a bovine renal epithelial Na$^+$ cannel subunit. *Am J Physiol* **269**: C641–C654.

Gadsby DC, Nairn AC. 1999. Control of CFTR channel gating by phosphorylation and nucleotide hydrolysis. *Physiol Rev* **79**: S77–S107.

Gentzsch M, Dang H, Dang Y, Garcia-Caballero A, Suchindran H, Boucher RC, Stutts MJ. 2010. The cystic fibrosis transmembrane conductance regulator impedes proteolytic stimulation of the epithelial Na$^+$ channel. *J Biol Chem* **285**: 32227–32232.

Goldman MJ, Anderson GM, Stolzenberg ED, Kari UP, Zaslof M, Wilson JM. 1997. Human β-defensin-1 is a salt-sensitive antibiotic in lung that is inactivated in cystic fibrosis. *Cell* **88**: 553–560.

Grubb BR, Boucher RC. 1997. Enhanced colonic Na$^+$ absorption in cystic fibrosis mice versus normal mice. *Am J Physiol* **272**: G393–G400.

Grubb BR, O'Neal WK, Ostrowski LE, Kreda SM, Button B, Boucher RC. 2012. Transgenic hCFTR expression fails to correct β-ENaC mouse lung disease. *Am J Physiol Lung Cell Mol Physiol* **302**: L238–L247.

Guggino WB. 1999. Cystic fibrosis and the salt controversy. *Cell* **96**: 607–610.

Hall RA, Ostedgaard LS, Premont RT, Blitzer JT, Rahman N, Welsh MJ, Lefkowitz RJ. 1998. A C-terminal motif found in the β2-adrenergic receptor, P2Y1 receptor and cystic fibrosis transmembrane conductance regulator determines binding to the Na$^+$/H$^+$ exchanger regulatory factor family of PDZ proteins. *Proc Natl Acad Sci* **95**: 8496–8501.

Hardcastle J, Taylor CJ, Hardcastle PT, Baxter PS, Goldhill J. 1990. Intestinal transport in cystic fibrosis (CF). *Acta Univ Carol (Praha)* **36**: 157–158.

Hardcastle J, Harwood MD, Taylor CJ. 2004. Small intestinal glucose absorption in cystic fibrosis: A study in human and transgenic ΔF508 cystic fibrosis mouse tissues. *J Pharm Pharmacol* **56**: 329–338.

Hughey RP, Bruns JB, Kinlough CL, Harkleroad KL, Tong Q, Carattino MD, Johnson JP, Stockand JD, Kleyman TR. 2004. Epithelial sodium channels are activated by furin-dependent proteolysis. *J Biol Chem* **279**: 18111–18114.

Hummler E, Horisberger JD. 1999. Genetic disorders of membrane transport. V. The epithelial sodium channel and its implication in human disease. *Am J Physiol* **276**: G567–G571.

Itani OA, Chen JH, Karp PH, Ernst S, Keshavjee S, Parekh K, Klesney-Tait J, Zabner J, Welsh MJ. 2011. Human cystic fibrosis airway epithelia have reduced Cl$^-$ conductance but not increased Na$^+$ conductance. *Proc Natl Acad Sci* **108**: 10260–10265.

Jasti J, Furukawa H, Gonzales EB, Gouaux E. 2007. Structure of acid-sensing ion channel 1 at 1.9 Å resolution and low pH. **449:** 316–323.

Jia Y, Mathews CJ, Hanrahan JW. 1997. Phosphorylation by protein kinase C is required for acute activation of cystic fibrosis transmembrane conductance regulator by protein kinase A. *J Biol Chem* **272:** 4978–4984.

Johnson W. 1956. Salivary electrolytes in fibrocystic disease. *Arch Dis Child* **31:** 477.

Joris L, Dab I, Quinton PM. 1993. Elemental composition of human airway surface fluid in healthy and diseased airways. *Am Rev Respir Dis* **148:** 1633–1637.

Kashlan OB, Adelman JL, Okumura S, Blobner BM, Zuzek Z, Hughey RP, Kleyman TR, Grabe M. 2011. Constraint-based, homology model of the extracellular domain of the epithelial Na^+ channel α subunit reveals a mechanism of channel activation by proteases. *J Biol Chem* **286:** 649–660.

Kellenberger S, Schild L. 2002. Epithelial sodium channel/degenerin family of ion channels: A variety of functions for a shared structure. *Physiol Rev* **82:** 735–767.

Kesimer M, Makhov AM, Griffith JD, Verdugo P, Sheehan JK. 2010. Unpacking a gel-forming mucin: A view of MUC5B organization after granular release. *Am J Physiol Lung Cell Mol Physiol* **298:** L15–L22.

Kilburn K. 1974. Functional morphology of the distal lung. *Int Rev Cytol* **37:** 153–270.

Kimura T, Kawabe H, Jiang C, Zhang W, Xiang YY, Lu C, Salter MW, Brose N, Lu WY, Rotin D. 2011. Deletion of the ubiquitin ligase Nedd4L in lung epithelia causes cystic fibrosis-like disease. *Proc Natl Acad Sci* **108:** 3216–3221.

Knowles MR, Boucher RC. 2002. Mucus clearance as a primary innate defense mechanism for mammalian airways. *J Clin Invest* **109:** 571–577.

Knowles M, Murray G, Shallal J, Askin F, Ranga V, Gatzy J, Boucher R. 1984. Bioelectric properties and ion flow across excised human bronchi. *J Appl Physiol* **56:** 868–877.

Knowles MR, Robinson JM, Wood RE, Pue CA, Mentz MW, Wager GC, Gatzy JT, Boucher RC. 1997. Ion composition of airway surface liquid of patients with cystic fibrosis as compared with normal and disease-control subjects. *J Clin Invest* **100:** 2588–2595.

Ko YH, Pedersen PL. 1995. The first nucleotide binding fold of the cystic fibrosis transmembrane conductance regulator can function as an active ATPase. *J Biol Chem* **270:** 22093–22096.

Kokko JP. 1973. Proximal tubule potential difference dependence on glucose, HCO_3^- and amino acids. *J Clin Invest* **52:** 1362–1367.

Kreda SM, Mall M, Mengos A, Rochelle L, Yankaskas J, Riordan JR, Boucher RC. 2005. Characterization of wild-type and ΔF508 cystic fibrosis transmembrane regulator in human respiratory epithelia. *Mol Biol Cell* **16:** 2154–2167.

Kunzelmann K, Mall M. 2001. Pharmacotherapy of the ion transport defect in cystic fibrosis. *Clin Exp Pharmacol Physiol* **28:** 857–867.

Kunzelmann K, Mall M. 2002. Electrolyte transport in the mammalian colon: Mechanisms and implications for disease. *Physiol Rev* **82:** 245–289.

Kunzelmann K, Schreiber R, Boucherot A. 2001. Mechanisms of the inhibition of epithelial Na^+ channels by CFTR and purinergic stimulation. *Kidney Int* **60:** 455–461.

Kunzelmann K, Bachhuber T, Regeer R, Markovich D, Sun J, Schreiber R. 2004. Purinergic inhibition of the epithelial Na^+ transport via hydrolysis of PIP2. *FASEB J* **19:** 142–143.

Luo J, Pato MD, Riordan JR, Hanrahan JW. 1998. Differential regulation of single CFTR channels by PP2C, PP2A, and other phosphatases. *Am J Physiol* **274:** C1397–C1410.

Ma HP, Saxena S, Warnock DG. 2002. Anionic phospholipids regulate native and expressed epithelial sodium channel (ENaC). *J Biol Chem* **277:** 7641–7644.

Mall MA. 2009. Role of the amiloride-sensitive epithelial Na^+ channel in the pathogenesis and as a therapeutic target for cystic fibrosis lung disease. *Exp Physiol* **94:** 171–174.

Mall M, Bleich M, Kuehr J, Brandis M, Greger R, Kunzelmann K. 1999. CFTR-mediated inhibition of epithelial Na^+ conductance in human colon is defective in cystic fibrosis. *Am J Physiol* **277:** G709–G716.

Matsui H, Davis CW, Tarran R, Boucher RC. 2000. Osmotic water permeabilities of cultured, well-differentiated normal and cystic fibrosis airway epithelia. *J Clin Invest* **105:** 1419–1427.

McPherson MA, Davies H, Mills CL, Pereira MMC, Goodchild MC, Dormer RL. 1992. Role of CFTR in salivary secretion. Elsevier, Amsterdam.

Mohammad-Panah R, Demolombe S, Riochet D, Leblais V, Loussouarn G, Pollard H, Baro I, Escande D. 1998. Hyperexpression of recombinant CFTR in heterologous cells alters its physiological properties. *Am J Physiol* **274:** C310–C318.

Mount DB, Romero MF. 2004. The SLC26 gene family of multifunctional anion exchangers. *Pflugers Arch* **447:** 710–721.

Myerburg MM, Butterworth MB, McKenna EE, Peters KW, Frizzell RA, Kleyman TR, Pilewski JM. 2006. Airway surface liquid volume regulates ENaC by altering the serine protease-protease inhibitor balance. *J Biol Chem* **281:** 27942–27949.

Myerburg MM, McKenna EE, Luke CJ, Frizzell RA, Kleyman TR, Pilewski JM. 2008. Prostasin expression is regulated by airway surface liquid volume and is increased in cystic fibrosis. *Am J Physiol Lung Cell Mol Physiol* **294:** L932–L941.

Myerburg MM, Harvey PR, Heidrich EM, Pilewski JM, Butterworth MB. 2010. Acute regulation of ENaC in airway epithelia by proteases and trafficking. *Am J Respir Cell Mol Biol* **43:** 712–719.

Olver RE, Davis B, Marin MG, Nadel JA. 1975. Active transport of Na^+ and Cl^- across the canine tracheal epithelium in vitro. *Am Rev Respir Dis* **112:** 811–815.

Ostedgaard LS, Meyerholz DK, Chen JH, Pezzulo AA, Karp PH, Rokhlina T, Ernst SE, Hanfland RA, Reznikov LR, Ludwig PS, et al. 2011. The ΔF508 mutation causes CFTR misprocessing and cystic fibrosis-like disease in pigs. *Sci Transl Med* **3:** 74ra24.

Palmer L, Patel A, Frindt G. 2012. Regulation and dysregulation of epithelial Na$^+$ channels. *Clin Exp Nephrol* **16:** 35–43.

Passero CJ, Mueller GM, Rondon-Berrios H, Tofovic SP, Hughey RP, Kleyman TR. 2008. Plasmin activates epithelial Na$^+$ channels by cleaving the γ subunit. *J Biol Chem* **283:** 36586–36591.

Picciotto MR, Cohn JA, Bertuzzi G, Greengard P, Nairn AC. 1992. Phosphorylation of the cystic fibrosis transmembrane conductance regulator. *J Biol Chem* **267:** 12742–12752.

Pilewski JM, Frizzell RA. 1999. Role of CFTR in airway disease. *Physiol Rev* **79:** S215–S255.

Prader A, Gautier E. 1955. Die Na$^-$ and K$^-$ Konzentration in gemischten Speichel: II Erhohte Werte bei der Pankreasfibrose. *Helv Paediat Acta* 10: 56.

Quinton PM. 1984. Exocrine glands. In *Cystic fibrosis* (ed. Taussig LM), pp. 338–375. Thieme-Stratton, New York.

Quinton PM. 1987. Physiology of sweat secretion. *Kidney Int Suppl* **21:** S102–S108.

Quinton PM. 1990. Cystic fibrosis: A disease in electrolyte transport. *FASEB J* **4:** 2709–2717.

Quinton PM. 1999. Physiological basis of cystic fibrosis: A historical perspective. *Physiol Rev* **79:** S3–S22.

Quinton PM. 2007. Cystic fibrosis: Lessons from the sweat gland. *Physiology (Bethesda)* **22:** 212–225.

Quinton PM, Reddy MM. 1991. Regulation of absorption in the human sweat duct. *Adv Exp Med Biol* **290:** 159–170.

Randak C, Neth P, Auerswald EA, Eckerskorn C, Assfalg-Machleidt I, Machleidt W. 1997. A recombinant polypeptide model of the second nucleotide-binding fold of the cystic fibrosis transmembrane conductance regulator functions as an active ATPase, GTPase and adenylate kinase. *FEBS Lett* **410:** 180–186.

Reddy MM, Quinton PM. 1987. Intracellular potentials of microperfused human sweat duct cells. *Pflugers Arch* **410:** 471–475.

Reddy MM, Quinton PM. 1989. Altered electrical potential profile of human reabsorptive sweat duct cells in cystic fibrosis. *Am J Physiol* **257:** C722–C726.

Reddy MM, Quinton PM. 1991. Intracellular potassium activity and the role of potassium in transepithelial salt transport in the human reabsorptive sweat duct. *J Membr Biol* **119:** 199–210.

Reddy MM, Quinton PM. 1992. cAMP activation of CF-affected Cl$^-$ conductance in both cell membranes of an absorptive epithelium. *J Membr Biol* **130:** 49–62.

Reddy MM, Quinton PM. 1994. Intracellular Cl activity: Evidence of dual mechanisms of cl absorption in sweat duct. *Am J Physiol* **267:** C1136–C1144.

Reddy MM, Quinton PM. 1996. Hydrolytic and nonhydrolytic interactions in the ATP regulation of CFTR Cl$^-$ conductance. *Am J Physiol* **271:** C35–C42.

Reddy MM, Quinton PM. 2002. Influence of cytoplasmic Cl$^-$ on the effect of CFTR activation on ENaC function. *Pediatr Pulmonology* **34** (Suppl)**:** 8–359.

Reddy MM, Quinton PM. 2003a. Functional interaction of CFTR and ENaC in sweat glands. *Pflugers Arch* **445:** 499–503.

Reddy MM, Quinton PM. 2003b. Control of dynamic CFTR selectivity by glutamate and ATP in epithelial cells. *Nature* **423:** 756–760.

Reddy MM, Light MJ, Quinton PM. 1999. Activation of the epithelial Na$^+$ channel (ENaC) requires CFTR Cl$^-$ channel function. *Nature* **402:** 301–304.

Reddy MM, Wang XF, Quinton PM. 2008. Effect of cytosolic pH on epithelial Na$^+$ channel in normal and cystic fibrosis sweat ducts. *J Membr Biol* **225:** 1–11.

Reuss L. 1985. Changes in cell volume measured with an electrophysiologic technique. *Proc Natl Acad Sci* **82:** 6014–6018.

Reuss L, Segal Y, Altenberg G. 1991. Regulation of ion transport across gallbladder epithelium. *Annu Rev Physiol* **53:** 361–373.

Riordan JR, Rommens JM, Kerem B, Alon N, Rozmahel R, Grzelczak Z, Zielenski J, Lok S, Plavsic N, Chou JL, et al. 1989. Identification of the cystic fibrosis gene: Cloning and characterization of complementary DNA. *Science* **245:** 1066–1073.

Rossier BC. 1996. The renal epithelial sodium channel: New insights in understanding hypertension. *Adv Nephrol Necker Hosp* **25:** 275–286.

Rotin D. 2000. Regulation of the epithelial sodium channel (ENaC) by accessory proteins. *Curr Opin Nephrol Hypertens* **9:** 529–534.

Russo MA, Hogenauer C, Coates SW Jr, Santa Ana CA, Porter JL, Rosenblatt RL, Emmett M, Fordtran JS. 2003. Abnormal passive chloride absorption in cystic fibrosis jejunum functionally opposes the classic chloride secretory defect. *J Clin Invest* **112:** 118–125.

Sato K, Sato F. 1984. Defective β adrenergic response of cystic fibrosis sweat glands in vivo and in vitro. *J Clin Invest* **73:** 1763–1771.

Schild L, Canessa CM, Shimkets RA, Gautschi I, Lifton RP, Rossier BC. 1995. A mutation in the epithelial sodium channel causing Liddle disease increases channel activity in the *Xenopus laevis* oocyte expression system. *Proc Natl Acad Sci* **92:** 5699–5703.

Schild L, Lu Y, Gautschi I, Schneeberger E, Lifton RP, Rossier BC. 1996. Identification of a PY motif in the epithelial Na channel subunits as a target sequence for mutations causing channel activation found in Liddle syndrome. *EMBO J* **15:** 2381–2387.

Schreiber R, Hopf A, Mall M, Greger R, Kunzelmann K. 1999. The first-nucleotide binding domain of the cystic-fibrosis transmembrane conductance regulator is important for inhibition of the epithelial Na$^+$ channel. *Proc Natl Acad Sci* **96:** 5310–5315.

Schwiebert EM, Benos DJ, Egan ME, Stutts MJ, Guggino WB. 1999. CFTR is a conductance regulator as well as a chloride channel. *Physiol Rev* **79:** S145–S166.

Sheppard DN, Welsh MJ. 1999. Structure and function of the CFTR chloride channel. *Physiol Rev* **79:** S23–S45.

Short DB, Trotter KW, Reczek D, Kreda SM, Bretscher A, Boucher RC, Stutts MJ, Milgram SL. 1998. An apical PDZ protein anchors the cystic fibrosis transmembrane conductance regulator to the cytoskeleton. *J Biol Chem* **273:** 19797–19801.

Smith JJ, Travis SM, Greenberg EM, Welsh MJ. 1996. Cystic fibrosis airway epithelia fail to kill bacteria because of abnormal airway surface fluid. *Cell* **85:** 229–236.

Song Y, Namkung W, Nielson DW, Lee JW, Finkbeiner WE, Verkman AS. 2009. Airway surface liquid depth measured in ex vivo fragments of pig and human trachea: Dependence on Na^+ and Cl^- channel function. *Am J Physiol Lung Cell Mol Physiol* **297:** L1131–L1140.

Staub O, Abriel H, Plant P, Ishikawa T, Kanelis V, Saleki R, Horisberger JD, Schild L, Rotin D. 2000. Regulation of the epithelial Na^+ channel by Nedd4 and ubiquitination. *Kidney Int* **57:** 809–815.

Stockand JD, Staruschenko A, Pochynyuk O, Booth RE, Silverthorn DU. 2008. Insight toward epithelial Na^+ channel mechanism revealed by the acid-sensing ion channel 1 structure. *IUBMB Life* **60:** 620–628.

Stutts MJ, Gatzy JT, Boucher RC. 1988. Activation of chloride conductance induced by potassium in tracheal epithelium. *Pflugers Arch* **411:** 252–258.

Stutts MJ, Canessa CM, Olsen JC, Hamrick M, Cohn JA, Rossier BC, Boucher RC. 1995. CFTR as a cAMP-dependent regulator of sodium channels. *Science* **269:** 847–850.

Svenningsen P, Uhrenholt TR, Palarasah Y, Skjødt K, Jensen BL, Skøtt O. 2009. Prostasin-dependent activation of epithelial Na^+ channels by low plasmin concentrations. *Am J Physiol Regul Integr Comp Physiol* **297:** R1733–R1741.

Tarran R, Grubb BR, Gatzy JT, Davis CW, Boucher RC. 2001. The relative roles of passive surface forces and active ion transport in the modulation of airway surface liquid volume and composition. *J Gen Physiol* **118:** 223–236.

Tarran R, Loewen ME, Paradiso AM, Olsen JC, Gray MA, Argent BE, Boucher RC, Gabriel SE. 2002. Regulation of murine airway surface liquid volume by CFTR and Ca^{2+}-activated Cl^- conductances. *J Gen Physiol* **120:** 407–418.

Tarran R, Button B, Boucher RC. 2005. Regulation of normal and cystic fibrosis airway surface liquid volume by phasic shear stress. *Annu Rev Physiol* **68:** 543–561.

Tarran R, Trout L, Donaldson SH, Boucher RC. 2006. Soluble mediators, not cilia, determine airway surface liquid volume in normal and cystic fibrosis. *J Gen Physiol* **127:** 591–604.

Taylor CJ, Baxter PS, Hardcastle J, Hardcastle PT. 1987. Absence of secretory response in jejunal biopsy samples from children with cystic fibrosis. *Lancet* **330:** 107–108.

Taylor CJ, Baxter PS, Hardcastle J, Hardcastle PT. 1988. Failure to induce secretion in jejunal biopsies from children with cystic fibrosis. *Gut* **29:** 957–962.

Teune TM, Timmers-Reker AJ, Bouquet J, Bijman J, De Jonge HR, Sinaasappel M. 1996. In vivo measurement of chloride and water secretion in the jejunum of cystic fibrosis patients. *Pediatr Res* **40:** 522–527.

Ussing HH, Erlij D, Lassen U. 1974. Transport pathways in biological membranes. *Annu Rev Physiol* **36:** 17–49.

Uyekubo SN, Fischer H, Maminishkis A, Illek B, Miller SS, Widdicombe JH. 1998. cAMP-dependent absorption of chloride across airway epithelium. *Am J Physiol* **275:** L1219–L1227.

Vaandrager AB, Ehlert EM, Jarchau T, Lohmann SM, de Jonge HR. 1996. N-terminal myristoylation is required for membrane localization of cGMP-dependent protein kinase type II. *J Biol Chem* **271:** 7025–7029.

Vaandrager AB, Smolenski A, Tilly BC, Houtsmuller AB, Ehlert EM, Bot AG, Edixhoven M, Boomaars WE, Lohmann SM, de Jonge HR. 1998. Membrane targeting of cGMP-dependent protein kinase is required for cystic fibrosis transmembrane conductance regulator Cl^- channel activation. *Proc Natl Acad Sci* **95:** 1466–1471.

Verkman AS. 2007. Role of aquaporins in lung liquid physiology. *Respir Physiol Neurobiol* **159:** 324–330.

Verkman AS, Matthay MA, Song Y. 2000. Aquaporin water channels and lung physiology. *Am J Physiol Lung Cell Mol Physiol* **278:** L867–L879.

Vuagniaux G, Vallet V, Jaeger NF, Pfister C, Bens M, Farman N, Courtois-Coutry N, Vandewalle A, Rossier BC, Hummler E. 2000. Activation of the amiloride-sensitive epithelial sodium channel by the serine protease mCAP1 expressed in a mouse cortical collecting duct cell line. *J Am Soc Nephrol* **11:** 828–834.

Wang J, Knight ZA, Fiedler D, Williams O, Shokat KM, Pearce D. 2008. Activity of the p110-α subunit of phosphatidylinositol-3-kinase is required for activation of epithelial sodium transport. *Am J Physiol Renal Physiol* **295:** F843–F850.

Welsh MJ. 1987. Electrolyte transport by airway epithelia. *Physiol Rev* **67:** 1143–1184.

Welsh MJ. 1996. Cystic fibrosis. In *Molecular biology of membrane transport disorders*, 2nd ed. (ed. Schultz SG), Chapter 30, pp. 605–623. Springer, New York.

Welsh MJ, Anderson MP, Rich DP, Berger HA, Denning GM, Ostedgaard LS, Sheppard DN, Cheng SH, Gregory RJ, Smith AE. 1992. Cystic fibrosis transmembrane conductance regulator: A chloride channel with novel regulation. *Neuron* **8:** 821–829.

Widdicombe JH, Widdicombe JG. 1995. Regulation of human airway surface liquid. *Respir Physiol* **99:** 3–12.

Willumsen NJ, Boucher RC. 1991. Sodium transport and intracellular sodium activity in cultured human nasal epithelium. *Am J Physiol* **261:** C319–C331.

Willumsen NJ, Davis CW, Boucher RC. 1989. Intracellular Cl^- activity and cellular Cl^- pathways in cultured human airway epithelium. *Am J Physiol* **256:** C1033–C1044.

Yue G, Malik B, Yue G, Eaton DC. 2002. Phosphatidylinositol 4,5-bisphosphate (PIP2) stimulates epithelial sodium channel activity in A6 cells. *J Biol Chem* **277:** 11965–11969.

Zabner J, Smith JJ, Karp PH, Widdicombe JH, Welsh MJ. 1998. Loss of CFTR chloride channels alters salt absorption by cystic fibrosis airway epithelia in vitro. *Mol Cell* **2:** 397–403.

Physiology of Epithelial Chloride and Fluid Secretion

Raymond A. Frizzell[1] and John W. Hanrahan[2]

[1]Department of Cell Biology and Physiology, University of Pittsburgh School of Medicine, Pittsburgh, Pennsylvania 15261

[2]Department of Physiology, McGill University, Montreal, Quebec H3G 1Y6, Canada

Correspondence: frizzell@pitt.edu

Epithelial salt and water secretion serves a variety of functions in different organ systems, such as the airways, intestines, pancreas, and salivary glands. In cystic fibrosis (CF), the volume and/or composition of secreted luminal fluids are compromised owing to mutations in the gene encoding CFTR, the apical membrane anion channel that is responsible for salt secretion in response to cAMP/PKA stimulation. This article examines CFTR and related cellular transport processes that underlie epithelial anion and fluid secretion, their regulation, and how these processes are altered in CF disease to account for organ-specific secretory phenotypes.

The formation and maintenance of the fluid compartment at the apical (luminal) surface of epithelia are critical for the normal functions of numerous organ systems. Examples include the airway surface liquid (ASL) layer of the conducting airways, which is vital to mucociliary clearance, generation of the liquid vehicle for secretion of pancreatic digestive enzymes and their delivery to the intestinal lumen, and in the intestine, maintenance of an appropriate level of luminal fluidity for digestion. In terms of volume, intestinal secretion is the greatest, estimated at $\sim$8 L/d (Barrett and Keely 2000). In cystic fibrosis (CF), the formation of luminal fluid, and often its composition, is impaired, and these deficits underlie the complex manifestations of organ pathophysiology that characterize this disease.

At the basis of these defects is CFTR, the apical membrane anion channel with the capacity for regulated secretion of Cl, HCO_3, and smaller amounts of other anions. Chloride secretion establishes an electrical driving force (lumen-negative) for trans-epithelial sodium secretion via the paracellular pathways, and together, their movements into the luminal compartment generate the osmotic driving force for water flow that yields an isotonic secretory product. The net result of a missing or defective CFTR channel is therefore disruption of secreted fluid volume and composition, and in the systems mentioned, defects in airway clearance of mucins and bacteria, pancreatic insufficiency, and intestinal obstruction (for organ system details, see below and "Role of CFTR in the Secretion of Other Anions").

Cite this article as *Cold Spring Harb Perspect Med* doi: 10.1101/cshperspect.a009563

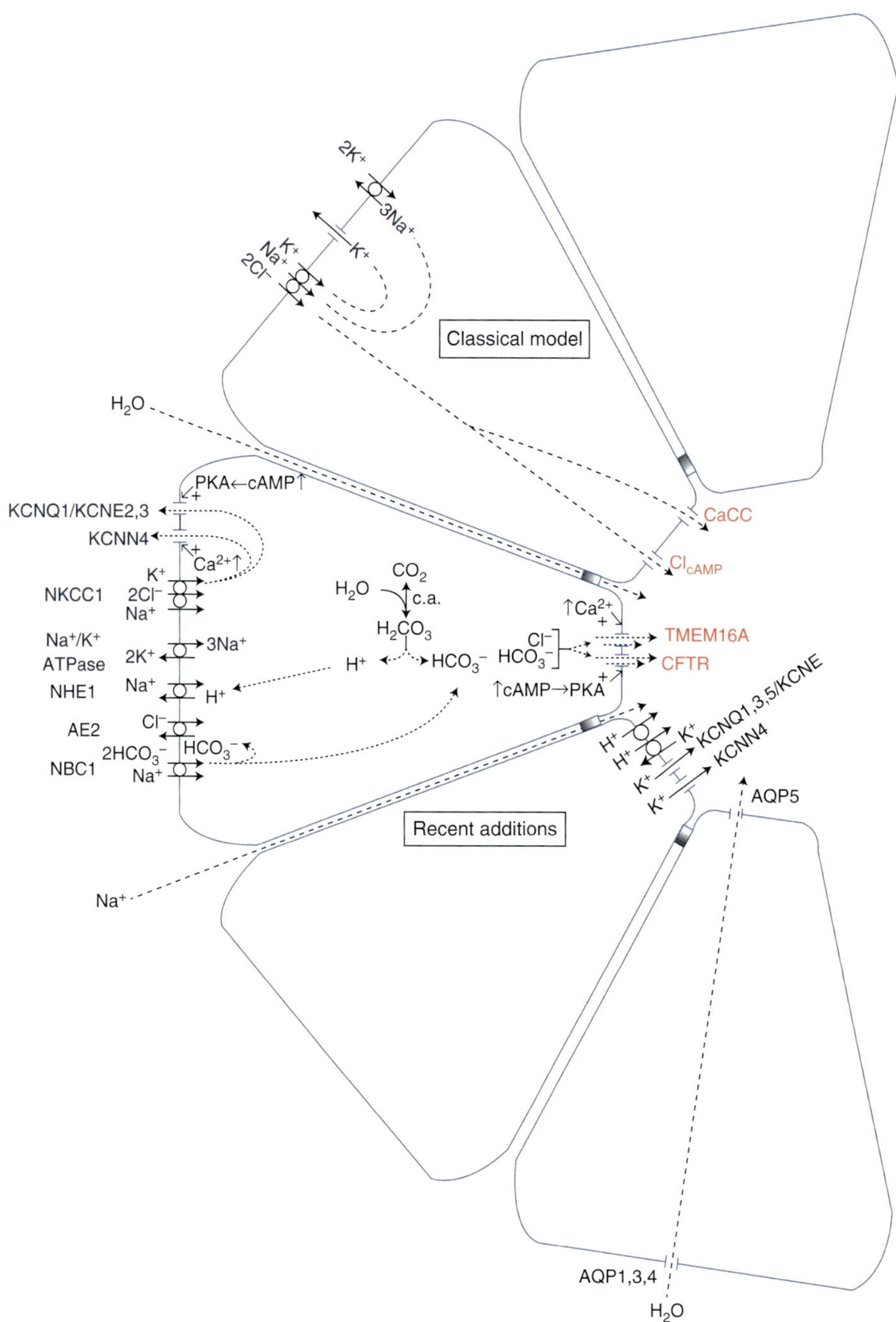

Figure 1. Schematic of the transport pathways in secretory epithelia. As shown in the "Classical model," Cl⁻ entry across the basolateral membrane is driven primarily via Na⁺ cotransport. Apical Cl⁻ efflux is mediated by cAMP- and Ca²⁺-activated anion conductances. The scheme labeled "Recent additions" identifies the numerous transporters and channels proposed to mediate ion and fluid secretion in various epithelia. Accordingly, basolateral Cl⁻ loading may occur via both NKCC1 and AE2. Cl⁻ and HCO₃⁻ exit involves apical CFTR, TMEM16A, and probably other Cl⁻ channels/transporters, such as members of the SLC26A family (data not shown). (*Legend continues on facing page.*)

Cite this article as *Cold Spring Harb Perspect Med* doi: 10.1101/cshperspect.a009563

This article examines the basic, cellular events that underlie epithelial anion and fluid secretion, including the component transport processes and their regulation. Newer findings linking other transport and channel functions to the activity or trafficking of CFTR will determine the tissue-specific secretory phenotype and how these physiological properties are altered in CF disease.

BASIC TRANSPORT COMPONENTS OF ANION SECRETION

Marine biology has made significant contributions to the study of epithelial ion transport. Key among them is a series of experiments performed in the late 1930s by the Danish physiologist and Nobel Laureate August Krogh, using the skin and gills of freshwater animals. This work led Krogh to the conclusion that metabolic energy was required to maintain the ionic gradients between the organism and its external environment that permit specialization, and he and his collaborators introduced the use of radioisotopes to define the underlying transport events. His student Hans Ussing combined isotopic ion fluxes and biophysical approaches to form our basic knowledge of epithelial polarity, active and passive transport, the paracellular pathway, and the origins of epithelial "bioelectric potentials" (Larsen 2002). Once Jens Christian Skou identified the Na,K-ATPase and linked it to cation transport (Skou 1998), the ensuing decades firmly established the central role of this active transport process as the energy source for the secondary active transport of many nutrients and ions. Ultimately, these principles were applied in the context of active anion secretion, leading to the classical cellular model, its basic principles being applicable to many secretory epithelia (Fig. 1).

Active Cl secretion and second-messenger pathways converged when the dramatic fluid secretion evoked by intestinal infection with *Vibrio cholerae* was linked with an active Cl secretory process, modeled in studies of rabbit intestine mounted in Ussing chambers and stimulated by cAMP or agents that increase its cellular levels (Kimberg et al. 1971). The basic transport principles were enumerated (Frizzell et al. 1979), and over the next decade, their molecular identities were established (see Fig. 1). A rise in cellular cAMP, which can be elicited by numerous primary agonists, evokes anion secretion by stimulating three basic transport processes: the apical anion channel CFTR, basolateral K channels of at least two types, and the basolateral Na-coupled Cl entry process, NKCC1 (Na/K/2Cl cotransporter 1).

Transport at the Apical Membrane

Apical CFTR

This section reviews the many facets of CFTR structure, function, and regulation. As noted, CFTR is in the large family of ATP-binding cassette (ABC) proteins, whose members are composed of two membrane-spanning domains (MSD1 and 2) and two nucleotide-binding domains (NBD1 and 2) arranged in series in the primary structure (Riordan et al. 1989). Uniquely, CFTR contains a central R (regulatory) region, which contains nine strong consensus sequences for phosphorylation by protein kinase A (PKA) and additional sites of phosphorylation by other kinases (e.g., PKC, AMPK). Another significant

---

Figure 1. (*Continued*) NBC1-mediated HCO_3^- entry during cAMP stimulation maintains intracellular pH and generates electrogenic trans-epithelial HCO_3^- secretion. NHE1 maintains intracellular pH during stimulation by Ca^{2+} secretagogues when cells are hyperpolarized and net Cl^- secretion is favored. There is evidence for apical H^+ secretion mediated by vacuolar proton pumps, H^+/K^+-ATPase, and also apical K channels, which contribute to apical hyperpolarization and thus enhance the driving force for anion secretion. Most trans-epithelial Na^+ flux occurs passively through the paracellular pathway driven by the lumen-negative voltage arising from electrogenic anion secretion. In both schemes, regulated apical anion conductance is rate-limiting at low/medium levels of anion secretion, but basolateral K conductance becomes increasingly limiting at higher secretory rates, explaining the additive nature of secretagogues that operate via different second-messenger pathways.

PKA phosphorylation site in the regulatory insertion of NBD1 is discussed below. As a primary event in secretion, protein kinase A (PKA) phosphorylates the R region, which then enables channel gating (opening and closing). Gating and Cl secretion are elicited by the binding and hydrolysis of ATP at the NBDs (see below for details). As discussed below, CFTR is not the sole pathway for apical Cl exit, but it is the predominant pathway for secretion in response to agonists that act via cAMP/PKA-dependent phosphorylation. In addition, CFTR exhibits a significant conductance to bicarbonate, which is reviewed below in "Anion Secretion by Alternate Apical Pathways."

Requirements at the Basolateral Membrane

Na/K Pump

The basolateral membrane Na/K pump (ATPase) is the primary, energy-using process that establishes the electrochemical gradients for secondary active anion secretion. It transports three Na ions out of the cell in exchange for two K ions pumped into the cell, with the energetic revenue derived from the conversion of one ATP to ADP (Skou 1998). Because the pump is itself electrogenic, it contributes to the inside-negative basolateral membrane potential, the magnitude of its contribution dependent on the parallel electrical resistance. The primary determinant of the membrane voltage, however, is the leakage of potassium out of the cell via a collection of basolateral (and apical in some cell types) K channels, which are primarily responsible for determining the membrane potential (see further discussion below). In most cells, the pump is a major consumer of metabolic energy, but in epithelia, it may account for two-thirds of the cell's energy expenditure. Airway cells in cystic fibrosis are reported to consume oxygen at a rate two to three times that observed in non-CF tissues, and they express more binding sites for ouabain (Stutts et al. 1986), findings that reflect the higher rate of pump expression and turnover due to enhanced Na absorption across CF airway epithelia ("Anion Secretion by Alternate Apical Pathways"). Nevertheless, recent

findings in an animal model of CF have challenged this view (Chen et al. 2010).

Basolateral K Channels

An increase in basolateral K conductance is necessary to recycle K brought into the cell by the Na/K pump and to maintain an electrical driving force for anion exit across the apical membrane. Owing to the large increase in Cl conductance associated with the activation of CFTR, or other apical anion channels, the apical membrane voltage approaches the Cl equilibrium potential. But in response to second-messenger pathways, basolateral K channels are also activated, to coordinate membrane repolarization with the activation of apical Cl conductance and permit the membrane potentials to repolarize toward the K equilibrium potential, establishing the electrical driving force for Cl exit across the apical membrane.

Activation of apical CFTR-mediated anion secretion by cAMP/PKA is paralleled by PKA phosphorylation and activation of a basolateral membrane K channel, KCNQ1 (aka LQT1). The requirement for basolateral K conductance activation is evidenced by the impact of blocking these pathways with the generalized K channel blocker barium. Barium inhibits anion secretion by depolarizing the membrane potentials toward electrochemical equilibrium for Cl at the apical membrane and thereby blocking Cl secretion (Smith and Frizzell 1984; Welsh and McCann 1985). In many epithelia, the basolateral K conductance is composed of at least two K channel types. KCNQ1 is activated primarily during cAMP/PKA-induced secretion, its single channel conductance is low ($\sim$3 pS), and it is inhibited selectively by the chromanol 293B and by HMR-1556 (Gerlach et al. 2001; Kunzelmann et al. 2001). The contribution of KCNQ1 to the total K conductance varies with secretory agonist type and is tissue/cell-type dependent (Liao et al. 2005).

On the other hand, secretion stimulated by a cellular Ca rise primarily activates the intermediate conductance, Ca-activated channel KCNN4 (Devor and Frizzell 1998; Kunzelmann et al. 2001). Nevertheless, it is difficult to generalize,

     Cite this article as *Cold Spring Harb Perspect Med* doi: 10.1101/cshperspect.a009563

because multiple channel isoforms and β subunits may be expressed in different cell types, and these can assemble as homo- or heterodimers with somewhat different properties. Two-pore K channels have also been found in intestinal secretory cells. These channels may also target to the apical membrane to mediate K secretion in some epithelia (e.g., airway) (Heitzmann and Warth 2008). It is interesting that a splice variant of the calcium-activated, intermediate conductance K channel KCNN4, which is usually considered basolateral, has also been shown at the apical membrane in colon, where it mediates K secretion and contributes to the driving force for Cl^- secretion (Joiner et al. 2003; Nanda Kumar et al. 2010). This may be a more general phenomenon than previously thought, because most epithelial secretions have elevated [K] compared with plasma (typically twofold to fourfold) and may also involve several members of the KCNQ channel family (Namkung et al. 2009).

Basolateral NKCC1

Chloride entry across the basolateral membranes of secretory epithelia is mediated primarily by this Na/K/2Cl cotransporter isoform. The driving force for cellular Cl entry and accumulation is supplied primarily by the inward chemical gradient of Na established by the pump; the basolateral membrane potential does not influence this electrically neutral process. Ultimately, the rate of basolateral Cl entry via NKCC1 determines the overall rate of Cl secretion, which requires coordination of transport events at the limiting membranes to maintain the steady state. The timing of NKCC1 activation generally follows that of CFTR and the K channels, resulting in some loss of cellular Cl and K and a resulting decrease in cell volume.

This is a key feature of NKCC1 regulation, because reduced cell Cl concentration and volume trigger the activation of the WNK kinase (McCormick and Ellison 2011), which phosphorylates the SPAK (surface presentation of antigens protein) and OSR1 (oxidative stress responsive 1) kinases, which, in turn, phosphorylate the cotransporter amino terminus to activate its turnover (Mercier-Zuber and

O'Shaughnessy 2011). NKCC1 was the first recognized binding partner of SPAK and OSR1. Phosphorylation of NKCC1 at three threonines (lying within amino acids 203–212) gives rise to the stimulation of NKCC1 turnover, and mutations at these sites reduce its response to low Cl or hyperosmotic-induced cell shrinkage (Richardson et al. 2008). The role of SPAK in this signaling cascade has been confirmed in knockout or knockin mice in which the expression of SPAK-bearing mutations at critical threonine residues blocks the ability of WNKs to activate NKCC1. It was also of interest that the expression level of NKCC1 was reduced in mice expressing mutant SPAK, suggesting that this kinase controls both the activity and the expression level of NKCC1. A current critical question in the field is the mechanism by which WNKs are activated by low Cl and cell shrinkage. An interaction with the cytoskeleton or cell adhesion molecules could perhaps serve as a transducer of its activity.

Basolateral Anion Exchange (AE)

Although early studies pointed to a mechanism in which secretion involved Na^+-coupled Cl^- entry at the basolateral membrane, a large fraction of the Cl and fluid secretion by salivary glands and other epithelia was found to be insensitive to the NKCC1 inhibitor bumetanide but was sensitive to inhibitors of carbonic anhydrase (acetazolamide and ethoxolamide) and to amiloride and dimethyl amiloride. In salivary glands (see Fig. 1), this component of secretion was attributed to the parallel operation of basolateral Cl/HCO_3 and Na/H exchangers (Novak and Young 1986; Pirani et al. 1987). Carbonic anhydrase was proposed to catalyze cellular HCO_3 synthesis, and because CO_2 hydration would generate carbonic acid and acid-load the cell, H extrusion, for example, by amiloride-sensitive Na/H exchange, was proposed to maintain intracellular pH and thus transepithelial Cl secretion. Basolateral Cl uptake by anion exchangers has been suggested in various epithelia including equine trachea (Tessier et al. 1990) and the Calu-3 cell line (Cuthbert et al. 2003; Krouse et al. 2004; Shan et al. 2011).

The molecular basis of the basolateral Cl⁻ entry process may be the Na-independent anion exchanger AE2 (SLC4A2), particularly in acid-secreting cells that must cope with an intracellular alkaline load. AE2 is expressed at high levels in parietal cells of the gastric epithelium, and null mice display a dramatic gastric epithelial phenotype with abnormal canaliculae and reduced HCl secretion (Gawenis et al. 2004). Other Cl-secreting epithelia are also affected, notably the proximal colon (Gawenis et al. 2010); the role of AE2 has been difficult to establish in most tissues because it is electrically silent, there are no specific inhibitors available, and AE2-null mice die soon after birth. AE2 is expressed in the basolateral membrane of Calu-3 cells (Loffing et al. 2000), and recent studies of a stable, AE2-deficient Calu-3 cell line under open-circuit conditions revealed that AE2 mediates most basolateral anion exchange (Huang et al. 2011) and accounts for up to 70% of the net Cl⁻ flux during cAMP-stimulated fluid secretion (Shan et al. 2011).

Essential Role of the Paracellular Pathway and Membrane Coupling

Secretory epithelia are polarized sheets of cells that require distinct sets of apical and basolateral membrane transporters to perform vectorial transport. Nevertheless, the apical and basolateral membranes are functionally coupled by the paracellular pathway, and this shunt for ions and fluid is essential in secretory epithelia, most of which are of moderate leakiness (intermediate conductance). Early evidence for its involvement in salivary gland secretion was provided by Lundberg (1957), although definitive demonstration of the paracellular shunt in epithelia did not come until much later (Frömter and Diamond 1972). The paracellular shunt facilitates secretion by causing hyperpolarization of the apical membrane away from the Cl⁻ equilibrium potential (E_{Cl}) during stimulation of apical Cl⁻ channels. According to the simplest equivalent circuit model for epithelia (Fig. 2):

$$V_a = \frac{E_a(R_b + R_s) + E_b R_a}{R_a + R_b + R_s}. \tag{1}$$

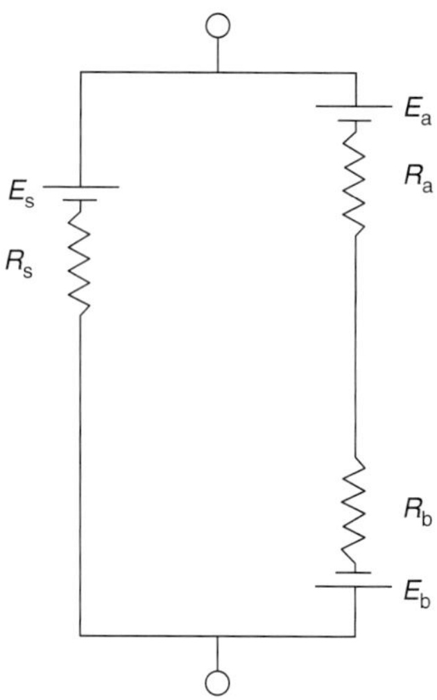

Figure 2. Simplest equivalent circuit that can describe ionic currents across epithelia. Net electromotive forces (EMFs) that result from ion concentration gradients across the apical membrane (E_a), basolateral membrane (E_b), and paracellular "shunt" pathway (E_s) are shown as batteries. Membrane conductances that mediate electrogenic (non-neutral) ion flows driven by those EMFs are depicted as resistors.

In Equation 1, V_a is the apical membrane potential; R_a, R_b, and R_s are the electrical resistances of the apical, basolateral, and shunt (paracellular) pathways, respectively; and E_a and E_b are the electromotive forces at the apical and basolateral membranes. The activation of large numbers of apical CaCC and CFTR Cl⁻ channels by secretagogues dramatically reduces apical membrane resistance. As R_a (and thus $E_b R_a$) falls, the ratio $(R_b + R_s)/(R_a + R_b + R_s)$ approaches 1 if R_s is high, and the apical membrane potential V_a approximates E_a, as in tight epithelia. Essentially, the high Cl⁻ conductance during stimulated secretion would clamp V_a at E_{Cl} ($=E_a$) if R_s were high, thereby eliminating the favorable driving force for Cl exit and secretion. However, when there is a large paracellular shunt and R_s is low compared with R_b, as in moderately leaky secretory epithelia, then V_a is influenced by the loop current, which flows through basolateral K⁺ channels and the shunt pathway, driven by E_K at the basolateral membrane. When this current flows through the apical membrane ($E_b R_a$), it causes hyperpolarization and promotes Cl⁻

Cite this article as *Cold Spring Harb Perspect Med* doi: 10.1101/cshperspect.a009563

secretion through an increase in net driving force ($V_a - E_{Cl}$) (Welsh et al. 1983; Willumsen et al. 1989). It is also obvious from Equation 1 that secretion is inefficient when Ca^{2+} and cAMP stimulate only apical channels, because the impact of the basolateral membrane and its driving force diminishes as apical channels become activated (i.e., when R_a and therefore $E_b R_a$ decrease). Therefore, it is not surprising that in most epithelia, activation of basolateral Ca- and/or cAMP-activated K channels (e.g., KCNN4 and KCNQ1/KCNE) by secretagogues is coordinated with that of apical Cl channels. In summary, secretory epithelia need to be somewhat leaky because the paracellular shunt pathway functionally couples the apical and basolateral membranes and maintains the apical driving force for Cl secretion. Indeed, an increase in the shunt can lead to excessive secretion as in the intestine, where a marked increase in paracellular permeability leads to secretion during T-cell-mediated diarrhea (Clayburgh et al. 2005).

ANION SECRETION BY ALTERNATE APICAL PATHWAYS

Calcium-Activated Chloride Channels

Calcium-activated Cl conductance (CaCC) is a major pathway for Cl transport in nearly all cells, but it was first characterized in detail in a secretory epithelium (Marty et al. 1984). It is activated when intracellular Ca is elevated to micromolar concentrations, and also by membrane depolarization at low cell [Ca], although it becomes almost voltage independent at high [Ca]. The single channel conductance estimated by noise analysis is $1-2$ pS, and the channels have a "weak field strength" selectivity sequence (Eisenman 1962; Wright and Diamond 1977; Evans and Marty 1986). Ca-activated Cl^- conductance was proposed to be the main apical Cl pathway during muscarinic stimulation of lacrimal glands (Marty et al. 1984), but this pathway has been identified in virtually all gland epithelial cells, where it is stimulated by a wide range of Ca-mobilizing agonists.

The search for the channel protein that mediates CaCC has uncovered many interesting candidates during the past 25 years. One of the early possibilities, human CLCA1, now appears to be a secreted protein rather than a Cl channel (Gibson et al. 2005), although it is nevertheless reported to modulate the conductance of endogenous Ca-activated Cl^- channels (Hamann et al. 2009). The putative high-conductance Ca-activated Cl channel tweety (hTTYH3) remains a candidate channel, although little is known regarding its physiological role (Suzuki and Mizuno 2004). A member of the ClC family, ClC3, has also been proposed to mediate CaCC but is predominantly intracellular and may function as a Cl/H exchanger. Human bestrophin 1, which is mutated in a juvenile form of macular degeneration called Best vitelliform macular dystrophy ("Best disease"), forms Cl- and HCO_3-permeable channels in the plasma membrane, but they have a biophysical signature that is distinct from that of epithelial CaCC (Hartzell et al. 2008).

The most compelling candidate to mediate CaCC is TMEM16A, which was identified almost simultaneously by three independent groups using different approaches: expression cloning (Schroeder et al. 2008), bioinformatics (Yang et al. 2008), and differential gene expression (Caputo et al. 2008). The channel has also been named "Anoctamin-1" because it is predicted to have eight transmembrane segments. The anion selectivity and pharmacology of TMEM16A are consistent with those of CaCC, and it is expressed in exocrine glands and, indeed, all epithelia that exhibit CaCC currents, including airways and renal tubules. Silencing its expression in cell cultures and tissues reduces CaCC activity (Yang et al. 2008), and knockout mice are deficient in CaCC activity in salivary glands and other tissues (Ousingsawat et al. 2009; Ferrera et al. 2010; Romanenko et al. 2010). Nevertheless, other channels may contribute to CaCC in airway and intestinal epithelia because basal secretion and up to 40% of the UTP-stimulated CaCC activity persists in TMEM16A knockout mice (Rock et al. 2009), and only a transiently activated CaCC current appears sensitive to RNAi knockdown and pharmacological inhibitors of TMEM16A (Namkung et al. 2011). Other CaCC channels might be expected under control conditions, because TMEM16A is normally expressed at low levels and strongly induced by proinflammatory cytokines (Caputo et al. 2008). This also raises the exciting prospect that more CaCCs remain to be discovered.

SLC26A Transporters and Channels

The contribution of the SLC26A family of anion transporters to HCO_3 secretion in several epithelia was established with the recognition that congenital chloride-losing diarrhea (CLD) was linked to mutations in the Cl/HCO_3 exchanger encoded by SLC26A3 (Hoglund et al. 1996). CLD patients exhibit diarrhea characterized by high Cl concentrations and metabolic alkalosis suggestive of defective Cl/HCO_3 exchange, and mice lacking Scl26a3 recapitulate this phenotype (Schweinfest et al. 2006). Some CLD-causing mutations lead to misfolding of the transporter and disrupt its apical trafficking (Dorwart et al. 2008a). Ten members of this family have been separated into three groups, based on their transport properties (Dorwart et al. 2008b). Group 1 contains the sulfate transporters A1 and A2; Group 2 the Cl/HCO_3 exchangers A3, A4, and A6; and Group 3, represented by A7 and A9, exhibits anion channel behavior. SLC26A5 is expressed in outer hair cells of the cochlea, and A8 is present in male germ cells. Many SCL26 transporter members have been associated with human diseases (Mount and Romero 2004).

In regard to epithelial HCO_3 secretion, differences in anion transport stoichiometry among Group 2 family members have been proposed to account for the ability of the pancreatic duct epithelium to generate a final $NaHCO_3$ concentration that is isotonic to the plasma. Specifically, A6, with a stoichiometry of $1Cl/2HCO_3$, appears to be required for the generation of high luminal HCO_3 concentrations as Cl concentrations fall along the pancreatic duct (Ko et al. 2004; Steward et al. 2005). Mice lacking Slc26a6 show impaired pancreatic duct and intestinal Cl/HCO3 exchange and impaired pancreatic fluid secretion (Wang et al. 2006; Ishiguro et al. 2007; Seidler et al. 2008). SLC26A6 may be responsible for generating other HCO_3-rich se-

cretory products, as observed in the salivary gland (Shcheynikov et al. 2008). The ability of intestinal A6 to transport oxalate is likely involved in protecting against renal stone formation (Aronson 2010).

The Group 3 anion channels include SLC26A7, which is reported to be relatively selective for Cl and to conduct Cl in a pH-dependent manner (Kim et al. 2005). However, other transport modes involving coupled HCO_3 exchange have also been proposed (Petrovic et al. 2003, 2004; Kim et al. 2005). A7 has been found in the basolateral membranes of gastric parietal cells, where it may contribute to Cl entry from the interstitium; it is found in several renal nephron segments. A7 may be located at the apical or basolateral membranes in various epithelia, and the factors that determine its targeting are not known. SLC26A9 is reported to lie at the apical membranes of ciliated airway epithelial cells and alveolar cells, and it is present also in the gastric mucosa (Xu et al. 2008). In primary cultures of human bronchial epithelium (HBE), SLC26A9 appears to account for the basal secretory current that remains after amiloride inhibition of Na absorption, and it contributes also to anion secretion in response to cAMP/PKA-mediated stimulation (Bertrand et al. 2009). Interestingly, A9 currents were inhibited by the CFTR blocker GlyH-101, but they showed less voltage-dependent block than CFTR, allowing these current components to be distinguished. As observed for other SLC26 transporters, the cAMP/PKA-stimulated currents observed during coexpression of A9 with CFTR exceeded those observed for either channel alone, indicative of a mutual stimulatory interaction (see below). Interestingly, A9 currents were absent from HBE or HEK cells expressing $\Delta F508$ CFTR, which may be consistent with the strong interaction between A9 and CFTR found in coimmunoprecipitation experiments.

Of great interest for the CF field is the regulation of SLC26A-mediated transport, because several members of this family have been linked functionally to the activation of CFTR (Ko et al. 2004; Dorwart et al. 2008b). As with CFTR, several SLC26A family members possess PDZ-domain-interacting motifs at their carboxyl termini. This motif allows them to associate with CFTR via scaffolding proteins such as NHERF 1 and 2 and CAP70 in macromolecular complexes that are likely to include cytoskeleton-interacting proteins (e.g., ezrin) (Short et al. 1998) and links to regulatory proteins such as PKA (Sun et al. 2000a,b). The intriguing studies of Ko et al. (2004) suggest that mutual stimulatory interactions between SLC26A3, A6, and CFTR involve regulated association between subdomains of these proteins, the R region of CFTR, and the carboxy-terminal STAS domain of A6, allowing their stimulatory reciprocity to be enhanced by PKA phosphorylation of the R region of CFTR (Ko et al. 2004). The interaction between STAS and the R region could influence CFTR gating by altering the dynamic interaction of the R region with NBD subdomains, identified by NMR (Baker et al. 2007; Kanelis et al. 2010). Thus, interactions between CFTR and different SLC26 transporters in different epithelia might explain how PKA-induced stimulation of CFTR can give rise to tissue-specific secretory products, as well as the diverse transport phenotypes observed when CFTR is absent or impaired in CF.

ROLE OF CFTR IN THE SECRETION OF OTHER ANIONS

Another important aspect of the physiology of anion secretion mediated by CFTR concerns the lack of strict anion selectivity that CFTR and other anion channels display, contrasting with their highly selective cation channel counterparts. In recent years, it has been recognized that two physiologically significant anions also permeate the CFTR channel—glutathione and thiocyanate.

Glutathione

CFTR is permeable to glutathione, and other ABC transporters have been shown also to transport glutathione (GSH), a major antioxidant that is present in cells at millimolar concentrations (Linsdell et al. 1997). GSH was 55% lower in airway surface liquid of CFTR-deficient cultures. The presence of airway bacteria has been shown to generate increases in airway surface liquid GSH, and in CFTR knockout mice, failure to secrete GSH has been linked to airway inflammation and oxidative stress (Day et al. 2004). Accordingly, in CF, deficient glutathione transport via CFTR has the potential to influence the redox state of both the luminal and cellular compartments of epithelial cells and lead to oxidative stress. These disturbances to the redox balance can evoke inflammatory signaling pathways and could be linked to apoptosis (Rahman and MacNee 1998; Haddad 2002; Rottner et al. 2009). Thus, altered glutathione levels can contribute to NF-κB activation and to the inflammation evoked by bacterial colonization.

Thiocyanate

Airway epithelia secrete thiocyanate anion (SCN) via CFTR, as a precursor for the production of the potent antibacterial compound hypothiocyanate anion (OSCN) (Conner et al. 2007). Early studies of CFTR's anion selectivity showed that SCN permeability exceeded that to Cl on a molar basis (Linsdell et al. 1997). Airway cells generate OSCN from H_2O_2 using membrane-tethered dual oxidases (Duox1 and 2), and H_2O_2 is metabolized by secreted lactoperoxidase to oxidize secreted SCN to the OSCN anion. SCN reaches concentrations of $\sim$0.5 mM in the ASL, which is at least $\sim$30 times its concentration in the serum. In vitro studies have implicated the Na-dependent iodide cotransporter, which is highly expressed in thyroid gland cells, in the accumulation of SCN across the basolateral membranes of airway cells for subsequent exit through CFTR. Accordingly, in CF, the loss of SCN transport would impair OSCN-dependent bacterial killing and contribute to bacterial colonization and disease pathogenesis.

It is interesting that therapy with hypertonic saline in CF (Elkins et al. 2006), which is thought to improve the hydration of the airway surface, has been reported to produce improvements in ASL levels of both SCN and GSH. Nebulization of hypertonic saline in mice increased ASL SCN and GSH, and these increases were smaller in CFTR knockout animals (Gould et al. 2010).

FLUID SECRETION LINKED TO SALT TRANSPORT

Which Transporters Drive Fluid Secretion and Mediate Water Flux?

Fluid secretion by airway submucosal gland acinar cells is driven by active Cl^- transport, whereas HCO_3^- net flux drives fluid transport across other epithelia, such as pancreatic ducts and regions of the GI tract (e.g., duodenum). The well-studied model human airway cell line Calu-3 does not transport Cl^- actively at detectable rates under short-circuit conditions (Devor et al. 1999); nevertheless, recent studies indicate that a small net Cl^- flux generates the osmotic driving force for fluid transport by Calu-3 cells under thin film conditions (Shan et al. 2011). This Cl^- transport requires basolateral anion exchange, and a similar mechanism may occur in salivary and airway glands and other tissues where a large component of the fluid secretion is insensitive to bumetanide. This HCO_3^--dependent Cl^- transport contrasts with the duodenum and especially the pancreatic duct, where fluid secretion is driven by HCO_3^- itself, and as discussed above, the luminal $NaHCO_3$ in the secretions reaches very high concentrations, for example, 140 mM (Steward et al. 2005). Water follows the net transport of solutes through both paracellular and trans-cellular pathways in secretory epithelia. In human airways, the trans-cellular pathway consists of apical AQP5 and predominantly basolateral AQP3 and AQP4 water channels (Kreda et al. 2001). Interestingly, the infection of mouse airways with adenovirus leads to reduced expression of AQP1 and AQP3 in endothelial and epithelial cells, respectively, which may contribute to pulmonary edema (Towne et al. 2000).

REGULATION

Evidence for Local Activation by cAMP/PKA

CFTR is hypothesized to be part of a signaling complex that includes scaffolds, adaptors, and many regulatory enzymes (kinases, phosphatases, cyclases, phosphodiesterases, cAMP exporters, etc.). Interactions with these proteins enables CFTR channel regulation to be spatially and temporally restricted (Huang et al. 2001). PKA is localized near CFTR by an A-kinase anchoring protein (AKAP) (Gray et al. 2000), probably ezrin, because it coimmunoprecipitates with CFTR, binds to a regulatory subunit of PKA, and is expressed at the apical membranes of Calu-3 and T84 cells (Sun et al. 2000b). CFTR channels can be activated by adding cAMP to excised membrane patches; therefore, the entire PKA holoenzyme must be tethered (Huang et al. 2000). This was also shown biochemically by coimmunoprecipitation of the catalytic and regulatory subunits of PKA with CFTR (Sun et al. 2000a,b) and by showing that endogenous PKA activity in the immunoprecipitates was sensitive to a peptide that disrupts AKAP interactions. Recent studies have shown that phosphodiesterase type 3A (PDE3A) is linked functionally to CFTR activation and that its inhibition produces a localized increase in cAMP, which also enhances the physical interaction of PDE3A with CFTR and augments CFTR activity (Penmatsa et al. 2010). This interaction depended on the cytoskeleton, and its disruption blocked compartmentalized cAMP signaling. PKC-ε and Src are also tethered in the putative CFTR complex by a protein called RACK1 (regulator of activated C kinase) (Yarwood et al. 1999; Chang et al. 2001).

Local Activation by cGMP/PKG

In intestinal epithelium, the cGMP-dependent kinase, type II (French et al. 1995), regulates CFTR activity independent of PKA and in response to luminal peptide hormones, the guanylins, and the diarrhea-producing, heat-stable enterotoxin, as shown by inhibitor and gene knockout studies (Vaandrager et al. 1998). The type II PKG is amino-terminally myristolated (Vaandrager et al. 1996) and is attached through this lipid anchor to the apical membranes of intestinal epithelial cells, where it is in a privileged position to elicit CFTR phosphorylation. Indeed, the kinase and CFTR coimmunoprecipitate. The association of PDE3A with CFTR, noted above, has implications not only for

 Cite this article as *Cold Spring Harb Perspect Med* doi: 10.1101/cshperspect.a009563

cAMP/PKA signaling but also for localized cGMP/PKG signaling. Because type 3A PDE degrades both cAMP and cGMP, increases in cGMP will block its degradation of cAMP and permit PKA-dependent CFTR activation in addition to that produced by the cGMP kinase.

Inactivation of CFTR

CFTR is deactivated by association with the phosphatases PP2C and PP2A (Zhu et al. 1999; Thelin et al. 2005). The cAMP signal is kept local and transient by the action of phosphodiesterase PDE4D, which interacts with both NHERF1 (see below) and RACK1 (Barnes et al. 2005). In the intestine, cAMP is also regulated locally by MRP4, a cAMP exporter that associates with CFTR (Li et al. 2007a). CFTR activity is downregulated during metabolic stress by the $\alpha 1$ and $\alpha 2$ catalytic subunits of AMP-activated protein kinase (AMPK), which binds to CFTR at its carboxyl terminus (Hallows et al. 2000). AMPK inhibits CFTR by phosphorylating an inhibitory site at position 768 within the R domain of CFTR. CFTR is also inhibited by binding of the t-SNARE syntaxin 1A, which disrupts the association of the amino terminus with the R domain (Naren et al. 2000). MUNC18 relieves this inhibition by binding to syntaxin 1A (Naren and Kirk 2000), although there is evidence that syntaxin 1A also inhibits CFTR trafficking to the membrane (Peters et al. 1999). The β-adrenergic receptor forms a macromolecular complex with CFTR in airway epithelial cells (Naren et al. 2003), and similar results have been obtained in mouse intestine (U Seidler, pers. comm.).

PDZ and WD-Domain Proteins Hold Things Together

At least two types of scaffold proteins have been identified that may help assemble CFTR regulatory complexes—PDZ domain proteins and WD-domain proteins. PDZ (postsynaptic density 95, Discs large, ZO-1) domain proteins bind to the carboxyl terminus of CFTR and help to stabilize it at the plasma membrane. Mutating threonine or leucine in the [1477]DTRL motif inhibits PDZ binding and leads to par-

tial redistribution of CFTR from the apical to the lateral plasma membrane (Short et al. 1998; Wang et al. 1998). However, removing the carboxyl terminus does not completely eliminate CFTR binding to PDZ domains because of the presence of an internal six-amino-acid binding site upstream in the carboxyl terminus of CFTR (LaRusch 2007). Sodium hydrogen exchange regulatory factor (NHERF1) is one such PDZ-domain protein that binds CFTR. It has two PDZ domains that can both bind CFTR with different affinities (see below). The carboxyl terminus of NHERF1 acts as a switch that enables PDZ2 to bind CFTR when the carboxyl terminus is phosphorylated by PKC or interacts with ezrin (Li et al. 2007b). PDZ2 also binds YAP65, which recruits the Src family tyrosine kinase p62[c-yes] to the apical membrane (Mohler et al. 1999), although the functional role of this interaction is not yet known.

A related PDZ-domain protein, NHERF2 (also called E3KARP), is expressed at the apical membrane of epithelia and can be immunoprecipitated with CFTR (Sun et al. 2000), as can CAP70 (NHERF3), a protein with four PDZ domains that was identified by affinity purification using the C-tail of CFTR (Wang et al. 2000). Adding CAP70 increases CFTR channel activity at low concentrations and decreases it at high concentrations, and similar results have been reported with fragments of NHERF1 (Raghuram et al. 2001). Another PDZ-domain protein that forms complexes with CFTR is CAL (CFTR-associated ligand), which has a single PDZ domain and two coiled-coil domains. CAL interacts with CFTR in the trans-Golgi network (TGN) and inhibits its surface expression (Cheng et al. 2010). The PDZ-domain protein Shank2E binds to the carboxyl terminus of CFTR and functions as a negative regulator. It has ankyrin repeats and is concentrated in actin complexes at the apical membrane of epithelial cells (McWilliams et al. 2008), where it may recruit the cyclic nucleotide phosphodiesterase PDE4D (Lee et al. 2008) and phospholipase C-$\beta 3$.

The best studied WD domain protein is RACK1 (receptor for activated C kinase). It has a β-propeller structure and sequence homology with the β subunit of heterotrimeric G-proteins, an ancient family of regulatory proteins with repeating units that usually end with Trp-Asp (WD). Based mainly on coimmunoprecipitation results, RACK1 can interact with NHERF1 (Liedtke and Wang 2006), PKC-ε (Liedtke et al. 2002), PDE4D, and Src (Chang et al. 2001; Cox et al. 2003).

It has been proposed that positive regulators such as NHERF1 and negative regulators such as CAL and Shank2 compete for CFTR (Kim et al. 2004). The CFTR affinities of NHERF1 (K_d = 200 and 1000 nM) and NHERF2 (K_d = 323 and 232 nM) are much higher than that of CAL (K_d > 600 μM), which may explain why CFTR normally traffics toward the plasma membrane rather than to the lysosome (Cushing et al. 2008). Finally, there is evidence for the interaction of CFTR with the actin cytoskeleton (Cantiello 2001), and the amino terminus of CFTR can also associate with the actin cytoskeleton through filamin (Thelin et al. 2007).

Cross Talk between Signaling Pathways

Although the apparent stimulation of CFTR by Ca-mobilizing agonists and synergistic regulation by signaling pathways has caused confusion, it is now clear that many observations can be explained by localized signaling domains near the plasma membrane and overlap between cAMP and Ca signaling. For example, ATP-induced Ca release from stores enhances store-operated Ca entry, and the subsequent local activation of Ca-stimulated adenylyl cyclase 1 or 8 and elevation of cAMP (Martin et al. 2009; Namkung et al. 2010). Even without triggering influx of extracellular Ca, release from stores can elevate cAMP through a store-operated cAMP signaling mechanism in which STIM1 activates adenylyl cyclases (Lefkimmiatis et al. 2009). These pathways also intersect at the level of Ca stores, where cAMP enhances IP3 receptor activation, thereby contributing to synergistic regulation of secretion by these two second messengers (Lee and Foskett 2010).

CF AFFECTS EPITHELIUM-SPECIFIC ANION SECRETORY FUNCTIONS

The impact of CFTR-mediated anion secretion in different organ systems is often modified or augmented by transporters/channels that lie in parallel with CFTR at the apical membrane, as discussed above. These pathways may either support its secretory function by adding to CFTR-mediated anion secretion or altering secretory composition (see alternate anion pathways, above), or they may oppose or balance net salt and water transport using an oppositely directed transport process, for example, Na absorption across airway epithelia. Organ-specific functions also arise from modification of the primary secretory product of exocrine gland acini by downstream, ductal transport events to provide a secretory product of volume and composition appropriate to the physiology of that tissue/organ. More detailed treatments of the functions of specific organ systems can be found in other chapters in this collection.

To generalize, CFTR-mediated anion secretion in some organs maintains the volume (and liquidity) of the luminal compartment and its contents, for example, airway, intestines. In the exocrine organs, for example, pancreas and salivary gland, CFTR-dependent anion secretion provides the vehicle for clearing secreted products like macromolecules to another location. It would be argued that anion secretion serves both functions in the surface and submucosal gland epithelia that line the airway. In contrast to these organs, the sweat gland is spared from similar pathologic consequences in CF, presumably because the secretory coil in humans is controlled primarily by CFTR-independent, cholinergic pathways and Ca-dependent channels.

Airway Epithelia

Anion secretion in the airways contributes to the formation and maintenance of a thin layer of liquid, the airway surface liquid (ASL), which consists of a lower-viscosity periciliary liquid (PCL) layer (referred to as "sol") adjacent to the apical membrane, in which the cilia beat, and a higher-viscosity gel layer where secreted mucins trap inhaled microorganisms and debris. Airway anion secretion is critically important in maintaining the volume and composition of the PCL. Opposing Na absorptive (ENaC) and Cl secretory (CFTR and CACC) transport processes are important in determining the depth of the PCL on which the efficiency of mucociliary clearance depends (see Fig. 3) (Matsui et al. 1998; Tarran 2004). Anion secretory processes are expressed in the ciliated surface airway epithelium, but a significant volume of liquid (and defense molecules) is generated also by the submucosal glands present in the more proximal airway. In the upper airways, gland secretion may contribute most of the secreted liquid, but adjustments performed by the surface epithelium via localized control mechanisms regulate an optimal PCL depth. It is clear, particularly from studies using primary cultures of upper airway cells, that the surface cells have the capacity to regulate PCL depth in the 6- to 10-μm range (Tarran 2004). The factors influencing ASL properties in the lower airways are much less clear.

In CF, the absence or defective function of CFTR at the apical membranes compromises the formation of a sufficient PCL, and ENaC-mediated Na absorption is enhanced. These defects result in the near absence of the low-viscosity fluid layer on which effective mucus clearance depends. This leads to the accumulation of mucus and bacteria, particularly *Pseudomonas aeruginosa* and other opportunistic infections that colonize airway surfaces and obstruct smaller airways and gland ducts caused by the loss of vehicle. This situation is further compromised by the loss of the innate immune substances present in their secretions.

Intestines

The secretion of salt and water by the small and large intestines is important for maintaining a semiliquid luminal environment that supports enzyme activities, the absorption of nutrients, and for optimal clearance of the luminal contents. The proximal small intestine is primarily involved in salt absorption and bicarbonate secretion, the latter to buffer gastric acid. Here, coupled Na/H and Cl/HCO3 exchangers account primarily for salt and water handling. In the more distal small intestine, CFTR is expressed primarily in the crypt cells as determined by in situ hybridization, immunofluorescence, and functional measurements,

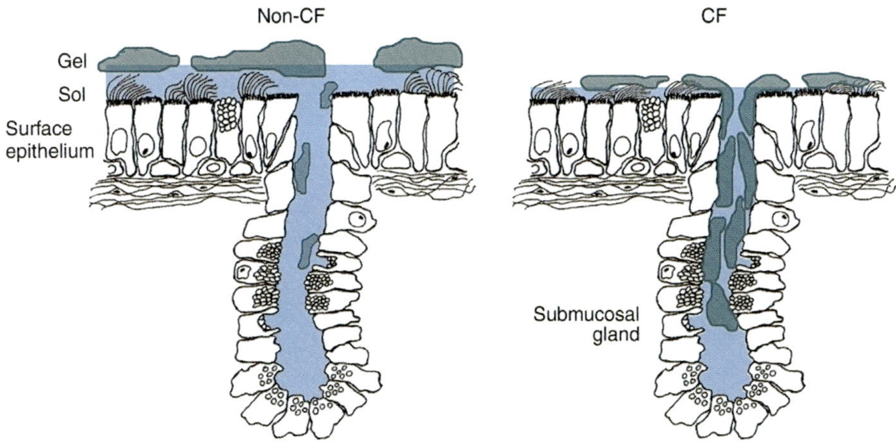

Figure 3. Cartoon illustrating the impact of impaired airway secretion on mucociliary clearance. Contributions to the volume and composition of the airway surface liquid (ASL) emerge from the surface epithelium and submucosal glands. The surface epithelium of non-CF airways adjusts the depth of the periciliary liquid layer (PCL) to a level that enables ciliary beating and effective mucociliary clearance. In CF, the absence of sufficient CFTR function compromises salt and water secretion, PCL regulation, and mucus clearance from the surface epithelium and gland ducts. Mucus accumulation leads to infection, inflammation, and bronchiectasis.

although high CFTR-expressing cells are evident on the villi (Ameen et al. 1995). In the large intestine, CFTR is localized to the crypt epithelium. The model of Figure 1 applies to CFTR-mediated anion secretion in the distal small intestine and colon, and the counterion Na moves through the paracellular pathway, as shown, in response to the favorable trans-epithelial lumen-negative voltage, as in airway cells.

In CF, insufficient anion secretion may lead to blockage of the intestinal lumen at birth by meconium as an early clinical symptom of the disease. Later in life, patients may experience chronic constipation, reflecting the continued lack of CFTR-mediated secretion. Mice without CFTR or expressing ΔF508 CFTR exhibit a more severe intestinal phenotype than CF patients, which is confirmed by compromised cAMP/PKA stimulation of Cl transport in trans-epithelial measurements in vitro (Clarke et al. 1992; Grubb and Gabriel 1997). On the other side of the coin, CFTR plays a key role in the amplified response observed in secretory diarrheas, such as those elicited by cholera toxin during infection with *Vibrio cholerae* or by heat-stable enterotoxin during infection with pathogenic *Escherichia coli* (Chao et al. 1994; Gabriel et al. 1994). Small-molecule blockers of CFTR may prove valuable in treating these conditions (e.g., Thiagarajah et al. 2004), because these toxins are ineffective in CF mice (Grubb and Gabriel 1997).

Pancreas and Salivary Gland

CFTR is expressed in the duct cells of the exocrine pancreas and salivary glands by in situ hybridization and antibody labeling (Marino et al. 1991; Trezise and Buchwald 1991). Evidence suggests also that CFTR is present in the salivary acini and in the intralobular cells of the pancreas, which extend into the acinar lumen. In these acinar regions, the expression level of CFTR appears to be lower than the corresponding ductal levels. The involvement of CFTR in formation of the primary secretory product by acinar or intralobular cells is supported by the actions of cAMP/PKA-dependent secretogogues, for example, secretin and adrenergic agonists. In the pancreatic duct, CFTR couples functionally to the anion exchangers SLC26A3 and SLC26A6 to generate NaHCO$_3$ secretion for alkalinizing the duodenal lumen (discussed under alternative anion transport, above). Together with the primary acinar secretion, these isotonic secretory products provide the vehicle for the delivery of pancreatic digestive enzymes to the duodenal lumen, and, physiologically, this process is similar for salivary clearance of mucins and enzymes (Zeng et al. 1997).

In CF, most but not all CFTR mutations produce pancreatic insufficiency, making pancreatic function a biomarker for disease severity (Wilschanski and Durie 2007). Patients could be grouped as pancreatic sufficient (PS) or insufficient (PI), based on whether they required replacement enzyme therapy, even before the CF gene was identified. Both PS and PI patients exhibit reduced fluid and anion secretion, with the secretory defect more pronounced in PI. Nevertheless, PS patients are prone to recurrent pancreatitis (see above). Failure of anion secretion results in duct or gland blockage by precipitated macromolecules, and the subsequent leakage of proteolytic enzymes, which destroys the exocrine parenchyma. Damage to the pancreas begins even in utero, when ductal obstruction leads to luminal dilation, atrophy, and fibrosis, and thus, cystic fibrosis of the pancreas. As patients age, pancreatic islets may also be lost with the development of CF-related diabetes.

CONCLUDING THOUGHTS

The loss of CFTR's ion channel function at the apical membranes of secretory epithelial cells leads to defects that are now conceptually understandable based on modifications of, or additions to, the basic model for secondary active Cl secretion that emerged in the 1970s (Fig. 1). Disease pathology within the affected organ systems depends on the contribution of CFTR function to salt and water secretion, relative to the activity of other pathways. Its impact in different organ systems relates also to the locus of CFTR expression within the tissue and the developmental consequences of deficient CFTR. For example, it is estimated that as little as

20% impairment in Cl secretion is sufficient to produce obstructive degeneration of the vas deferens (Marcorelles et al. 2011), making its development particularly sensitive to CFTR expression. In the pancreas, $\sim$20% of normal CFTR function is required for pancreatic-sufficient CF disease. In the sweat gland, a 50% impairment of CFTR activity is not detectable as a change in sweat Cl, and relatively little CFTR activity produces a significant drop in sweat Cl. A broad variation in CFTR function is found amidst the more than 1700 reported mutations and polymorphisms in CFTR, and these can provide an indication of the relative amount of CFTR function needed for therapeutic benefit. It becomes increasingly clear that the core defect in CF disease is anion transport, making CFTR the prime target for therapeutic development.

ACKNOWLEDGMENTS

The work of the authors is supported by grants from the NIH-NIDDK and from the Cystic Fibrosis Foundation to R.A.F.; and from Cystic Fibrosis Canada and the CIHR to J.W.H. We also thank Natalie Schweninger and Xiaoyan Gong for help with the manuscript.

REFERENCES

Ameen NA, Ardito T, Kashgarian M, Marino CR. 1995. A unique subset of rat and human intestinal villus cells express the cystic fibrosis transmembrane conductance regulator. *Gastroenterology* **108:** 1016–1023.

Aronson PS. 2010. Role of SLC26A6-mediated Cl-oxalate exchange in renal physiology and pathophysiology. *J Nephrol* **23**(Suppl 16): S158–S164.

Baker JM, Hudson RP, Kanelis V, Choy WY, Thibodeau PH, Thomas PJ, Forman-Kay JD. 2007. CFTR regulatory region interacts with NBD1 predominantly via multiple transient helices. *Nat Struct Mol Biol* **14:** 738–745.

Barnes AP, Livera G, Huang P, Sun C, O'Neal WK, Conti M, Stutts MJ, Milgram SL. 2005. Phosphodiesterase 4D forms a cAMP diffusion barrier at the apical membrane of the airway epithelium. *J Biol Chem* **280:** 7997–8003.

Barrett KE, Keely SJ. 2000. Chloride secretion by the intestinal epithelium: Molecular basis and regulatory aspects. *Annu Rev Physiol* **62:** 535–572.

Bertrand CA, Zhang R, Pilewski JM, Frizzell RA. 2009. SLC26A9 is a constitutively active, CFTR-regulated anion conductance in human bronchial epithelia. *J Gen Physiol* **133:** 421–438.

Cantiello HF. 2001. Role of actin filament organization in CFTR activation. *Pflugers Arch* **443** (Suppl 1): S75–S80.

Caputo A, Caci E, Ferrera L, Pedemonte N, Barsanti C, Sondo E, Pfeffer U, Ravazzolo R, Zegarra-Moran O, Galietta LJ. 2008. TMEM16A, a membrane protein associated with calcium-dependent chloride channel activity. *Science* **322:** 590–594.

Chang BY, Chiang M, Cartwright CA. 2001. The interaction of Src and RACK1 is enhanced by activation of protein kinase C and tyrosine phosphorylation of RACK1. *J Biol Chem* **276:** 20346–20356.

Chao AC, de Sauvage FJ, Dong YJ, Wagner JA, Goeddel DV, Gardner P. 1994. Activation of intestinal CFTR Cl⁻ channel by heat-stable enterotoxin and guanylin via cAMP-dependent protein kinase. *EMBO J* **13:** 1065–1072.

Chen JH, Stoltz DA, Karp PH, Ernst SE, Pezzulo AA, Moninger TO, Rector MV, Reznikov LR, Launspach JL, Chaloner K, et al. 2010. Loss of anion transport without increased sodium absorption characterizes newborn porcine cystic fibrosis airway epithelia. *Cell* **143:** 911–923.

Cheng J, Moyer BD, Milewski M, Loffing J, Ikeda M, Mickle JE, Cutting GR, Li M, Stanton BA, Guggino WB. 2002. A Golgi-associated PDZ domain protein modulates cystic fibrosis transmembrane regulator plasma membrane expression. *J Biol Chem* **227:** 3520–3529.

Cheng J, Cebotaru V, Cebotaru L, Guggino WB. 2010. Syntaxin 6 and CAL mediate the degradation of the cystic fibrosis transmembrane conductance regulator. *Mol Biol Cell* **21:** 1178–1187.

Clarke LL, Grubb BR, Gabriel SE, Smithies O, Koller BH, Boucher RC. 1992. Defective epithelial chloride transport in a gene-targeted mouse model of cystic fibrosis. *Science* **257:** 1125–1128.

Clayburgh DR, Barrett TA, Tang Y, Meddings JB, Van Eldik LJ, Watterson DM, Clarke LL, Mrsny RJ, Turner JR. 2005. Epithelial myosin light chain kinase-dependent barrier dysfunction mediates T cell activation-induced diarrhea in vivo. *J Clin Invest* **115:** 2702–2715.

Conner GE, Wijkstrom-Frei C, Randell SH, Fernandez VE, Salathe M. 2007. The lactoperoxidase system links anion transport to host defense in cystic fibrosis. *FEBS Lett* **581:** 271–278.

Cox EA, Bennin D, Doan AT, O'Toole T, Huttenlocher A. 2003. RACK1 regulates integrin-mediated adhesion, protrusion, and chemotactic cell migration via its Src binding site. *Mol Biol Cell* **14:** 658–669.

Cushing PR, Fellows A, Villone D, Boisguérin P, Madden DR. 2008. The relative binding affinities of PDZ partners for CFTR: A biochemical basis for efficient endocytic recycling. *Biochemistry* **47:** 10084–10098.

Cuthbert AW, Supuran CT, MacVinish LJ. 2003. Bicarbonate-dependent chloride secretion in Calu-3 epithelia in response to 7,8-benzoquinoline. *J Physiol* **551:** 79–92.

Day BJ, van Heeckeren AM, Min E, Velsor LW. 2004. Role for cystic fibrosis transmembrane conductance regulator protein in a glutathione response to bronchopulmonary pseudomonas infection. *Infect Immun* **72:** 2045–2051.

Devor DC, Frizzell RA. 1998. Modulation of K⁺ channels by arachidonic acid in T84 cells. I. Inhibition of the Ca²⁺-dependent K⁺ channel. *Am J Physiol* **274:** C138–C148.

Devor DC, Singh AK, Lambert LC, DeLuca A, Frizzell RA, Bridges RJ. 1999. Bicarbonate and chloride secretion in

Calu-3 human airway epithelial cells. *J Gen Physiol* **113**: 743–760.

Dorwart MR, Shcheynikov N, Baker JM, Forman-Kay JD, Muallem S, Thomas PJ. 2008a. Congenital chloride-losing diarrhea causing mutations in the STAS domain result in misfolding and mistrafficking of SLC26A3. *J Biol Chem* **283**: 8711–8722.

Dorwart MR, Shcheynikov N, Yang D, Muallem S. 2008b. The solute carrier 26 family of proteins in epithelial ion transport. *Physiology (Bethesda)* **23**: 104–114.

Eisenman G. 1962. Cation selective glass electrodes and their mode of operation. *Biophys J* **2**: 259–323.

Elkins MR, Robinson M, Rose BR, Harbour C, Moriarty CP, Marks GB, Belousova EG, Xuan W, Bye PT. 2006. A controlled trial of long-term inhaled hypertonic saline in patients with cystic fibrosis. *N Engl J Med* **354**: 229–240.

Evans MG, Marty A. 1986. Calcium-dependent chloride currents in isolated cells from rat lacrimal glands. *J Physiol* **378**: 437–460.

Ferrera L, Caputo A, Galietta LJ. 2010. TMEM16A protein: A new identity for Ca^{2+}-dependent Cl^- channels. *Physiology (Bethesda)* **25**: 357–363.

French PJ, Bijman J, Edixhoven M, Vaandrager AB, Scholte BJ, Lohmann SM, Nairn AC, de Jonge HR. 1995. Isotype-specific activation of cystic fibrosis transmembrane conductance regulator-chloride channels by cGMP-dependent protein kinase II. *J Biol Chem* **270**: 26626–26631.

Frizzell RA, Field M, Schultz SG. 1979. Sodium-coupled chloride transport by epithelial tissues. *Am J Physiol* **236**: F1–F8.

Frömter E, Diamond J. 1972. Route of passive ion permeation in epithelia. *Nat New Biol* **235**: 9–13.

Gabriel SE, Brigman KN, Koller BH, Boucher RC, Stutts MJ. 1994. Cystic fibrosis heterozygote resistance to cholera toxin in the cystic fibrosis mouse model. *Science* **266**: 107–109.

Gawenis LR, Ledoussal C, Judd LM, Prasad V, Alper SL, Stuart-Tilley A, Woo AL, Grisham C, Sanford LP, Doetschman T, et al. 2004. Mice with a targeted disruption of the AE2$^{Cl^-/HCO_3^-}$ exchanger are achlorhydric. *J Biol Chem* **279**: 30531–30539.

Gawenis LR, Bradford EM, Alper SL, Prasad V, Schull GE. 2010. AE2$^{Cl^-/HCO_3^-}$ exchanger is required for normal cAMP-stimulated anion secretion in murine proximal colon. *Am J Physiol Gastrointest Liver Physiol* **298**: G493–G503.

Gerlach U, Brendel J, Lang HJ, Paulus EF, Weidmann K, Bruggemann A, Busch AE, Suessbrich H, Bleich M, Greger R. 2001. Synthesis and activity of novel and selective I_{Ks}-channel blockers. *J Med Chem* **44**: 3831–3837.

Gibson A, Lewis AP, Affleck K, Aitken AJ, Meldrum E, Thompson N. 2005. hCLCA1 and mCLCL3 are secreted non-integral membrane proteins and therefore are not ion channels. *J Biol Chem* **280**: 27205–27212.

Gould NS, Gauthier S, Kariya CT, Min E, Huang J, Brian DJ. 2010. Hypertonic saline increases lung epithelial lining fluid glutathione and thiocyanate: Two protective CFTR-dependent thiols against oxidative injury. *Respir Res* **11**: 119.

Gray PC, Scott JD, Catterall WA. 2000. Regulation of ion channels by cAMP-dependent protein kinase and A-kinase anchoring proteins. *Curr Opin Neurobiol* **8**: 330–334.

Grubb BR, Gabriel SE. 1997. Intestinal physiology and pathology in gene-targeted mouse models of cystic fibrosis. *Am J Physiol* **273**: G258–G266.

Haddad JJ. 2002. Redox regulation of pro-inflammatory cytokines and IκB-α/NF-κB nuclear translocation and activation. *Biochem Biophys Res Commun* **296**: 847–856.

Hallows KR, Raghuram V, Kemp BE, Witters LA, Foskett JK. 2000. Inhibition of cystic fibrosis transmembrane conductance regulator by novel interaction with the metabolic sensor AMP-activated protein kinase. *J Clin Invest* **105**: 1711–1721.

Hamann M, Gibson A, Davies N, Jowett A, Walhin JP, Partington L, Affleck K, Trezise D, Main M. 2009. Human ClCa1 modulates anionic conduction of calcium-dependent chloride currents. *J Physiol* **587**: 2255–2274.

Hartzell HC, Qu Z, Yu K, Xiao Q, Chien LT. 2008. Molecular physiology of bestrophins:multifunctional membrane proteins linked to Best disease and other retinopathies. *Physiol Rev* **88**: 639–672.

Heitzmann D, Warth R. 2008. Physiology and pathophysiology of potassium channels in gastrointestinal epithelia. *Physiol Rev* **88**: 1119–1182.

Hoglund P, Haila S, Socha J, Tomaszewski L, Saarialho-Kere U, Karjalainen-Lindsberg ML, Airola K, Holmberg C, de la Chapelle A, Kere J. 1996. Mutations of the Down-regulated in adenoma (DRA) gene cause congenital chloride diarrhoea. *Nat Genet* **14**: 316–319.

Huang P, Trotter K, Boucher RC, Milgram SL, Stutts MJ. 2000. PKA holoenzyme is functionally coupled to CFTR by AKAPs. *Am J Physiol Cell Physiol* **278**: C417–C422.

Huang P, Lazarowski ER, Tarran R, Milgram SL, Boucher RC, Stutts MJ. 2001. Compartmentalized autocrine signaling to cystic fibrosis transmembrane conductance regulator at the apical membrane of airway epithelial cells. *Proc Natl Acad Sci* **98**: 14120–14125.

Huang J, Shan J, Alper SL, Hanrahan JW. 2011. Secretion by the human airway epithelial cell line Calu-3 depends on basolateral chloride loading by AE2. *Proc Physiol Soc* **23**: 160.

Ishiguro H, Namkung W, Yamamoto A, Wang Z, Worrell RT, Xu J, Lee MG, Soleimani M. 2007. Effect of Slc26a6 deletion on apical Cl^-/HCO_3^- exchanger activity and cAMP-stimulated bicarbonate secretion in pancreatic duct. *Am J Physiol Gastrointest Liver Physiol* **292**: G447–G455.

Joiner WJ, Basavappa S, Vidyasagar S, Nehrke K, Krishnan S, Binder HJ, Boupaep EL, Rajendran VM. 2003. Active K^+ secretion through multiple K_{Ca}-type channels and regulation by IK_{Ca} channels in rat proximal colon. *Am J Physiol Gastrointest Liver Physiol* **285**: G185–G196.

Kanelis V, Hudson RP, Thibodeau PH, Thomas PJ, Forman-Kay JD. 2010. NMR evidence for differential phosphorylation-dependent interactions in WT and ΔF508 CFTR. *EMBO J* **29**: 263–277.

Kim JY, Han W, Namkung W, Lee JH, Kim KH, Shin H, Kim E, Lee MG. 2004. Inhibitory regulation of cystic fibrosis transmembrane conductance regulator anion-transporting activities by Shank2. *J Biol Chem* **279**: 10389–10396.

Kim KH, Shcheynikov N, Wang Y, Muallem S. 2005. SLC26A7 is a Cl⁻ channel regulated by intracellular pH. *J Biol Chem* **280:** 6463–6470.

Kimberg DV, Field M, Johnson J, Henderson A, Gershon E. 1971. Stimulation of intestinal mucosal adenyl cyclase by cholera enterotoxin and prostaglandins. *J Clin Invest* **50:** 1218–1230.

Ko SB, Zeng W, Dorwart MR, Luo X, Kim KH, Millen L, Goto H, Naruse S, Soyombo A, Thomas PJ, et al. 2004. Gating of CFTR by the STAS domain of SLC26 transporters. *Nat Cell Biol* **6:** 343–350.

Kreda SM, Gynn MC, Fenstermacher DA, Boucher RC, Gabriel SE. 2001. Expression and localization of epithelial aquaporins in the adult human lung. *Am J Respir Cell Mol Biol* **24:** 224–234.

Krouse ME, Talbott JF, Lee MM, Joo NS, Wine JJ. 2004. Acid and base secretion in the Calu-3 model of human serous cells. *Am J Physiol Lung Cell Mol Physiol* **287:** L1274–L1283.

Kunzelmann K, Hubner M, Schreiber R, Levy-Holzman R, Garty H, Bleich M, Warth R, Slavik M, von Hahn T, Greger R. 2001. Cloning and function of the rat colonic epithelial K⁺ channel KVLQT1. *J Membr Biol* **179:** 155–164.

Larsen EH. 2002. Hans H. Ussing—Scientific work: Contemporary significance and perspectives. *Biochim Biophys Acta* **1566:** 2–15.

LaRusch JA. 2007. "A novel internal binding motif in the CFTR C-terminus enhances EBP50 multimerization and facilitates endocytic recycling." PhD thesis, The Johns Hopkins University, Baltimore, MD.

Lee RJ, Foskett JK. 2010. cAMP-activated Ca²⁺ signalling is required for CFTR-mediated serous cell fluid secretion in porcine and human airways. *J Clin Invest* **120:** 3137–3148.

Lee JH, Richter W, Namkung W, Kim KH, Kim E, Conti M, Lee MG. 2008. Dynamic regulation of cystic fibrosis transmembrane conductance regulator by competitive interactions of molecular adaptors. *J Biol Chem* **282:** 10414–10422.

Lefkimmiatis K, Srikanthan M, Maiellaro I, Moyer MP, Curci S, Hofer AM. 2009. Store-operated cyclic AMP signalling mediated by STIM1. *Nat Cell Biol* **11:** 433–442.

Li C, Krishnamurthy PC, Penmatsa H, Marrs KL, Wang XQ, Zaccolo M, Jalink K, Li M, Nelson DJ, Schuetz JD, et al. 2007a. Spatiotemporal coupling of cAMP transporter to CFTR chloride channel function in the gut epithelia. *Cell* **131:** 940–951.

Li J, Poulikakos PI, Dai Z, Testa JR, Callaway DJE, Bu Z. 2007b. Protein kinase C phosphorylation disrupts Na⁺/H⁺ exchanger regulatory factor 1 autoinhibition and promotes cystic fibrosis transmembrane conductance regulator macromolecular assembly. *J Biol Chem* **282:** 27086–27099.

Liao T, Wang L, Halm ST, Lu L, Fyffe RE, Halm DR. 2005. K+ channel KVLQT1 located in the basolateral membrane of distal colonic epithelium is not essential for activating Cl⁻ secretion. *Am J Physiol Cell Physiol* **289:** C564–C575.

Liedtke CM, Wang X. 2006. The N-terminus of the WD5 repeat of human RACK1 binds to airway epithelial NHERF1. *Biochemistry* **45:** 10270–10277.

Liedtke CM, Yun CHC, Kyle N, Wang D. 2002. PKC-ε dependent regulation of CFTR involves binding to RACK1, a receptor for activated C kinase, and RACK1 binding to NHERF1. *J Biol Chem* **277:** 22925–22933.

Linsdell P, Tabcharani JA, Rommens JM, Hou YX, Chang XB, Tsui LC, Riordan JR, Hanrahan JW. 1997. Permeability of wild-type and mutant cystic fibrosis transmembrane conductance regulator chloride channels to polyatomic anions. *J Gen Physiol* **110:** 355–364.

Loffing J, Moyer BD, Reynolds D, Shmukler BE, Alper SL, Stanton BA. 2000. Functional and molecular characterization of an anion exchanger in airway serous epithelial cells. *Am J Physiol Cell Physiol* **279:** C1016–C1023.

Lundberg A. 1957. Secretory potentials in the sublingual gland of the cat. *Acta Physiol Scand* **40:** 21–34.

Marcorelles P, Gillet D, Friocourt G, Ledé F, Samaison L, Huguen G, Ferec C. 2011. Cystic fibrosis transmembrane conductance regulator protein expression in the male excretory duct system during development. *Hum Pathol* **43:** 390–397.

Marino CR, Matovcik LM, Gorelick FS, Cohn JA. 1991. Localization of the cystic fibrosis transmembrane conductance regulator in pancreas. *J Clin Invest* **88:** 712–716.

Martin AC, Willoughby D, Ciruela A, Ayling LJ, Pagano M, Wachten S, Tengholm A, Cooper DM. 2009. Capacitative Ca²⁺ entry via Orai1 and stromal interacting molecule 1 (STIM1) regulates adenylyl cyclase type 8. *Mol Pharmacol* **75:** 830–842.

Marty A, Tan YP, Trautmann A. 1984. Three types of calcium-dependent channel in rat lacrimal glands. *J Physiol* **357:** 293–325.

Matsui H, Randell SH, Peretti SW, Davis CW, Boucher RC. 1998. Coordinated clearance of periciliary liquid and mucus from airway surfaces. *J Clin Invest* **102:** 1125–1131.

McCormick JA, Ellison DH. 2011. The WNKs: Atypical protein kinases with pleiotropic actions. *Physiol Rev* **91:** 177–219.

McWilliams RR, Gidey E, Fouassier L, Weed SA, Doctor RB. 2008. Characterization of an ankyrin repeat-containing Shank2 isoform (Shank2E) in liver epithelial cells. *Biochem J* **380:** 181–191.

Mercier-Zuber A, O'Shaughnessy KM. 2011. Role of SPAK and OSR1 signalling in the regulation of NaCl cotransporters. *Curr Opin Nephrol Hypertens* **20:** 534–540.

Mohler PJ, Kreda SM, Boucher RC, Sudol M, Stutts MJ, Milgram SL. 1999. Yes-associated protein 65 localizes p62^c-Yes to the apical compartment of airway epithelia by association with EBP50. *J Cell Biol* **147:** 879–890.

Mount DB, Romero MF. 2004. The SLC26 gene family of multifunctional anion exchangers. *Pflugers Arch* **447:** 710–721.

Namkung W, Song Y, Mills AD, Padmawar P, Finkbeiner WE, Verkman AS. 2009. In situ measurement of airway surface liquid [K+] using a ratioable K+-selective fluorescent dye. *J Biol Chem* **284:** 15916–15926.

Namkung W, Finkbeiner WE, Verkman AS. 2010. CFTR-adenylyl cyclase 1 association responsible for UTP activation of CFTR in well-differentiated primary human bronchial cell cultures. *Mol Biol Cell* **21:** 2639–2648.

Namkung W, Phuan P-W, Verkman AS. 2011. TMEM16A inhibitors reveal TMEM16A as a minor component of calcium-activated chloride channel conductance in airway and intestinal epithelial cells. *J Biol Chem* **286:** 2365–2374.

Nanda Kumar NS, Sing SK, Rajendran VM. 2010. Mucosal potassium efflux mediated via Kcnn4 channels provides the driving force for electrogenic anion secretion in colon. *Am J Physiol Gastrointest Liver Physiol* **299:** G707–G714.

Naren AP, Kirk KL. 2000. CFTR chloride channels: Binding partners and regulatory networks. *News Physiol Sci* **15:** 57–61.

Naren AP, Di A, Cormet-Boyaka E, Boyaka PN, McGhee JR, Zhou W, Akagawa K, Fujiwara T, Thome U, Engelhardt JF, et al. 2000. Syntaxin 1A is expressed in airway epithelial cells, where it modulates CFTR Cl⁻ currents. *J Clin Invest* **105:** 377–386.

Naren AP, Cobb B, Li C, Roy K, Nelson D, Heda GD, Liao J, Kirk KL, Sorscher EJ, Hanrahan JW, et al. 2003. A macromolecular complex of β_2 adrenergic receptor, CFTR, and ezrin/radixin/moesin-binding phosphoprotein 50 is regulated by PKA. *Proc Natl Acad Sci* **100:** 342–346.

Novak I, Young JA. 1986. Two independent anion transport systems in rabbit mandibular salivary glands. *Pflügers Arch* **407:** 649–656.

Ousingsawat J, Martins JR, Schreiber R, Rock JR, Harfe BD, Kunzelmann K. 2009. Loss of TMEM16A causes a defect in epithelial Ca^{2+}-dependent chloride transport. *J Biol Chem* **284:** 28698–28703.

Penmatsa H, Zhang W, Yarlagadda S, Li C, Conoley VG, Yue J, Bahouth SW, Buddington RK, Zhang G, Nelson DJ. 2010. Compartmentalized cyclic adenosine $3',5'$-monophosphate at the plasma membrane clusters PDE3A and cystic fibrosis transmembrane conductance regulator into microdomains. *Mol Biol Cell* **21:** 1097–1110.

Peters KW, Qi J, Watkins SC, Frizzell RA. 1999. Syntaxin 1A inhibits regulated CFTR trafficking in *Xenopus* oocytes. *Am J Physiol Cell Physiol* **277:** C174–C180.

Petrovic S, Ma L, Wang Z, Soleimani M. 2003. Identification of an apical Cl⁻/HCO₃⁻ exchanger in rat kidney proximal tubule. *Am J Physiol Cell Physiol* **285:** C608–C617.

Petrovic S, Barone S, Xu J, Conforti L, Ma L, Kujala M, Kere J, Soleimani M. 2004. SLC26A7: A basolateral Cl⁻/HCO₃⁻ exchanger specific to intercalated cells of the outer medullary collecting duct. *Am J Physiol Renal Physiol* **286:** F161–F169.

Pirani D, Evans LA, Cook DI, Young JA. 1987. Intracellular pH in the rat mandibular salivary gland: The role of Na⁻H and Cl⁻HCO₃ antiports in secretion. *Pflugers Arch* **408:** 178–184.

Raghuram V, Mak D-OD, Foskett JK. 2001. Regulation of cystic fibrosis transmembrane conductance regulator single-channel gating by bivalent PDZ-domain-mediated interaction. *Proc Natl Acad Sci* **98:** 1300–1305.

Rahman I, MacNee W. 1998. Role of transcription factors in inflammatory lung diseases. *Thorax* **53:** 601–612.

Richardson C, Rafiqi FH, Karlsson HK, Moleleki N, Vandewalle A, Campbell DG, Morrice NA, Alessi DR. 2008. Activation of the thiazide-sensitive Na+-Cl⁻ cotransporter by the WNK-regulated kinases SPAK and OSR1. *J Cell Sci* **121:** 675–684.

Riordan JR, Rommens JM, Kerem B, Alon N, Rozmahel R, Grzelczak Z, Zielenski J, Lok S, Plavsic N, Chou JL, et al. 1989. Identification of the cystic fibrosis gene: Cloning and characterization of complementary DNA. *Science* **245:** 1066–1073.

Rock JR, O'Neal WK, Gabriel SE, Randell SH, Harfe BD, Boucher RC, Grubb BR. 2009. Transmembrane protein 16A (TMEM16A) is a Ca^{2+}-regulated Cl⁻ secretory channel in mouse airways. *J Biol Chem* **284:** 14875–14880.

Romanenko VG, Catalán MA, Brown DA, Putzier I, Hartzell HC, Marmorstein AD, Gonzalez-Begne M, Rock JR, Harfe BD, Melvin JE. 2010. Tmem16A encodes the Ca^{2+}-activated Cl⁻ channel in mouse submandibular salivary gland acinar cells. *J Biol Chem* **285:** 12990–13001.

Rottner M, Freyssinet JM, Martinez MC. 2009. Mechanisms of the noxious inflammatory cycle in cystic fibrosis. *Respir Res* **10:** 23.

Schroeder BC, Cheng T, Jan YN, Jan LY. 2008. Expression cloning of TMEM16A as a calcium-activated chloride channel subunit. *Cell* **134:** 1019–1029.

Schweinfest CW, Spyropoulos DD, Henderson KW, Kim JH, Chapman JM, Barone S, Worrell RT, Wang Z, Soleimani M. 2006. slc26a3 (dra)-deficient mice display chloride-losing diarrhea, enhanced colonic proliferation, and distinct up-regulation of ion transporters in the colon. *J Biol Chem* **281:** 37962–37971.

Seidler U, Rottinghaus I, Hillesheim J, Chen M, Riederer B, Krabbenhoft A, Engelhardt R, Wiemann M, Wang Z, Barone S, et al. 2008. Sodium and chloride absorptive defects in the small intestine in Slc26a6 null mice. *Pflugers Arch* **455:** 757–766.

Shan J, Huang J, Liao J, Robert R, Hanrahan JW. 2011. Anion secretion by a model epithelium: More lessons from Calu-3. *Acta Physiol* **202:** 523–531.

Shcheynikov N, Yang D, Wang Y, Zeng W, Karniski LP, So I, Wall SM, Muallem S. 2008. The Slc26a4 transporter functions as an electroneutral Cl⁻/I⁻/HCO₃⁻ exchanger: Role of Slc26a4 and Slc26a6 in I⁻ and HCO₃⁻ secretion and in regulation of CFTR in the parotid duct. *J Physiol* **586:** 3813–3824.

Short DB, Trotter KW, Reczek D, Kreda SM, Bretscher A, Boucher RC, Stutts MJ, Milgram SL. 1998. An apical PDZ protein anchors the cystic fibrosis transmembrane conductance regulator to the cytoskeleton. *J Biol Chem* **273:** 19797–19801.

Skou JC. 1998. Nobel Lecture. The identification of the sodium pump. *Biosci Rep* **18:** 155–169.

Smith PL, Frizzell RA. 1984. Chloride secretion by canine tracheal epithelium: IV. Basolateral membrane K permeability parallels secretion rate. *J Membr Biol* **77:** 187–199.

Steward MC, Ishiguro H, Case RM. 2005. Mechanisms of bicarbonate secretion in the pancreatic duct. *Annu Rev Physiol* **67:** 377–409.

Stutts MJ, Knowles MR, Gatzy JT, Boucher RC. 1986. Oxygen consumption and ouabain binding sites in cystic fibrosis nasal epithelium. *Pediatr Res* **20:** 1316–1320.

Sun F, Hug MJ, Lewarchik CM, Yun CH, Bradbury NA, Frizzell RA. 2000a. E3KARP mediates the association of ezrin and protein kinase A with the cystic fibrosis transmembrane conductance regulator in airway cells. *J Biol Chem* **275:** 29539–29546.

Sun F, Hug MJ, Bradbury NA, Frizzell RA. 2000b. Protein kinase A associates with cystic fibrosis transmembrane conductance regulator via an interaction with ezrin. *J Biol Chem* **275:** 14360–14366.

Suzuki M, Mizuno A. 2004. A novel human Cl⁻ channel family related to *Drosophila flightless* locus. *J Biol Chem* **279:** 22461–22468.

Tarran R. 2004. Regulation of airway surface liquid volume and mucus transport by active ion transport. *Proc Am Thorac Soc* **1:** 42–46.

Tessier GJ, Traynor TR, Kannan MS, O'Grady SM. 1990. Mechanisms of sodium and chloride transport across equine tracheal epithelium. *Am J Physiol Lung Cell Mol Physiol* **259:** L459–L467.

Thelin WR, Kesimer M, Tarran R, Kreda SM, Grubb BR, Sheehan JK, Stutts MJ, Milgram SL. 2005. The cystic fibrosis transmembrane conductance regulator is regulated by a direct interaction with the protein phosphatase 2A. *J Biol Chem* **280:** 41512–41520.

Thelin WR, Chen Y, Gentzsch M, Kreda SM, Sallee JL, Scarlett CO, Borchers CH, Jacobson K, Stutts MJ, Milgram SL. 2007. Direct interaction with filamins modulates the stability and plasma membrane expression of CFTR. *J Clin Invest* **117:** 364–374.

Thiagarajah JR, Broadbent T, Hsieh E, Verkman AS. 2004. Prevention of toxin-induced intestinal ion and fluid secretion by a small-molecule CFTR inhibitor. *Gastroenterology* **126:** 511–519.

Towne JE, Harrod KS, Krane CM, Menon AG. 2000. Decreased expression of aquaporin (AQP)1 and AQP5 in mouse lung after acute viral infection. *Am J Respir Cell Mol Biol* **22:** 34–44.

Trezise AE, Buchwald M. 1991. In vivo cell-specific expression of the cystic fibrosis transmembrane conductance regulator. *Nature* **353:** 434–437.

Vaandrager AB, Ehlert EM, Jarchau T, Lohmann SM, de Jonge HR. 1996. N-terminal myristoylation is required for membrane localization of cGMP-dependent protein kinase type II. *J Biol Chem* **271:** 7025–7029.

Vaandrager AB, Smolenski A, Tilly BC, Houtsmuller AB, Ehlert EM, Bot AG, Edixhoven M, Boomaars WE, Lohmann SM, de Jonge HR. 1998. Membrane targeting of cGMP-dependent protein kinase is required for cystic fibrosis transmembrane conductance regulator Cl- channel activation. *Proc Natl Acad Sci* **95:** 1466–1471.

Wang S, Raab RW, Schatz PJ, Guggino WB, Li M. 1998. Peptide binding consensus of the NHE-RF-PDZ1 do-main matches the C-terminal sequence of cystic fibrosis transmembrane conductance regulator (CFTR). *FEBS Lett* **427:** 103–108.

Wang S, Yue H, Derin RB, Guggino WB, Li M. 2000. Accessory protein facilitated CFTR–CFTR interaction, a molecular mechanism to potentiate the chloride channel activity. *Cell* **103:** 169–179.

Wang Y, Soyombo AA, Shcheynikov N, Zeng W, Dorwart M, Marino CR, Thomas PJ, Muallem S. 2006. Slc26a6 regulates CFTR activity in vivo to determine pancreatic duct HCO₃⁻ secretion: Relevance to cystic fibrosis. *EMBO J* **25:** 5049–5057.

Welsh MJ, McCann JD. 1985. Intracellular calcium regulates basolateral potassium channels in a chloride-secreting epithelium. *Proc Natl Acad Sci* **82:** 8823–8826.

Welsh MJ, Smith PL, Frizzell RA. 1983. Chloride secretion by canine tracheal epithelium: III. Membrane resistances and electromotive forces. *J Membr Biol* **71:** 209–218.

Willumsen NJ, Davis CW, Boucher RC. 1989. Intracellular Cl⁻ activity and cellular Cl⁻ pathways in cultured human airway epithelium. *Am J Physiol* **256:** C1033–C1044.

Wilschanski M, Durie PR. 2007. Patterns of GI disease in adulthood associated with mutations in the CFTR gene. *Gut* **56:** 1153–1163.

Wright EM, Diamond JM. 1977. Anion selectivity in biological systems. *Physiol Rev* **57:** 109–157.

Xu J, Song P, Miller ML, Borgese F, Barone S, Riederer B, Wang Z, Alper SL, Forte JG, Shull GE, et al. 2008. Deletion of the chloride transporter Slc26a9 causes loss of tubulovesicles in parietal cells and impairs acid secretion in the stomach. *Proc Natl Acad Sci* **105:** 17955–17960.

Yang YD, Cho H, Koo JY, Tak MH, Cho Y, Shim W-S, Park SP, Lee J, Lee B, Kim B-M, et al. 2008. TMEM16A confers receptor-activated calcium-dependent chloride conductance. *Nature* **455:** 1210–1215.

Yarwood SJ, Steele MR, Scotland G, Houslay MD, Bolger GB. 1999. The RACK1 signaling scaffold protein selectively interacts with the cAMP-specific phosphodiesterase PDE4D5 isoform. *J Biol Chem* **274:** 14909–14917.

Zeng W, Lee MG, Yan M, Diaz J, Benjamin I, Marino CR, Kopito R, Freedman S, Cotton C, Muallem S, et al. 1997. Immuno and functional characterization of CFTR in submandibular and pancreatic acinar and duct cells. *Am J Physiol* **273:** C442–C455.

Zhu T, Dahan D, Evaglelidis A, Zheng S-X, Luo J, Hanrahan JW. 1999. Association of cystic fibrosis transmembrane conductance regulator and protein phosphatase 2C. *J Biol Chem* **274:** 29102–29107.

Mechanisms of Bicarbonate Secretion: Lessons from the Airways

Robert J. Bridges

Department of Physiology and Biophysics, Rosalind Franklin University of Medicine and Sciences, North Chicago, Illinois 60064

Correspondence: bob.bridges@rosalindfranklin.edu

Early studies showed that airway cells secrete HCO_3^- in response to cAMP-mediated agonists and HCO_3^- secretion was impaired in cystic fibrosis (CF). Studies with Calu-3 cells, an airway serous model with high expression of CFTR, also show the secretion of HCO_3^- when cells are stimulated with cAMP-mediated agonists. Activation of basolateral membrane hIK-1 K^+ channels inhibits HCO_3^- secretion and stimulates Cl^- secretion. CFTR mediates the exit of both HCO_3^- and Cl^- across the apical membrane. Entry of HCO_3^- on a basolateral membrane NBC or Cl^- on the NKCC determines which anion is secreted. Switching between these two secreted anions is determined by the activity of hIK-1 K^+ channels.

The recognition that HCO_3^- secretion is impaired in CF patients dates back to the studies of Hadorn and coworkers in the 1960s (Hadorn et al. 1968). These investigators showed that pancreatic HCO_3^- and fluid secretion were diminished in CF patients. Moreover, CF patients were refractory to secretin, a cAMP-mediated secretory agonist in the pancreas. These studies were confirmed by Gaskin et al. (1982) and Kopelman et al. (1985, 1988). Extensive transport studies of the exocrine pancreas have shown that HCO_3^- is the primary anion secreted by the ductal cells, the predominate site of CFTR expression (Marino et al. 1991). The human pancreas can secrete a fluid of 130 mM HCO_3^- (Schultz 1987). Secreted HCO_3^- electrically draws Na^+ into the lumen and H_2O follows osmotically. The secreted fluid and electrolytes serve to flush the digestive enzymes from the acini and ducts of the pancreas. Thus, impaired HCO_3^- secretion results in poor clearance of the digestive enzymes, and their premature activation eventuates in the destruction of the pancreas in CF. We surmise that a similar sequela follows from impaired HCO_3^- secretion in the submucosal glands and airways of CF patients. Indeed, several recent studies from Wine and coworkers have shown cAMP-stimulated fluid secretion is impaired from CFTR-deficient submucosal glands (Joo et al. 2006). Verkman and coworkers have also shown impaired fluid secretion from the submucosal glands of CF patients and shown that the secreted fluid is hyperviscous and acidic compared with glands from non-CF patients (Salinas et al. 2005; Song et al. 2006). Analogous to the pancreas, the submucosal glands secrete mucins, protease inhibitors, antibiotic peptides, and enzymes that must be flushed from the glands onto the airway surface epithelium (Basbaum et al. 1990). Moreover, the physical properties of mucus are intrinsically dependent on the composition of the fluid. Most notably, alterations in ionic strength, divalent cation concentration, and pH have profound effects on the viscoelastic properties of mucins (Forstner et al. 1976; List et al. 1978; Tam et al. 1981; Lin et al. 1993). In the pancreas, the pH of the ductal fluid plays a critical role in regulating the activity of the exocytosed digestive enzymes. In contrast, very little is known regarding the electrolyte composition and pH of the submucosal gland fluid and the role it might play in the 1000-fold expansion that a mucin granule undergoes upon release and degranulation (Yeates et al. 1997; Verdugo and Hauser 2012). In addition, the surface epithelium must maintain a periciliary fluid of appropriate volume and composition to ensure proper mucociliary clearance (Randell and Boucher 2006; Boucher 2007). Adversely affected mucus leads to impaired mucociliary clearance from the submucosal glands and airway surface. The uncleared mucus then becomes a sink for bacterial binding, infection, and inflammation, thereby perpetuating a vicious cycle leading to further mucus secretion (Quinton 1999). This sequence of events is not restricted to the 40,000 individuals suffering from CF, but also occurs in more than 10 million patients suffering from COPD (chronic obstructive pulmonary disease) (Celli et al. 1995; O'Byrine et al. 1999). Thus, impaired fluid secretion by the submucosal glands or surface epithelium hinders clearance from the glands and airway surface. Until recently, Cl^- was considered to be the secreted anion responsible for fluid secretion in the airways. However, recent studies suggest that HCO_3^- secretion importantly contributes to the airway surface and submucosal gland microenvironments.

AIRWAY CELLS SECRETE BICARBONATE

Several early studies indicated that the short-circuit current (I_{SC}), a measure of net electrolyte transport, across airway epithelia was not fully accounted for by the net movements of Na$^+$ and Cl$^-$. Instead, 35%–45% of the I_{SC} had to be attributed to an additional ion species. Ion substitution studies revealed that HCO_3^- but not Cl$^-$ was required to observe a cAMP-stimulated increase in I_{SC} in airway monolayers (Al-Bazzaz et al. 1979, 1981; Welsh 1983). These early observations prompted Smith and Welsh (1992) to investigate HCO_3^- secretion in normal and CF-cultured airway epithelia. The results of this investigation revealed that cAMP-stimulated HCO_3^- secretion across normal but not CF airway epithelia. These investigators went on to conclude that HCO_3^- exit at the apical membrane is through the Cl$^-$ channel, which is defectively regulated in CF epithelia. In addition, they suggested the possibility that a defect in HCO_3^- secretion may contribute to the pathophysiology of CF pulmonary disease.

The intracellular microelectrode studies of Willumsen and Boucher (Willumsen et al. 1992), although designed for a different intent, provided important insight regarding the driving forces acting on HCO_3^- and the question of why normal, but not, CF epithelia can secrete HCO_3^-. These investigators observed that the intracellular pHs of normal and CF airway cells were equal (normal = 7.15 $\pm$ 0.02 vs. CF = 7.11 $\pm$ 0.05), but that the apical membrane potentials (V_{ap}) were different and of opposite polarity (normal = -19 ± 2 mV vs. CF = 3 $\pm$ 5 mV). This difference has been confirmed in several additional studies by Boucher and co-workers (Boucher et al. 1988; Willumsen et al. 1989a,b). Thus, at an extracellular HCO_3^- concentration of 25 mM, an extracellular and intracellular pCO$_2$ of 40 mm Hg, and a pH of 7.15, there is an outwardly directed driving force for HCO_3^- secretion of 6 mV across the apical membrane of normal cells. In contrast, in CF cells, there is an inwardly directed driving force for HCO_3^- absorption of 16 mV across the apical membrane. Therefore, provided there is a conductive pathway to mediate the movement of HCO_3^- across the apical membrane, normal cells will *secrete* HCO_3^- and CF cells will *absorb* HCO_3^-. Although it is often suggested that impaired Cl$^-$ secretion must be corrected in CF, it is noteworthy, given the above V_{ap}s and the usual extracellular and intracellular Cl$^-$

 Cite this article as *Cold Spring Harb Perspect Med* doi: 10.1101/cshperspect.a015016

concentrations, that the net driving force acting on Cl^- is in the absorptive direction for both normal (9 mV) and CF (31 mV) cells. Indeed, only after Na^+ transport is down-regulated or inhibited with amiloride is it possible to observe Cl^- secretion in normal cells. It is generally accepted that the difference in V_{ap} between normal and CF cells is due to the loss of CFTR channels and the higher Na^+ permeability in CF cells (Boucher et al. 1986, 1988; Boucher 1994a,b). Besides Cl^-, CFTR also conducts HCO_3^-, as shown in the studies of Gray et al. (1990) and Linsdell and coworkers (Linsdell et al. 1997; Tang et al. 2009; Li et al. 2011) and more recently by Ishiguro et al. (2009). Therefore, normal airway cells can secrete HCO_3^- through CFTR, whereas CF cells show a HCO_3^- impermeability. This reasoning leads us to assert that, in addition to abnormal Cl^- and Na^+ transport, HCO_3^- transport is also dysfunctional in CF airway epithelia and that HCO_3^- secretion and not Cl^- secretion is critical for normal surface airway epithelial function.

STUDIES WITH Calu-3 CELLS

A second line of evidence that HCO_3^- secretion in the airways may be more important than previously appreciated comes from the studies of Wine and coworkers (Lee et al. 1998) and the results on Calu-3 cells from my own group (Devor et al. 1999). Calu-3 cells are a human airway serous cell line developed by Wine and Widdicombe and coworkers (Shen et al. 1994). The Calu-3 cells were selected from among 12 lung adenocarcinomas as a cell line consistent with cell biological and electrophysiological characteristics of airway serous cells. Calu-3 cells form confluent monolayers with transepithelial resistances of several hundred ohm-centimeters squared (Ω cm^2), express high levels of CFTR, and respond to both cAMP- and Ca^{2+}-mediated agonists with changes in net transepithelial ion transport as measured by I_{SC} (Finkbeiner et al. 1993; Shen et al. 1994). In addition, the Calu-3 cells produce several serous cell-associated proteins including lysozyme, lactoferrin, serine leukoprotease inhibitor (SLP1), secretory component, and mucins (MUC1 and MUC2)

(Finkbeiner et al. 1993). The Calu-3 cells are now used by many laboratories as a serous cell model (Kelley et al. 1995; Grygorczyk et al. 1997; Liedtke et al. 1998; Al-Nakkash et al. 1999; Berger et al. 1999; Duszyk et al. 1999; Illek et al. 1999; Ito et al. 1999; Waters et al. 1999). Our studies on Calu-3 cells showed that they secrete HCO_3^- rather than Cl^- in response to cAMP-mediated agonists. In effect, the Calu-3 cells function as one would expect for a pancreatic ductal cell; however, a cell line of the latter is thus far unavailable. Calu-3 cells express high levels of CFTR, as do serous cells of the submucosal glands of the native epithelium (Puchelle et al. 1992; Engelhardt et al. 1994) and pancreatic ductal cells (Marino et al. 1991). In addition, Calu-3 cells express Na:HCO_3^- cotransporter isoforms pNBC1 (SLC4A4, NBCe1) and NBC4 (SLC4A5, NBCe2) in the basolateral membrane (Kreindler et al. 2006). Studies from Case and coworkers on pancreatic ducts (Ishiguro et al. 1996a,b) and the studies of my laboratory on Calu-3 cells (Devor et al. 1999) suggest that HCO_3^- entry across the basolateral membrane is mediated by an NBC. If we assume that the transport phenotype expressed by Calu-3 cells accurately reflects native serous cells, then HCO_3^- secretion must be important in the physiology of submucosal glands. In agreement with this hypothesis, the studies of Ballard and coworkers have shown that inhibitors of HCO_3^- secretion, acetazolamide, a carbonic anhydrase inhibitor, and DIDS, an inhibitor of Cl^-:HCO_3^- exchangers as well as NBCs, caused mucus obstruction of the submucosal glands in secretagogue-stimulated porcine distal bronchi (Inglis et al. 1997, 1998; Trout et al. 1998). Furthermore, bumetanide, an inhibitor of the Na^+:-K^+:2 Cl^- cotransporter (NKCC1) and thereby Cl^- secretion failed to inhibit cAMP-induced gland fluid secretion (Corrales et al. 1984). Consistent with these results, the forskolin-stimulated increase in I_{SC} in Calu-3 cells is insensitive to bumetanide but is inhibited by acetazolamide and DNDS, another inhibitor of NBCs (Devor et al. 1999). Collectively, these studies lead us to conclude that HCO_3^- secretion is important in the physiology of submucosal glands and the airway surface epithelium.

MODEL FOR ANION SECRETION IN AIRWAY CELLS

Our model for anion secretion in Calu-3 cells is illustrated in Figure 1. Forskolin-stimulated Calu-3 cells secrete HCO_3^- by an electrogenic mechanism, that is, Cl^--independent, serosal Na^+-dependent, serosal bumetanide-insensitive and inhibited by serosal disulfonic stilbene (DNDS) as judged by transepithelial currents, isotope fluxes, and the results of ion substitution, pharmacology, and pH studies (Devor et al. 1999). However, Calu-3 cells are not limited to the secretion of HCO_3^-. Instead, when stimulated by 1-EBIO (1-ethyl-2-benzimidazolinone), an activator of the basolateral membrane, Ca^+-activated CTX-sensitive, K^+ channels (hIK1, KCNN4), Calu-3 cells secrete Cl^- by an electrogenic bumetanide-sensitive mechanism, and HCO_3^- secretion is diminished. Moreover, when stimulated by both forskolin and 1-EBIO, the secretion of HCO_3^- is diminished and Cl^- secretion dominates. To account for these results, we proposed the above model of anion secretion whereby CFTR serves as the cAMP/PKA-activated anion channel for both Cl^- and HCO_3^- exit across the apical membrane. Activation of CFTR alone tends to bring V_{ap} to the equilibrium potential for Cl^- (E_{Cl}), a value greater than the equilibrium potential for HCO_3^- (E_{HCO3}), and thereby provides the driving force for HCO_3^- exit across the apical membrane. Stimulation by cAMP (forskolin) alone leaves the basolateral membrane potential (V_{bl}) less hyperpolarized than the reversal potential of the DNDS-sensitive NBC (E_{revNBC}), and HCO_3^- is secreted. Subsequent activation of hIk1 by 1-EBIO or cholinergic agonists hyperpolarizes V_{bl} so that $V_{bl} > E_{revNBC}$, which inhibits HCO_3^- uptake by an NBC but provides a driving force for Cl^- secretion as V_{ap} becomes $> E_{Cl}$.

The initial results that led to the discovery that Calu-3 cells secrete HCO_3^- in response to cAMP stimulation and Cl^- when hIK1 channels are activated by 1-EBIO are illustrated in Figure 2 (Devor et al. 1999). Calu-3 cells were grown on collagen-coated filters and studied by standard short-circuit current (I_{SC}) methods. In standard bath solutions of NaCl and $NaHCO_3$, the Calu-3 cells display a basal I_{SC} of 13 ± 0.8 μA cm^{-2} and a transepithelial resistance (R_t) of 353 $\pm 14\,\Omega$ cm^2 ($n = 216$ filters). Stimulation

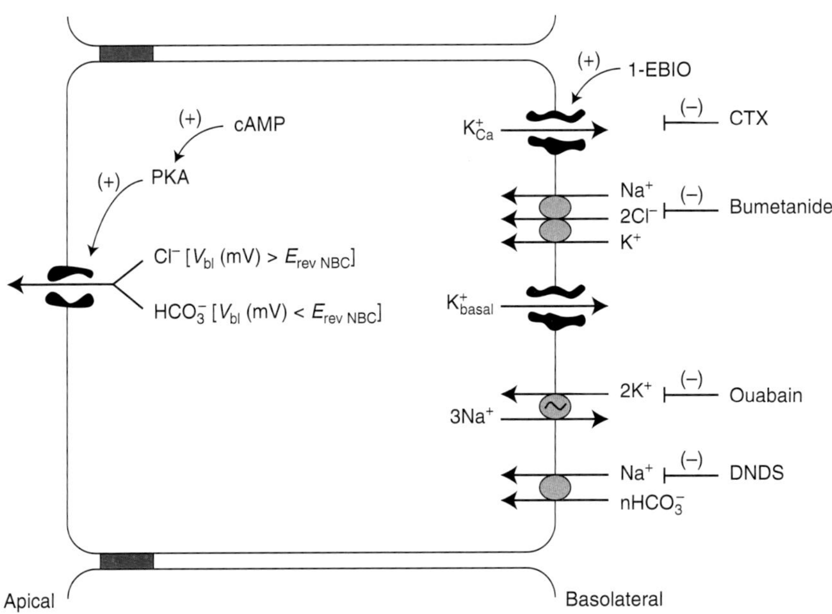

Figure 1. Model of anion secretion in Calu-3 cells.

Cite this article as *Cold Spring Harb Perspect Med* doi: 10.1101/cshperspect.a015016

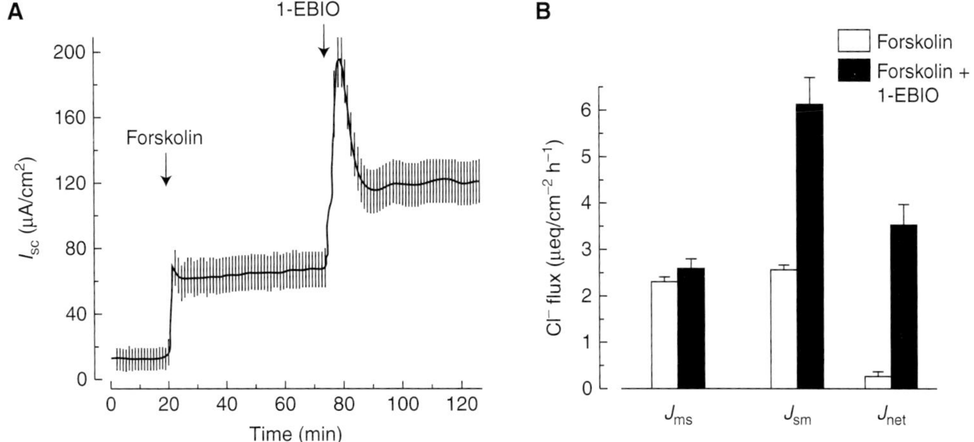

Figure 2. Effect of forskolin (2 μM) and 1-EBIO (1 mM) on I_{sc} and Cl^- fluxes across Calu-3 cells. Ion fluxes are given as absolute values. Forskolin caused a fivefold increase in each of the unidirectional fluxes without any net secretion of Cl^-. Control data are not shown. (Figure is from Devor et al. 1999; reprinted, with permission, from the author.)

with forskolin (2 μM) induced a damped oscillatory response that became stable and sustained after 5–10 min at a plateau value of 66 $\pm$ 4 μA cm^{-2} and at a reduced R_T of 189 $\pm$ 7 Ω cm^{-2} ($n = 109$). The subsequent addition of 1-EBIO (1 mM) caused a further increase in I_{SC} to 114 $\pm$ 5 μA cm^{-2} without any further significant decrease in R_t (173 $\pm$ 9 Ω cm^2, $n = 36$). Unidirectional ion fluxes revealed that forskolin caused a fivefold increase in both the mucosal-to-serosal and serosal-to-mucosal fluxes of Cl^-, but no net secretion of Cl^- (Fig. 2B). However, when the forskolin-stimulated monolayers were further stimulated with 1-EBIO, there was a further increase in the serosal-to-mucosal flux of Cl^-, and the net secretion of Cl^- was nearly equal to the I_{SC}. Moreover, whereas the forskolin-stimulated I_{SC} and Cl^- fluxes were insensitive to bumetanide, bumetanide completely inhibited the I_{SC} and the net secretion of Cl^- in forskolin plus 1-EBIO–treated monolayers (results not shown). Additional ion substitution studies revealed that the forskolin-stimulated I_{SC} was Cl^- independent, HCO_3^- dependent, and serosal Na^+ dependent. Pharmacological studies revealed that forskolin-stimulated I_{SC} was DNDS and acetazolamide sensitive. Collectively, these studies led us to conclude that Calu-3 cells secrete HCO_3^- in response to cAMP stim-

ulation. This conclusion was strongly supported by the observation that forskolin stimulation caused an alkalinization of the mucosal bathing solution of Calu-3 cells studied under open circuit conditions. In agreement with our estimates, Wine and coworkers (Irokawa et al. 2004), using a novel chamber to collect the fluid secreted by Calu-3 cells, showed that forskolin stimulated the secretion of a solution of $\sim$80 mM HCO_3^-. Moreover, when forskolin plus 1-EBIO was used to stimulate the cells, the fluid was no longer alkaline, and the HCO_3^- concentration fell to 17 mM (Irokawa et al. 2004).

CFTR CL$^-$-DEPENDENT SECRETION

The studies with 1-EBIO reveal that Calu-3 cells are not limited to HCO_3^- secretion but can also secrete Cl^-. We used 1-EBIO in these studies because it causes a sustained activation of the basolateral membrane Ca^{2+}-activated, CTX-sensitive K^+ channels (hIK1) (Devor et al. 1996a,b; Syme et al. 2000). Although relatively high concentrations of 1-EBIO are required (1 mM), subsequent studies with newly synthesized derivatives that we have prepared with 100-fold higher affinities produce the same effects (Singh et al. 1999). Wine and coworkers (Lee et al. 1998) also observed similar results using thapsigargin to elevate intracellular Ca^{2+} and activate hIK channels. These channels would normally be activated by Ca^{2+}-mediated agonists such as acetylcholine or substance-P. However, the former does not cause sustained increases in intracellular Ca^{2+} and I_{SC}. In addition to causing Cl^- secretion, the activation of hIK1 channels diminishes HCO_3^- secretion. These results combined with the observations that forskolin-stimulated HCO_3^- secretion required serosal Na^+ and was inhibited by DNDS led us to propose that the influx of HCO_3^- across the basolateral membrane was mediated by an electrogenic NBC that carries two or more HCO_3^- ions for each Na^+ ion. Activation of basolateral membrane K^+ channels tends to hyperpolarize V_{bl} and inhibit NBC-dependent HCO_3^- influx. The hyperpolarization of V_{bl} also activates the basolateral membrane NaK2Cl-cotransporter (Haas and Forbush 2000), allowing for the influx of Cl^- and its net secretion.

Therefore, the activation and inactivation of basolateral membrane K^+ channels provides a means of switching between HCO_3^- and Cl^- secretion across airway epithelia as illustrated in Figure 1.

MICROELECTRODE AND IMPEDANCE ANALYSIS STUDIES

To obtain a better understanding of the conductances and driving forces involved in these different modes of anion secretion in Calu-3 cells, we performed microelectrode and impedance analysis experiments (Tamada et al. 2001). The results of these experiments are summarized in cell models shown in Figure 3. We were able to maintain microelectrode impalements for 10–30 min on a routine basis in Calu-3 cells, which allowed us to monitor the same cell under control, forskolin, and forskolin plus 1-EBIO–stimulated conditions. Cells were

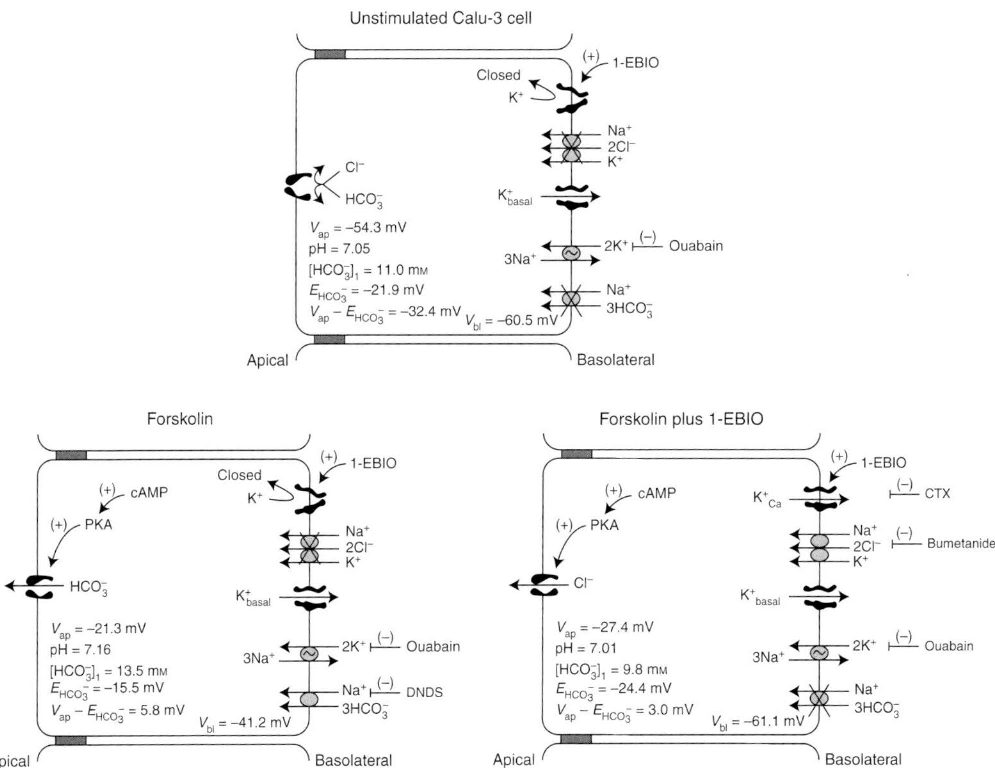

Figure 3. Cell models of unstimulated control cells, forskolin-stimulated cells secreting HCO_3^-, and forskolin plus 1-EBIO cells secreting Cl^-. See text for details.

studied under open circuit conditions, and the transepithelial voltage (V_t) and resistance (R_t) as well as the V_{ap} were measured. The apical fractional resistance (F_{Rap}) was calculated from the ΔV_t and the $\Delta V_{ap}/\Delta V_t$ ratio in response to a 50-μA transepithelial pulse and the V_{bl} from the V_t and V_{ap} values. Forskolin caused a hyperpolarization of V_t and decreased R_t similar to the changes seen in the short-circuit current experiments yielding equivalent current changes (I_{eq}) similar to the previously reported I_{SC} values of $\sim$65 μA/cm^2. Forskolin depolarized both V_{ap} and V_{bl}. V_{ap} decreased from a control value in unstimulated cells of -54.3 mV to -21.3 mV and V_{bl} from -60.5 mV to -41.2 mV. Consistent with the activation of an apical membrane conductance, the F_{Rap} decreased from 0.55 to 0.074 with forskolin stimulation. Double barrel voltage and pH electrodes were used to obtain estimates of the intracellular pH (pH_i). Forskolin increased pH_i from a control value of 7.02 to 7.16. Assuming a pCO$_2$ of 40 mm Hg, the intracellular HCO$_3^-$ concentration increased from 11 mM in control cells to 13.5 mM in forskolin-stimulated cells. Thus, forskolin stimulated a driving force of 5.8 mV for HCO$_3^-$ exit from the cell. Consistent with the dramatic decrease in the F_{Rap}, impedance analysis showed that the apical membrane resistance fell to 14 Ω cm^2 in forskolin-stimulated cells corresponding to an apical membrane conductance of 71 mS/cm^2. With a driving force of 5.8 mV, only 12 mS/cm^2 of the apical membrane conductance of 71 mS/cm^2 is needed to account for the observed HCO$_3^-$ current of 65 μA/cm^2. The conductance ratio of 12 mS/cm^2 to 71 mS/cm^2 is 0.17, and this too is consistent with the reported HCO$_3^-$ conductance compared with Cl$^-$ of the apical membrane of Calu-3 cells (Illek et al. 1999) as well as CFTR (Linsdell et al. 1997; Tang et al. 2009; Man-song et al. 2010). It must also be noted that a driving force of 5.8 mV does not explain the reported 80 mM concentration of HCO$_3^-$ in the thin film experiments (Devor et al. 1999; Irokawa et al. 2004).

It is noteworthy that a driving force of 32.4 mV exists for HCO$_3^-$ exit in unstimulated control cells, and yet not until CFTR is activated is HCO$_3^-$ secreted, and this secretion is driven by a much lower driving force (5.8 mV). Thus, if alternative HCO$_3^-$ transporters are present in the apical membrane, they are not active until the cells are stimulated with forskolin. In addition, the intracellular HCO$_3^-$ concentration is actually increased in forskolin-stimulated cells, indicating that the secreted HCO$_3^-$ is replenished. We surmise that HCO3$^-$ entry via a basolateral membrane NBC is activated by depolarization of the basolateral membrane. Given a reversal potential for an NBC with a stoichiometry of 1 Na$^+$ to 2 HCO$_3^-$ of nearly -90 mV, we suggested that the NBC that mediates HCO$_3^-$ entry has a stoichiometry of 1 Na$^+$ to 3 HCO$_3^-$ with a reversal potential of approximately -60 mV. Depolarization of V_{bl} from -61.5 mV in unstimulated cells to -44 mV would activate an NBC with a 1:3 stoichiometry. Addition of DNDS to the basolateral side caused a depolarization of 8.5 mV in forskolin-stimulated cells but not in unstimulated cells. Replacement of HCO$_3^-$ with HEPES also depolarized V_{bl} by 8.5 mV (T Tamada and RJ Bridges, unpubl.). These results are consistent with the activation of an electrogenic, DNDS-sensitive basolateral membrane NBC in forskolin-stimulated cells. Confirmation of the exact stoichiometry of the NBC will require measurements of the intracellular Na$^+$ activity.

The microelectrode and impedance studies revealed that 1-EBIO hyperpolarized V_{bl} and V_{ap} when added to forskolin-stimulated cells. V_{bl} hyperpolarized from -41.2 mV to -61.1 mV and V_{ap} from -21.3 mV to -27.4 mV when 1-EBIO was added to the forskolin-stimulated cells. The hyperpolarization caused by 1-EBIO was reversed by the addition of hIK1 blockers (T Tamada and RJ Bridges, unpubl.). Impedance analysis confirmed that 1-EBIO activated a basolateral membrane conductance, and the reversal of this conductance increased with hIK-1 blockers. 1-EBIO also decreased pH_i from 7.16 to 7.01, and intracellular HCO$_3^-$ from 13.5 to 9.8 mM. V_{bl} was no longer sensitive to DNDS or to serosal HCO$_3^-$ replacement with HEPES. However, bumetanide decreased V_t, and as noted above, the current was almost fully accounted for by Cl$^-$ secretion in forskolin plus 1-EBIO–stimulated cells. These results are consistent with the activation of basolateral membrane hIK-1 potassium channels. For an NBC with a 1:3 Na^+-to-HCO_3^- stoichiometry, the hyperpolarization of V_{bl} to -61.1 exceeds the reversal potential of the NBC and thereby would allow HCO_3^- exit from the cell on the NBC. Consequently, both pH_i and intracellular HCO_3^- decrease. Even though the pH_i and intracellular HCO_3^- have decreased, a driving force of 3.0 mV still exists for HCO_3^- exit across the apical membrane. If cell metabolism produces sufficient HCO_3^-, then HCO_3^- secretion should persist. Indeed, acetazolamide causes a small decrease in the I_{SC} in forskolin plus 1-EBIO-stimulated cells.

1-EBIO also appears to activate the basolateral membrane NKCC cotransporter. Activation of the NKCC may result from the hyperpolarization of V_{bl} as well as cell shrinkage due to the loss of intracellular Cl^- and K^+ (Haas and Forbush 2000). Impedance analysis and cell height measurements confirm that Calu-3 cells shrink when 1-EBIO is added to forskolin-stimulated cells (RJ Bridges and W van Driessche, unpubl.). Addition of 1-EBIO also hyperpolarizes V_{ap} in forskolin-stimulated cells. If intracellular Cl^- were to remain unchanged in forskolin- and forskolin plus 1-EBIO-stimulated cells, a driving force of 6.1 mV would exist for Cl^- exit across the apical membrane. A 6.1-mV driving force across an apical membrane conductance of 71 mS/cm^2 should result in a current of 433 μA/cm^2. Because we observe a Cl^- current of only $\sim$140 μA/cm^2, we predict that the intracellular Cl^- concentration must decrease by 8 mM when 1-EBIO is added to the forskolin-stimulated cells. Unfortunately, we were unsuccessful in our attempts to fabricate a double barrel voltage and chloride-sensitive microelectrode to test this hypothesis.

SUMMARY

Calu-3 cells, an airway serous cell model with high levels of CFTR expression, secrete HCO_3^- in response to a cAMP-mediated agonist but can be stimulated to secrete Cl^- with a basolateral membrane K^+ channel-activating agonist such as 1-EBIO. Activation of CFTR by for-

skolin results in a very high apical membrane conductance and a depolarization of V_{ap} to a value that approaches E_{Cl}. With V_{ap} at E_{Cl}, there is a net driving force for HCO_3^- exit from the cell across the apical membrane. Because of this driving force and the very high apical membrane conductance, HCO_3^- can be secreted from the cells despite the lower conductance of CFTR for HCO_3^- compared with Cl^-. The very high apical membrane anion conductance also results in a depolarization of V_{bl} and the activation of a basolateral membrane NBC. Secreted HCO_3^- is supplied by the basolateral entry of HCO_3^- by this NBC. Thus, the activation of CFTR sets the V_{ap} at E_{Cl}, provides the driving force for HCO_3^- exit, and results in the activation of a basolateral membrane NBC. Activation of basolateral membrane K^+ channels hyperpolarizes V_{bl} and V_{ap}. The hyperpolarization of V_{ap} provides a driving force for Cl^- secretion. The hyperpolarization of V_{bl} results in an inhibition of the NBC and the activation of the NKCC. Therefore, the switch from HCO_3^- secretion to Cl^- secretion is mediated by the activity of the basolateral K^+ channel and which of the basolateral membrane cotransporters is active, NBC or the NKCC. Depending on which of these cotransporters is active, CFTR will secret either HCO_3^- or Cl^-.

ACKNOWLEDGMENTS

I thank my collaborators Dr. Dan Devor, Dr. Tsutomu Tamada, Dr. Martin Hug, Dr. Jim Kreindler, Dr. Willy van Driessche, and Dr. Raymond Frizzell for their many helpful discussions and contributions during the course of these studies. The skillful technical assistance of Mathew Green and Ashwini Mokashi as well as the secretarial assistance of Kim Hankin are acknowledged. Special thanks go to Dr. Paul Quinton, who encouraged me to prepare this article and waited very patiently for its completion.

REFERENCES

Reference is also in this collection.

Al-Bazzaz FJ. 1981. Role of cyclic AMP in regulation of chloride secretion by canine tracheal mucosa. *Am Rev Respir Dis* **123:** 295–298.

Al-Bazzaz FJ, Al Awqati Q. 1979. Interaction between sodium and chloride transport in canine tracheal mucosa. *J App Physiol* **46:** 111–119.

Al-Nakkash L, Hwang TC. 1999. Activation of wild-type and ΔF508-CFTR by phosphodiesterase inhibitors through cAMP-dependent and -independent mechanisms. *Pflugers Arch* **437:** 553–561.

Basbaum CB, Jany B, Finkbeiner WE. 1990. The serous cell. *Annu Rev Physiol* **52:** 97–113.

Berger JT, Voynow JA, Peters KW, Rose MC. 1999. Respiratory carcinoma cell lines. *MUC*genes and glycoconjugates. *Am J Respir Cell Mol Biol* **20:** 500–510.

Boucher RC. 1994a. Human airway ion transport. Part one. *Am J Respir Crit Care Med* **150:** 271–281.

Boucher RC. 1994b. Human airway ion transport. Part two. *Am J Respir Crit Care Med* **150:** 581–593.

Boucher RC. 2007. Airway surface dehydration in cystic fibrosis: Pathogenesis and therapy. *Annu Rev Med* **58:** 157–170.

Boucher RC, Stutts MJ, Knowles MR, Cantley L, Gatzy JT. 1986. Na$^+$ transport in cystic fibrosis respiratory epithelia. Abnormal basal rate and response to adenylate cyclase activation. *J Clin Invest* **78:** 1245–1252.

Boucher RC, Cotton CU, Gatzy JT, Knowles MR, Yankaskas JR. 1988. Evidence for reduced Cl$^-$ and increased Na$^+$ permeability in cystic fibrosis human primary cell cultures. *J Physiol* **405:** 77–103.

Celli BR, Snider GL, Heffner J, Tiep B, Ziment I, Make B, Braman S, Olsen G, Phillips Y. 1995. Standards for the diagnosis and care of patients with chronic obstructive pulmonary disease. *Am J Respir Crit Care Med* **152:** S77–S120.

Corrales RJ, Nadel JA, Widdicombe JH. 1984. Source of the fluid component of secretions from tracheal submucosal glands in cats. *J Appl Physiol* **56:** 1076–1082.

Devor DC, Singh AK, Frizzell RA, Bridges RJ. 1996a. Modulation of Cl$^-$ secretion by benzimidazolones. I. Direct activation of a Ca^{2+}-dependent K$^+$ channel. *Am J Physiol* **271:** L775–L784.

Devor DC, Singh AK, Bridges RJ, Frizzell RA. 1996b. Modulation of Cl$^-$ secretion by benzimidazolones. II. Coordinate regulation of apical GCl and basolateral GK. *Am J Physiol* **271:** L785–L795.

Devor DC, Singh AK, Lambert LC, DeLuca A, Frizzell RA, Bridges RJ. 1999. Bicarbonate and chloride secretion in Calu-3 human airway epithelial cells. *J Gen Physiol* **113:** 743–760.

Duszyk M, Shu Y, Sawicki G, Radomski A, Man SF, Radomski MW. 1999. Inhibition of matrix metalloproteinase MMP-2 activates chloride current in human airway epithelial cells. *Can J Physiol Pharmacol* **77:** 529–535.

Engelhardt JF, Zepeda M, Cohn JA, Yankaskas JR, Wilson JM. 1994. Expression of the cystic fibrosis gene in adult human lung. *J Clin Invest* **93:** 737–749.

Finkbeiner WE, Carrier SD, Teresi CE. 1993. Reverse transcription-polymerase chain reaction (RT-PCR) phenotypic analysis of cell cultures of human tracheal epithelium, tracheobronchial glands, and lung carcinomas. *Am J Respir Cell Mol Biol* **9:** 547–556.

Forstner JF, Jabbal I, Findlay BP, Forstner GG. 1976. Interaction of mucins with calcium, H$^+$ ion and albumin. *Mod Probl Paediatr* **19:** 54–65.

Gaskin KJ, Durie PR, Corey M, Wei P, Forstner GG. 1982. Evidence for a primary defect of pancreatic HCO$_3^-$ secretion in cystic fibrosis. *Pediatr Res* **16:** 554–557.

Gray MA, Pollard CE, Harris A, Coleman L, Greenwell JR, Argent BE. 1990. Anion selectivity and block of the small-conductance chloride channel on pancreatic duct cells. *Am J Physiol* **259:** C752–C761.

Grygorczyk R, Hanrahan JW. 1997. CFTR-independent ATP release from epithelial cells triggered by mechanical stimuli. *Am J Physiol* **272:** C1058–C1066.

Haas M, Forbush B III. 2000. The Na-K-Cl cotransporter of secretory epithelia. *Annu Rev Physiol* **62:** 515–534.

Hadorn B, Johansen PG, Anderson CM. 1968. Pancreozymin secretin test of exocrine pancreatic function in cystic fibrosis and the significance of the result for the pathogenesis of the disease. *Can Med Assoc J* **98:** 377–385.

Illek B, Tam AW, Fischer H, Machen TE. 1999. Anion selectivity of apical membrane conductance of Calu-3 human airway epithelium. *Pflugers Arch* **437:** 812–822.

Inglis SK, Corboz MR, Taylor AE, Ballard ST. 1997. Effect of anion transport inhibition on mucus secretion by airway submucosal glands. *Am J Physiol* **272:** L372–L377.

Inglis SK, Corboz MR, Ballard ST. 1998. Effect of anion secretion inhibitors on mucin content of airway submucosal gland ducts. *Am J Physiol* **274:** L762–L766.

Irokawa T, Krouse ME, Joo NS, Wu JV, Wine JJ. 2004. A "virtual gland" method for quantifying epithelial fluid secretion. *Am J Physiol Lung Cell Mol Physiol* **287:** L784–L793.

Ishiguro H, Steward MC, Wilson RW, Case RM. 1996a. Bicarbonate secretion in interlobular ducts from guinea-pig pancreas. *J Physiol* **495:** 179–191.

Ishiguro H, Steward MC, Lindsay AR, Case RM. 1996b. Accumulation of intracellular HCO$_3^-$ by Na$^+$–HCO$_3^-$ cotransport in interlobular ducts from guinea-pig pancreas. *J Physiol* **495:** 169–178.

Ishiguro H, Steward MC, Naruse S, Ko SBH, Goto H, Case RM, Kondo T, Yamamoto A. 2009. CFTR functions as a bicarbonate channel in pancreatic duct cells. *J Gen Physiol* **133:** 315–326.

Ito Y, Kume H, Yamaki K, Takagi K. 1999. Tetracyclines reduce Na$^+$/K$^+$ pump capacity in Calu-3 human airway cells. *Biochem Biophys Res Commun* **260:** 13–16.

Joo NS, Irokawa T, Robbins RC, Wine JJ. 2006. Hyposecretion, not hyperabsorption, is the basic defect of cystic fibrosis airway glands. *J Biol Chem* **281:** 7392–7398.

Kelley TJ, al-Nakkash L, Drumm ML. 1995. CFTR-mediated chloride permeability is regulated by type III phosphodiesterases in airway epithelial cells. *Am J Respir Cell Mol Biol* **13:** 657–664.

Kopelman H, Durie P, Gaskin K, Weizman Z, Forstner G. 1985. Pancreatic fluid secretion and protein hyperconcentration in cystic fibrosis. *N Engl J Med* **312:** 329–334.

Kopelman H, Corey M, Gaskin K, Durie P, Weizman Z, Forstner G. 1988. Impaired chloride secretion, as well as bicarbonate secretion, underlies the fluid secretory defect in the cystic fibrosis pancreas. *Gastroenterology* **95:** 349–355.

Kreindler JL, Peters KW, Frizzell RA, Bridges RJ. 2006. Identification and membrane localization of electrogenic sodium bicarbonate cotransporters in Calu-3 cells. *Biochim Biophys Acta* **1762:** 704–710.

Lee MC, Penland CM, Widdicombe JH, Wine JJ. 1998. Evidence that Calu-3 human airway cells secrete bicarbonate. *Am J Physiol* **274:** L450–L453.

Li M-S, Holstead RG, Wang W, Linsdell P. 2011. Regulation of CFTR chloride channel macroscopic conductance by extracellular bicarbonate. *Am J Physiol Cell Physiol* **300:** C65–C74.

Liedtke CM, Cole TS. 1998. Antisense oligonucleotide to PKC-ε alters cAMP-dependent stimulation of CFTR in Calu-3 cells. *Am J Physiol* **275:** C1357–C1364.

Lin SY, Amidon GL, Weiner ND, Goldberg AH. 1993. Viscoelasticity of anionic polymers and their mucociliary transport on the frog palate. *Pharm Res* **10:** 411–417.

Linsdell P, Tabcharani JA, Hanrahan JW. 1997. Multi-ion mechanism for ion permeation and block in the cystic fibrosis transmembrane conductance regulator chloride channel. *J Gen Physiol* **110:** 365–377.

List SJ, Findlay BP, Forstner GG, Forstner JF. 1978. Enhancement of the viscosity of mucin by serum albumin. *Biochem J* **175:** 565–571.

Marino CR, Matovcik LM, Gorelick FS, Cohn JA. 1991. Localization of the cystic fibrosis transmembrane conductance regulator in pancreas. *J Clin Invest* **88:** 712–716.

O'Byrine PM, Postma DS. 1999. The many faces of airway inflammation: Asthma and chronic obstructive pulmonary disease. *Am J Respir Crit Care Med* **159:** S41–S66.

Puchelle E, Gaillard D, Ploton D, Hinnrasky J, Fuchey C, Boutterin MC, Jacquot J, Dreyer D, Pavirani A, Dalemans W. 1992. Differential localization of the cystic fibrosis transmembrane conductance regulator in normal and cystic fibrosis airway epithelium. *Am J Respir Cell Mol Biol* **7:** 485–491.

Quinton PM. 1999. Physiological basis of cystic fibrosis: A historical perspective. *Physiol Rev* **79:** S3–S22.

Randell SH, Boucher RC. 2006. Effective mucus clearance is essential for respiratory health. *Am J Respir Cell Mol Biol* **35:** 20–28.

Salinas D, Haggie PM, Thiagarajah JR, Song Y, Rosbe K, Finkbeiner WE, Nielson DW, Verkman AS. 2005. Submucosal gland dysfunction as a primary defect in cystic fibrosis. *FASEB J* **19:** 431–433.

Schultz I. 1987. Electrolyte and fluid secretion in the exocrine pancreas. In *Physiology of the gastrointestinal tract* (ed. Johnson LR), Vol. 2, pp. 1147–1172. Raven, New York.

Shen BQ, Finkbeiner WE, Wine JJ, Mrsny RJ, Widdicombe JH. 1994. Calu-3: A human airway epithelial cell line that shows cAMP-dependent Cl⁻ secretion. *Am J Physiol* **266:** L493–L501.

Singh S, Syme CA, Singh AK, Devor DC, Bridges RJ. 1999. Development of benzimidazolones as chloride secretory agonists. *Ped Pulmonl Suppl* **19:** 190.

Smith JJ, Welsh MJ. 1992. cAMP stimulates bicarbonate secretion across normal, but not cystic fibrosis airway epithelia. *J Clin Invest* **89:** 1148–1153.

Song Y, Salinas D, Nielson DW, Verkman AS. 2006. Hyperacidity of secreted fluid from submucosal glands in early cystic fibrosis. *Am J Physiol Cell Physiol* **290:** C741–C749.

Syme CA, Gerlach AC, Singh AK, Devor DC. 2000. Pharmacological activation of cloned intermediate- and small-conductance Ca^{2+}-activated K^+ channels. *Am J Physiol Cell Physiol* **278:** C570–C581.

Tam PY, Verdugo P. 1981. Control of mucus hydration as a Donnan equilibrium process. *Nature* **292:** 340–342.

Tamada T, Hug MJ, Frizzell RA, Bridges RJ. 2001. Microelectrode and impedance analysis of anion secretion in Calu-3 cells. *JOP* **2:** 219–228.

Tang L, Fatehi M, Linsdell P. 2009. Mechanism of direct bicarbonate transport by the CFTR anion channel. *J Cyst Fibros* **8:** 115–121.

Trout L, King M, Feng W, Inglis SK, Ballard ST. 1998. Inhibition of airway liquid secretion and its effect on the physical properties of airway mucus. *Am J Physiol* **274:** L258–L263.

* Verdugo P. 2012. Supramolecular dynamics of mucus. *Cold Spring Harb Perspect Med* **2:** a009597.

Waters CM, Savla U. 1999. Keratinocyte growth factor accelerates wound closure in airway epithelium during cyclic mechanical strain. *J Cell Physiol* **181:** 424–432.

Welsh MJ. 1983. Inhibition of chloride secretion by furosemide in canine tracheal epithelium. *J Membr Biol* **71:** 219–226.

Willumsen NJ, Boucher RC. 1992. Intracellular pH and its relationship to regulation of ion transport in normal and cystic fibrosis human nasal epithelia. *J Physiol* **455:** 247–269.

Willumsen NJ, Davis CW, Boucher RC. 1989a. Intracellular Cl⁻ activity and cellular Cl⁻ pathways in cultured human airway epithelium. *Am J Physiol* **256:** C1033–C1044.

Willumsen NJ, Davis CW, Boucher RC. 1989b. Cellular Cl⁻ transport in cultured cystic fibrosis airway epithelium. *Am J Physiol* **256:** C1045–C1053.

Yeates DB, Besseris GJ, Wong LB. 1997. Physiochemical properties of mucus and its propulsion. In *The lung, scientific foundations*, 2nd ed. (ed. Crystal RG), pp. 487–515. Lippincott Williams & Wilkins, New York.

Transepithelial Bicarbonate Secretion: Lessons from the Pancreas

Hyun Woo Park and Min Goo Lee

Department of Pharmacology, Brain Korea 21 Project for Medical Science, Yonsei University College of Medicine, Seoul 120-752, Korea

Correspondence: mlee@yuhs.ac

Many cystic fibrosis transmembrane conductance regulator (CFTR)-expressing epithelia secrete bicarbonate (HCO_3^-)-containing fluids. Recent evidence suggests that defects in epithelial bicarbonate secretion are directly involved in the pathogenesis of cystic fibrosis, in particular by building up hyperviscous mucus in the ductal structures of the lung and pancreas. Pancreatic juice is one of the representative fluids that contain a very high concentration of bicarbonate among bodily fluids that are secreted from CFTR-expressing epithelia. We introduce up-to-date knowledge on the basic principles of transepithelial bicarbonate transport by showing the mechanisms involved in pancreatic bicarbonate secretion. The model of pancreatic bicarbonate secretion described herein may also apply to other exocrine epithelia. As a central regulator of bicarbonate transport at the apical membrane, CFTR plays an essential role in both direct and indirect bicarbonate secretion. The major role of CFTR in bicarbonate secretion would be variable depending on the tissue and cell type. For example, in epithelial cells that produce a low concentration of bicarbonate-containing fluid (up to 80 mm), either CFTR-dependent Cl^-/HCO_3^- exchange or CFTR anion channel with low bicarbonate permeability would be sufficient to generate such fluid. However, in cells that secrete high-bicarbonate-containing fluids, a highly selective CFTR bicarbonate channel activity is required. Therefore, understanding the molecular mechanism of transepithelial bicarbonate transport and the role of CFTR in each specific epithelium will provide therapeutic strategies to recover from epithelial defects induced by hyposecretion of bicarbonate in cystic fibrosis.

Epithelial cells in respiratory, gastrointestinal, and genitourinary systems secrete bicarbonate (HCO_3^-)-containing fluids, which include saliva, pancreatic juice, intestinal fluids, airway surface fluid, and fluids secreted by reproductive organs. Bicarbonate is an essential ingredient in these fluids and plays critical roles. For example, bicarbonate is the biological pH buffer that guards against toxic intracellular and extracellular fluctuations in pH (Roos and Boron 1981). Bicarbonate in pancreatic juice and duodenal fluids neutralizes gastric acid and provides an optimal pH environment for digestive enzymes to function properly in the duodenum (Lee and Muallem 2008). In addition, as a moderate chaotropic ion, bicarbonate facilitates the solubilization of macromolecules such as mucins (Hatefi and Hanstein 1969). Recent studies suggest that the abnormal bicarbonate secretion observed in cystic fibrosis (CF) leads to altered mucin hydration and solubilization (Quinton 2010), resulting in hyperviscous mucus that blocks ductal structures of the lung and pancreas (Quinton 2001, 2008).

Pancreatic juice is one of the representative fluids that contain a very high concentration of bicarbonate among bodily fluids secreted from exocrine epithelia. At pH 7.4 and 5% CO_2, the bicarbonate equilibrium concentration is $\sim$25 mM according to the Henderson–Hasselbalch equation. In humans and several other species, such as dogs, cats, pigs, and guinea pigs, the pancreas is capable of generating $\sim$140 mM HCO_3^--containing pancreatic fluid upon stimulation, which is at least a fivefold higher concentration than the plasma. (Domschke et al. 1977; Lee and Muallem 2008). Therefore, pancreatic bicarbonate secretion has attracted attention as a typical model to gain insight into the bicarbonate transport mechanism in diverse epithelial cells. How exocrine glands secrete copious amounts of fluid and bicarbonate has long been a puzzle. The discovery of acidic pancreatic juice from patients with CF was an important advance in understanding the physiological mechanisms of pancreatic bicarbonate secretion (Johansen et al. 1968). In addition, significant progress has been made during the last 25 years with the identification of the molecular nature of many epithelial ion transporters and channels including the cystic fibrosis transmembrane conductance regulator (CFTR), which is mutated in patients with CF (Kerem et al. 1989). We introduce the basic principles of transepithelial bicarbonate transport, in particular in pancreatic duct cells, and the role of CFTR in this process. Additional information on pancreatic bicarbonate secretion can be found in Lee and Muallem (2008) and Lee et al. (2012).

TRANSPORTERS INVOLVED IN TRANSEPITHELIAL BICARBONATE TRANSPORT

Overview

Transepithelial bicarbonate secretion is mediated by a coordinated function of transporters in epithelial cells, whereby transporters in the basolateral membrane absorb bicarbonate from the blood and those in the apical membrane secrete bicarbonate to the luminal space of hollow viscus or exocrine ducts. Recent progress in molecular and physiological techniques revealed the molecular identity and function of epithelial ion transporters at the basolateral and apical membranes (Lee et al. 2012). Major discoveries include the identification of anion channels and transporters in the apical membrane, such as CFTR and Cl^-/HCO_3^- exchangers belonging to the SLC26 family (Ko et al. 2004; Dorwart et al. 2008), in combination with the basolateral bicarbonate uptake mechanisms, such as the Na^+-HCO_3^- cotransporter (NBC) (Zhao et al. 1994; Lee et al. 2000). This article describes major ion transporters expressed at the basolateral and apical membranes in bicarbonate-secreting epithelial cells and their roles in transepithelial bicarbonate secretion. The basic characteristics of basolateral and apical transporters are illustrated in Figure 1.

Transporters in the Basolateral Membrane

Na^+/K^+ ATPase Pump

The Na^+/K^+ ATPase pump is expressed in the basolateral membrane of the epithelial cells that actively secrete fluid and electrolytes such as pancreatic duct cells. The Na^+/K^+ ATPase exchanges three Na_i^+ for two K_o^+ with energy generated by ATP hydrolysis (Morth et al. 2011). The Na^+/K^+ ATPase pump in conjunction with the basolateral K^+ channels converts the chemical energy of ATP into osmotic energy in the form of the Na^+ and K^+ gradients and into electrical energy of a negative membrane potential. These osmotic and electrical energies fuel the fluid and electrolyte transport in epithelial monolayers. The Na^+ gradient is used for cytosolic

Cite this article as *Cold Spring Harb Perspect Med* doi: 10.1101/cshperspect.a009571

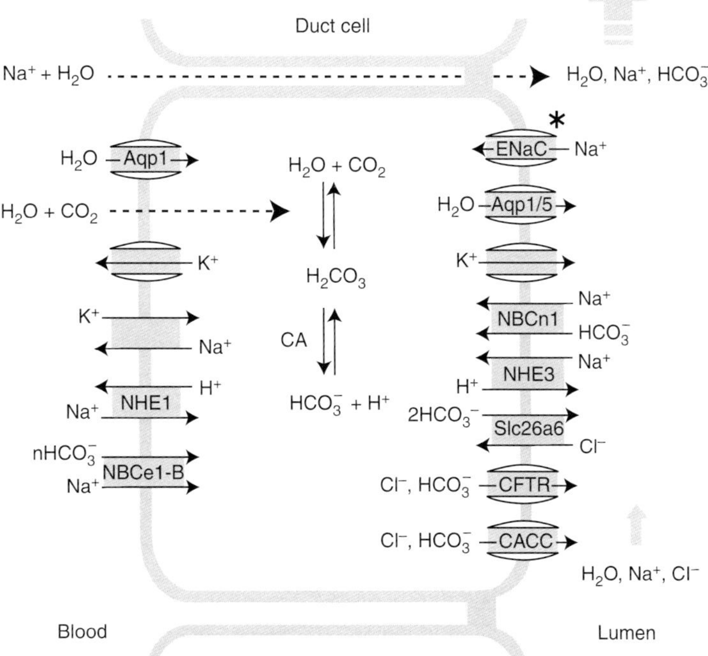

Figure 1. Ion transporters involved in transepithelial bicarbonate transport. Major transporters in the basolateral and luminal membranes of bicarbonate-secreting epithelial cells are illustrated. Bicarbonate uptake through the basolateral membrane is achieved by the NBCe1-B and the combinatorial function of NHE1 and CAs. Apical HCO_3^- secretion is mostly mediated by the CFTR anion channel and the Cl^-/HCO_3^- exchanger Slc26a6 in pancreatic duct cells. Some epithelial cells also express (1) bicarbonate-reabsorbing mechanisms such as NHE3 and NBCn1-A; (2) ENaC, which mediates electrogenic Na^+ absorption; and (3) K^+ channels that secrete K^+ to the luminal fluids in the apical membrane. Overall fluid secretion is driven by HCO_3^- secretion, and Na^+ and water follow via a paracellular route. Water can also travel through a transcellular route via aquaporins.

bicarbonate accumulation by Na^+-HCO_3^- co-transporters (NBCs) and Na^+/H^+ exchangers (NHEs) in the basolateral membrane, and the negative membrane potential facilitates bicarbonate efflux via the electrogenic anion channels and transporters in the apical membrane.

Na^+/H^+ Exchanger (NHE), V-Type H^+ Pump, and H^+/K^+ ATPase Pump

Transepithelial bicarbonate secretion requires bicarbonate entry through the basolateral membrane to maintain adequate bicarbonate concentration in the cytoplasm. This can be achieved by the function of H^+ extrusion mechanisms in the basolateral membrane in conjunction with cytosolic carbonic anhydrases (CAs)

that eventually produce bicarbonate from membrane-diffused CO_2 and water molecules (Fig. 1). The NHEs are electroneutral 1 $Na^+/1$ H^+ exchangers and exchange Na_o^+ for H_i^+ in physiological ion gradients. The mammalian NHE gene family (SLC9A) is composed of three gene clusters: (1) five plasma membrane-type Na^+-selective NHEs (NHE1–NHE5); (2) four organellar cation nonselective NHEs (NHE6–NHE9); and (3) two distantly related NHE-like genes, termed Na^+/H^+ antiporter 1 (NHA1) and NHA2 (Lee et al. 2012). The ubiquitous housekeeping NHE1 is essential for pH$_i$ homeostasis, and it is localized at the basolateral membrane in epithelial cells. Bicarbonate secretion through apical transporters will decrease the bicarbonate concentrations and induce intracellular acidification. In acidic pH_i, NHE1 is activated, and it extrudes H_i^+ in exchange for Na_o^+. This, in turn, facilitates the production and accumulation of bicarbonate inside epithelial cells. Similarly, H^+ extrusion through a vesicular V-type H^+-ATPase pump or H^+/K^+ ATPase pumps at the basolateral membrane can accumulate bicarbonate in epithelial cells. It has been shown that the V-type H^+-ATPase pump is expressed in the basolateral membrane of pig pancreatic duct cells (Villanger et al. 1995) and that the gastric and nongastric types of H^+/K^+ ATPase pumps are expressed in the basolateral membrane of rat pancreatic duct cells (Novak et al. 2011).

However, the ability of these basolateral H^+ extrusion mechanisms to accumulate bicarbonate seems to be limited and does not fully meet the required bicarbonate uptake through the basolateral membrane in cells that secrete high bicarbonate-containing fluids, such as human and guinea pig pancreatic duct cells. For example, NHE1 is activated below a pH_i of 7.0, indicating that it can retain intracellular bicarbonate concentrations only ~ 10 mM at 5% CO_2. In this case, the maximum bicarbonate concentration in pancreatic juice would be only 100 mM even when the hypothetical bicarbonate channel in the apical membrane is maximally activated (membrane potential at -60 mV). Several mathematical models and experimental data suggest that pancreatic duct cells maintain a 20 mM intracellular bicarbonate concentration (pH_i 7.3) that is suitable to generate >140 mM bicarbonate-containing fluid (Ishiguro et al. 1996a; Sohma et al. 2000; Whitcomb and Ermentrout 2004). In addition, inhibitors of NHE1 and the V-type H^+-ATPase pump produced no or variable effects in reducing pancreatic bicarbonate secretion (Lee and Muallem 2008). Therefore, a more direct bicarbonate uptake mechanism is required in the basolateral membrane of pancreatic duct cells.

Na^+-HCO_3^- Cotransporter (NBC)

Evidence suggests that NBC activity at the basolateral membrane is the major route for the basolateral bicarbonate uptake in the pancreatic duct cells (Ishiguro et al. 1996a). The basolateral NBC isoform was cloned from the pancreas and named pNBC1 (Abuladze et al. 1998). Subsequently, pNBC1 was renamed NBCe1-B, which is an electrogenic transporter with 1 Na^+:2 HCO_3^- stoichiometry in pancreatic duct cells. The stoichiometry of electrogenic NBC1 (NBCe1) is an important parameter that determines the direction of bicarbonate movement. The stoichiometry of NBCe1 appears to be dependent on the cell type and PKA-dependent phosphorylation status (Gross et al. 2003). For example, the kidney type NBC1 (kNBC1, NBCe1-A), another variant transcribed from the same gene, seems to have 1 Na^+:3 HCO_3^- stoichiometry in the basolateral membrane of proximal tubule, where it mediates transepithelial $NaHCO_3$ absorption (hence, outward movement of bicarbonate at the basolateral membrane). In this case, the electrorepulsive force of 3 HCO_3^- overcomes the inward movement of 1 Na^+ molecule. The pancreatic electrogenic NBCe1-B uses the Na^+ gradient more efficiently to accumulate cytosolic bicarbonate than the electroneutral NHE1, because NBCe1-B transports two bicarbonate molecules into the cells using the electrochemical energy of one Na^+ molecule. In fact, it has been shown in guinea pig pancreatic duct cells, that NBCe1-B mediates the bulk of basolateral bicarbonate entry during stimulated secretion (Ishiguro et al. 1996a,b). Recent studies suggested that IRBIT (inositol-1,4,5-triphosphate [IP_3] receptor-binding protein released with IP_3) is an important regulator of NBCe1-B in pancreas (Shirakabe et al. 2006).

K^+ Channel

The negative membrane potential generated by K^+ channels in the basolateral membrane of epithelial cells provides the driving force for Cl^- and bicarbonate to exit through the apical membrane, which is a key step that precedes fluid and electrolyte secretion in all secretory epithelial cells. In general, secretory epithelial cells express two important K^+ channels in the basolateral membrane, a Ca^{2+}- and voltage-activated K^+ channel of a large conductance (Maruyama et al. 1983) and a time- and

Cite this article as *Cold Spring Harb Perspect Med* doi: 10.1101/cshperspect.a009571

voltage-independent K$^+$ channel of intermediate conductance (Hayashi et al. 1996; Nehrke et al. 2003). The molecular identity of the channels was subsequently determined as the MaxiK channels encoded by *KCNMA1* (Nehrke et al. 2003; Romanenko et al. 2006) and the IK1 channels encoded by *KCNN4* (Begenisich et al. 2004; Hayashi et al. 2004), respectively. In the pancreatic duct, MaxiK seems to be the major channel maintaining a negative membrane potential during stimulated bicarbonate secretion (Gray et al. 1990). However, MaxiK channels do not appear to contribute to resting membrane potential, possibly because they have a very low open probability during the unstimulated state. The IK1 channel is a potential basolateral K$^+$ channel responsible for the resting K$^+$ permeability (Novak and Greger 1988).

Na$^+$/K$^+$/2 Cl$^-$ Cotransporter (NKCC) and Cl$^-$/HCO$_3^-$ Exchanger (Anion Exchanger, AE)

Epithelial cells that secrete Cl$^-$-rich fluids, such as acinar cells in the exocrine pancreas and salivary glands, express the Na$^+$/K$^+$/2 Cl$^-$ cotransporter NKCC1 in the basolateral membrane. Owing to the electroneutrality of its transport process, NKCC1 maintains intracellular Cl$^-$ concentrations above the electrochemical equilibrium. This high intracellular Cl$^-$ concentration together with the negative membrane potential generated by the basolateral K$^+$ channel provides the driving force for fluid and electrolyte secretion to the luminal space when the apical Cl$^-$ channel is opened by physiological stimuli. However, as we discuss below, a high intracellular Cl$^-$ concentration is unfavorable to secrete a high concentration of bicarbonate via CFTR or Cl$^-$/HCO$_3^-$ exchangers at the apical membrane. Therefore, basolateral NKCC is absent in epithelial cells that produce fluids containing an extremely high concentration of bicarbonate, such as human and guinea pig pancreatic duct cells.

The Cl$^-$/HCO$_3^-$ exchanger AE2 is found in the basolateral membrane of almost all epithelial cells. AE2 is activated by alkaline pH$_i$ to extrude excessive cytosolic bases (Olsnes et al. 1986). In physiological ion gradients, the baso-lateral AE accumulates Cl$^-$ inside the cells and dissipates accumulated intracellular bicarbonate. Therefore, although basolateral AE2 is required for a housekeeping function of cells preventing overt intracellular alkalinization, its activation would inhibit apical bicarbonate secretion in epithelial cells. In fact, mathematical models suggest that inhibition of basolateral AE activity is required for the high bicarbonate secretion in pancreatic duct cells (Sohma et al. 1996, 2000; Whitcomb and Ermentrout 2004).

Transporters in the Apical Membrane

Cystic Fibrosis Transmembrane Conductance Regulator (CFTR)

CFTR (ABCC7) belongs to the "C" branch of the ATP-binding cassette (ABC) transporter superfamily. Most ABC transporters function as membrane pumps, which transport their substrates against the electrochemical gradient using energy generated from ATP hydrolysis (Deeley et al. 2006). Unlike other ABC transporters, CFTR is an anion channel that permits diffusion of substrate ion molecules down to the preexisting electrochemical gradient. In expression cloning, CFTR functions as a cAMP-activated Cl$^-$ channel that has a small conductance (5–10 pS) and a linear current–voltage relationship (Tabcharani et al. 1991).

CFTR is expressed in the apical membrane of secretory epithelial cells. CFTR Cl$^-$ channels have a limited permeability to bicarbonate in typical physiologic conditions. At normal intracellular ($>$20 m$_M$) and extracellular ($>$100 m$_M$) Cl$^-$ concentrations, the P$_{HCO3}$/P$_{Cl}$ of CFTR is 0.2–0.5 (Poulsen et al. 1994; Linsdell et al. 1997; Shcheynikov et al. 2004). Importantly, CFTR bicarbonate permeability is dynamically regulated by intracellular Cl$^-$ concentration-sensitive kinases (Park et al. 2010). Recently, two related kinase families—with-no-lysine (WNK) kinases and sterile 20 (STE20)-like kinases—have emerged as osmotic sensors that modulate diverse ion transporters (Anselmo et al. 2006; Richardson and Alessi 2008). In general, osmotic stress such as a decrease in the intracellular Cl$^-$ concentration, [Cl$^-$]$_i$, activates WNK kinases, including WNK1, which subsequently phosphorylate and activate downstream STE20-like kinases, especially oxidative stress-responsive kinase 1 (OSR1) and STE20/SPS1-related proline/alanine-rich kinase (SPAK) (Richardson and Alessi 2008). In pancreatic duct cells, CFTR activation greatly reduces $[Cl^-]_i$, which triggers the activation of WNK1-SPAK/OSR1 kinase cascade. Activation of the WNK1-SPAK/OSR1 pathway results in a dramatic increase in CFTR bicarbonate permeability, making CFTR primarily a bicarbonate-selective channel (Park et al. 2010). A bicarbonate channel forms an electrodiffusive bicarbonate efflux pathway that is essential for the generation of pancreatic juice containing bicarbonate at concentrations exceeding 140 mM.

In addition to Cl^- and bicarbonate channel activity, CFTR has been suggested to be a central regulator of fluid and electrolyte secretion in many epithelia by regulating other membrane transporters (Lee et al. 2012). CFTR exists in a macromolecular complex at the apical membrane of secretory epithelia. The carboxyl terminus of CFTR forms a PDZ ligand that binds to PDZ domain-containing scaffolds (Short et al. 1998; Wang et al. 1998). In addition, CFTR interacts with SNARE proteins, AKAPs, kinases, and phosphatases (Guggino 2004). In these complexes, CFTR directly or indirectly regulates the activity of several transporters. Functional interactions with CFTR were reported for epithelial Na^+ channels (ENaCs), outwardly rectifying Cl^- channels, Ca^{2+}-activated Cl^- channels, ROMK2 and KvLQT1 K^+ channels, SLC26 transporters, NHE3, NBCn1-A (NBC3), and aquaporins (AQPs) (Kunzelmann 2001; Lee and Muallem 2008).

Cl^-/HCO_3^- Exchanger (Anion Exchanger, AE)

Many exocrine glands are composed of two types of cells: the acinar cells initially secreting Cl^--rich fluids, and the duct cells modifying the ionic composition of the fluids. Pancreatic duct cells absorb luminal Cl^- and secrete the bulk of bicarbonate and fluids. Although the CFTR anion conductive pathway could directly produce bicarbonate-containing fluids, thermodynamically, an apical Cl^-/HCO_3^- exchanger can much more efficiently secrete bicarbonate into the lumen when the luminal space contains Cl^-. In fact, it has been proposed that the Cl^-/HCO_3^- exchanger in the apical membrane mediates bicarbonate secretion in cooperation with an apical Cl^- channel in pancreatic ducts (Steward et al. 2005; Lee and Muallem 2008). In this case, apical Cl^- channels facilitate bicarbonate secretion by recycling Cl^- to the lumen, which maintains the luminal Cl^- concentration for continuous apical Cl^-/HCO_3^- exchange.

The first family of Cl^-/HCO_3^- exchangers to be considered is members of the solute-linked carrier 4 (SLC4) family (Lee et al. 2012). The pancreatic duct cells express the SLC4 transporter AE2 (SLC4A2), but it is localized on the basolateral membrane (Roussa et al. 1999, 2001). The second type of Cl^-/HCO_3^- exchangers comprise transporters belonging to the SLC26 transporter family, which consists of 10 members and transports diverse anions, such as chloride, bicarbonate, oxalate, and sulfate (Ohana et al. 2009). Among the members, SLC26A3, SLC26A4, and SLC26A6 function as $Cl^-/HCO_3^-/I^-$ exchangers (Ko et al. 2002; Xie et al. 2002). Slc26a6 appears to be the major Cl^-/HCO_3^- exchanger in the apical membrane of rat pancreatic duct cells and has electrogenic 1 $Cl^-/2\ HCO_3^-$ exchange activity (Wang et al. 2006; Shcheynikov et al. 2008; Stewart et al. 2009). As discussed below, this electrogenicity of SLC26A6 further contributes to the outward transport of bicarbonate at physiological negative membrane potential. An important feature of the SLC26 transporters and CFTR is their mutual regulation. Thus, the STAS domain of SLC26 transporters interacts with the R domain of CFTR. In addition, the two transporters are connected by PDZ-based adaptors. These interactions are required for activation of both the SLC26 transporters and CFTR (Ko et al. 2004).

Ca^{2+}-Activated Cl^- Channel (CaCC)

CaCC activity is present in the apical membrane of almost all secretory epithelial cells (Gray et al. 1989, 1994; Zeng et al. 1997; Venglovecz et al. 2008). The biophysical property of the channel has been characterized to be a voltage- and Ca^{2+}-activated, time-dependent outwardly rectifying channel (Melvin et al. 2005; Kunzelmann et al. 2009). Recently, members of the anoctamin (ANO; also known as TMEM16) family, in particular ANO1/TMEM16A and ANO2/TMEM16B, were shown to function as CaCCs in several epithelial and neuronal tissues (Caputo et al. 2008; Schroeder et al. 2008; Yang et al. 2008; Stephan et al. 2009; Romanenko et al. 2010). ANO1 is expressed at high levels in the apical membranes of salivary glands (Schroeder et al. 2008; Yang et al. 2008) and pancreatic acinar cells (Huang et al. 2009). However, the molecular identity of CaCC in the pancreatic duct cell is still unknown. The discovery of the ANO/TMEM16 family as the CaCC in several epithelial cells suggests that the ductal CaCC is likely a member of this family. Whether the other ANO/TMEM16 isoforms mediate CaCC activity in pancreatic duct cells awaits further studies (Lee et al. 2012). An interesting possibility is that CaCCs may replace the anion channel function of CFTR in bicarbonate secretion (Zsembery et al. 2000). The P_{HCO3}/P_{Cl} of heterologously expressed ANO1 is 0.1–0.5, indicating that CaCC has a limited ability to produce bicarbonate secretion at typical physiological conditions. However, the P_{HCO3}/P_{Cl} of ANO1 seems to be dynamically regulated by $[Ca^{2+}]_i$ (MG Lee, unpubl.), suggesting that CaCC may play a role in epithelial bicarbonate secretion under certain specific conditions.

K^+ Channel

Some secretory epithelial cells secrete K^+ into the lumen. For example, the salivary gland duct cells absorb Na^+ and secrete K^+ into salivary fluid. Deletion of *Kcnma1*, which encodes MaxiK, impairs salivary K^+ secretion, suggesting MaxiK to be the channel responsible for K^+ efflux in salivary glands (Nakamoto et al. 2008). A recent study showed that MaxiK channels are also expressed at the apical membrane of pancreatic duct cells in guinea pigs (Venglovecz et al. 2011), which may contribute to the potentiation of secretin and CCK (or other Ca^{2+} agonists such as cholinergic) response in pancre-

atic secretion (Gray et al. 1990). However, the physiological role of the apical K^+ channel in pancreatic duct cells is not fully understood at present, because unlike saliva, the major cation in pancreatic juice is Na^+ and pancreatic duct cells do not secrete the bulk of K^+.

Epithelial Na^+ Channel (ENaC)

Some surface epithelial cells of respiratory and digestive tracts and duct cells of salivary glands express ENaC at the apical membrane, which mediates electrogenic Na^+ uptake in these cells (Cook et al. 2002; Catalan et al. 2010). However, pancreatic duct cells do not express ENaC, because they do not absorb Na^+. Simple absorption of Na^+ without secreting K^+ is unfavorable for bicarbonate and fluid secretion because it will evoke a lumen-negative transepithelial potential and net fluid absorption. In the lungs, CFTR is proposed to inhibit ENaC channel function, and deletion of CFTR increases ENaC activity and fluid absorption, which causes a contraction in airway surface fluids (Berdiev et al. 2009). However, in salivary glands, deletion of CFTR inhibited Na^+ absorption by ENaC and greatly diminished the expression of α-ENaC (Catalan et al. 2010). Similarly, ENaC activity is markedly reduced in the sweat ducts of CF patients (Reddy et al. 1999). It is not known whether these tissue-specific regulations of ENaC by CFTR affect transepithelial bicarbonate secretion in ENaC-expressing epithelial cells.

NHE and NBC

Many Na^+- and fluid-absorbing epithelial cells express NHE in the apical membrane, which mediates electroneutral Na^+ uptake (Cook et al. 2002; Catalan et al. 2010). NHE2 and NHE3 are the major NHE isotypes expressed in the apical membrane of epithelial cells. NHE activity in the apical membrane will nullify bicarbonate secretion by donating protons to the lumen. Interestingly, NHEs are expressed in the apical membrane of cells in the distal parts of large-sized salivary and pancreatic ducts, and they may reabsorb bicarbonate when bicarbonate secretion is not needed in these cells (Marteau et al. 1995;

Park et al. 1999; Lee et al. 2000; Luo et al. 2001; Bobulescu et al. 2005). At resting state or low flow rates, pancreatic juice is acidic and contains a high amount of CO_2, indicating an active H^+ secretion by the apical transporters (Gerolami et al. 1989; Marteau et al. 1993). This seems to be mainly mediated by NHE3 in the large-sized ducts (Ahn et al. 2001). The salivary and pancreatic ducts also express the electroneutral NBCn1-A (NBC3) in the apical membrane (Park et al. 2002), which absorbs Na^+ and bicarbonate from luminal fluids. Similar to the case of NHE3, NBCn1-A appears to have a bicarbonate-salvaging function in the resting state (Park et al. 2002).

Other Factors

Aquaporin (AQP)

Transcellular ion secretion results in osmotic water flow. Although the membrane lipid bilayer and paracellular junctions are partially permeable to water, transepithelial water flow is facilitated by the water channel AQP. The AQP family consists of 13 members, all of which can function as water channels, but some can transport additional molecules such as glycerol, urea, ions, and CO_2 (Verkman 2008). In epithelia, the AQPs show highly restricted and cell-specific expression patterns. For example, AQP1 is expressed in both the apical and basolateral domains (Burghardt et al. 2003), whereas AQP5 is expressed in the apical membrane of salivary gland and pancreatic duct cells (Delporte and Steinfeld 2006).

Carbonic Anydrase (CA)

CAs comprise a class of proteins essential for all bicarbonate-related transport functions. The global CA inhibitor acetazolamide significantly inhibits pancreatic fluid and bicarbonate secretion (Pak et al. 1966; Dyck et al. 1972). Further studies revealed that CAs are present at the bicarbonate transporting complex to facilitate bicarbonate transport via membrane transporters, such as NBCs and SLC26 transporters (McMurtrie et al. 2004).

PDZ-Based Adaptor

Transepithelial bicarbonate transport is achieved by the cooperative operation of several membrane proteins in the basolateral and apical membranes. A mechanism that enhances the efficiency of this close cooperation is protein complex formation by adaptor proteins containing PDZ domains (Fig. 2). The PDZ domain was identified as a conserved domain in three proteins: PSD-95, Discs-large, and ZO-1. The PDZ domain is a protein–protein interaction module consisting of 80–90 amino acids that typically binds to target proteins harboring specific carboxy-terminal sequences called PDZ-binding motifs. Although they were initially identified in neuronal tissues, subsequent studies revealed that PDZ-based adaptor proteins are also expressed in diverse epithelia and play critical roles in transepithelial fluid and electrolyte transport (Gee et al. 2009).

The most well-known PDZ-based adaptors in epithelia are the Na^+/H^+ exchanger regulatory factor (NHERF) proteins. NHERF1 (EBP50), NHERF2 (E3KARP), and NHERF3 (CAP70, PDZK1) facilitate the PKA-dependent phosphorylation and membrane trafficking of CFTR and NHE3 in epithelial cells of respiratory and digestive systems, including pancreatic duct cells. These adaptors assemble a large protein complex in the apical membrane of secretory epithelia, where CFTR functions as a central regulator. For example, the mutual regulation of CFTR and SLC26 transporters is enhanced by protein–protein interactions through PDZ-based scaffolds (Ko et al. 2002). In addition, the apical NHE3 and NBCn1-A are associated with CFTR via PDZ-based adaptors, such as NHERF1 and NHERF2, and their activity is regulated by CFTR in the protein complex (Ahn et al. 2001; Park et al. 2002). Assembling a large protein complex greatly enhances signaling efficiency in confined regions of the apical membrane. Upon stimulation with cAMP, bicarbonate-secreting transporters, such as CFTR and apical Cl^-/HCO_3^- exchangers, are activated, and at the same time, bicarbonate-absorbing transporters, such as NHE3 and NBCn1-A, are inhibited to efficiently

    Cite this article as *Cold Spring Harb Perspect Med* doi: 10.1101/cshperspect.a009571

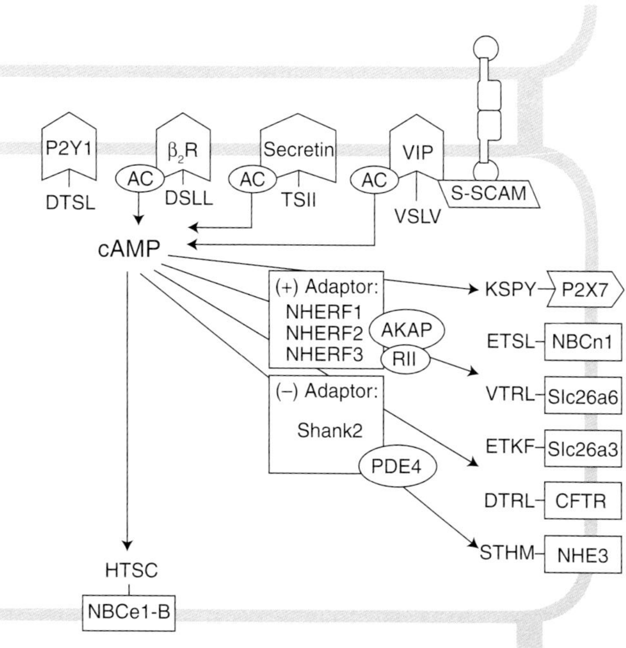

Figure 2. PDZ-based protein–protein interaction in bicarbonate-transporting epithelia. Many membrane receptors and transporters participating in the bicarbonate homeostasis in epithelial cells have a PDZ-binding motif (-X-T/S-X-hydrophobic amino acid) on their carboxyl terminus. The carboxy-terminal sequences are based on the human clones. Excluding purinergic receptors, most of the proteins are associated with cAMP-dependent processes. AC, Adenylyl cyclase; AKAP, cAMP-dependent protein kinase-anchoring protein; P2, purinergic receptor; RII, regulatory subunit of protein kinase A type II. (Other abbreviations are the same as described in the text.)

mediate bicarbonate and fluid secretion (Lee et al. 2012).

In addition to the NHERFs, several other scaffolds with PDZ domains, such as Shank2, S-SCAM, SAP97, and PSD-95, are expressed in various epithelial cells and participate in the regulation of transepithelial fluid and ion transport (Kim et al. 2004; Gee et al. 2009). For example, Shank2 is expressed at the apical region of transporting epithelial cells and modulates the activity of CFTR and NHE3 (Kim et al. 2004; Han et al. 2006). In contrast with NHERF proteins that deliver cAMP/PKA signals, Shank2 mediates an inhibitory effect on the cAMP/PKA pathway. Shank2 associates with phosphodiesterase 4D, which hydrolyzes cAMP, hence lowering local cAMP concentrations in the apical microdomains (Lee et al. 2007). Therefore, the activity of CFTR is dynamically regulated by a competitive balance between CFTR-activating

(NHERFs) and CFTR-inactivating (Shank2) PDZ-domain interactions (Lee et al. 2012).

Many G-protein-coupled receptors (GPCRs) in epithelial cells also have PDZ ligands at their carboxyl termini (Fig. 2). This recruits the GPCRs to the transporting complex, resulting in polarized GPCR expression and delivery of the second messengers to the specific intracellular microdomains. A good example for such an arrangement is the vasoactive intestinal polypeptide (VIP) receptor VPAC₁, which binds to the PDZ-based scaffold S-SCAM (Gee et al. 2009). S-SCAM associates with E-cadherin, a key protein at the adherens junction, and recruits VPAC₁ to the junctional area near the apical pole. Confined localization of VPAC₁ at the junctional area generates a localized cAMP signal close to the apical effectors such as CFTR. This, in turn, enables efficient bicarbonate and fluid secretion in epithelial cells in response to

VIP with minimal effects on the cell interior (Gee et al. 2009; Lee et al. 2012).

MODEL FOR PANCREATIC BICARBONATE SECRETION

Studies on the mechanism and regulation of epithelial bicarbonate secretion have been concentrated in the pancreatic ducts, because the pancreatic duct epithelial cells secrete copious amounts of bicarbonate. Argent and Case (1994) proposed the first model to explain pancreatic bicarbonate secretion. In this model, basolateral NHE1 in cooperation with cytosolic CAs provides a bicarbonate entry mechanism. Bicarbonate is then secreted by Cl^-/HCO_3^- exchange at the apical membrane. During this process, the apical Cl^-/HCO_3^- exchange absorbs the Cl^-, which is then recycled by exiting through CFTR (Argent and Case 1994; Steward et al. 2005). Subsequently, this model was revised by several major findings. For example, the major basolateral bicarbonate influx mechanism was identified as NBCe1-B, which functions as a $1 Na^+/2 HCO_3^-$ cotransporter (Zhao et al. 1994; Ishiguro et al. 1996a; Abuladze et al. 1998). In addition, studies in perfused guinea pig pancreatic ducts and mathematical modeling suggested that a bicarbonate channel activity is required to set the final bicarbonate concentration in the pancreatic juice to 140 mM (Sohma et al. 2000; Whitcomb and Ermentrout 2004).

The pancreatic duct absorbs Cl^- and secretes bicarbonate to generate pancreatic juice containing $\sim$20 mM Cl^- and 140 mM HCO_3^-. The loss of pancreatic bicarbonate secretion in patients with CF (Johansen et al. 1968) indicates that CFTR plays a critical role in bicarbonate secretion. Interestingly, the activity of the apical Cl^-/HCO_3^- exchanger is dependent on the expression of CFTR (Lee et al. 1999a,b). These results strengthened the idea that Cl^-/HCO_3^- exchange mediates pancreatic bicarbonate secretion. However, a limitation of this model has been realized by many people that a classical 1:1 electroneutral Cl^-/HCO_3^- exchanger is able to secrete only a maximum of 80 mM HCO_3^- when intracellular concentrations of Cl^- and bicarbonate are estimated to be equal.

In comparison to these electroneutral transporters, electrogenic bicarbonate transporters are capable of secreting higher concentrations of bicarbonate when the electrorepulsive force generated by the negative membrane potential is coupled to the efflux of bicarbonate. Interestingly, recent studies have suggested that the apical Cl^-/HCO_3^- exchangers may be electrogenic with distinct $Cl^-:HCO_3^-$ stoichiometry (Ko et al. 2004). In case of an electrogenic $1 Cl^-/2 HCO_3^-$, it can accumulate greater amounts of bicarbonate in pancreatic juice compared with an electroneutral exchanger (Steward et al. 2005). Nevertheless, this transporter cannot fully account for the bicarbonate-driven fluid secretion in pancreatic duct cells because a Cl^-/HCO_3^- exchanger with 1:2 stoichiometry is still insufficient to attain 140 mM HCO_3^- in the lumen (Steward et al. 2005). More importantly, the driving force for bicarbonate secretion by the Cl^-/HCO_3^- exchanger are greatly weakened when the luminal Cl^- concentration decreases in the subsequent step of pancreatic secretion, because luminal Cl^- must be required for the Cl^-/HCO_3^- exchanger-mediated bicarbonate secretion. However, it has been shown that a significant fraction of pancreatic bicarbonate secretion is retained even in the absence of luminal Cl^- (Ishiguro et al. 1998, 2009). This implies that an unknown mechanism is likely to be responsible for the ductal bicarbonate secretion, especially at the subsequent stage when the bicarbonate concentration mediated by Cl^-/HCO_3^- exchange approaches equilibrium.

Bicarbonate channels are strong candidates for the transporter responsible for high-concentration bicarbonate secretion (Steward et al. 2005; Lee and Muallem 2008). Having a bicarbonate-selective channel in the apical membrane of a duct cell makes it theoretically possible to secrete up to 200 mM HCO_3^- if cells maintain a membrane potential of -60 mV. It has been previously suggested that CFTR functions as a bicarbonate channel in pancreatic duct cells under some specific conditions (O'Reilly et al. 2000; Reddy and Quinton 2003; Shcheynikov et al. 2004; Whitcomb and Ermentrout 2004; Ishiguro et al. 2009). However, the P_{HCO3}/P_{Cl} of CFTR was reported to be $\sim$0.2–0.5

     Cite this article as *Cold Spring Harb Perspect Med* doi: 10.1101/cshperspect.a009571

(O'Reilly et al. 2000; Sohma et al. 2000). With this permeability ratio, the CFTR anion channel would secrete Cl^- much faster than it would secrete bicarbonate. Thus, CFTR is unable to secrete sufficient amounts of bicarbonate when cells retain a significant intracellular concentration of Cl^-. In one of the previous models, the basolateral membrane was set to be absolutely impermeable to Cl^- but allowed to uptake bicarbonate continuously via basolateral pNBC (Whitcomb and Ermentrout 2004). This caused an extreme reduction in $[Cl^-]_i$ close to 0 mM by CFTR activation, which secretes Cl^- to the lumen. Under these circumstances, the CFTR anion channel secretes isotonic bicarbonate, because of the absence of Cl^- inside the cells. As exemplified in this model, $[Cl^-]_i$ is a critical parameter that determines the anion composition in fluids secreted by epithelial cells, and a decrease in $[Cl^-]_i$ can greatly increase bicarbonate secretion via apical anion channels and transporters. In fact, experimental data in guinea pig ducts and human pancreatic duct cells revealed that $[Cl^-]_i$ in duct cells is $\sim$20 mM at the resting state, and this level is reduced when CFTR is stimulated with cAMP (Ishiguro et al. 2002; Park et al. 2010). However, the minimum $[Cl^-]_i$ that can be induced by a maximum cAMP stimulation is 5 mM in a practical situation, perhaps because of a small Cl^- leak via basolateral Cl^-/HCO_3^- exchangers or Cl^- channels. At this $[Cl^-]_i$, known anion transporters in the apical membrane are still unable to accumulate bicarbonate at concentrations exceeding 140 mM in pancreatic juice. At 20 mM $[HCO_3^-]_i$ and 5 mM $[Cl^-]_i$, the maximum bicarbonate equilibrium concentration in pancreatic juice that can be attained by the $1Cl^-/2HCO_3^-$ exchanger is 128 mM (Steward et al. 2005), and that attained by the CFTR anion channel with a P_{HCO3}/P_{Cl} of 0.2 is 100 mM (Whitcomb and Ermentrout 2004).

Recently, these limitations were complemented by a unique mode of CFTR regulation whereby the P_{HCO3}/P_{Cl} of CFTR is dynamically modulated by the $[Cl^-]_i$-sensitive WNK1-SPAK/OSR1 kinase pathway (Park et al. 2010). In pancreatic duct cells, CFTR activation reduces $[Cl^-]_i$, which subsequently activates the WNK1-SPAK/OSR1 kinase cascade. Activation of the WNK1-SPAK/OSR1 pathway significantly increases the bicarbonate permeability of CFTR, making CFTR into a bicarbonate-selective channel with a P_{HCO3}/P_{Cl} of $\sim$1.5. This CFTR-derived bicarbonate channel would form an electrodiffusive bicarbonate efflux pathway, which can effectively generate pancreatic juice containing $>$140 mM bicarbonate. Interestingly, SPAK and OSR1 activation simultaneously inhibits the CFTR-dependent Cl^-/HCO_3^- exchange activity. It is conceivable that continuous activation of apical 1 Cl^-/2 HCO_3^- exchange would actually absorb bicarbonate from the lumen by its reversed activity when the luminal bicarbonate concentration becomes $>$128 mM. Therefore, inhibition of the apical Cl^-/HCO_3^- exchange is required to prevent the reverse mode of Cl^-/HCO_3^- exchange activity and ultimately achieve high-bicarbonate-containing pancreatic juice. In fact, the requirement of inhibiting the apical Cl^-/HCO_3^- exchange during stimulated secretion had been previously assumed by the mathematical models on high pancreatic bicarbonate secretion (Sohma et al. 2000; Steward et al. 2005).

An integrated model for pancreatic bicarbonate secretion based on the aforementioned findings is illustrated in Figure 3. In response to a meal, secretin hormone and VIP released from the vagal nerve endings stimulate pancreatic duct cells and generate cAMP signals (Lee and Muallem 2008). This initially activates bicarbonate secretion mediated by CFTR-dependent Cl^-/HCO_3^- exchange. The continuous increase in luminal bicarbonate and consequent reduction in luminal Cl^- causes a decrease in $[Cl^-]_i$ to close to an electrical equilibrium (10-fold lower than luminal $[Cl^-]$ at a membrane potential of -60 mV) via apical CFTR Cl^- channel activity. The low $[Cl^-]_i$, in turn, activates WNK1 and the downstream STE20-like kinases, such as SPAK and OSR1. Activation of WNK1-SPAK/OSR1 exerts two critical effects on the apical ion-transporting proteins. First, activation of WNK1-SPAK/OSR1 generates an electrogenic pathway for bicarbonate secretion by increasing the bicarbonate permeability of CFTR, which is essential for the secretion of

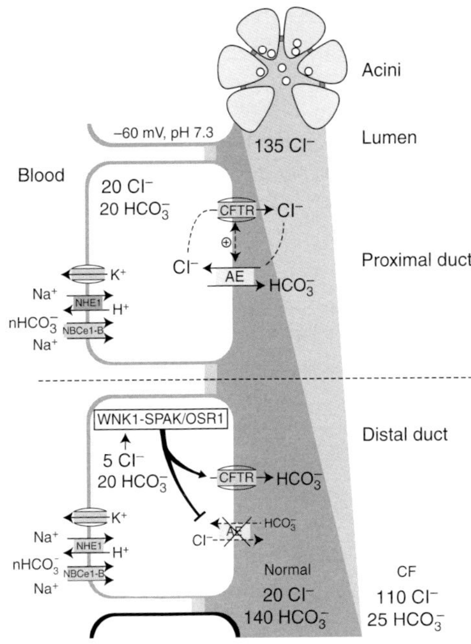

Figure 3. A model of pancreatic fluid and bicarbonate secretion. In the proximal pancreatic duct, cAMP signals activate the CFTR-dependent Cl^-/HCO_3^- exchange at the apical membrane, which enables the duct to absorb part of the Cl^- and secrete as much as 80–100 mM HCO_3^- along with a large volume of fluid into the pancreatic juice. As the fluid arrives at the more distal portions of the duct, the reduced luminal Cl^- and activated CFTR lower the intracellular Cl^- concentration $[Cl^-]_i$ to <10 mM. The low $[Cl^-]_i$ activates WNK1, which phosphorylates SPAK/OSR1, which, in turn, acts on CFTR by converting it into a bicarbonate-selective channel. In contrast, the WNK1-SPAK/OSR1 pathway concurrently inhibits the function of apical Cl^-/HCO_3^- exchange to prevent bicarbonate reabsorption.

pancreatic juice containing bicarbonate at a concentration exceeding 140 mM. Second, WNK1-SPAK/OSR1 activation inhibits apical Cl^-/HCO_3^- exchange activity that may reabsorb bicarbonate from the high-bicarbonate-containing pancreatic juice.

CONCLUSION

The model in Figure 3 describes the mechanisms underlying fluid and bicarbonate secretion in the pancreatic duct, which may also apply to other exocrine epithelia. Mechanisms of

basolateral bicarbonate uptake are now relatively well established. The H^+ extrusion mechanisms including NHE1 expressed in the basolateral membrane of almost all epithelial cells contribute to the cellular accumulation of bicarbonate. In some cells specialized for bicarbonate secretion, such as pancreatic duct cells, NBC mediates the bulk of basolateral bicarbonate uptake. In contrast, the bicarbonate exit pathways in the apical membrane are diverse and depend on the specific condition of each type of epithelia. In epithelial cells that secrete a low concentration of bicarbonate-containing fluids (up to 80 mM), either Cl^-/HCO_3^- exchange or CFTR anion channel with low bicarbonate permeability is sufficient to perform this task. Expression patterns of ion transporters at the apical membrane and the presence of Cl^- in the lumen will determine the route of apical bicarbonate exit. In cells that secrete high-bicarbonate-containing fluids, a CFTR bicarbonate channel activity that is activated by low $[Cl^-]_i$ and WNK1-SPAK/OSR1 kinases is required. It is important to mention that many pathological conditions in the respiratory tract and intestinal lumen can induce hypo- or hyperosmotic stress, which also activates the WNK1-SPAK/OSR1 kinases. Therefore, CFTR bicarbonate channel activity may play a role in epithelial defense against various noxious stimuli in these organs. For example, bicarbonate is an important ingredient to maintain appropriate viscosity and proper function of mucin molecules, which play a critical role in mucosal immunity and the epithelial defense system (Lee and Muallem 2008; Quinton 2008). The fact that CFTR functions not only as a Cl^- channel but also a bicarbonate channel will greatly influence the arena of CFTR research and drug development toward CF patients. In addition, models of epithelial bicarbonate secretion will continue to progress as our knowledge on established pathways expands and novel mechanisms are further elucidated.

ACKNOWLEDGMENTS

We thank Dong-Su Jang for editorial assistance. M.G.L. is supported by grant 2012-0005644 from the National Research Foundation; the

 Cite this article as *Cold Spring Harb Perspect Med* doi: 10.1101/cshperspect.a009571

Ministry of Education, Science, and Technology, Korea; and grant A111218-11-PG03 from the National Project for Personalized Genomic Medicine, Korea Health 21 R&D Project, Ministry of Health & Welfare, Korea.

REFERENCES

Abuladze N, Lee I, Newman D, Hwang J, Boorer K, Pushkin A, Kurtz I. 1998. Molecular cloning, chromosomal localization, tissue distribution, and functional expression of the human pancreatic sodium bicarbonate cotransporter. *J Biol Chem* **273:** 17689–17695.

Ahn W, Kim KH, Lee JA, Kim JY, Choi JY, Moe OW, Milgram SL, Muallem S, Lee MG. 2001. Regulatory interaction between the cystic fibrosis transmembrane conductance regulator and HCO_3^- salvage mechanisms in model systems and the mouse pancreatic duct. *J Biol Chem* **276:** 17236–17243.

Anselmo AN, Earnest S, Chen W, Juang YC, Kim SC, Zhao Y, Cobb MH. 2006. WNK1 and OSR1 regulate the Na^+, K^+, $2Cl^-$ cotransporter in HeLa cells. *Proc Natl Acad Sci* **103:** 10883–10888.

Argent B, Case R. 1994. Pancreatic ducts. Cellular mechanism and control of bicarbonate secretion. In *Physiology of the gastrointestinal tract* (ed. Johnson LR), pp. 1473–1497. Raven, New York.

Begenisich T, Nakamoto T, Ovitt CE, Nehrke K, Brugnara C, Alper SL, Melvin JE. 2004. Physiological roles of the intermediate conductance, Ca^{2+}-activated potassium channel Kcnn4. *J Biol Chem* **279:** 47681–47687.

Berdiev BK, Qadri YJ, Benos DJ. 2009. Assessment of the CFTR and ENaC association. *Mol Biosyst* **5:** 123–127.

Bobulescu IA, Di Sole F, Moe OW. 2005. Na^+/H^+ exchangers: Physiology and link to hypertension and organ ischemia. *Curr Opin Nephrol Hypertens* **14:** 485–494.

Burghardt B, Elkaer ML, Kwon TH, Racz GZ, Varga G, Steward MC, Nielsen S. 2003. Distribution of aquaporin water channels AQP1 and AQP5 in the ductal system of the human pancreas. *Gut* **52:** 1008–1016.

Caputo A, Caci E, Ferrera L, Pedemonte N, Barsanti C, Sondo E, Pfeffer U, Ravazzolo R, Zegarra-Moran O, Galietta LJ. 2008. TMEM16A, a membrane protein associated with calcium-dependent chloride channel activity. *Science* **322:** 590–594.

Catalan MA, Nakamoto T, Gonzalez-Begne M, Camden JM, Wall SM, Clarke LL, Melvin JE. 2010. Cftr and ENaC ion channels mediate NaCl absorption in the mouse submandibular gland. *J Physiol* **588:** 713–724.

Cook DI, Dinudom A, Komwatana P, Kumar S, Young JA. 2002. Patch-clamp studies on epithelial sodium channels in salivary duct cells. *Cell Biochem Biophys* **36:** 105–113.

Deeley RG, Westlake C, Cole SP. 2006. Transmembrane transport of endo- and xenobiotics by mammalian ATP-binding cassette multidrug resistance proteins. *Physiol Rev* **86:** 849–899.

Delporte C, Steinfeld S. 2006. Distribution and roles of aquaporins in salivary glands. *Biochim Biophys Acta* **1758:** 1061–1070.

Domschke S, Domschke W, Rosch W, Konturek SJ, Sprugel W, Mitznegg P, Wunsch E, Demling L. 1977. Inhibition by somatostatin of secretin-stimulated pancreatic secretion in man: A study with pure pancreatic juice. *Scand J Gastroenterol* **12:** 59–63.

Dorwart MR, Shcheynikov N, Yang D, Muallem S. 2008. The solute carrier 26 family of proteins in epithelial ion transport. *Physiology (Bethesda)* **23:** 104–114.

Dyck WP, Hightower NC, Janowitz HD. 1972. Effect of acetazolamide on human pancreatic secretion. *Gastroenterology* **62:** 547–552.

Gee HY, Kim YW, Jo MJ, Namkung W, Kim JY, Park HW, Kim KS, Kim H, Baba A, Yang J, et al. 2009. Synaptic scaffolding molecule binds to and regulates vasoactive intestinal polypeptide type-1 receptor in epithelial cells. *Gastroenterology* **137:** 607–617.

Gerolami A, Marteau C, Matteo A, Sahel J, Portugal H, Pauli AM, Pastor J, Sarles H. 1989. Calcium carbonate saturation in human pancreatic juice: Possible role of ductal H^+ secretion. *Gastroenterology* **96:** 881–884.

Gray MA, Harris A, Coleman L, Greenwell JR, Argent BE. 1989. Two types of chloride channel on duct cells cultured from human fetal pancreas. *Am J Physiol* **257:** C240–C251.

Gray MA, Greenwell JR, Garton AJ, Argent BE. 1990. Regulation of maxi-K^+ channels on pancreatic duct cells by cyclic AMP-dependent phosphorylation. *J Membr Biol* **115:** 203–215.

Gray MA, Winpenny JP, Porteous DJ, Dorin JR, Argent BE. 1994. CFTR and calcium-activated chloride currents in pancreatic duct cells of a transgenic CF mouse. *Am J Physiol* **266:** C213–C221.

Gross E, Fedotoff O, Pushkin A, Abuladze N, Newman D, Kurtz I. 2003. Phosphorylation-induced modulation of pNBC1 function: Distinct roles for the amino- and carboxy-termini. *J Physiol* **549:** 673–682.

Guggino WB. 2004. The cystic fibrosis transmembrane regulator forms macromolecular complexes with PDZ domain scaffold proteins. *Proc Am Thorac Soc* **1:** 28–32.

Han W, Kim KH, Jo MJ, Lee JH, Yang J, Doctor RB, Moe OW, Lee J, Kim E, Lee MG. 2006. Shank2 associates with and regulates Na^+/H^+ exchanger 3. *J Biol Chem* **281:** 1461–1469.

Hatefi Y, Hanstein WG. 1969. Solubilization of particulate proteins and nonelectrolytes by chaotropic agents. *Proc Natl Acad Sci* **62:** 1129–1136.

Hayashi T, Young JA, Cook DI. 1996. The Ach-evoked Ca^{2+}-activated K^+ current in mouse mandibular secretory cells. Single channel studies. *J Membr Biol* **151:** 19–27.

Hayashi M, Kunii C, Takahata T, Ishikawa T. 2004. ATP-dependent regulation of SK4/IK1-like currents in rat submandibular acinar cells: Possible role of cAMP-dependent protein kinase. *Am J Physiol Cell Physiol* **286:** C635–C646.

Huang F, Rock JR, Harfe BD, Cheng T, Huang X, Jan YN, Jan LY. 2009. Studies on expression and function of the TMEM16A calcium-activated chloride channel. *Proc Natl Acad Sci* **106:** 21413–21418.

Ishiguro H, Steward MC, Lindsay AR, Case RM. 1996a. Accumulation of intracellular HCO_3^- by Na^+-HCO_3^-

cotransport in interlobular ducts from guinea-pig pancreas. *J Physiol* **495:** 169–178.

Ishiguro H, Steward MC, Wilson RW, Case RM. 1996b. Bicarbonate secretion in interlobular ducts from guinea-pig pancreas. *J Physiol* **495:** 179–191.

Ishiguro H, Naruse S, Steward MC, Kitagawa M, Ko SB, Hayakawa T, Case RM. 1998. Fluid secretion in interlobular ducts isolated from guinea-pig pancreas. *J Physiol* **511:** 407–422.

Ishiguro H, Naruse S, Kitagawa M, Mabuchi T, Kondo T, Hayakawa T, Case RM, Steward MC. 2002. Chloride transport in microperfused interlobular ducts isolated from guinea-pig pancreas. *J Physiol* **539:** 175–189.

Ishiguro H, Steward MC, Naruse S, Ko SB, Goto H, Case RM, Kondo T, Yamamoto A. 2009. CFTR functions as a bicarbonate channel in pancreatic duct cells. *J Gen Physiol* **133:** 315–326.

Johansen PG, Anderson CM, Hadorn B. 1968. Cystic fibrosis of the pancreas. A generalised disturbance of water and electrolyte movement in exocrine tissues. *Lancet* **1:** 455–460.

Kerem B, Rommens JM, Buchanan JA, Markiewicz D, Cox TK, Chakravarti A, Buchwald M, Tsui LC. 1989. Identification of the cystic fibrosis gene: Genetic analysis. *Science* **245:** 1073–1080.

Kim JY, Han W, Namkung W, Lee JH, Kim KH, Shin H, Kim E, Lee MG. 2004. Inhibitory regulation of cystic fibrosis transmembrane conductance regulator anion-transporting activities by Shank2. *J Biol Chem* **279:** 10389–10396.

Ko SB, Shcheynikov N, Choi JY, Luo X, Ishibashi K, Thomas PJ, Kim JY, Kim KH, Lee MG, Naruse S, et al. 2002. A molecular mechanism for aberrant CFTR-dependent HCO_3^- transport in cystic fibrosis. *EMBO J* **21:** 5662–5672.

Ko SB, Zeng W, Dorwart MR, Luo X, Kim KH, Millen L, Goto H, Naruse S, Soyombo A, Thomas PJ, et al. 2004. Gating of CFTR by the STAS domain of SLC26 transporters. *Nat Cell Biol* **6:** 343–350.

Kunzelmann K. 2001. CFTR: Interacting with everything? *News Physiol Sci* **16:** 167–170.

Kunzelmann K, Kongsuphol P, Aldehni F, Tian Y, Ousingsawat J, Warth R, Schreiber R. 2009. Bestrophin and TMEM16-Ca^{2+} activated Cl^- channels with different functions. *Cell Calcium* **46:** 233–241.

Lee MG, Muallem S. 2008. Physiology of duct cell secretion. In *Pancreas: An integrated textbook of basic science, medicine, and surgery* (ed. Beger H, et al.), pp. 78–90. Blackwell, Oxford.

Lee MG, Choi JY, Luo X, Strickland E, Thomas PJ, Muallem S. 1999a. Cystic fibrosis transmembrane conductance regulator regulates luminal Cl^-/HCO_3^- exchange in mouse submandibular and pancreatic ducts. *J Biol Chem* **274:** 14670–14677.

Lee MG, Wigley WC, Zeng W, Noel LE, Marino CR, Thomas PJ, Muallem S. 1999b. Regulation of Cl^-/HCO_3^- exchange by cystic fibrosis transmembrane conductance regulator expressed in NIH 3T3 and HEK 293 cells. *J Biol Chem* **274:** 3414–3421.

Lee MG, Ahn W, Choi JY, Luo X, Seo JT, Schultheis PJ, Shull GE, Kim KH, Muallem S. 2000. Na^+-dependent transporters mediate HCO_3^- salvage across the luminal membrane of the main pancreatic duct. *J Clin Invest* **105:** 1651–1658.

Lee JH, Richter W, Namkung W, Kim KH, Kim E, Conti M, Lee MG. 2007. Dynamic regulation of cystic fibrosis transmembrane conductance regulator by competitive interactions of molecular adaptors. *J Biol Chem* **282:** 10414–10422.

Lee MG, Ohana E, Park HW, Yang D, Muallem S. 2012. Molecular mechanism of pancreatic and salivary gland fluid and HCO_3^- secretion. *Physiol Rev* **92:** 39–74.

Linsdell P, Tabcharani JA, Rommens JM, Hou YX, Chang XB, Tsui LC, Riordan JR, Hanrahan JW. 1997. Permeability of wild-type and mutant cystic fibrosis transmembrane conductance regulator chloride channels to polyatomic anions. *J Gen Physiol* **110:** 355–364.

Luo X, Choi JY, Ko SB, Pushkin A, Kurtz I, Ahn W, Lee MG, Muallem S. 2001. HCO_3^- salvage mechanisms in the submandibular gland acinar and duct cells. *J Biol Chem* **276:** 9808–9816.

Marteau C, Blanc G, Devaux MA, Portugal H, Gerolami A. 1993. Influence of pancreatic ducts on saturation of juice with calcium carbonate in dogs. *Dig Dis Sci* **38:** 2090–2097.

Marteau C, Silviani V, Ducroc R, Crotte C, Gerolami A. 1995. Evidence for apical Na^+/H^+ exchanger in bovine main pancreatic duct. *Dig Dis Sci* **40:** 2336–2340.

Maruyama Y, Petersen OH, Flanagan P, Pearson GT. 1983. Quantification of Ca^{2+}-activated K^+ channels under hormonal control in pig pancreas acinar cells. *Nature* **305:** 228–232.

McMurtrie HL, Cleary HJ, Alvarez BV, Loiselle FB, Sterling D, Morgan PE, Johnson DE, Casey JR. 2004. The bicarbonate transport metabolon. *J Enzyme Inhib Med Chem* **19:** 231–236.

Melvin JE, Yule D, Shuttleworth T, Begenisich T. 2005. Regulation of fluid and electrolyte secretion in salivary gland acinar cells. *Annu Rev Physiol* **67:** 445–469.

Morth JP, Pedersen BP, Buch-Pedersen MJ, Andersen JP, Vilsen B, Palmgren MG, Nissen P. 2011. A structural overview of the plasma membrane Na^+,K^+-ATPase and H^+-ATPase ion pumps. *Nat Rev Mol Cell Biol* **12:** 60–70.

Nakamoto T, Romanenko VG, Takahashi A, Begenisich T, Melvin JE. 2008. Apical maxi-K (KCa1.1) channels mediate K^+ secretion by the mouse submandibular exocrine gland. *Am J Physiol Cell Physiol* **294:** C810–C819.

Nehrke K, Quinn CC, Begenisich T. 2003. Molecular identification of Ca^{2+}-activated K^+ channels in parotid acinar cells. *Am J Physiol Cell Physiol* **284:** C535–C546.

Novak I, Greger R. 1988. Electrophysiological study of transport systems in isolated perfused pancreatic ducts: Properties of the basolateral membrane. *Pflugers Arch* **411:** 58–68.

Novak I, Wang J, Henriksen KL, Haanes KA, Krabbe S, Nitschke R, Hede SE. 2011. Pancreatic bicarbonate secretion involves two proton pumps. *J Biol Chem* **286:** 280–289.

Ohana E, Yang D, Shcheynikov N, Muallem S. 2009. Diverse transport modes by the solute carrier 26 family of anion transporters. *J Physiol* **587:** 2179–2185.

Cite this article as *Cold Spring Harb Perspect Med* doi: 10.1101/cshperspect.a009571

Olsnes S, Tonnessen TI, Sandvig K. 1986. pH-regulated anion antiport in nucleated mammalian cells. *J Cell Biol* **102:** 967–971.

O'Reilly CM, Winpenny JP, Argent BE, Gray MA. 2000. Cystic fibrosis transmembrane conductance regulator currents in guinea pig pancreatic duct cells: Inhibition by bicarbonate ions. *Gastroenterology* **118:** 1187–1196.

Pak BH, Hong SS, Pak HK, Hong SK. 1966. Effects of acetazolamide and acid–base changes on biliary and pancreatic secretion. *Am J Physiol* **210:** 624–628.

Park K, Olschowka JA, Richardson LA, Bookstein C, Chang EB, Melvin JE. 1999. Expression of multiple Na^+/H^+ exchanger isoforms in rat parotid acinar and ductal cells. *Am J Physiol* **276:** G470–G478.

Park M, Ko SB, Choi JY, Muallem G, Thomas PJ, Pushkin A, Lee MS, Kim JY, Lee MG, Muallem S, et al. 2002. The cystic fibrosis transmembrane conductance regulator interacts with and regulates the activity of the HCO_3^- salvage transporter human Na^+-HCO_3^- cotransport isoform 3. *J Biol Chem* **277:** 50503–50509.

Park HW, Nam JH, Kim JY, Namkung W, Yoon JS, Lee JS, Kim KS, Venglovecz V, Gray MA, Kim KH, et al. 2010. Dynamic regulation of CFTR bicarbonate permeability by $[Cl^-]_i$ and its role in pancreatic bicarbonate secretion. *Gastroenterology* **139:** 620–631.

Poulsen JH, Fischer H, Illek B, Machen TE. 1994. Bicarbonate conductance and pH regulatory capability of cystic fibrosis transmembrane conductance regulator. *Proc Natl Acad Sci* **91:** 5340–5344.

Quinton PM. 2001. The neglected ion: HCO_3^-. *Nat Med* **7:** 292–293.

Quinton PM. 2008. Cystic fibrosis: Impaired bicarbonate secretion and mucoviscidosis. *Lancet* **372:** 415–417.

Quinton PM. 2010. Role of epithelial HCO_3 transport in mucin secretion: Lessons from cystic fibrosis. *Am J Physiol Cell Physiol* **299:** C1222–C1233.

Reddy MM, Quinton PM. 2003. Control of dynamic CFTR selectivity by glutamate and ATP in epithelial cells. *Nature* **423:** 756–760.

Reddy MM, Light MJ, Quinton PM. 1999. Activation of the epithelial Na^+ channel (ENaC) requires CFTR Cl^- channel function. *Nature* **402:** 301–304.

Richardson C, Alessi DR. 2008. The regulation of salt transport and blood pressure by the WNK-SPAK/OSR1 signalling pathway. *J Cell Sci* **121:** 3293–3304.

Romanenko V, Nakamoto T, Srivastava A, Melvin JE, Begenisich T. 2006. Molecular identification and physiological roles of parotid acinar cell maxi-K channels. *J Biol Chem* **281:** 27964–27972.

Romanenko VG, Catalan MA, Brown DA, Putzier I, Hartzell HC, Marmorstein AD, Gonzalez-Begne M, Rock JR, Harfe BD, Melvin JE. 2010. Tmem16A encodes the Ca^{2+}-activated Cl^- channel in mouse submandibular salivary gland acinar cells. *J Biol Chem* **285:** 12990–13001.

Roos A, Boron WF. 1981. Intracellular pH. *Physiol Rev* **61:** 296–434.

Roussa E, Romero MF, Schmitt BM, Boron WF, Alper SL, Thevenod F. 1999. Immunolocalization of anion exchanger AE2 and Na^+-HCO_3^- cotransporter in rat parotid and submandibular glands. *Am J Physiol* **277:** G1288–G1296.

Roussa E, Alper SL, Thevenod F. 2001. Immunolocalization of anion exchanger AE2, Na^+/H^+ exchangers NHE1 and NHE4, and vacuolar type H^+-ATPase in rat pancreas. *J Histochem Cytochem* **49:** 463–474.

Schroeder BC, Cheng T, Jan YN, Jan LY. 2008. Expression cloning of TMEM16A as a calcium-activated chloride channel subunit. *Cell* **134:** 1019–1029.

Shcheynikov N, Kim KH, Kim KM, Dorwart MR, Ko SB, Goto H, Naruse S, Thomas PJ, Muallem S. 2004. Dynamic control of cystic fibrosis transmembrane conductance regulator Cl^-/HCO_3^- selectivity by external Cl^-. *J Biol Chem* **279:** 21857–21865.

Shcheynikov N, Yang D, Wang Y, Zeng W, Karniski LP, So I, Wall SM, Muallem S. 2008. The Slc26a4 transporter functions as an electroneutral $Cl^-/I^-/HCO_3^-$ exchanger: Role of Slc26a4 and Slc26a6 in I^- and HCO_3^- secretion and in regulation of CFTR in the parotid duct. *J Physiol* **586:** 3813–3824.

Shirakabe K, Priori G, Yamada H, Ando H, Horita S, Fujita T, Fujimoto I, Mizutani A, Seki G, Mikoshiba K. 2006. IRBIT, an inositol 1,4,5-trisphosphate receptor-binding protein, specifically binds to and activates pancreas-type Na^+/HCO_3^- cotransporter 1 (pNBC1). *Proc Natl Acad Sci* **103:** 9542–9547.

Short DB, Trotter KW, Reczek D, Kreda SM, Bretscher A, Boucher RC, Stutts MJ, Milgram SL. 1998. An apical PDZ protein anchors the cystic fibrosis transmembrane conductance regulator to the cytoskeleton. *J Biol Chem* **273:** 19797–19801.

Sohma Y, Gray MA, Imai Y, Argent BE. 1996. A mathematical model of the pancreatic ductal epithelium. *J Membr Biol* **154:** 53–67.

Sohma Y, Gray MA, Imai Y, Argent BE. 2000. HCO_3^- transport in a mathematical model of the pancreatic ductal epithelium. *J Membr Biol* **176:** 77–100.

Stephan AB, Shum EY, Hirsh S, Cygnar KD, Reisert J, Zhao H. 2009. ANO2 is the cilial calcium-activated chloride channel that may mediate olfactory amplification. *Proc Natl Acad Sci* **106:** 11776–11781.

Steward MC, Ishiguro H, Case RM. 2005. Mechanisms of bicarbonate secretion in the pancreatic duct. *Annu Rev Physiol* **67:** 377–409.

Stewart AK, Yamamoto A, Nakakuki M, Kondo T, Alper SL, Ishiguro H. 2009. Functional coupling of apical Cl^-/HCO_3^- exchange with CFTR in stimulated HCO_3^- secretion by guinea pig interlobular pancreatic duct. *Am J Physiol Gastrointest Liver Physiol* **296:** G1307–G1317.

Tabcharani JA, Chang XB, Riordan JR, Hanrahan JW. 1991. Phosphorylation-regulated Cl^- channel in CHO cells stably expressing the cystic fibrosis gene. *Nature* **352:** 628–631.

Venglovecz V, Rakonczay Z Jr, Ozsvari B, Takacs T, Lonovics J, Varro A, Gray MA, Argent BE, Hegyi P. 2008. Effects of bile acids on pancreatic ductal bicarbonate secretion in guinea pig. *Gut* **57:** 1102–1112.

Venglovecz V, Hegyi P, Rakonczay Z Jr, Tiszlavicz L, Nardi A, Grunnet M, Gray MA. 2011. Pathophysiological relevance of apical large-conductance Ca^{2+}-activated potassium channels in pancreatic duct epithelial cells. *Gut* **60:** 361–369.

Verkman AS. 2008. Mammalian aquaporins: Diverse physiological roles and potential clinical significance. *Expert Rev Mol Med* **10:** e13.

Villanger O, Veel T, Raeder MG. 1995. Secretin causes H^+/HCO_3^- secretion from pig pancreatic ductules by vacuolar-type H^+-adenosine triphosphatase. *Gastroenterology* **108:** 850–859.

Wang S, Raab RW, Schatz PJ, Guggino WB, Li M. 1998. Peptide binding consensus of the NHE-RF-PDZ1 domain matches the C-terminal sequence of cystic fibrosis transmembrane conductance regulator (CFTR). *FEBS Lett* **427:** 103–108.

Wang Y, Soyombo AA, Shcheynikov N, Zeng W, Dorwart M, Marino CR, Thomas PJ, Muallem S. 2006. Slc26a6 regulates CFTR activity in vivo to determine pancreatic duct HCO_3^- secretion: Relevance to cystic fibrosis. *EMBO J* **25:** 5049–5057.

Whitcomb DC, Ermentrout GB. 2004. A mathematical model of the pancreatic duct cell generating high bicarbonate concentrations in pancreatic juice. *Pancreas* **29:** e30–e40.

Xie Q, Welch R, Mercado A, Romero MF, Mount DB. 2002. Molecular characterization of the murine Slc26a6 anion exchanger: Functional comparison with Slc26a1. *Am J Physiol Renal Physiol* **283:** F826–F838.

Yang YD, Cho H, Koo JY, Tak MH, Cho Y, Shim WS, Park SP, Lee J, Lee B, Kim BM, et al. 2008. TMEM16A confers receptor-activated calcium-dependent chloride conductance. *Nature* **455:** 1210–1215.

Zeng W, Lee MG, Muallem S. 1997. Membrane-specific regulation of Cl^- channels by purinergic receptors in rat submandibular gland acinar and duct cells. *J Biol Chem* **272:** 32956–32965.

Zhao H, Star RA, Muallem S. 1994. Membrane localization of H^+ and HCO_3^- transporters in the rat pancreatic duct. *J Gen Physiol* **104:** 57–85.

Zsembery A, Strazzabosco M, Graf J. 2000. Ca^{2+}-activated Cl^- channels can substitute for CFTR in stimulation of pancreatic duct bicarbonate secretion. *FASEB J* **14:** 2345–2356.

CFTR, Mucins, and Mucus Obstruction in Cystic Fibrosis

Silvia M. Kreda[1], C. William Davis[1,2], and Mary Callaghan Rose[3,4]

[1]Cystic Fibrosis/Pulmonary Research and Treatment Center, University of North Carolina, Chapel Hill, North Carolina 27517-7248

[2]Department of Cell and Molecular Physiology, University of North Carolina, Chapel Hill, North Carolina 27517-7248

[3]Center for Genetic Medicine Research, Children's National Medical Center, George Washington University School of Medicine and Health Sciences, Washington, D.C. 20010

[4]Department of Integrative Systems Biology and Department of Pediatrics, George Washington University School of Medicine and Health Sciences, Washington, D.C. 20010

Correspondence: mrose@childrensnational.org

Mucus pathology in cystic fibrosis (CF) has been known for as long as the disease has been recognized and is sometimes called mucoviscidosis. The disease is marked by mucus hyperproduction and plugging in many organs, which are usually most fatal in the airways of CF patients, once the problem of meconium ileus at birth is resolved. After the CF gene, *CFTR*, was cloned and its protein product identified as a cAMP-regulated Cl^- channel, causal mechanisms underlying the strong mucus phenotype of the disease became obscure. Here we focus on mucin genes and polymeric mucin glycoproteins, examining their regulation and potential relationships to a dysfunctional cystic fibrosis transmembrane conductance regulator (CFTR). Detailed examination of CFTR expression in organs and different cell types indicates that changes in CFTR expression do not always correlate with the severity of CF disease or mucus accumulation. Thus, the mucus hyperproduction that typifies CF does not appear to be a direct cause of a defective CFTR but, rather, to be a downstream consequence. In organs like the lung, up-regulation of mucin gene expression by inflammation results from chronic infection; however, in other instances and organs, the inflammation may have a non-infectious origin. The mucus plugging phenotype of the β-subunit of the epithelial Na^+ channel (βENaC)-overexpressing mouse is proving to be an archetypal example of this kind of inflammation, with a dehydrated airway surface/concentrated mucus gel apparently providing the inflammatory stimulus. Data indicate that the luminal HCO_3^- deficiency recently described for CF epithelia may also provide such a stimulus, perhaps by causing a mal-maturation of mucins as they are released onto luminal surfaces. In any event, the path between CFTR dysfunction and mucus hyperproduction has proven tortuous, and its unraveling continues to offer its own twists and turns, along with fascinating glimpses into biology.

Mucus has long been recognized to reside on the moist, external and internal surfaces of the body. In the early 1930s, for instance, long before he became famous for the codiscovery of penicillin, Howard Florey wrote extensively on mucus secretion in the intestine and airways, as well as on the functions of mucus (Florey 1930; Goldsworthy and Florey 1930; Florey et al. 1932)[5]; his observations strike a surprisingly contemporary chord! The concept of mucus as a barrier against harsh environments or pathogens, however, is a more recent concept that was described first for the stomach (Forte and Forte 1970) and the cervix (Enhorning et al. 1970) in 1970. Since that time, our appreciation of this barrier function offered by mucus has been refined and generalized (Cone 2009) and is presently being appreciated at the molecular level (Pickles 2004; Linden et al. 2008). Also more recently appreciated is the role that the breakdown of the mucus barrier plays in inflammatory diseases (Rhodes 1989; Knowles and Boucher 2002; Johansson et al. 2010). Studies of the genetic disease cystic fibrosis (CF) have been central to much of our current understanding of the function of the mucus barrier in health and its dysfunction in disease, especially with respect to chronic infection and inflammation of the airways in the lungs of CF patients.

Mucus on most epithelial surfaces, such as airways and ocular surfaces, resides as a single layer of polymeric mucin gel (Hollingsworth and Swanson 2004), a few tens of micrometers thick, overlying the epithelial glycocalyx, which includes a class of mucins tethered to the apical plasma membrane (Hattrup and Gendler 2008; Govindarajan and Gipson 2010). In the stomach and colon, however, there are two layers of mucus, one that is adherent, forms from mucins released by goblet cells, and is impermeable to bacteria. The second is more luminal and non-adherent, forms from the enzymatic processing of mucins that are continuously released into the adherent layer, and harbors bacteria (Allen et al. 1984; Taylor et al. 2004; Johansson et al. 2011). Under normal conditions, the mucus barrier on epithelia functions as part of the innate immune system; however, under conditions of inflammation, mucus production is accelerated as part of the body's response to infection or other insults, and the resulting hyperproduction can be deleterious to the health of the patient. This phenomenon of mucus hyperproduction and mucus plugging in the airways is especially true for CF patients, as we consider below.

CYSTIC FIBROSIS, MUCUS, AND MUCINS

Overview

Cystic fibrosis is often referred to as "mucoviscidosis" in early descriptions of the disease (Farber et al. 1943) because copious amounts of viscid mucus were observed in the gastrointestinal and respiratory tracts of children when the disease was first reported in 1938 (Anderson 1938). Mucus is a viscoelastic material that covers and protects the apical surfaces of the respiratory, gastrointestinal, and reproductive epithelial tracts and is a composite of components secreted apically (luminally) by epithelial and glandular cells in the mucosal epithelium (Rose 2006). Mucin glycoproteins (mucins) are major macromolecular components of lung mucus (Rose et al. 1987; Kesimer et al. 2008). Mucins (MUCs) are heavily O-glycosylated (Fig. 1A,B) and are now identified by the *MUC* genes that encode their protein backbones. Because these are well described (Rose and Voynow 2006; Thornton et al. 2007; Hattrup and Gendler 2008), they are only briefly reviewed below. The potential involvement in CF of polymeric (Fig. 1A) (Rose 1988; Rose and Voynow 2006) and membrane-tethered mucins (Fig. 1B) (Hattrup and Gendler 2008) has been reviewed. We review more recently published information on mucins and mucin gene regulation and integrate it into our evolving understanding of the contribution of mucins to CF

---

[5]Many of the manuscripts of Florey and other authors publishing in the 1800s and early 1900s are now available through the archives of PubMed Central, part of the National Library of Medicine. The archive can be searched through PubMed (available at http://www.ncbi.nlm.nih.gov/sites/entrez), but, because the older references in the PMC database may not be included in the main, PubMed database, one must select the PMC database to access it.

 Cite this article as *Cold Spring Harb Perspect Med* doi: 10.1101/cshperspect.a009589

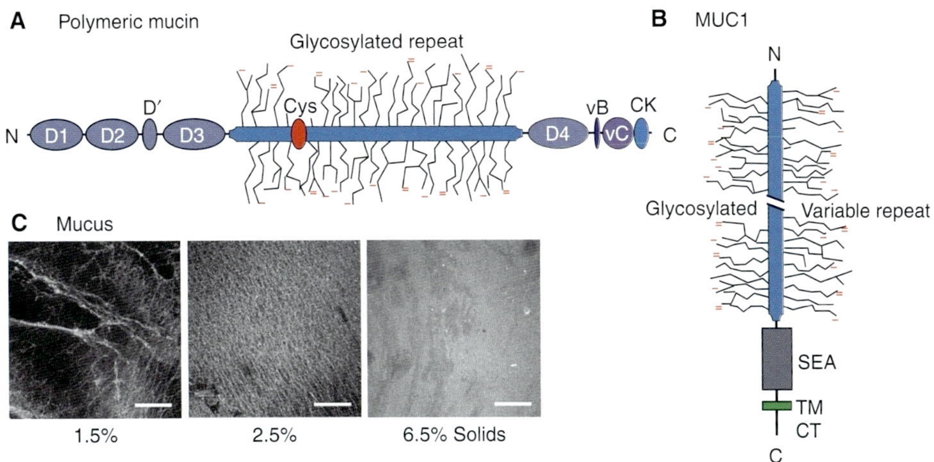

Figure 1. Mucin domain maps and mucus. (*A*) Domain map for a generic, polymeric mucin (after Fig. 1a in Thornton et al. 2007). Indicated are the vWF vB, vC, and D domains, a Cys-rich domain, and the carboxy-terminal cystine knot (CK), in addition to the central glycosylated TR domains. The map most closely represents MUC2; unlike MUC2, MUC5AC and MUC5B have multiple Cys-rich domains within the glycosylated repeat domain (Rose and Voynow 2006). (*B*) Domain map for a tethered mucin, MUC1 (after Fig. 1A in Hattrup and Gendler 2008). Indicated are the cytoplasmic tail (CT), transmembrane (TM), and SEA domains, as well as the glycosylated TR domains, which for MUC1 can have considerable variation in the number of TRs (Hattrup and Gendler 2008). (*C*) Mucus harvested from HBE cell cultures, stained with fluorescent wheat germ agglutinin, and concentrated to 2.5% or 6.5% solids, from a control of 1.5%. The highest concentration (6.5% solids) is representative of CF sputum. (Panel *C* is from Matsui et al. 2005; reprinted, with express permission, from the author.)

disease, especially with regard to the localization and expression of the cystic fibrosis transmembrane regulator (*CFTR*) gene, the causative gene of CF disease.

The initial pathology in CF is manifested in mucosal epithelia and represents the basis for the long-standing concept of a fundamental mucus abnormality in CF patients (Fig. 1C). This is supported by ultrasound studies in the gastrointestinal tract, wherein 90% of CF fetuses manifest inspissated meconium in their distal ileum during the 17–19-wk gestation period (Duchatel et al. 1993). Although this is generally reabsorbed, a small percentage of CF infants present with mucin-containing meconium ileus at birth (Schachter and Dixon 1965). Interestingly, CFTR-null ($Cftr^{-/-}$) mice, pigs, and ferrets, that is, gene-targeted animal models that that do not express the *Cftr* gene, have meconium ileus at birth (Snouwaert et al. 1992; Rogers et al. 2008; Sun et al. 2010; Klymiuk et al. 2011). This mucus overproduction, which reflects

markedly increased levels of Muc1 mRNA and a moderate increase of Muc1 mucin, is obviated in CF mice that do not express the Muc1 gene (Parmley and Gendler 1998), suggesting that the membrane-tethered mucin Muc1 plays an important role in mucus obstruction in the CF colon.

Inspissated material is also found in the pancreas and bile ducts in CF patients and CF animal models. Although these tissues are not typically described as mucosal epithelia, the pancreas expresses MUC6 mRNA and the gallbladder expresses both MUC6 and MUC5B mRNA (Reid et al. 1997b; Hollingsworth 1999). Inspissated material in the pancreatic ducts of CF patients contains MUC6 mucin (Reid et al. 1997b; Hollingsworth 1999), whereas CF mice show higher Muc6 mRNA and protein levels than wild-type mice in their pancreatic ducts (Gouyer et al. 2010). These data suggest that the absence of functional CFTR/Cftr protein may impact regulation or secretion of MUC6/Muc6 mucin in the gastrointestinal tract and contribute to the formation of materials that block pancreatic acini and ducts in CF. Gallstones are frequent in CF patients and generally contain "black" pigment (i.e., Ca bilirubinate) with an appreciable cholesterol admixture. CF mice exhibit similar pathophysiological changes in their gallbladder bile (Freudenberg et al. 2010). Whether or not mucins are part of the gallstone admixture in CF bile is not yet known, but the biliary tract can express and secrete mucins in disease conditions (Gouyer et al. 2010).

In contrast to the gastrointestinal tract, there are no marked morphological abnormalities in the airways of CF fetuses or neonates, and mucus obstruction in CF airways is not observed prenatally or at birth (Zuelzer and Newton 1949; Sturgess and Imrie 1982). Postnatally, dilated acinar and duct lumens in submucosal glands are observed in CF airways early in life with the earliest consistent pathological lesion and evidence of mucus obstruction being observed in the bronchioles (Sturgess 1982). Nevertheless, pulmonary complications (reflecting predisposition to infection and recurring cycles of infection) result in airway mucus obstruction and progressive lung disease, which are the major cause of morbidity and mortality in CF patients (Boat et al. 1989; Welsh et al. 1995). The existence of mucus plugs containing mucins, bacteria, and neutrophils that block the lower airways of CF patients and of sputum with similar characteristics continue to support the concept of "abnormal" lung mucus and/or mucins in CF. Further exacerbating the problem of stagnant mucus in CF airways, the airway surface liquid of CF patients has decreased bactericidal activity (Smith et al. 1996), which may be due to the decreased lactoferrin activity observed in CF sputum (Rogan et al. 2004). In addition, CFTR transport of glutathione and its thiocyanate conjugates is defective in CF airways, inactivating the oxidative antimicrobial system in CF mucus (Childers et al. 2007; Moskwa et al. 2007).

Interestingly, genetic knockout of *Cftr* in mice does not cause lung disease, although the gastrointestinal phenotype of CF disease, meconium ileus, is duplicated in the CFTR-null mouse quite well (Snouwaert et al. 1992). The lack of CF lung disease in the mouse model reflects the dominant expression of a Ca^{2+}-activated Cl^- channel and poor expression of CFTR in the lungs. The mouse model that produces the best CF-like lung phenotype is the transgenic overexpression, selectively in Clara cells, of βENaC, the β-subunit of the epithelial Na^+ channel (Mall et al. 2004a; Zhou et al. 2011). The increased liquid absorption in the lungs of the βENaC Tg mouse appears to drive an inflammatory process that results in mucus metaplasia and overproduction, and mucus plugging (Livraghi et al. 2009). We pursue this "non-infectious inflammation" phenomenon below, after consideration of infectious inflammation.

Mucin Glycoproteins

Mucins are large glycoproteins with a carbohydrate content that accounts for 50%–90% of their molecular mass. They are characterized by a high number of *O*-glycans and an extensive number of tandem repeats (TRs) in their protein backbones that are high in threonine and/or serine and proline, as well as domains specific to individual mucins (Fig. 1A,B). TR domains are a characteristic feature that distinguishes mucins from mucin-like glycoproteins, especially membrane-bound glycoproteins/receptors that have extracellular regions high in serine, threonine, and proline. Mucins are classified by their MUC protein backbones, which are encoded by one of 18 *MUC* genes, as membrane-tethered (MUC1, MUC3A, MUC3B, MUC4, MUC11, MUC12, MUC13, MUC16, MUC17, MUC20), secreted, polymeric, and cysteine rich (MUC2, MUC5AC, MUC5B, MUC6, MUC19) or secreted and non-cysteine rich (MUC7, MUC8, MUC9) mucins (Rose and Voynow 2006). Models of secreted mucins and the membrane-tethered MUC1 mucin are shown in Figure 1, A and B. MUC1 is typically expressed in the apical membrane of epithelial cells. Mucins show somewhat restricted tissue and cell specificity, which is often altered in disease states, especially cancer. The gel-forming mucins are synthesized in goblet and mucosal cells in epithelial tracts (Table 1) and stored in secretory granules until stimulated for luminal

Table 1. Human tissues (normal) expressing polymeric mucin glycoproteins

Tissue	MUC2	MUC5AC	MUC5B	MUC6	MUC19	References
Salivary glands (sublingual)			X		[a]	Zalewska et al. 2000
Nasal glands			X			Martinez-Anton et al. 2006
Nasal septum and nasopharynx		X				Martinez-Anton et al. 2006
Airway superficial epithelium (trachea and bronchii)[b]		X				Rose and Voynow 2006; Thornton et al. 2007
Airway submucosal glands			X			Rose and Voynow 2006; Thornton et al. 2007
Stomach		X		X		Corfield et al. 2000; Linden et al. 2008
Small and large intestine	X					Corfield et al. 2000; Linden et al. 2008
Gallbladder and hepatobilliary ducts			X	X		Sasaki et al. 2007
Pancreatic ducts				X		Reid et al. 1997b; Hollingsworth 1999
Cervix		X	X	X (low)		Andersch-Bjorkman et al. 2007
Conjunctiva	X (low)	X	[c]			Spurr-Michaud et al. 2007

[a]MUC19 mucin glycoprotein is expressed in saliva from rats, horses, pigs, and cows, but was not detected in human saliva (Rousseau et al. 2008).

[b]Muc5b is expressed in Clara cells of wild-type (WT) mouse airways under control conditions, and during development and in mucus metaplasia (Zhu et al. 2008; Roy et al. 2011). Muc5ac is expressed during mucus metaplasia (Zuhdi Alimam et al. 2000; Zhu et al. 2008); however, the human MUC gene products expressed in healthy and inflamed human small airways (bronchioles) remain to be identified.

[c]MUC5B appears to be expressed in lacrimal glands but was not detected in tears (Spurr-Michaud et al. 2007).

release by secretagogues (Kim et al. 2003; Davis and Dickey 2008).

Mucus and Mucins in Healthy and CF Lungs

Mucus and mucins provide a physiological barrier to environmental toxins and pathogens (Hollingsworth and Swanson 2004; Linden et al. 2008; Cone 2009) and are part of the first line of innate immune responses in the conducting airway epithelium (Knowles and Boucher 2002). Polymeric mucins are the major macromolecular components of lung mucus and form viscoelastic gels by interacting with other mucins and/or proteins (Fig. 1C) (Litt et al. 1974; Rose et al. 1979, 1987). MUC5AC and MUC5B are the predominant mucins in lung secretions. In healthy lungs, MUC5AC mRNA expression is restricted to goblet cells in the conducting airway epithelium. MUC5B mRNA is expressed in mucosal cells of the submucosal glands, but is also expressed in goblet cells in the second trimester during gestation (Reid et al. 1997a) and recently has been observed at the protein level in adult lungs (Fig. 2, left panel). Although altered localization of cellular expression of secretory mucin genes has been reported—MUC5B mRNA in goblet cells in patients with various obstructive lung diseases (Chen et al. 2001a) and MUC5AC protein in glandular cells of COPD patients (Caramori et al. 2009)—this has not been observed in CF lungs. However, in neonatal wild-type and CF piglets, both Muc5ac and Muc5b mucins are expressed in goblet cells in the conducting airway epithelium, whereas Muc5b, but not Muc5ac, is expressed in glandular cells. Interestingly, the submucosal glands of CF piglets have

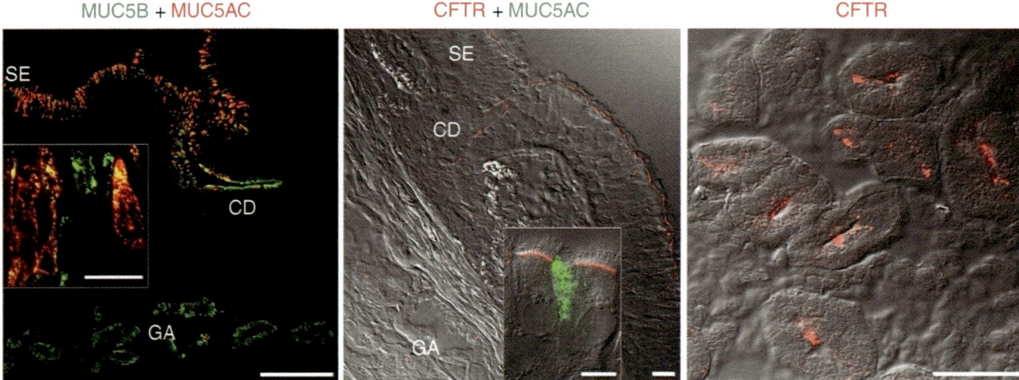

Figure 2. Immunolocalization of CFTR, MUC5AC, and MUC5B in normal human airway epithelium. (*Left* panel) Confocal microscopy overlay image of a human frozen bronchial section from a normal lung donor immunostained with antibodies against MUC5AC (clone 45M1) and MUC5B (VNTR region). MUC5AC expression is restricted to goblet cells of the surface epithelium, whereas MUC5B is expressed in the goblet cells of the surface epithelium and predominantly in submucosal glands both in the ciliated ducts (CD) and acini (GA) in a gradient pattern with the highest levels in the acini. Scale bar, 200 μm. (*Inset*) Goblet cells of the surface epithelium show that MUC5B is present in the surface epithelium and can be coexpressed with MUC5AC. Scale bar, 20 μm. (*Center* panel) Confocal microscopy image of a frozen section of human bronchus immunostained with a CFTR antibody. Images represent an overlay of the differential interference contrast (DIC) (gray) and CFTR immunofluorescence (red) confocal channels, and indicate that CFTR protein is mainly localized in the apical membrane of ciliated cells of the surface epithelium and ciliated ducts; CFTR immunostaining is negligible in serous cells of the gland acini. Scale bar, 20 μm. (Only ∼20% of normal bronchial specimens display CFTR immunostaining in glandular serous cells.) (*Inset*) Confocal image of coimmunostaining of CFTR and MUC5AC in primary cultures of human bronchial epithelium indicating that CFTR is not immunolocalized in goblet cells. Scale bar, 10 μm. (Images from Kreda et al. 2005; reproduced, with permission, from ASCB/ *Molecular Biology of the Cell*.) (*Right* panel) Confocal microscopy images of CFTR immunolocalization in the gland acini of human nasal epithelium. CFTR is localized in the apical membrane of serous cells in ∼50% of normal nasal specimens (Kreda et al. 2005). Scale bar, 20 μm. SE, surface epithelium; CD, ciliated duct; GA, gland acini.

reduced Muc5b immunostaining relative to gland volume, suggesting that mucous cell development is reduced and that airway changes that begin during fetal life may contribute to CF pathogenesis and clinical disease during postnatal life (Meyerholz et al. 2010).

MUC5AC and MUC5B mucins are overproduced in CF lung secretions (Kirkham et al. 2002), especially following exacerbations (Henke et al. 2007). An earlier report that MUC5AC and MUC5B mucin levels are decreased in CF sputum (Henke et al. 2004) highlights the challenges inherent in analyzing CF lung sputum. Early studies in the field identified differences in levels of mucins and the presence of DNA and inflammatory markers in CF sputum (Boat et al. 1976). Biochemical studies show that CF lung mucins in sputum are heterogeneous in size, likely reflect-

ing digestion of mucins by proteinases in CF sputum (Rose et al. 1987). MUC5AC mucin is degraded by neutrophil elastase (Voynow et al. 1999; Henke et al. 2011), as is MUC5B mucin (Henke et al. 2011). MUC2 mucin or peptides are barely detectable in normal (Hovenberg et al. 1996a,b) or CF sputum (Davies et al. 1999; Kirkham et al. 2002), although MUC2 mRNA is well expressed in CF bronchial explants and is up-regulated by *Pseudomonas aeruginosa* lipopolysaccharide (Li et al. 1997) and other inflammatory mediators. Taken together, these data indicate that inflammatory mediators and bacterial byproducts in CF lungs contribute to the overproduction of mucins in vivo, as they do in vitro (for review, see Rose and Voynow 2006).

Knowledge of CFTR genotypes has taught us important lessons about CF (Dorfman et al.

2010), but not enough to completely predict phenotype and disease severity, which has resulted in searches for genetic modifiers in CF patients. A few mucin genes have now been evaluated with regard to their variable number of TR lengths (Guo et al. 2011). A significant association of a 6.4-kb TR in the *MUC5AC* gene is associated with severity of CF lung disease, suggesting that this polymeric mucin gene may be a genetic modifier for CF lung disease. No strong associations were found for *MUC1*, *MUC2*, or *MUC7* mucin genes. Whether the genes that encode the polymeric MUC5B or the membrane-tethered MUC4 and MUC16 mucins are also genetic modifiers of CF lung disease remains to be established.

IMPACT OF CFTR EXPRESSION ON MUCUS PRODUCTION IN CF

Overview

One of the most important aspects of understanding the pathogenesis and pathophysiology of CF with regard to the mucus problem in CF involves understanding CFTR expression at the cellular and tissue levels. The organs that show pathophysiology in CF patients are those of epithelial origin, especially mucosal epithelia (Quinton 1990; Mall et al. 2004b; Kreda et al. 2005). As detailed below, CFTR expression and function have been described in epithelial cells of upper and lower respiratory tracts (Trezise and Buchwald 1991; Engelhardt et al. 1992, 1994; Trezise et al. 1993; Kalin et al. 1999; Penque et al. 2000; Carvalho-Oliveira et al. 2004; Kreda et al. 2005); the gastrointestinal tract (Trezise and Buchwald 1991; Tizzano et al. 1993; Manson et al. 1997; Kalin et al. 1999; Mall et al. 2004b), including pancreatic glands (Crawford et al. 1991; Marino et al. 1991; Trezise and Buchwald 1991; Tizzano et al. 1993; Trezise et al. 1993), liver and gallbladder (Tizzano et al. 1993; Yang et al. 1993); salivary glands (Trezise and Buchwald 1991; Best and Quinton 2005); and male and female reproductive tracts (Trezise and Buchwald 1991; Tizzano et al. 1993, 1994b; Trezise et al. 1993). However, CFTR is also expressed and functional in epithelia that

lack mucus secreting cells—for example, sweat ducts (Quinton 1983, 1990; Cohn et al. 1991; Kartner et al. 1992; Claass et al. 2000; Kreda and Gentzsch 2011) and kidney (Jouret and Devuyst 2009)—but show ion transport abnormalities in CF. Furthermore, CFTR expression has been detected in non-epithelial tissues such as heart (Davies et al. 2004), brain (Mulberg et al. 1998; Robert et al. 2004), smooth muscle (Robert et al. 2004), and lymphocytes (Yoshimura et al. 1991), and the contribution of these sites to epithelial organ damage and CF disease, although underinvestigated, may be significant (Tirouvanziam et al. 2002; Bruscia et al. 2009; Hodges and Drumm 2009).

Although epithelial and non-epithelial tissues show CFTR-associated chloride and HCO_3^- transport abnormalities in CF patients, the most severe lesions responsible for CF morbidity are the result of mucus-based obstructions (often chronically infected) in the lumens of organs that support epithelial gel-forming mucin production (i.e., airways, intestine, and reproductive organs) (Boat et al. 1989; Tizzano and Buchwald 1995; Welsh et al. 1995). That some mucus-producing organs do not show obstructive or severe complications in CF (e.g., salivary and lacrimal glands) (Tizzano and Buchwald 1995; Castagna et al. 2001) has obscured the linkage of CFTR expression/deficiency with mucus production/obstruction, and ultimately, with organ failure. As a consequence, several hypotheses have arisen to explain the relationship between CFTR deficiency and mucus obstruction across different organs. The hypotheses fall into two broad categories— one represented by the concept that mucus obstruction and organ failure is a direct consequence of epithelial CFTR deficiency, the other that mucus obstruction and organ failure occur secondarily to epithelial CFTR deficiency. The hypothesis that CFTR is involved directly in mucus production (i.e., mucin synthesis, post-translation modifications, trafficking, secretion, and/or postsecretion modifications) holds that CFTR deficiency results in mucins with abnormal physicochemical qualities, favoring luminal accumulation, abnormal adhesion, and infection (Barasch et al. 1991; Barasch and al-Awqati

1993; Zhang et al. 1995; Xia et al. 2005; Quinton 2010). However, this hypothesis has never been unequivocally confirmed, as detailed below.

The second possibility, that CFTR is indirectly related to mucus obstruction, has given rise to several hypotheses. One is the "dehydration hypothesis," in which mutated CFTR fails to contribute to the ionic drive needed for appropriate hydration of the mucus layer, resulting in thick mucus with concentrated mucins that collapses onto the underlying epithelium as a first step toward chronic infection and organ disease (Matsui et al. 1998; Boucher 2007; Chen et al. 2010). Another is the "innate immune hypothesis" wherein mutated CFTR results in abnormal mucus with diminished expression/ activity of epithelial antimicrobial molecules important for homeostasis following inflammation and infection, thereby contributing to chronic infection and organ disease (Smith et al. 1996; Zabner et al. 1998). When focused on altered secretion of innate immune proteins by the submucosal glands, this hypothesis is termed "the serous cell malfunction hypothesis" because airway serous cells require CFTR for the secretion of the antimicrobial-rich fluid elaborated by submucosal glands in response to irritants. In CF, fluid secretion by these cells is greatly diminished, resulting in airway mucus that is thicker (Wine and Joo 2004; Ballard and Spadafora 2007).

CFTR Expression

One of the most controversial aspects of CFTR studies has been the elucidation of CFTR expression in non-CF and CF organs. The rationale for these studies has been that determining the localization and levels of CFTR in different organs will help to elucidate the pathogenesis of CF and thus better direct the design of effective therapies. For the last 20 years, many laboratories using different experimental approaches (i.e., expression and localization of CFTR mRNA and protein and functional assessment of ion channel activity) have pursued studies producing some common findings, as well as some more controversial results. In analyzing the published data, it is necessary to take into

account the experimental design and the reagents used, in particular the CFTR antibodies used to produce CFTR expression data in native tissues by means of immunostaining and western blotting. For example, a few studies, which produced controversial expression data in native tissues, used CFTR antibodies that were subsequently reported to perform differently according to tissue quality and histological techniques (Claass et al. 2000; Kreda and Gentzsch 2011). Not all antibodies are created equal, and rigorous controls and independent techniques should be used in parallel to detect low levels of CFTR protein expression in native tissues. Currently, highly sensitive and specific CFTR antibodies are readily available through the North America CF Foundation, and detection techniques are much improved, which has allowed scientists in the field to confirm or challenge expression data reported earlier. In the next sections, some of the key findings on CFTR mRNA and protein expression in the most severely affected organs are discussed from the perspective of the "mucus problem."

CFTR Expression in Goblet Cells of the Respiratory and Gastrointestinal Tracts

Notably, the most severe and life-threatening complications of CF disease are associated with aberrant mucus that obstructs and infects the airway and intestinal lumens, but the functional relationships between CFTR and goblet cells in those organs have not been completely defined. With the exception of a few immunostaining studies reporting CFTR expression in goblet cells of airway and intestinal tissues (Jacquot et al. 1993; Dray-Charier et al. 1995; Kalin et al. 1999), most studies fail to detect significant levels of expression (or channel activity) in goblet cells. As we review below, CFTR expression/activity, instead, is mainly localized apically not to MUC5AC-expressing goblet cells, but to the neighboring epithelial cells (Fig. 2, center panel) known to be involved in ion and fluid transport. Thus, aberrant mucus production appears to be secondary to defective, CFTR-driven ion composition and/or fluid volume in CF epithelia (Kreda et al. 2010; Quinton

Cite this article as *Cold Spring Harb Perspect Med* doi: 10.1101/cshperspect.a009589

2010; Wine et al. 2011). Yet, evidence of chemical differences (e.g., terminal glycosylation with regard to sulfation, sialylation, fucosylation, and carbohydrate composition) in CF mucins (Cheng et al. 1989; Barasch et al. 1991; Barasch and al-Awqati 1993; Zhang et al. 1995; Xia et al. 2005) might implicate a direct role for CFTR in the posttranslational processing of mucins (see section below: Mucin Glycosylation and CF). Because the glycosylstransferases requisite for *O*-glycosylation are predominant in goblet and mucous cells, it would imply that CFTR activity is present in intracellular compartments and is necessary for proper mucin biosynthesis. Emerging technical advances in the field of mucin chemistry and goblet cell and mucin granule isolation will allow us to address these important questions in the near future; however, as the following subsections indicate, the available evidence suggests that CFTR is absent from mucin-secreting cells (Fig. 2, center and right panels), with the exception of pancreatic and hepatobiliary ductal epithelia whose simple, cuboidal epithelial cells express CFTR and are responsible for both transepithelial ion and fluid transport and mucin secretion (Hollingsworth 1999; Kuver et al. 2000; Sasaki et al. 2007).

CFTR Expression in the Upper and Lower Respiratory Tracts

CFTR transcripts are expressed during development in the pseudoglandular epithelium of the human fetal primordial lung (Tizzano et al. 1994a). Subsequently during fetal development, CFTR expression decreases in the alveoli and is gradually restricted to the surface epithelium in the large and small airways, and not detected in fetal submucosal glands (Tizzano et al. 1993, 1994a; Trezise et al. 1993). In contrast, in the neonatal period, CFTR mRNA expression decreases in the surface of the conductive airway epithelia but appears gradually in the submucosal glands (Tizzano et al. 1993, 1994a; Trezise et al. 1993). In the adult lung, although the level of CFTR mRNA expression is lower than in the fetal lung, mRNA is observed in the surface epithelium of nose, trachea, bronchi, and proximal bronchioles, while being scarce in the distal bronchioles and alveoli (Engelhardt et al. 1994; Kreda et al. 2005). In the airway submucosal glands, CFTR mRNA expression appears to decrease distally toward the acini (Engelhardt et al. 1994; Kreda et al. 2005). Interestingly, this pattern of CFTR mRNA expression appears to be at odds with CF pathogenesis. For example, CFTR expression is highest during fetal life, but CF lungs are mostly normal at birth; similarly, CF lung disease develops in the infant and is the primary cause of mortality of young and adult CF patients (Boat et al. 1989), despite low levels of CFTR expression at these ages. Moreover, submucosal gland hyperplasia is one of the earliest pathogenic changes, yet CFTR expression in the glands is seen only after birth (Tizzano et al. 1994a).

Immunolocalization of CFTR Protein Does Not Correlate with CFTR mRNA in Lung Tissues

Localization of CFTR protein is hindered by the capability of CFTR antibodies to detect low levels of protein expressed in adult human lung tissues. Consequently, CFTR expression data, by means of immunostaining, can vary according to the CFTR antibodies used and do not always reflect mRNA expression (Kreda and Gentzsch 2011). For example, an early study by Engelhardt and collaborators, using a polyclonal antibody that recognizes the carboxyl terminus of CFTR, describes CFTR protein expression being associated mainly with the apical membrane of the serous cells in the submucosal glands, although mRNA expression was high in the superficial and gland ductal epithelia (Engelhardt et al. 1992, 1994). In contrast, a subsequent study by Kreda and collaborators, using very sensitive monoclonal antibodies recognizing the NBD2 domain of CFTR (Mall et al. 2004b), describes CFTR protein expression mainly in the apical membrane of all ciliated cells in the epithelium lining the nose and all airway regions, including submucosal gland ducts; however, CFTR protein was not obviously expressed in the acinar cells of bronchial submucosal glands (Kreda et al. 2005). Interestingly, CFTR was detected in the glandular serous cells in ~20% of normal bronchial specimens (Fig. 2, center panel) and in 50% of normal nasal specimens (Fig. 2, right panel) (Kreda et al. 2005). Despite both studies (Engelhardt et al. 1992; Kreda et al. 2005) showing a similar pattern of localization of CFTR transcripts in the human lung, the CFTR protein immuno-localization data are very different. The distribution of immunostaining signal described by Kreda et al. (2005) corresponds with the pattern of mRNA localization in the lung, suggesting that CFTR expression is important in the surface epithelia and that it decreases distally toward the acinar structures in the submucosal glands. However, in the airway glands, CFTR activity appears to be more important in the acinus than in the proximal ducts (Joo et al. 2002; Ballard and Inglis 2004; Ballard and Spadafora 2007; Wu et al. 2007b; Choi et al. 2009; Wine et al. 2011). Thus, one conclusion of these studies is that expression of CFTR gene products may not always predict accurately how critical CFTR activity might affect the physiological functions of a particular tissue or organ. Additionally, these data stress the need to validate immunostaining findings with alternative approaches to identify CFTR expression (i.e., western blotting and channel activity).

At the cellular level, most localization studies identify CFTR protein expression in the apical plasma membrane of ciliated cells in nasal and airway epithelia (Puchelle et al. 1992; Brezillon et al. 1995; Kalin et al. 1999; Penque et al. 2000; Carvalho-Oliveira et al. 2004; Kreda et al. 2005; Kreda and Gentzsch 2011) and the acinar serous cells of submucosal glands (Fig. 2) (Engelhardt et al. 1992; Kalin et al. 1999; Kreda et al. 2005), and more recently, in isolated type II alveolar cells (Brochiero et al. 2004; Bove et al. 2010). Although a few studies indicated CFTR localization in airway goblet cells (Jacquot et al. 1993; Kalin et al. 1999), most laboratories did not detect CFTR expression in mucous/goblet cells in human lung tissues (Engelhardt et al. 1992, 1994; Puchelle et al. 1992; Brezillon et al. 1995; Kalin et al. 1999; Kreda et al. 2005; Kreda and Gentzsch 2011), in primary cultures of airway epithelial cells (Kreda et al. 2005; Cholon et al. 2010), or in mucin-secreting cell lines

derived from airway epithelia, for example, SPOC1 cells (Abdullah et al. 1997) and Calu-3 cells (Kreda et al. 2007). Calu-3 is an adenocarcinoma cell line that differentiates into two distinctive cell types: one that expresses high levels of functional CFTR in the apical membrane, and another that expresses MUC5AC mucins (Kreda et al. 2007). CFTR was not detected in mucin granule membranes isolated from Calu-3 cells (Kreda et al. 2010; SM Kreda and J Sesma, unpubl.), even though Calu-3 cultures do express high levels of CFTR (Shen et al. 1994; Kreda et al. 2007). Thus, CFTR expression and channel activity in mucosal epithelia appear to be associated with cell types involved in ion/fluid transport rather than with mucin-producing cells in the human lung.

CFTR Expression in Human Gastrointestinal Organs

Unlike the lung, the pattern of CFTR mRNA expression in the gastrointestinal track is similar in the fetus and adult, suggesting that CFTR performs similar functions in the fetal and adult gastrointestinal tracts (Crawford et al. 1991; Trezise et al. 1993; Strong et al. 1994) and is consistent with the fact that severe gastrointestinal disease is the initial pathological feature in CF (Park and Grand 1981; Boat et al. 1989). Already during fetal development, and continuing through life, high levels of CFTR mRNA expression are observed in the epithelium of all gastrointestinal regions (Tizzano et al. 1993; Strong et al. 1994). CFTR mRNA expression is highest in the duodenum and decreases along the small intestine. It is lowest in the colon and presents as a decreasing gradient of expression along the crypt–villous axis, with CFTR mRNA being lowest in the luminal half of the villus in all intestinal regions (Tizzano et al. 1993; Trezise et al. 1993; Strong et al. 1994). High levels of CFTR mRNA are also observed in the Brunner's glands (Strong et al. 1994) and in the ductal and centroacinar cells at later development stages and in the adult (Tizzano et al. 1993; Trezise et al. 1993; Strong et al. 1994). In the liver, CFTR mRNA is localized to the epithelia of bile ducts and ductules, whereas in the gallbladder, high

     Cite this article as *Cold Spring Harb Perspect Med* doi: 10.1101/cshperspect.a009589

levels of CFTR mRNA are observed in the epithelium, with no expression detected in the hepatocytes or bile canaliculi (Tizzano et al. 1993; Trezise et al. 1993; Strong et al. 1994).

Importantly, the pattern of CFTR mRNA expression is indicative of the pathological changes observed in the gastrointestinal track of affected CF individuals (Crawford et al. 1991; Tizzano et al. 1993; Trezise et al. 1993; Strong et al. 1994). Meconium ileus and intestinal mucus-based obstruction are among the earliest manifestations of CF (Duchatel et al. 1993) and may reflect a failure in utero to proteolytically digest swallowed and sloughed proteins, which may include mucins. Distal intestinal mucus-based obstruction occurs in >20% of adult patients. Ductal obstruction and dilation in the Brunner's glands, the crypts of Lieberkuhn, and the pancreas have been observed in affected fetuses, while pancreatic insufficiency requires oral enzyme supplementation in ~85% of CF patients. Similarly, liver disease characterized by excessive accumulation of mucus in bile ducts and ductules is observed already early in life and evolves in focal biliary fibrosis and cirrhosis, affecting 20%–50% of all CF patients (Park and Grand 1981; Boat et al. 1989; Tizzano et al. 1993; Yang et al. 1993).

Immunolocalization of CFTR Protein Correlates with CFTR mRNA Findings in Gastrointestinal Tissues

As in the lung, the quality of the antibodies and techniques used must be taken into account in analysis of CFTR immunolocalization data in gastrointestinal tissues (Fig. 2). In the human stomach, CFTR immunolocalizes to the epithelium but at low levels (SM Kreda, unpubl.). In the intestine, CFTR immunolocalizes specifically to the apical membrane of enterocytes in the crypts and villi of the small intestine (Mall et al. 2004b; Kreda and Gentzsch 2011) and the crypts of the jejunum and colon (Crawford et al. 1991; Mall et al. 2004b); CFTR protein immunostaining is more prominent in the base decreasing toward the tip of the crypts (Mall et al. 2004b; Kreda and Gentzsch 2011), following the pattern of CFTR mRNA expression. In the pancreas, liver, and gallbladder, CFTR immunostaining also reflects the pattern of mRNA expression, with CFTR protein being detected chiefly in the apical membrane of non-mucous cells (Crawford et al. 1991; Marino et al. 1991; Cohn et al. 1993; Yang et al. 1993). Although goblet cell hyperplasia is present in all gastrointestinal regions and mucus obstruction is a feature of gastrointestinal CF disease (Park and Grand 1981; Boat et al. 1989; Tizzano et al. 1993), most studies have failed to detect CFTR immunostaining associated with goblet cells in the GI epithelia, even using epitope retrieval and other signal-enhancing techniques (Mall et al. 2004b; Kreda and Gentzsch 2011). A few studies reported CFTR immunostaining associated specifically with intracellular organelles of goblet cells, in all intestinal regions (Kalin et al. 1999) and in the gallbladder (Kuver et al. 2000); however, the antibodies used in these studies have been reported to produce unexpected results in human tissues (Claass et al. 2000). Moreover, CFTR is immunolocalized to the apical membrane of enterocytes in murine intestine, and no CFTR expression was observed in goblet cells in the proximal and distal intestine (Jakab et al. 2011). The selective expression of CFTR in the enterocytes supports a role of CFTR in regulating ion and fluid transport in the human and rodent intestine (Garcia et al. 2009).

CFTR Expression in the Reproductive Organs

Virtually all CF males are azoospermic because of atrophy and obstruction of the epididymis, vas deferens, and seminal vesicles. These pathological changes have often been interpreted as a secondary lesion to mucus-based obstruction of the epididymis and vas deferens ductal system (Tizzano et al. 1994b). Accordingly, CFTR mRNA is expressed mainly in the ductal epithelial cells of the epididymis in the human male fetus, newborn, and infant (Trezise and Buchwald 1991; Tizzano et al. 1993, 1994b). In adults, the epithelium of the epididymis and vas deferens shows a high level of CFTR mRNA expression, whereas low mRNA expression is observed in testis, prostate, or seminal vesicles throughout life (Tizzano et al. 1994b).

CFTR immunostaining reflects mRNA expression and is observed at the luminal surface of epithelial cells of the efferent ducts and the epididymis (Hihnala et al. 2006; Kujala et al. 2007). In the seminal vesicles and prostate, no immunostaining is observed in the glandular epithelium, but strong labeling for CFTR is present on the luminal surface of the ductal epithelium (Hihnala et al. 2006; Kujala et al. 2007). CFTR is also immunolocalized to the equatorial segment of human and mouse spermatozoids (Hihnala et al. 2006; Kujala et al. 2007; Xu et al. 2007). More studies are needed to understand the link between CFTR deficiency and the pathological changes in the CF male reproductive tract.

Females with CF have reduced fertility, but the underlying causes are not well understood. Infertility has been proposed to be mainly a consequence of abnormally thick, dense cervical mucus that presents a barrier for sperm penetration and transport to the egg (Brugman and Taussig 1984; Tizzano et al. 1994b; Hodges et al. 2008). However, altered ion and fluid transport (due to CFTR deficiency) throughout the female reproductive track, endocrine abnormalities, and menstrual irregularities may also account for infertility in women and mouse models with CF (Brugman and Taussig 1984; Tizzano et al. 1994b; Hodges et al. 2008; Chan et al. 2009). Expression of CFTR mRNA is evident in the female fetus after the third trimester in the epithelium of the uterine cervix and the Fallopian tube (Tizzano et al. 1993). In newborns and infants, CFTR mRNA expression is found at high, moderate, and low levels in the epithelium and glands of the cervix, the Fallopian tubes, and the endometrium, respectively; whereas in adults, a high level of mRNA expression is detected in the cervix and to a lesser degree in both endometrium and Fallopian tubes; no expression was observed in ovaries at any age (Tizzano et al. 1994b). By immunostaining, CFTR has been observed in the apical surface of uterine and oviduct epithelial cells involved in fluid secretion (Rochwerger and Buchwald 1993). In the Fallopian tubes, CFTR is localized specifically to the apical membrane of epithelial cells (Ajonuma et al. 2005). CFTR immuno-

staining is also present in the apical membrane of fluid-secreting cells of the endocervix (SM Kreda, unpubl.). The levels of CFTR mRNA expression change during the menstrual and (estrous) cycle in the female reproductive track (Hodges et al. 2008; Chan et al. 2009). Whether these changes correlate with the fluctuations in levels of specific mucins and alterations in mucin glycosylation and physical properties that occur during estrous has not yet been evaluated as far as we are aware, but may be informative in understanding the impact of CFTR on mucins and mucosal components in an epithelial tract that is not typically infected in CF.

CFTR Expression in the Organs of CF Patients

The pattern of expression of mRNA transcripts for ΔF508 CFTR (the most frequent mutation in the Caucasian population) and wild-type CFTR is similar in human tissues, indicating that there is no significant defect at the level of mRNA expression in (ΔF508) CF tissues (Trapnell et al. 1991; Engelhardt et al. 1992). However, the protein expression and localization of mutated CFTR, in particular ΔF508 CFTR, have produced different results depending on the antibodies used. For example, most CFTR antibodies recognize wild-type CFTR protein at the apical membrane of epithelial cells in sweat ducts from normal but not from ΔF508 homozygous individuals (Quinton 1990; Cohn et al. 1991; Kartner et al. 1992; Kalin et al. 1999; Claass et al. 2000; Kreda and Gentzsch 2011), in agreement with the concept that the absence of protein in the apical membrane of epithelial cells in ΔF508 homozygous tissues reflects lack of fully glycosylated CFTR protein and channel activity in those tissues (Cheng et al. 1990; Quinton 1990; Cohn et al. 1991; Marino et al. 1991; Puchelle et al. 1992; Zeng et al. 1997; Mall et al. 2004b; Kreda et al. 2005; Kreda and Gentzsch 2011). In contrast, a few immunostaining studies reported that localization of human ΔF508 CFTR did not differ from wild-type CFTR in airway and intestinal tissues (Dray-Charier et al. 1995; Dupuit et al. 1995; Kalin et al. 1999; Penque et al. 2000). However, with

     Cite this article as *Cold Spring Harb Perspect Med* doi: 10.1101/cshperspect.a009589

the availability of CFTR antibodies of high sensitivity and specificity and sensitive confocal microscopy techniques, these issues have been revisited. In these studies, ΔF508 CFTR protein was not detected in the apical membrane of epithelial cells in native tissues and primary cell cultures from human nose and lung and intestine (Mall et al. 2004b; Kreda et al. 2005; Kreda and Gentzsch 2011), as originally observed in the sweat duct of CF patients (see above). These observations reflect the loss of processing and trafficking of CFTR that occurs in mutated CFTR.

MUCIN OVERPRODUCTION IN CF

Overview

Several processes can account for mucin overproduction and increased mucin levels in the human respiratory tract, including hypersecretion of mucins, proteolytic cleavage of membrane-tethered mucins, up-regulation of *MUC* gene expression, and epithelial remodeling by goblet cell hyperplasia/metaplasia and/or glandular hyperplasia (for review, see Rose and Voynow 2006). A network of genes that transcriptionally mediate goblet cell metaplasia/hyperplasia in murine and human epithelium is emerging (Chen et al. 2009). In CF lung epithelial, patches of goblet cell hyperplasia were observed (Voynow et al. 2005). In contrast, goblet cell hyperplasia is not observed in CF bronchial biopsies, although an increase in goblet cell size in CF tissue, as well as a fourfold increase in submucosal gland volume, was reported (Hays and Fahy 2006). It is worth noting, however, that goblet cell size is not necessarily a good marker of mucin production, because a cell in which mucin synthesis is elevated could be secreting mucins at an equal rate. In fact, at this point, we have no idea of the cellular mechanisms regulating the quantity of mucin stores in goblet cells.

In the murine conducting airway epithelium, allergenic sensitization and challenge or the Th2 cytokine IL13 induce mucus cell metaplasia, and thus expression of Muc5ac in goblet cells (Zuhdi Alimam et al. 2000). It is now clearly evident that murine lung secretory cells express Muc5b at baseline conditions and transdifferentiate to goblet cells that express Muc5b and Muc5ac in response to allergenic sensitization and challenge (Zhu et al. 2008). Based on these studies and studies on mucin gene expression during development of mouse lungs (Roy et al. 2011) and mucin gene knockout studies (C Evans, unpubl.), emerging paradigms are that Muc5b is an intrinsic component of homeostatic mucosal defense in murine lungs and that both *Muc5ac* and *Muc5b* genes can be up-regulated in vivo by mediators activated during inflammation and infection. The relevance of these concepts to human lung diseases, especially CF, will unfold further as studies are performed in *Cftr* ferrets and pigs, which have abundant submucosal glands and thus more closely resemble human airways.

A suggestion that appears to arise periodically is that it is the exocytic secretion of mucins that is affected in CF. The existing data, however, do not support this notion. Basal- and agonist-stimulated mucin secretion are not different in primary cultures of airway epithelial cells from non-CF and CF individuals (Fig. 3A) (Lethem et al. 1993) or from non-CF sources treated with CFTR inh172, a potent CFTR inhibitor (Fig. 3B) (Hodges and Drumm 2009). Another suggestion, discussed briefly below (see section on Mucin Secretion and Regulation), in favor for several years, held that a dysfunctional CFTR was responsible for changes in glycosylation, sulfation, and sialylation of CF mucins. However, this would require expression of CFTR in goblet cells, which is not supported by current data, as addressed above. The current paradigm is that the various processes that drive mucin overproduction and hypersecretion are directed to a large extent by inflammation and are predominantly impacted by mechanisms that drive mucin gene regulation and mucin secretion. These are independent processes, although the terms "mucin production" and "mucin secretion" are still sometimes used interchangeably in the literature. Understanding how mucin gene regulation and secretion are modulated during inflammation in CF airways is a fundamental question that remains to be answered. Thus, recent information on mucin gene regulation and

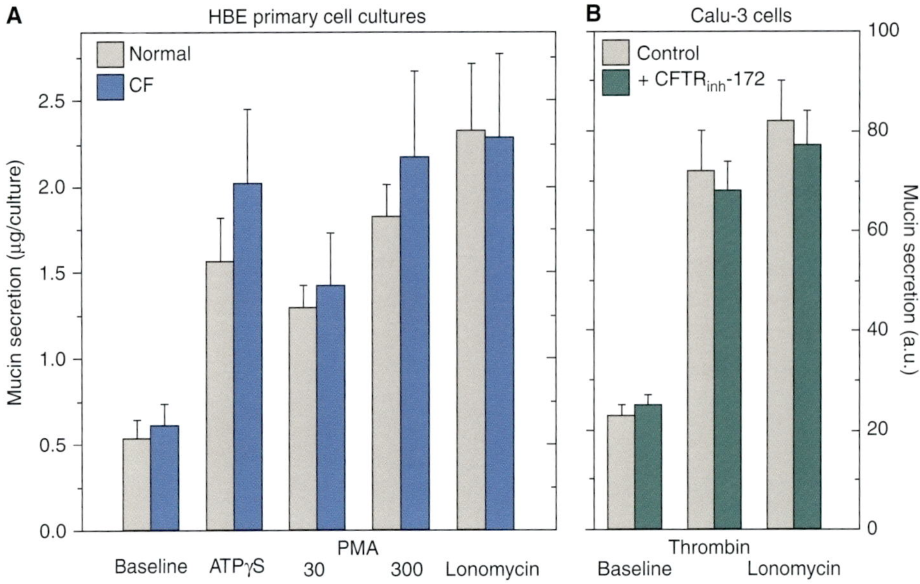

Figure 3. Mucin secretion in human bronchial epithelial cell cultures from healthy (normal) and CF individuals (*A*), and from Calu-3 cells exposed to $CFTR_{inh}$-172 (*B*). Mucin secretion, assessed as described in Abdullah et al. (2012) and Kreda et al. (2007), was not affected in either case. The data are expressed as the mean $\pm$ S.E.: HBE cell culture data (LH Abdullah and CW Davis, unpubl.) ($n = 6$); Calu-3 cell data (S Kreda, unpubl.) ($n = 3$). a.u., arbitrary units.

mucin secretion potentially relevant to our understanding of CF lung disease is reviewed below.

Mucin Gene Regulation

Multiple stimuli (bacterial, viral, inflammatory) induce overproduction of polymeric mucins by up-regulating expression of polymeric mucin genes. This regulation is complex and can be both cell and mucin gene specific. Transcriptional up-regulation of *MUC* gene expression is activated following binding of mediators to membrane receptors on secretory cells, resulting in activation of various signal transduction pathways and transcription factors that then translocate to the nucleus and bind to *cis*-sites on *MUC* gene promoters. Typically, studies initially use lung epithelial cancer cell lines (NCI-H292 and A549) and/or immortalized lung epithelial cell lines, for example, HBE1 cells (Yankaskas et al. 1993) or 16HBE cells (SV-40 virus-transformed, immortalized HBE cell lines) (Bruscia et al. 2002). To assess the rele-

vance of observations, findings are then evaluated in primary differentiated human bronchial epithelial (HBE) or human nasal epithelial (HNE) cells.[6] These latter cells differentiate to a polarized epithelium with ciliated, goblet, and basal cells when grown under air–liquid interface conditions and are model systems that morphologically recapitulate the conducting epithelium of the respiratory tract (Wu et al. 1990; Kondo et al. 1993; Gruenert et al. 1995). Some mechanisms present in lung cancer cell lines are lacking in HBE cells, and instances are pointed out below. This reinforces the importance of validating results obtained in cancer or immortalized cell lines with differentiated HBE cells or animal models that reflect human airways when investigating therapies for mucus overproduction. In the sections below, the experimental cell types used are stated, rather than the generic term "lung epithelial cells."

---

[6]Here these model systems are referred to as differentiated HBE or HNE cells.

Most of the literature with regard to mediators that increase *MUC2, MUC5AC,* and *MUC5B* gene expression from 1995 to 2005 was recently reviewed (Rose and Voynow 2006), and schematics identifying the impact of these pathways on specific *cis*-sites in the promoters of *MUC2, MUC5AC,* and *MUC5B* genes (Thai et al. 2008) or the *MUC5AC* gene (Voynow and Rubin 2009) have been published. This section attempts to update and integrate information recently reported and focuses on regulation of the *MUC5AC* (Fig. 4) and *MUC5B*

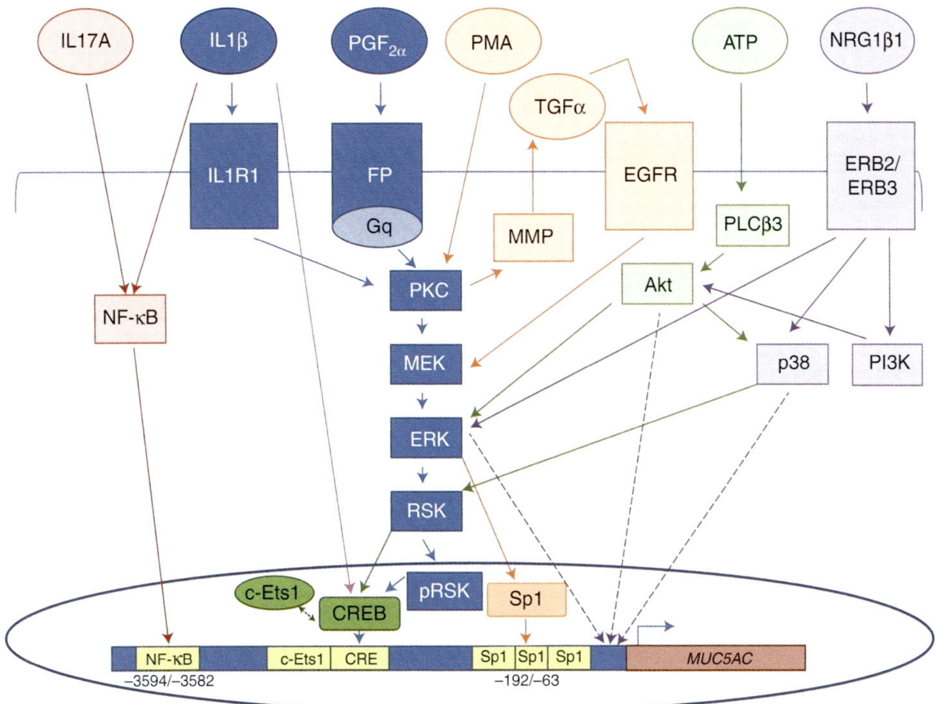

Figure 4. *MUC5AC* gene expression is transcriptionally regulated by various mediators and pathways. The figure depicts mediators, receptors, and *cis*-sites reported since earlier reviews on regulation of *MUC5AC* gene expression (Rose and Voynow 2006; Thai et al. 2008; Voynow and Rubin 2009). Each reference is color-coded by arrows. (Red) IL1β and IL17A induce *MUC5AC* expression by activation of the NF-κB *cis*-site at −3594/−3582 bp. Direct binding of the NF-κB p50 subunit was observed in HBE1 cells (Fujisawa et al. 2009). (Blue) PGF$_{2\alpha}$ signals through the prostaglandin F receptor (FP), which is coupled to the Gq protein to activate PKC, which then signals through an MEK, ERK, RSK pathway. RSK translocates to the nucleus after phosphorylation and activates CREB to bind to the CRE *cis*-site (−878/−871) to up-regulate *MUC5AC* expression in differentiated HBE cells and NCI-H292 cells. A model summarizing the mechanisms whereby prostaglandins and IL1β induce *MUC5AC* overexpression has been reported (Chung et al. 2009). (Pink) CREB binds maximally to the CRE *cis*-site in the *MUC5AC* promoter 2 h after IL1β exposure to A549 or differentiated HBE cells (YA Chen, AM Watson, LM Garvin, et al., unpubl.). (Orange) PMA activates the PKC isoforms δ and θ, resulting in activation of matrix metalloproteinase (MMP) and secretion of TGFα, which binds to its cognate receptor EGFR. EGFR activates the MEK/ERK1/2 pathway to activate Sp1, which binds to three Sp1 sites (−192/−63) in the *MUC5AC* promoter. Similar results were obtained in H292 cells (Hewson et al. 2004) and in differentiated HBE and HBE1 cells (Yuan-Chen et al. 2007). (Green) ATP exposure to H292 cells induces *MUC5AC* expression by activating PLCβ3, which, in turn, activates Akt to signal through ERK or p38 to activate RSK, thereby activating CREB and c-Ets1 binding to the CRE (−889/−869) and c-Ets1 (−938/−930) sites in the *MUC5AC* promoter (Song et al. 2008). (Purple) Neuregulin 1β1 (NRG1β1) signals through the ERB2/ERB3 heterodimer receptor to activate three different pathways: p38, ERK1/2, and PI3K, which signals through Akt to up-regulate *MUC5AC* gene expression in differentiated HBE cells (Kettle et al. 2010).

(Fig. 5) genes, which account for mucin over-production in human lungs.

Transcriptional Regulation of Secretory Mucin Genes

Ligand binding to cognate membrane receptors—for example, the epidermal growth factor receptor (EGFR, i.e., ErbB1, a member of the ErbB family of receptor tyrosine kinases), IL1R1, FP (Prostaglandin F receptor)—activates specific signal transduction pathways many of which converge or interact (Figs. 4 and 5). This results in nuclear translocation and/or phosphorylation of specific transcription factors, for example, CREB (cyclic AMP response element-binding), NF-κB (nuclear factor κ-light-chain-enhancer of activated B-cells), or Sp1 (specificity protein 1), which bind to specific *cis*-sites on polymeric *MUC* gene promoters. A predominant mechanism for regulating *MUC2* and *MUC5AC* gene expression is via activation of EGFR, which results in activation of extracellular signal-related kinases (ERK)1,2 (Takeyama et al. 1999; Lemjabbar and Basbaum 2002). Recent data show that the TGFα-induced up-regulation of *MUC5AC* gene expression in NCI-H292 cells is mediated via the EGFR-

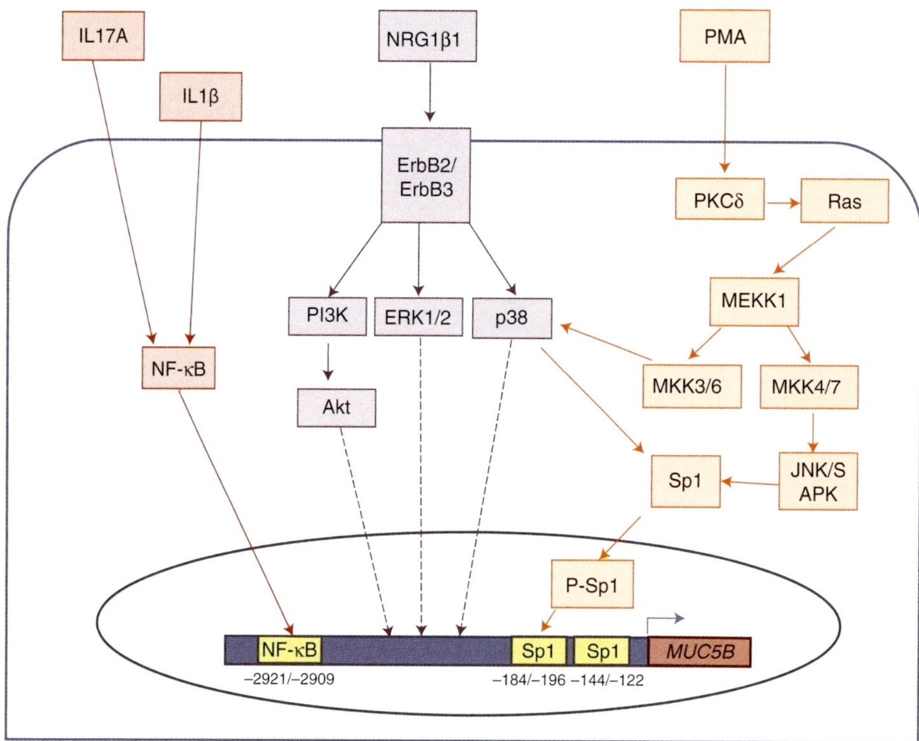

Figure 5. *MUC5B* gene expression is transcriptionally regulated by various mediators and pathways. (Red) IL1β and IL17A induce *MUC5B* gene expression by activation of the NF-κB *cis*-site at −2921/−2909 bp in the *MUC5B* promoter. Direct binding of the NF-κB p50 subunit was observed by ChIP in HBE1 cells (Fujisawa et al. 2011). (Orange) PMA activates PKCδ, which signals through Ras to activate MEKK1, which then signals through either a p38 pathway or the JNK pathway to phosphorylate Sp1, which binds to the Sp1 site at −184/−196. The Sp1 *cis*-site (−144/−122) functions to maintain basal regulation of *MUC5B* in differentiated HBE cells and in HBE1 cells (Wu et al. 2007a). (Purple) NRG1β1 increases *MUC5B* gene expression in differentiated HBE cells by signaling through the ErbB2/ErbB3 heterodimer receptor, which then signals through three different pathways including p38, ERK1/2, and PI3k. PI3k goes on to activate Akt (Kettle et al. 2010).

Cite this article as *Cold Spring Harb Perspect Med* doi: 10.1101/cshperspect.a009589

activated ERK1,2 pathway and also by ERK activation and secretion of CCL20. The latter mechanism activates the G-protein-coupled receptor CCR6, resulting in metalloprotease cleavage of EGFR pro-ligands and activation of *MUC5AC* gene expression. However, this is not a functional mechanism in primary differentiated HBE cells because they do not express CCR6 in vitro (Kim et al. 2011).

Neuregulin (NRG) 1β1, a member of the neuregulin growth factor family, binds to ERB2/ERB3 receptors, but not ERB4 or EGFR, in differentiated HBE cells, resulting in up-regulation of both the *MUC5AC* and *MUC5B* genes by activation of PI3K, p38, and Akt (Kettle et al. 2010). The activated transcription factors and *cis*-sites on the *MUC5AC* and *MUC5B* promoters to which they bind were not identified. Cigarette smoke also up-regulates expression of *MUC5AC* in 16HBE cells through the neuregulin 1β/ERB3 pathway (Yu et al. 2011). In contrast, in differentiated HBE cells, cigarette smoke up-regulates monocyte chemotactic protein-1, which increases expression of both *MUC5AC* and *MUC5B* genes via a cascade initiated by ligand/receptor (CCR2B) interactions and activation of PLCβ/PKC/MAPK (Monzon et al. 2011). PGF$_{2\alpha}$ binds to FP, which couples to the Gq protein to activate PKC, resulting in signaling through the MEK/ERK/RSK pathway and activation of CREB to induce *MUC5AC* expression in differentiated HBE cells (Chung et al. 2009). Double-stranded RNA (Zhu et al. 2009) and rhinovirus (RV) (Hewson et al. 2010) up-regulate *MUC5AC* expression in NCI-H292 cells through serially linked signaling cascades that result in activation of the EGFR-ERK pathway. This may reflect the exaggerated EGFR response in H292 cells (Kim et al. 2011) as RV up-regulates *MUC5AC* expression via a Src/MEK/NF-κB pathway in primary tracheal cells (Inoue et al. 2006). Understanding how RV up-regulates secretory mucin gene expression may prove important in understanding the response of lung epithelial cells to virally induced exacerbations, which are clinically relevant in CF and asthma.

Bacterial products produced by the Gram-negative bacterium *Pseudomonas aeruginosa* predominate in CF airways and can directly mediate lung mucin overproduction. *P. aeruginosa* lipopolysaccharide (LPS) was the first mediator shown to transcriptionally up-regulate expression of two secretory mucin genes—*MUC2* (Li et al. 1997) and *MUC5AC* (Li et al. 1998)—in lung epithelial cells. Recently, the redox-active virulence factor pyocyanin produced by *Pseudomonas* PAO1 and other strains has been shown to increase MUC5AC and MUC2 mRNA and protein abundance in NCI-H292, but not in BEAS-2B, IB3-1, or 16BHE14o-, cells. PAO1 strains that express pyocyanin but lack genes that encode LPS, flagellum, or pilus increase MUC5AC and MUC2 mRNA levels (Rada et al. 2011), raising the possibility that commercial preparations of LPS used in earlier studies contained pyocyanin and that pyocyanin, rather than LPS or other bacterial components, is responsible for inducing *MUC2* and *MUC5AC* gene expression during bacterial challenges. Pyocyanin-activated intracellular reactive oxygen species (ROS) and expression array analyses of NCI-H292 cells showed that it also induced expression of mediators potentially relevant to CF lung disease, including IL8, IL1β, and TNFα (Rada et al. 2011). ROS production has also been shown to increase expression of MUC5AC mRNA and protein, as well as the pro-inflammatory mediators IL-8 and IL1β (but not TNFα), in NCI-H292 cells following exposure to *Naegleria fowleri*, an amoeboflagellate that invades the olfactory mucosa to access the central nervous system (Cervantes-Sandoval et al. 2009).

Neutrophils, the predominant inflammatory cells in CF airways, secrete neutrophil elastase, which was initially shown to increase *MUC5AC* gene expression both posttranscriptionally in A549 and differentiated HBE cells (Voynow et al. 1999) and transcriptionally in H292 cells (Kohri et al. 2002). Defensin human neutrophil peptide-1 (HNP-1), an antimicrobial peptide secreted by neutrophils and present at high concentrations in the airway fluid of CF patients, also increases levels of *MUC5AC* mRNA and protein in NCI-H292 cells, and like LPS, induces phosphorylation of ERK1/2 (Ishimoto et al. 2009).

Interleukin (IL) 17 cytokines and receptors are of considerable interest in CF because they modulate neutrophil recruitment to the lungs and host defense mechanisms against Gram-negative bacteria like *Pseudomonas*. IL17 cytokines are elevated in CF sputum (McAllister et al. 2005; Decraene et al. 2010) and in neutrophils in CF lungs (Brodlie et al. 2011). IL17 cytokines signal through IL17R, which is localized to basal cells in the differentiated HBE system (McAllister et al. 2005). Because IL17R is not expressed in secretory cells in the differentiated HBE system (McAllister et al. 2005), the mechanism whereby IL17 signaling is activated in secretory cells remains unclear. Nevertheless, IL17A increases *MUC5AC* (Fujisawa et al. 2009) and *MUC5B* (Fujisawa et al. 2011) gene expression in differentiated HBE cells through activation of NF-κB and binding of the NF-κB subunit p50 at position $-3594/-3582$ in the *MUC5AC* promoter and at $2921/-2909$ in the *MUC5B* promoter. The Th2 cytokines IL4, IL9, and IL13 do not increase MUC5AC mRNA expression in CF sinus mucosal tissues following a 24-h exposure, although they do increase expression of hCLCA1 (Hauber et al. 2010), a secretory granule membrane protein. These data support the concept that the TH2 cytokines IL4 and IL13 do not directly mediate *MUC5AC* mucin gene expression but, rather, impact the pathways that lead to formation of secretory granules in goblet cells, indirectly resulting in increased MUC5AC levels in respiratory tract tissues.

The CREB transcription factor is emerging as a major player in up-regulating *MUC5AC* gene expression (Fig. 4), which may have important clinical implications for the management of airway mucus overproduction in CF. There is a CRE *cis*-site ($-878/-871$ nt) in the *MUC5AC* promoter and IL1β activates binding of CREB to that site in NCI-H292 and differentiated HNE cells (Song et al. 2003), as well as in A549 and differentiated HBE cells (YA Chen, AM Watson, LM Garvin, et al., unpubl.). In contrast, one study shows that HBE1 cells exposed to IL1β result in binding of the NF-κB subunit p50 to a NF-κB *cis*-site ($-3594/-3582$ nt) in the *MUC5AC* promoter (Fujisawa et al. 2009). Binding of CREB to the CRE *cis*-site is also activated by the prostaglandins PGE2 (Gray et al. 2004) and PGF$_{2\alpha}$ (Chung et al. 2009) in NCI-

H292 and differentiated HBE cells. ATP exposure to NCI-H292 cells induces signal transduction pathways that activate CREB binding to the CRE site and also activate E26 transformation-specific (Ets)1 binding to a nearby c-Ets1 site in the *MUC5AC* promoter (Song et al. 2008).

Posttranscriptional Regulation of Secretory Mucin Genes

Some mediators (acrolein, TNFα, neutrophil elastase) regulate secretory mucin genes posttranscriptionally (for review, see Rose and Voynow 2006). IL8, which is present at high levels in CF airways, does not transcriptionally regulate secretory mucin gene expression but, rather, increases MUC5AC and MUC5B mRNA abundance in A549, H292, and differentiated HBE cells posttranscriptionally. IL8 increases the half-life of mucin transcripts and induces binding of RNA-binding proteins to specific sites in the 3′-untranslated sequence of MUC5AC (Bautista et al. 2009).

Repression of *MUC5AC* Gene Expression

Pharmacological agents like glucocorticoids and macrolides that are used to treat lung diseases also reduce mucin gene expression. Studies to investigate mechanisms, especially under conditions in which lung epithelial cells are exposed to inflammatory mediators, are underway in several laboratories. The glucocorticoid dexamethasone (Dex) transcriptionally represses *MUC5AC* gene expression following binding of Dex-activated GR to two GRE *cis*-sites on the *MUC5AC* promoter in A549 (Chen et al. 2006) and in differentiated HBE cells (Chen et al. 2012) and is mediated by histone deacetylase 2 (Chen et al. 2012).

The macrolide antibiotic azithromycin (AZT), which at low doses improves clinical outcome in CF patients, alters the gene profile of differentiated HBE cells by significantly increasing the expression of lipid/cholesterol genes and decreasing the expression of cell cycle/mitosis genes. Interestingly, AZT pre-treatment of cells exposed to inflammatory stimuli does not prevent up-regulation of most inflammatory mediators, but prevents up-regulation of the *MUC5AC* gene (Ribeiro et al. 2009), suggesting that AZT may directly impact expression of that mucin gene. In H292 cells, AZT and clarithromycin have inhibitory effects on *MUC5AC* mRNA levels induced by HNP-1 or lipopolysaccharide (LPS), whereas telithromycin, a semisynthetic erythromycin, has an inhibitory effect on MUC5AC production induced by LPS, but not by HNP-1. All three compounds have inhibitory effects on ERK1/2 phosphorylation induced by LPS, but not by HNP-1, indicating that these compounds interfere with different intracellular signal transduction pathways (Ishimoto et al. 2009).

Summary: Mucin Gene Regulation

The *MUC5AC* gene is typically expressed in goblet cells in the conducting airway epithelium, which are well positioned to respond to environmental and pathogenic airborne challenges. The *MUC5B* gene is typically expressed in glandular mucosal cells in healthy lungs and also in secretory cells in the conducting airway epithelium in diseases conditions. Both mucin genes are up-regulated by inflammatory mediators that impact diverse pathways and activate various transcription factors in lung epithelial cells. Although inroads into understanding the complexity of how *MUC5AC* gene expression is up-regulated and repressed, information on regulation of *MUC5B* gene expression is more limited. Accumulating data show that regulation of both *MUC5AC* and *MUC5B* genes is modulated by diverse pathways in lung epithelial cells and that mechanisms in lung cancer cell lines, immortalized cells, and primary differentiated HBE cells are often, but not always, similar. Increased knowledge of the transcription factors and functional *cis*-sites in the 5′-flanking region and of the RNA-binding proteins and 3′-untranslated sites in polymeric *MUC* genes that are responsive to inflammatory mediators and to pharmacological agents will require additional studies to address their therapeutic impact. Further areas ripe for investigation include posttranscriptional and epigenetic regulation of mucin gene expression, as well as pharmacological modes of repressing mucin gene expression.

Non-Infectious Inflammation Also Impacts Mucus Overproduction

Inflammation is commonly thought of as being caused by infection and, as presented above, clearly impacts mucus overproduction. Other processes, for example, goblet cell and glandular hyperplasia, contribute to mucus overproduction (Rose and Voynow 2006). However, there are clearly non-infectious causes of mucus overproduction, as exemplified by the selective transgenic overexpression of βENaC in Clara cells in murine lungs using the CC10 promoter (Mall et al. 2004a; Zhou et al. 2011). Chronic inhibition of Na^+ in the mice by delivering amiloride in an aerosol blocks the mucus phenotype (Zhou et al. 2008), suggesting that it is not the overexpression of ENaC, per se, that drives the inflammation. Although there are few signs of lung infection in βENaC Tg mice, some bacteria can be detected as expected because microbiome studies show a natural flora in the lungs. However, raising the mice in a germ-free (gnotobiotic), LPS-free facility yields mice with the same mucus plugging phenotype (Livraghi-Butrico et al. 2012). Hence, the inflammation and mucus metaplasia/overproduction appear to be caused by hyperabsorption of airway surface liquid and resulting concentration of the mucus gel rather than being secondary to bacterial infection. The implication of this observation is that the epithelium initiates an inflammatory attack against what it senses on the luminal surface as an abnormal mucus gel.

Notably, this hypothesis offers an explanation for the mucus meta/hyperplasia and plugging observed in many organs of CF patients that are not affected by bacterial infections (Quinton 2010). The cause of mucus plugging in many CF organs is uncertain, especially those that do not express ENaC and therefore do not experience the degree of dehydration that occurs in the airways (Donaldson and Boucher 2007; Bridges 2012). A hypothesis gaining widespread interest revolves around a deficiency in HCO_3^- caused by functional impairment of mutated CFTR (Donaldson and Boucher 2007) and a resulting, profound failure of mucins to mature properly following their exocytic release, resulting in the formation of a dense, sticky mucus (Quinton 2010). This abnormal mucus is speculated then to trigger an inflammatory response in the epithelium, possibly similar to that occurring in the βENaC Tg mouse.

MUCIN BIOSYNTHESIS, GLYCOSYLATION, AND SECRETION

Overview

The emerging paradigm with regard to CF and mucins, as presented above, is that CFTR is not expressed in goblet or mucosal cells and thus does not directly impact fundamental processes in mucin production and secretion. Nevertheless, mucins are clearly overproduced in CF and affect the morbidity and mortality of CF patients. Therefore, the current status of mucin biosynthesis, glycosylation, and secretion, as well as of mucus dynamics, and their contribution to CF diseases are reviewed below.

Mucin Biosynthesis and Assembly

Historically, mucin biosynthesis in intestinal goblet cells played an important role in unraveling the cell biology of protein synthesis and secretion; for example, the Golgi apparatus was originally identified as the site of protein glycosylation (Peterson and LeBlong 1964) using ^{3}H-glucose labeling of what we now know to be MUC2. Mucins are similar to other proteins and glycoproteins in their synthesis, contrasting mainly by virtue of their immense size and extensive glycosylation (for a recent review, see Thornton et al. 2007). They are secreted or localized to plasma membranes, thus their primary functions are centered on the extracellular environment, and their protein backbones are synthesized by ribosomes associated with the endoplasmic reticulum. The polymeric mucin proteins are dimerized in the endoplasmic reticulum lumen via their carboxy-terminal cysteine knot domains (Fig. 1A) (Dekker and Strous 1990; Bell et al. 2001). A specific protein

disulfide isomerase, AGR2, has recently been shown to be essential for intestinal mouse Muc2 biosynthesis, because knockout of AGR2 leads to the loss of mucin secretion and colitis (Park et al. 2009). After transit to the Golgi apparatus, the mucin proteins (apomucins) are glycosylated, multimerized in the case of the polymeric species, and sorted at the *trans*-Golgi network into the regulated (polymeric mucins, MUC7) or constitutive (tethered mucins) limbs of the secretory pathway (see Davis and Dickey 2008). Recent studies with MUC5AC have revealed that GalNAc residues are transferred to serine and threonine residues in the MUC5AC apomucin in the Golgi and that stepwise synthesis and elongation of the final *O*-glycan structures are elaborated quickly in the HT-29 gastric cell line (Sheehan et al. 2004). Fully glycosylated lung mucins are filamentous (Rose et al. 1984), and mucin dimers of the polymeric mucins are on the order of 1 μm in length, and 5 MDa in molecular mass (Thornton et al. 2007). The importance of carboxy-terminal dimerization and amino-terminal multimerization of polymeric mucins was recognized with the sequencing of MUC2 and the identification, respectively, of the cystine knot and D domains, both having high homology to the respective domains of von Willebrand factor (vWF) (Gum et al. 1992; Desseyn et al. 2000). Present efforts on the assembly of polymeric mucins are using the recently revealed multimerization mechanism for vWF as a model (Huang et al. 2008). Given the scale of complexity and sizes of these massive molecules, the recent estimate for the full biosynthesis of a mucin polymer molecule in 2–4 h (Sheehan et al. 2004) is truly a remarkable feat of biology!

Mucin Secretion and Regulation

The sequence of signaling events regulating the exocytosis of mucin granules in goblet cells is complex and in a recent review (Davis and Dickey 2008) has been shown to share similarities with other secretory cells, for example, neurons, neuroendocrine cells, and pancreatic acinar cells (Burgoyne and Morgan 2003). Briefly, mucin granules are released via Ca^{2+}-

  Cite this article as *Cold Spring Harb Perspect Med* doi: 10.1101/cshperspect.a009589

dependent exocytosis following interaction of an agonist with the goblet cell, with the best described secretagogues being ATP and UTP, agonists of the P2Y2 receptor, coupled to $G\alpha q/11$ and intracellular Ca^{2+} store mobilization (Kreda et al. 2007; Davis and Dickey 2008). However, other Ca^{2+}-dependent agonists like serine proteases acting via protease-activated receptors have been recently identified as important mucin secretagogues in airway epithelia (Kreda et al. 2010). In the airways, cAMP-mediated agonists (e.g., isoproterenol) that activate CFTR do not stimulate mucin secretion from goblet cells (Davis and Dickey 2008), reinforcing the concept that the activation or inhibition of CFTR has no effect on mucin secretion (see Fig. 3) (Kreda et al. 2007), although cAMP does mediate mucin secretion in intestinal goblet cells (Bradbury 2000; Garcia et al. 2009). Interestingly, mucin secretagogues may also up-regulate mucin gene expression in vitro. For example, UTP, but not ATP, has been shown to up-regulate expression of the *MUC5AC* and *MUC5B* mucin genes in differentiated HBE cells (Chen et al. 2001b), but similar results have not been obtained in cancer cell lines. ATP has been reported to up-regulate *MUC5AC* gene expression in NCI-H292 cells in one study (Song et al. 2008), but another study showed that ATP does not increase MUC5AC protein levels in H292 cells (Rada et al. 2011). In differentiated HBE cells. neutrophil elastase functions both as a mucin secretagogue (Park et al. 2005) and an up-regulator of *MUC5AC* gene expression (Fischer and Voynow 2002; Voynow et al. 2004).

Recently, activations of the small G-protein RhoA and its downstream effector Rho kinase (ROCK) have been identified as critical steps in mucin granule exocytosis (Kreda et al. 2010). The main downstream effector of ROCK in airway goblet cells appears to be myosin light chain kinase (MLCK) (Seminario-Vidal et al. 2009; Kreda et al. 2010), which is essential in actomyosin cytoskeleton remodeling during mucin granule exocytosis (Davis and Dickey 2008; Kreda et al. 2010). However, other ROCK targets involved in agonist-stimulated granule secretion have been described; for example,

ROCK has been described to regulate MARCKS by phosphorylation (Sasaki 2003). MARCKS is also activated downstream from PKC and has been reported to regulate mucin granule exocytosis (Li et al. 2001; Singer et al. 2004).

The last steps of exocytosis are effected by the proteins that tether and dock the granule at the plasma membrane before exocytic fusion. The proteins that regulate this process have been well reviewed (Burgoyne and Morgan 2003; Davis and Dickey 2008). In regulated exocytosis of mucin granules, this interaction requires activation of Munc13 in a process termed "priming," by which Munc13 opens the conformation of the soluble NSF attachment protein (SNAP) receptors (SNAREs) to allow the assembly of the "core complex" or "SNARE core" (Davis and Dickey 2008). Genetic ablation of the Munc13-2 isoform in mice produces a defect in mucin secretion that is reflected, in part, by an accumulation of mucin granules within the goblet cells (Zhu et al. 2008). The SNARE complex, required for rapid fusion of the granule and cell membranes, is formed by the interaction of one R-SNARE, or VAMP (vesicle-associated membrane protein), and two or three Q-SNARES, whose identities in goblet cells remain undefined (Burgoyne and Morgan 2003; Davis and Dickey 2008). Syntaxins 2, 3, and 11 are candidate Q-SNARES for mucin granule exocytosis, and we have shown that syntaxin 3 is specifically localized apically in airway goblet cells (Kreda et al. 2007). Similarly, SNAP23 or SNAP25 has been postulated as the other Q-SNARE in the exocytotic core of mucin granules (Davis and Dickey 2008; Evans and Koo 2009), and deep-sequencing mRNA data showed that SNAP23, but not SNAP25, is highly expressed in airway epithelial cells (Jones et al. 2011). VAMP8 appears to be the R-SNARE in the exocytotic core of goblet cells because it is more than 10 times more highly expressed than other VAMPs in airway epithelial cells, and specific reduction of VAMP8 expression drastically decreases basal and agonist-mediated mucin granule exocytosis in airway goblet cells (Jones et al. 2011). Thus, a probable conformation of the mucin granule SNARE complex has VAMP8 in the granule membrane and SNAP23 and syntaxin 3 in the plasma membrane of airway goblet cells. The core complex activity is acutely regulated in response to second messengers Ca^{2+} and DAG, and the low-affinity, fast calcium-sensing synaptotagmin-2 is required for this activity in airway goblet cells as it is in neurons (Davis and Dickey 2008), because knockout of the isoform in mice produces a severe defect in ATP-induced mucin secretion in the airways (Tuvim et al. 2009).

Identification of other signaling components of the mucin secretory pathway (Davis and Dickey 2008) will be possible by genetic and pharmacological manipulations and high-throughput analysis of expression data, as well as high-resolution video-microscopy techniques. Importantly, some of these signaling components might be good targets for therapeutic inhibition of mucin hypersecretion in CF-affected organs, possibly as a prelude to intervention with anti-inflammatory or other therapeutic agents.

Mucin Glycosylation and CF

The O-glycans attached to serines and threonines in MUC protein backbones account for 50%–90% of the molecular mass of mucins (Fig. 1A,B). Historically, aberrant glycosylation of mucins in CF patients figured large before the cloning of CFTR in 1989, because there were several reports of higher levels of acidic mucins, along with increases in sulfation and either increases or decreases in the fucosylation and sialylation of mucins (Roussel et al. 1975; Chace et al. 1983; Cheng et al. 1989; also, for review, see Rose 1988; Rose and Voynow 2006). With the identification of CFTR and the suggestion that its effective absence from organelle and plasma membranes in CF cells was responsible for alkalization and dysfunction along the secretory pathway (Barasch et al. 1991; Barasch and al-Awqati 1993), a possible mechanism was suggested. After many years of controversy, which led to increasingly sophisticated studies, the issue appears to have been resolved with the weight of the evidence indicating the physical absence of CFTR in the Golgi apparatus (see above), and no apparent role for CFTR in the

maintenance of Golgi pH (Seksek et al. 1996; Gibson et al. 2000; Chandy et al. 2001; Haggie and Verkman 2009) or in the glycosylation of mucins and other glycoproteins (Jiang et al. 1997; Holmen et al. 2004; Leir et al. 2005). Biochemical analyses of O-glycans indicate that changes in glycosylation of CF mucins appear to be a secondary effect of the disease that is largely due to inflammation (Davril et al. 1999; Roussel and Lamblin 2003; Schulz et al. 2007). Consistent with this notion, recent observations indicate that cytokines and other inflammatory mediators not only induce increases in mucin gene expression (as reviewed above) but also increase the expression of glycosyltransferases and sulfotransferases within the bronchial mucosa (Delmotte et al. 2002; Groux-Degroote et al. 2008). Possibly because of the large number and diversity of glycosyltransferases (Brockhausen et al. 2009), their roles in mucin synthesis and mucus biology are vastly underappreciated. The number of investigators in this field is small, but they are persistent, and a basic understanding of glycosyltransferase cell biology and physiology as it relates to mucin O-glycosylation is gradually emerging (Ten Hagen et al. 2003; Cheng and Radhakrishnan 2011; Gerken et al. 2011). This information will impact our understanding of the molecular roles of O-glycans in specific mucins and their functions in protecting the epithelium against pathogens in health and disease, including CF.

Dynamics of Mucus Formation and Function

The world of mucus is dynamic. These dynamics are perhaps most obvious in the airways, where beating cilia drive a sheet of mucus cephalad, continuously, as reported by Florey et al. (1932). Recent, closer observation by modern video-microscopy has revealed that polymeric mucins, MUC5B in particular, form strands and clumps that bind transiently to the tips of beating cilia and appear to be transported by being passed along from one ciliated cell to the next (Sears et al. 2011). These observations suggest a much closer and dynamic interaction between mucin molecules and ciliary surfaces than previously suspected, and imply transient molecular binding interactions forming between MUC5B and ciliary surfaces.

More subtly, mucus on mucosal surfaces is constantly being formed and degraded. Clara cells in the mouse airways do not react with periodic acid Schiff (PAS) stain, the archetypal mucin stain (Spicer et al. 1971). However, in the Munc13-2-null mouse, which has a defect in basal mucin secretion, the cells are PAS$^+$, indicating an apparent accumulation that occurs as a result of hindered release of mucins (Zhu et al. 2008). Similarly, mucins are released continuously in the intestinal tract, forming the adherent mucus layer in the colon and stomach, then becoming part of the "floppy" layer as cross-links responsible for the mesh-like arrangement in adherent layers are cleaved (Johansson et al. 2011; Ambort et al. 2012).

Even more subtly, mucus dynamics extends to the birth of mucins following their exocytic release from goblet and mucous cells. It has long been appreciated that mucins expand enormously as they hydrate following exocytic release from goblet cells, a process that was thought to be due to ionic repulsion of the fixed negative charges on mucin O-glycans, because Ca^{2+} packaged with the mucins was exchanged for Na^+ on the luminal surface (Verdugo 1991, 2012) A recent analysis of salivary MUC5B mucin implies an even more ordered packing of mucins in secretory granules than previously imagined. In these studies (Kesimer et al. 2009a), salivary MUC5B was subjected to density gradient centrifugation, which revealed three morphotypes of mucin by electron microscopy. One, a "compact form" that comprised <10% of the total MUC5B in the sample, migrated to the bottom of the gradient. The polymer molecules had roughly circular profiles and cross-linked in such a way that the amino- and carboxy-terminal domains, arranged as nodes ~10 nm in diameter, were in the central part of the complex, with the glycosylated domains of the mucin arrayed in the outer parts. Present in the middle of the gradient was a "moderately compact form," one in which the mucin appeared to be in the process of expanding from the compact form. And, at the top of the gradi-

ent, representing the bulk of the total MUC5B, was the "linearized form," in which the mucins assumed open, linear, non-cross-linked confirmations. The hypothesis offered, presently under evaluation, is that the "compact form" represents a mucin molecule right after exocytic release into the luminal environment, and that it undergoes a maturation process, which may involve enzymatic cleavage, following which it assumes the "linearized form"; following this it is fully rendered into the mucus gel.

CONCLUSIONS AND EMERGING DIRECTIONS

The mechanism of mucin secretion in CF mucosal cells is normal and does not appear to be linked in any direct way to CFTR. Similarly, the mucus hyperproduction that typifies CF does not appear to be a direct cause of a defective CFTR, a major reason being the lack of CFTR expression in most mucin-secreting (goblet, mucous) cells. The overwhelming weight of the evidence reviewed above favors an indirect connection between CFTR expression and mucin overproduction in CF. The downstream consequences of mutant CFTR, for example, the lack of functional CFTR at the plasma membrane, clearly affect mucin overproduction in individuals with CF in ways that are not yet well understood. This is most apparent in CF lungs, where chronic inflammation and infection sustain conditions wherein mucin genes are up-regulated, resulting in increased production and secretion of polymeric mucins. Although recent studies from several laboratories have reinforced the concept that diverse stimuli modulate expression of the *MUC5AC* and *MUC5B* mucin genes in lung epithelial cells, regulation of these genes is complex. Additional studies are required for a full understanding of how these genes are modulated in CF and non-CF cells before therapeutic interventions can be maximized. In addition to inflammation that arises from infection, other pathways may stem from luminal environments gone awry. The βENaC-overexpressing mouse mucus plugging phenotype is proving to be an archetypal example of this kind of inflammation, with a dehydrated airway surface/concentrated mucus gel apparently providing the inflammatory stimulus. We speculate that the luminal HCO_3^- deficiency presently being described for CF epithelia may also provide such a stimulus, by causing a mal-maturation of mucins as they are released onto luminal surfaces. The molecular mechanisms triggered by these stimuli have not been identified, but they are presently under study in multiple laboratories. Additionally, innate immune defense proteins secreted by lung cells that express CFTR may be altered in their levels or properties in CF patients as a direct consequence of CFTR, thereby contributing to the pathogenesis of CF lung diseases.

We continue to be amazed at the complex, pervasive pathways of dysfunction manifested in numerous cells and organs that result from gene defects in a single gene, *CFTR*, that underlies CF disease, especially mucosal epithelia. Thirty years ago, CF was thought to be a disease of mucus. Twenty years ago, the identification of CFTR as a Cl^- channel caused the field to shift its attention dramatically away from mucus and mucins. Ten years ago, we began to appreciate that tertiary or higher-order effects of the defect in CFTR were driving mucus dysfunction and hyperproduction in CF. Finally, today, we may be approaching a full understanding of the train of molecular events that lead from mutations in CFTR to mucus overproduction, possibly via defects induced in the postsecretory maturation of mucins. In light of the numerous systems, organs, and tissues that are affected, CF is a systems biology disease, although it is the lungs that account for morbidity and mortality. Given the diversity of CFTR mutations, the challenge of circumventing airway mucus obstruction may be amenable in the future to a personalized-medicine approach to therapy.

ACKNOWLEDGMENTS

We are grateful to the CF and non-CF patient volunteers, globally, for tissue specimen donations, for the development of the knowledge expressed in this work would not have been possible without this material. Dr. Hiro Matsui kindly gave permission to use his images of concentrated mucus (Fig. 1C), Dr. Lubna Abdulla graciously contributed the data shown in Figure 3A, and Lindsay Garvin, PhD candidate, generated Figures 4 and 5. This work is supported by several grants from the Cystic Fibrosis Foundation (to S.M.K., C.W.D., and M.C.R.); grants from Cystic Fibrosis Foundation Therapeutics (Davis 1997); the Mary Lynn Richardson Fund (to S.M.K.); and the National Institutes of Health grants HL34322 (to S.M.K.), HL34582 (to S.M.K.), HL51818 (to S.M.K.), HL34322 (Davis 1997), HL063756 (Davis 1997), HL60280 (Davis 1997), HL 97000 (Davis 1997), HL 33152 (to M.C.R.), and AI087717 (to M.C.R.).

REFERENCES

*Reference is also in this collection.

Abdullah LH, Conway JD, Cohn JA, Davis CW. 1997. Protein kinase C and Ca^{2+} activation of mucin secretion in airway goblet cells. *Am J Physiol* **273:** L201–L210.

Abdullah LH, Wolber C, Kesimer M, Sheehan JK, Davis CW. 2012. Studying mucin secretion from human bronchial epithelial cell primary cultures. *Methods Mol Biol* **842:** 259–277.

Ajonuma LC, Ng EH, Chow PH, Hung CY, Tsang LL, Cheung AN, Brito-Jones C, Lok IH, Haines J, Chan HC. 2005. Increased cystic fibrosis transmembrane conductance regulator (CFTR) expression in the human hydrosalpinx. *Hum Reprod* **20:** 1228–1234.

Allen A, Hutton DA, Pearson JP, Sellers LA. 1984. Mucus glycoprotein structure, gel formation and gastrointestinal mucus function. *Ciba Found Symp* **109:** 137–156.

* Ambort D, Johansson MEV, Gustafsson JK, Ermund A, Hansson GC. 2012. Perspectives on mucus properties and formation—Lessons from the biochemical world. *Cold Spring Harb Perspect Med* **2:** a014159.

Andersch-Bjorkman Y, Thomsson KA, Holmen Larsson JM, Ekerhovd E, Hansson GC. 2007. Large scale identification of proteins, mucins, and their O-glycosylation in the endocervical mucus during the menstrual cycle. *Mol Cell Proteomics* **6:** 708–716.

Anderson DH. 1938. Cystic fibrosis of the pancreas and its relation to celiac disease: A clinical and pathological study. *Am J Dis Child* **56:** 344–399.

Ballard ST, Inglis SK. 2004. Liquid secretion properties of airway submucosal glands. *J Physiol* **556:** 1–10.

Ballard ST, Spadafora D. 2007. Fluid secretion by submucosal glands of the tracheobronchial airways. *Respir Physiol Neurobiol* **159:** 271–277.

Barasch J, al-Awqati Q. 1993. Defective acidification of the biosynthetic pathway in cystic fibrosis. *J Cell Sci Suppl* **17:** 229–233.

Barasch J, Kiss B, Prince A, Saiman L, Gruenert D, al-Awqati Q. 1991. Defective acidification of intracellular organelles in cystic fibrosis. *Nature* **352:** 70–73.

Bautista MV, Chen Y, Ivanova VS, Rahimi MK, Watson AM, Rose MC. 2009. IL-8 regulates mucin gene expression at the posttranscriptional level in lung epithelial cells. *J Immunol* **183:** 2159–2166.

Bell SL, Xu G, Forstner JF. 2001. Role of the cystine-knot motif at the C-terminus of rat mucin protein Muc2 in dimer formation and secretion. *Biochem J* **357:** 203–209.

Best JA, Quinton PM. 2005. Salivary secretion assay for drug efficacy for cystic fibrosis in mice. *Exp Physiol* **90:** 189–193.

Boat TF, Cheng PW, Wood RE. 1976. Tracheobronchial mucus secretion in vivo and in vitro by epithelial tissues from cystic fibrosis and control subjects. *Mod Probl Paediatr* **19:** 141–152.

Boat TF, Welsh MJ, Beaudet AL. 1989. Cystic fibrosis. In *The metabolic basis of inherited disease* (ed. Scriver CR, et al.), pp. 2649–2680. McGraw-Hill, New York.

Boucher RC. 2007. Cystic fibrosis: A disease of vulnerability to airway surface dehydration. *Trends Mol Med* **13:** 231–240.

Bove PF, Grubb BR, Okada SF, Ribeiro CM, Rogers TD, Randell SH, O'Neal WK, Boucher RC. 2010. Human alveolar type II cells secrete and absorb liquid in response to local nucleotide signaling. *J Biol Chem* **285:** 34939–34949.

Bradbury NA. 2000. Protein kinase-A-mediated secretion of mucin from human colonic epithelial cells. *J Cell Physiol* **185:** 408–415.

Brezillon S, Dupuit F, Hinnrasky J, Marchand V, Kalin N, Tummler B, Puchelle E. 1995. Decreased expression of the CFTR protein in remodeled human nasal epithelium from non-cystic fibrosis patients. *Lab Invest* **72:** 191–200.

* Bridges RJ. 2012. Mechanisms of bicarbonate secretion: Lessons from the airways. *Cold Spring Harb Perspect Med* **2:** a015016.

Brochiero E, Dagenais A, Prive A, Berthiaume Y, Grygorczyk R. 2004. Evidence of a functional CFTR Cl⁻ channel in adult alveolar epithelial cells. *Am J Physiol Lung Cell Mol Physiol* **287:** L382–L392.

Brockhausen I, Schachter H, Stanley P. 2009. O-GalNAc glycans. In *Essentials of glycobiology*, 2nd ed. (ed. Varki A, et al.), Chapter 9. Cold Spring Harbor Laboratory Press, Cold Spring Harbor, NY.

Brodlie M, McKean MC, Johnson GE, Anderson AE, Hilkens CM, Fisher AJ, Corris PA, Lordan JL, Ward C. 2011. Raised interleukin-17 is immunolocalised to neutrophils in cystic fibrosis lung disease. *Eur Respir J* **37:** 1378–1385.

Brugman SM, Taussig LM. 1984. The reproductive system. In *Cystic fibrosis* (ed. Taussig LM), pp. 323–337. Thieme-Stratton, New York.

Bruscia E, Sangiuolo F, Sinibaldi P, Goncz KK, Novelli G, Gruenert DC. 2002. Isolation of CF cell lines corrected at ΔF508-CFTR locus by SFHR-mediated targeting. *Gene Ther* **9:** 683–685.

Bruscia EM, Zhang PX, Ferreira E, Caputo C, Emerson JW, Tuck D, Krause DS, Egan ME. 2009. Macrophages directly contribute to the exaggerated inflammatory response in cystic fibrosis transmembrane conductance regulator⁻/⁻ mice. *Am J Respir Cell Mol Biol* **40:** 295–304.

Burgoyne RD, Morgan A. 2003. Secretory granule exocytosis. *Physiol Rev* **83:** 581–632.

Caramori G, Casolari P, Di GC, Saetta M, Baraldo S, Boschetto P, Ito K, Fabbri LM, Barnes PJ, Adcock IM, et al. 2009. MUC5AC expression is increased in bronchial submucosal glands of stable COPD patients. *Histopathology* **55:** 321–331.

Carvalho-Oliveira I, Efthymiadou A, Malho R, Nogueira P, Tzetis M, Kanavakis E, Amaral MD, Penque D. 2004. CFTR localization in native airway cells and cell lines expressing wild-type or F508del-CFTR by a panel of different antibodies. *J Histochem Cytochem* **52:** 193–203.

Castagna I, Roszkowska AM, Fama F, Sinicropi S, Ferreri G. 2001. The eye in cystic fibrosis. *Eur J Ophthalmol* **11:** 9–14.

Cervantes-Sandoval I, Serrano-Luna JJ, Meza-Cervantez P, Arroyo R, Tsutsumi V, Shibayama M. 2009. *Naegleria fowleri* induces MUC5AC and pro-inflammatory cytokines in human epithelial cells via ROS production and EGFR activation. *Microbiology* **155:** 3739–3747.

Chace KV, Leahy DS, Martin R, Carubelli R, Flux M, Sachdev GP. 1983. Respiratory mucous secretions in patients with cystic fibrosis: Relationship between levels of highly sulfated mucin component and severity of the disease. *Clin Chim Acta* **132:** 143–155.

Chan HC, Ruan YC, He Q, Chen MH, Chen H, Xu WM, Chen WY, Xie C, Zhang XH, Zhou Z. 2009. The cystic fibrosis transmembrane conductance regulator in reproductive health and disease. *J Physiol* **587:** 2187–2195.

Chandy G, Grabe M, Moore HP, Machen TE. 2001. Proton leak and CFTR in regulation of Golgi pH in respiratory epithelial cells. *Am J Physiol Cell Physiol* **281:** C908–C921.

Chen Y, Zhao YH, Di YP, Wu R. 2001a. Characterization of human mucin 5B gene expression in airway epithelium and the genomic clone of the amino-terminal and 5′-flanking region. *Am J Respir Cell Mol Biol* **25:** 542–553.

Chen Y, Zhao YH, Wu R. 2001b. Differential regulation of airway mucin gene expression and mucin secretion by extracellular nucleotide triphosphates. *Am J Respir Cell Mol Biol* **25:** 409–417.

Chen Y, Nickola TJ, DiFronzo NL, Colberg-Poley AM, Rose MC. 2006. Dexamethasone-mediated repression of *MUC5AC* gene expression in human lung epithelial cells. *Am J Respir Cell Mol Biol* **34:** 338–347.

Chen G, Korfhagen TR, Xu Y, Kitzmiller J, Wert SE, Maeda Y, Gregorieff A, Clevers H, Whitsett JA. 2009. SPDEF is required for mouse pulmonary goblet cell differentiation and regulates a network of genes associated with mucus production. *J Clin Invest* **119:** 2914–2924.

Chen EY, Yang N, Quinton PM, Chin WC. 2010. A new role for bicarbonate in mucus formation. *Am J Physiol Lung Cell Mol Physiol* **299:** L542–L549.

Chen YA, Watson AM, Williamson CD, Rahimi MK, Liang C, Colberg-Poley AM, Rose MC. 2012. Glucocorticoid receptor and HDAC2 mediate the dexamethasone-induced repression of the *MUC5AC* gene (submitted).

Cheng PW, Radhakrishnan P. 2011. Mucin O-glycan branching enzymes: Structure, function, and gene regulation. *Adv Exp Med Biol* **705:** 465–492.

Cheng PW, Boat TF, Cranfill K, Yankaskas JR, Boucher RC. 1989. Increased sulfation of glycoconjugates by cultured nasal epithelial cells from patients with cystic fibrosis. *J Clin Invest* **84:** 68–72.

Cheng SH, Gregory RJ, Marshall J, Paul S, Souza DW, White GA, O'Riordan CR, Smith AE. 1990. Defective intracellular transport and processing of CFTR is the molecular basis of most cystic fibrosis. *Cell* **63:** 827–834.

Childers M, Eckel G, Himmel A, Caldwell J. 2007. A new model of cystic fibrosis pathology: Lack of transport of glutathione and its thiocyanate conjugates. *Med Hypotheses* **68:** 101–112.

Choi JY, Khansaheb M, Joo NS, Krouse ME, Robbins RC, Weill D, Wine JJ. 2009. Substance P stimulates human airway submucosal gland secretion mainly via a CFTR-dependent process. *J Clin Invest* **119:** 1189–1200.

Cholon DM, O'Neal WK, Randell SH, Riordan JR, Gentzsch M. 2010. Modulation of endocytic trafficking and apical stability of CFTR in primary human airway epithelial cultures. *Am J Physiol Lung Cell Mol Physiol* **298:** L304–L314.

Chung WC, Ryu SH, Sun H, Zeldin DC, Koo JS. 2009. CREB mediates prostaglandin F2α-induced MUC5AC overexpression. *J Immunol* **182:** 2349–2356.

Claass A, Sommer M, de JH, Kalin N, Tummler B. 2000. Applicability of different antibodies for immunohistochemical localization of CFTR in sweat glands from healthy controls and from patients with cystic fibrosis. *J Histochem Cytochem* **48:** 831–837.

Cohn JA, Melhus O, Page LJ, Dittrich KL, Vigna SR. 1991. CFTR: Development of high-affinity antibodies and localization in sweat gland. *Biochem Biophys Res Commun* **181:** 36–43.

Cohn JA, Strong TV, Picciotto MR, Nairn AC, Collins FS, Fitz JG. 1993. Localization of the cystic fibrosis transmembrane conductance regulator in human bile duct epithelial cells. *Gastroenterology* **105:** 1857–1864.

Cone RA. 2009. Barrier properties of mucus. *Adv Drug Deliv Rev* **61:** 75–85.

Corfield AP, Myerscough N, Longman R, Sylvester P, Arul S, Pignatelli M. 2000. Mucins and mucosal protection in the gastrointestinal tract: New prospects for mucins in the pathology of gastrointestinal disease. *Gut* **47:** 589–594.

Crawford I, Maloney PC, Zeitlin PL, Guggino WB, Hyde SC, Turley H, Gatter KC, Harris A, Higgins CF. 1991. Immunocytochemical localization of the cystic fibrosis gene product CFTR. *Proc Natl Acad Sci* **88:** 9262–9266.

Davies JR, Svitacheva N, Lannefors L, Kornfalt R, Carlstedt I. 1999. Identification of MUC5B, MUC5AC and small amounts of MUC2 mucins in cystic fibrosis airway secretions. *Biochem J* **344:** 321–330.

Davies WL, Vandenberg JI, Sayeed RA, Trezise AE. 2004. Cardiac expression of the cystic fibrosis transmembrane conductance regulator involves novel exon 1 usage to produce a unique amino-terminal protein. *J Biol Chem* **279:** 15877–15887.

Davis CW. 1997. Goblet cells: Physiology and pharmacology. In *Airway mucus: Basic mechanisms and clinical perspectives* (ed. Rogers DF, Lethem MI), pp. 150–177. Berkhauser, Basel, Switzerland.

Davis CW, Dickey BF. 2008. Regulated airway goblet cell mucin secretion. *Annu Rev Physiol* **70:** 487–512.

Davril M, Degroote S, Humbert P, Galabert C, Dumur V, Lafitte JJ, Lamblin G, Roussel P. 1999. The sialylation of bronchial mucins secreted by patients suffering from cystic fibrosis or from chronic bronchitis is related to the severity of airway infection. *Glycobiology* **9:** 311–321.

Decraene A, Willems-Widyastuti A, Kasran A, De Boeck K, Bullens DM, Dupont LJ. 2010. Elevated expression of both mRNA and protein levels of IL-17A in sputum of stable cystic fibrosis patients. *Respir Res* **11:** 177.

Dekker J, Strous GJ. 1990. Covalent oligomerization of rat gastric mucin occurs in the rough endoplasmic reticulum, is *N*-glycosylation-dependent, and precedes initial *O*-glycosylation. *J Biol Chem* **265:** 18116–18122.

Delmotte P, Degroote S, Lafitte JJ, Lamblin G, Perini JM, Roussel P. 2002. Tumor necrosis factor α increases the expression of glycosyltransferases and sulfotransferases responsible for the biosynthesis of sialylated and/or sulfated Lewis x epitopes in the human bronchial mucosa. *J Biol Chem* **277:** 424–431.

Desseyn JL, Aubert JP, Porchet N, Laine A. 2000. Evolution of the large secreted gel-forming mucins. *Mol Biol Evol* **17:** 1175–1184.

Donaldson SH, Boucher RC. 2007. Sodium channels and cystic fibrosis. *Chest* **132:** 1631–1636.

Dorfman R, Nalpathamkalam T, Taylor C, Gonska T, Keenan K, Yuan XW, Corey M, Tsui LC, Zielenski J, Durie P. 2010. Do common in silico tools predict the clinical consequences of amino-acid substitutions in the CFTR gene? *Clin Genet* **77:** 464–473.

Dray-Charier N, Paul A, Veissiere D, Mergey M, Scoazec JY, Capeau J, Brahimi-Horn C, Housset C. 1995. Expression of cystic fibrosis transmembrane conductance regulator in human gallbladder epithelial cells. *Lab Invest* **73:** 828–836.

Duchatel F, Muller F, Oury JF, Mennesson B, Boue J, Boue A. 1993. Prenatal diagnosis of cystic fibrosis: Ultrasonography of the gallbladder at 17–19 weeks of gestation. *Fetal Diagn Ther* **8:** 28–36.

Dupuit F, Kalin N, Brezillon S, Hinnrasky J, Tummler B, Puchelle E. 1995. CFTR and differentiation markers expression in non-CF and ΔF 508 homozygous CF nasal epithelium. *J Clin Invest* **96:** 1601–1611.

Engelhardt JF, Yankaskas JR, Ernst SA, Yang Y, Marino CR, Boucher RC, Cohn JA, Wilson JM. 1992. Submucosal glands are the predominant site of CFTR expression in the human bronchus. *Nat Genet* **2:** 240–248.

Engelhardt JF, Zepeda M, Cohn JA, Yankaskas JR, Wilson JM. 1994. Expression of the cystic fibrosis gene in adult human lung. *J Clin Invest* **93:** 737–749.

Enhorning G, Huldt L, Melen B. 1970. Ability of cervical mucus to act as a barrier against bacteria. *Am J Obstet Gynecol* **108:** 532–537.

Evans CM, Koo JS. 2009. Airway mucus: The good, the bad, the sticky. *Pharmacol Ther* **121:** 332–348.

Farber S, Shwachman H, Maddock CL. 1943. Pancreatic function and disease in early life. I. Pancreatic enzyme activity and the celiac syndrome. *J Clin Invest* **22:** 827–838.

Fischer BM, Voynow JA. 2002. Neutrophil elastase induces *MUC5AC* gene expression in airway epithelium via a pathway involving reactive oxygen species. *Am J Respir Cell Mol Biol* **26:** 447–452.

Florey H. 1930. The secretion of mucus by the colon. *Br J Exp Pathol* **11:** 348–361.

Florey H, Carleton HM, Wells AQ. 1932. Mucus secretion in the trachea. *Br J Exp Pathol* **13:** 269–284.

Forte TM, Forte JG. 1970. Histochemical staining and characterization of glycoproteins in acid-secreting cells of frog stomach. *J Cell Biol* **47:** 437–452.

Freudenberg F, Leonard MR, Liu SA, Glickman JN, Carey MC. 2010. Pathophysiological preconditions promoting mixed "black" pigment plus cholesterol gallstones in a ΔF508 mouse model of cystic fibrosis. *Am J Physiol Gastrointest Liver Physiol* **299:** G205–G214.

Fujisawa T, Velichko S, Thai P, Hung LY, Huang F, Wu R. 2009. Regulation of airway MUC5AC expression by IL-1β and IL-17A; the NF-κB paradigm. *J Immunol* **183:** 6236–6243.

Fujisawa T, Chang MM, Velichko S, Thai P, Hung LY, Huang F, Phuong N, Chen Y, Wu R. 2011. NF-κB mediates IL-1β- and IL-17A-induced MUC5B expression in airway epithelial cells. *Am J Respir Cell Mol Biol* **45:** 246–252.

Garcia MA, Yang N, Quinton PM. 2009. Normal mouse intestinal mucus release requires cystic fibrosis transmembrane regulator-dependent bicarbonate secretion. *J Clin Invest* **119:** 2613–2622.

Gerken TA, Jamison O, Perrine CL, Collette JC, Moinova H, Ravi L, Markowitz SD, Shen W, Patel H, Tabak LA. 2011. Emerging paradigms for the initiation of mucin-type protein *O*-glycosylation by the polypeptide GalNAc transferase family of glycosyltransferases. *J Biol Chem* **286:** 14493–14507.

Gibson GA, Hill WG, Weisz OA. 2000. Evidence against the acidification hypothesis in cystic fibrosis. *Am J Physiol Cell Physiol* **279:** C1088–C1099.

Goldsworthy NE, Florey H. 1930. Some properties of mucus, with special reference to its antibacterial functions. *Br J Exp Pathol* **11:** 192–208.

Gouyer V, Leir SH, Tetaert D, Liu Y, Gottrand F, Harris A, Desseyn JL. 2010. The characterization of the first antimouse Muc6 antibody shows an increased expression of the mucin in pancreatic tissue of Cftr-knockout mice. *Histochem Cell Biol* **133:** 517–525.

Govindarajan B, Gipson IK. 2010. Membrane-tethered mucins have multiple functions on the ocular surface. *Exp Eye Res* **90:** 655–663.

Gray T, Nettesheim P, Loftin C, Koo JS, Bonner J, Peddada S, Langenbach R. 2004. Interleukin-1β-induced mucin production in human airway epithelium is mediated by cyclooxygenase-2, prostaglandin E2 receptors, and cyclic AMP-protein kinase A signaling. *Mol Pharmacol* **66:** 337–346.

Groux-Degroote S, Krzewinski-Recchi MA, Cazet A, Vincent A, Lehoux S, Lafitte JJ, Van S, Delannoy P. 2008. IL-6 and IL-8 increase the expression of glycosyltransferases and sulfotransferases involved in the biosynthesis of sialylated and/or sulfated Lewis x epitopes in the human bronchial mucosa. *Biochem J* **410:** 213–223.

Gruenert DC, Finkbeiner WE, Widdicombe JH. 1995. Culture and transformation of human airway epithelial cells. *Am J Physiol* **268:** L347–L360.

Gum JRJr, Hicks JW, Toribara NW, Rothe EM, Lagace RE, Kim YS. 1992. The human MUC2 intestinal mucin has cysteine-rich subdomains located both upstream and downstream of its central repetitive region. *J Biol Chem* **267:** 21375–21383.

Guo X, Pace RG, Stonebraker JR, Commander CW, Dang AT, Drumm ML, Harris A, Zou F, Swallow DM, Wright FA, et al. 2011. Mucin variable number tandem repeat polymorphisms and severity of cystic fibrosis lung disease: Significant association with *MUC5AC*. *PLoS ONE* **6:** e25452.

Haggie PM, Verkman AS. 2009. Defective organellar acidification as a cause of cystic fibrosis lung disease: Reexamination of a recurring hypothesis. *Am J Physiol Lung Cell Mol Physiol* **296:** L859–L867.

Hattrup CL, Gendler SJ. 2008. Structure and function of the cell surface (tethered) mucins. *Annu Rev Physiol* **70:** 431–457.

Hauber HP, Lavigne F, Hung HL, Levitt RC, Hamid Q. 2010. Effect of Th2 type cytokines on hCLCA1 and mucus expression in cystic fibrosis airways. *J Cyst Fibros* **9:** 277–279.

Hays SR, Fahy JV. 2006. Characterizing mucous cell remodeling in cystic fibrosis: Relationship to neutrophils. *Am J Respir Crit Care Med* **174:** 1018–1024.

Henke MO, Renner A, Huber RM, Seeds MC, Rubin BK. 2004. MUC5AC and MUC5B mucins are decreased in cystic fibrosis airway secretions. *Am J Respir Cell Mol Biol* **31:** 86–91.

Henke MO, John G, Germann M, Lindemann H, Rubin BK. 2007. MUC5AC and MUC5B mucins increase in cystic fibrosis airway secretions during pulmonary exacerbation. *Am J Respir Crit Care Med* **175:** 816–821.

Henke MO, John G, Rheineck C, Chillappagari S, Naehrlich L, Rubin BK. 2011. Serine proteases degrade airway mucins in cystic fibrosis. *Infect Immun* **79:** 3438–3444.

Hewson CA, Edbrooke MR, Johnston SL. 2004. PMA induces the MUC5AC respiratory mucin in human bronchial epithelial cells, via PKC, EGF/TGF-α, Ras/Raf, MEK, ERK and Sp1-dependent mechanisms. *J Mol Biol* **344:** 683–695.

Hewson CA, Haas JJ, Bartlett NW, Message SD, Laza-Stanca V, Kebadze T, Caramori G, Zhu J, Edbrooke MR, Stanciu LA, et al. 2010. Rhinovirus induces MUC5AC in a human infection model and in vitro via NF-κB and EGFR pathways. *Eur Respir J* **36:** 1425–1435.

Hihnala S, Kujala M, Toppari J, Kere J, Holmberg C, Hoglund P. 2006. Expression of SLC26A3, CFTR and NHE3 in the human male reproductive tract: Role in male subfertility caused by congenital chloride diarrhoea. *Mol Hum Reprod* **12:** 107–111.

Hodges CA, Drumm ML. 2009. CF: Tissue by tissue. *Pediatr Pulmonol* **32:** 188–189.

Hodges CA, Palmert MR, Drumm ML. 2008. Infertility in females with cystic fibrosis is multifactorial: Evidence from mouse models. *Endocrinology* **149:** 2790–2797.

Hollingsworth MA. 1999. Proteins expressed by pancreatic duct cells and their relatives. *Ann NY Acad Sci* **880:** 38–49.

Hollingsworth MA, Swanson BJ. 2004. Mucins in cancer: Protection and control of the cell surface. *Nat Rev Cancer* **4:** 45–60.

Holmen JM, Karlsson NG, Abdullah LH, Randell SH, Sheehan JK, Hansson GC, Davis CW. 2004. Mucins and their O-glycans from human bronchial epithelial cell cultures. *Am J Physiol Lung Cell Mol Physiol* **287:** L824–L834.

Hovenberg HW, Davies JR, Carlstedt I. 1996a. Different mucins are produced by the surface epithelium and the submucosa in human trachea: Identification of MUC5AC as a major mucin from the goblet cells. *Biochem J* **318:** 319–324.

Hovenberg HW, Davies JR, Herrmann A, Linden CJ, Carlstedt I. 1996b. MUC5AC, but not MUC2, is a prominent mucin in respiratory secretions. *Glycoconj J* **13:** 839–847.

Huang RH, Wang Y, Roth R, Yu X, Purvis AR, Heuser JE, Egelman EH, Sadler JE. 2008. Assembly of Weibel-Palade body-like tubules from N-terminal domains of von Willebrand factor. *Proc Natl Acad Sci* **105:** 482–487.

Inoue D, Yamaya M, Kubo H, Sasaki T, Hosoda M, Numasaki M, Tomioka Y, Yasuda H, Sekizawa K, Nishimura H, et al. 2006. Mechanisms of mucin production by rhinovirus infection in cultured human airway epithelial cells. *Respir Physiol Neurobiol* **154:** 484–499.

Ishimoto H, Mukae H, Sakamoto N, Amenomori M, Kitazaki T, Imamura Y, Fujita H, Ishii H, Nakayama S, Yanagihara K, et al. 2009. Different effects of telithromycin on MUC5AC production induced by human neutrophil peptide-1 or lipopolysaccharide in NCI-H292 cells compared with azithromycin and clarithromycin. *J Antimicrob Chemother* **63:** 109–114.

Jacquot J, Puchelle E, Hinnrasky J, Fuchey C, Bettinger C, Spilmont C, Bonnet N, Dieterle A, Dreyer D, Pavirani A. 1993. Localization of the cystic fibrosis transmembrane conductance regulator in airway secretory glands. *Eur Respir J* **6:** 169–176.

Jakab RL, Collaco AM, Ameen NA. 2011. Physiological relevance of cell-specific distribution patterns of CFTR, NKCC1, NBCe1, and NHE3 along the crypt–villus axis in the intestine. *Am J Physiol Gastrointest Liver Physiol* **300:** G82–G98.

Jiang X, Hill WG, Pilewski JM, Weisz OA. 1997. Glycosylation differences between a cystic fibrosis and rescued airway cell line are not CFTR dependent. *Am J Physiol* **273:** L913–L920.

Johansson ME, Gustafsson JK, Sjoberg KE, Petersson J, Holm L, Sjovall H, Hansson GC. 2010. Bacteria penetrate the inner mucus layer before inflammation in the dextran sulfate colitis model. *PLoS ONE* **5:** e12238.

Johansson ME, Ambort D, Pelaseyed T, Schutte A, Gustafsson JK, Ermund A, Subramani DB, Holmen-Larsson JM, Thomsson KA, Bergstrom JH, et al. 2011. Composition and functional role of the mucus layers in the intestine. *Cell Mol Life Sci* **68:** 3635–3641.

Jones LC, Moussa L, Fulcher LM, Zhu Y, Hudson EJ, O'Neal WK, Randell SH, Lazarowski ER, Boucher RC, Kreda SM. 2011. VAMP8 is a vesicle SNARE that regulates mucin secretion in airway goblet cells. *J Physiol* **590:** 545–562.

Joo NS, Irokawa T, Wu JV, Robbins RC, Whyte RI, Wine JJ. 2002. Absent secretion to vasoactive intestinal peptide in cystic fibrosis airway glands. *J Biol Chem* **277:** 50710–50715.

Jouret F, Devuyst O. 2009. CFTR and defective endocytosis: New insights in the renal phenotype of cystic fibrosis. *Pflugers Arch* **457:** 1227–1236.

Kalin N, Claass A, Sommer M, Puchelle E, Tummler B. 1999. ΔF508 CFTR protein expression in tissues from patients with cystic fibrosis. *J Clin Invest* **103:** 1379–1389.

Kartner N, Augustinas O, Jensen TJ, Naismith AL, Riordan JR. 1992. Mislocalization of ΔF508 CFTR in cystic fibrosis sweat gland. *Nat Genet* **1:** 321–327.

Kesimer M, Kirkham S, Pickles RJ, Henderson AS, Alexis NE, DeMaria G, Knight D, Thornton DJ, Sheehan JK. 2008. Tracheobronchial air–liquid interface cell culture: A model for innate mucosal defense of the upper airways? *Am J Physiol Lung Cell Mol Physiol* **296:** L92–L100.

Kesimer M, Makhov AM, Griffith JD, Verdugo P, Sheehan JK. 2009a. Unpacking a gel forming mucin: A view of MUC5B organization after granular release. *Am J Physiol Lung Cell Mol Physiol* **298:** L15–L22.

Kettle R, Simmons J, Schindler F, Jones P, Dicker T, Dubois G, Giddings J, Van HG, Jones CE. 2010. Regulation of neuregulin 1β1-induced MUC5AC and MUC5B expression in human airway epithelium. *Am J Respir Cell Mol Biol* **42:** 472–481.

Kim KC, Hisatsune A, Kim DJ, Miyata T. 2003. Pharmacology of airway goblet cell mucin release. *J Pharmacol Sci* **92:** 301–307.

Kim S, Lewis C, Nadel JA. 2011. CCL20/CCR6 feedback exaggerates epidermal growth factor receptor-dependent MUC5AC mucin production in human airway epithelial (NCI-H292) cells. *J Immunol* **186:** 3392–3400.

Kirkham S, Sheehan JK, Knight D, Richardson PS, Thornton DJ. 2002. Heterogeneity of airways mucus: Variations in the amounts and glycoforms of the major oligomeric mucins MUC5AC and MUC5B. *Biochem J* **361:** 537–546.

Klymiuk N, Mundhenk L, Kraehe K, Wuensch A, Plog S, Emrich D, Langenmayer MC, Stehr M, Holzinger A, Kroner C, et al. 2011. Sequential targeting of CFTR by BAC vectors generates a novel pig model of cystic fibrosis. *J Mol Med (Berl)* **90:** 597–608.

Knowles MR, Boucher RC. 2002. Mucus clearance as a primary innate defense mechanism for mammalian airways. *J Clin Invest* **109:** 571–577.

Kohri K, Ueki IF, Nadel JA. 2002. Neutrophil elastase induces mucin production by ligand-dependent epidermal growth factor receptor activation. *Am J Physiol Lung Cell Mol Physiol* **283:** L531–L540.

Kondo M, Finkbeiner WE, Widdicombe JH. 1993. Cultures of bovine tracheal epithelium with differentiated ultrastructure and ion transport. *In Vitro Cell Dev Biol* **29A:** 19–24.

Kreda SM, Gentzsch M. 2011. Imaging CFTR protein localization in cultured cells and tissues. *Methods Mol Biol* **742:** 15–33.

Kreda SM, Mall M, Mengos A, Rochelle L, Yankaskas J, Riordan JR, Boucher RC. 2005. Characterization of wild-type and ΔF508 cystic fibrosis transmembrane regulator in human respiratory epithelia. *Mol Biol Cell* **16:** 2154–2167.

Kreda SM, Okada SF, van Heusden CA, O'Neal W, Gabriel S, Abdullah L, Davis CW, Boucher RC, Lazarowski ER. 2007. Coordinated release of nucleotides and mucin from human airway epithelial Calu-3 cells. *J Physiol* **584:** 245–259.

Kreda SM, Seminario-Vidal L, van Heusden CA, O'Neal W, Jones L, Boucher RC, Lazarowski ER. 2010. Receptor-promoted exocytosis of airway epithelial mucin granules containing a spectrum of adenine nucleotides. *J Physiol* **588:** 2255–2267.

Kujala M, Hihnala S, Tienari J, Kaunisto K, Hastbacka J, Holmberg C, Kere J, Hoglund P. 2007. Expression of ion transport-associated proteins in human efferent and epididymal ducts. *Reproduction* **133:** 775–784.

Kuver R, Klinkspoor JH, Osborne WR, Lee SP. 2000. Mucous granule exocytosis and CFTR expression in gallbladder epithelium. *Glycobiology* **10:** 149–157.

Leir SH, Parry S, Palmai-Pallag T, Evans J, Morris HR, Dell A, Harris A. 2005. Mucin glycosylation and sulphation in airway epithelial cells is not influenced by cystic fibrosis transmembrane conductance regulator expression. *Am J Respir Cell Mol Biol* **32:** 453–461.

Lemjabbar H, Basbaum C. 2002. Platelet-activating factor receptor and ADAM10 mediate responses to *Staphylococcus aureus* in epithelial cells. *Nat Med* **8:** 41–46.

Lethem MI, Dowell ML, Van Scott M, Yankaskas JR, Egan T, Boucher RC, Davis CW. 1993. Nucleotide regulation of goblet cells in human airway epithelial explants: Normal exocytosis in cystic fibrosis. *Am J Respir Cell Mol Biol* **9:** 315–322.

Li JD, Dohrman AF, Gallup M, Miyata S, Gum JR, Kim YS, Nadel JA, Prince A, Basbaum CB. 1997. Transcriptional activation of mucin by *Pseudomonas aeruginosa* lipopolysaccharide in the pathogenesis of cystic fibrosis lung disease. *Proc Natl Acad Sci* **94:** 967–972.

Li D, Gallup M, Fan N, Szymkowski DE, Basbaum CB. 1998. Cloning of the amino-terminal and 5′-flanking region of the human *MUC5AC* mucin gene and transcriptional upregulation by bacterial exoproducts. *J Biol Chem* **273:** 6812–6820.

Li Y, Martin LD, Spizz G, Adler KB. 2001. MARCKS protein is a key molecule regulating mucin secretion by human airway epithelial cells in vitro. *J Biol Chem* **276:** 40982–40990.

Linden SK, Sutton P, Karlsson NG, Korolik V, McGuckin MA. 2008. Mucins in the mucosal barrier to infection. *Mucosal Immunol* **1:** 183–197.

Litt M, Khan MA, Chakrin LW, Wardell JR Jr, Christian P. 1974. The viscoelasticity of fractionated canine tracheal mucus. *Biorheology* **11:** 111–117.

Livraghi A, Grubb BR, Hudson EJ, Wilkinson KJ, Sheehan JK, Mall MA, O'Neal WK, Boucher RC, Randell SH. 2009. Airway and lung pathology due to mucosal surface dehydration in β-epithelial Na$^+$ channel-overexpressing mice: Role of TNF-α and IL-4Rα signaling, influence of neonatal development, and limited efficacy of glucocorticoid treatment. *J Immunol* **182:** 4357–4367.

Livraghi-Butrico A, Kelly EJ, Klem ER, Dang H, Wolfgang MC, Boucher RC, Randell SH, O'Neal WK. 2012. Mucus clearance and MyD88-dependent immunity modulate lung susceptibility to spontaneous bacterial infection and inflammation. *Mucosal Immunol* doi: 10.1038/mi.2012.17.

Mall M, Grubb BR, Harkema JR, O'Neal WK, Boucher RC. 2004a. Increased airway epithelial Na$^+$ absorption produces cystic fibrosis-like lung disease in mice. *Nat Med* **10:** 487–493.

Mall M, Kreda SM, Mengos A, Jensen TJ, Hirtz S, Seydewitz HH, Yankaskas J, Kunzelmann K, Riordan JR, Boucher RC. 2004b. The ΔF508 mutation results in loss of CFTR function and mature protein in native human colon. *Gastroenterology* **126:** 32–41.

Manson AL, Trezise AE, MacVinish LJ, Kasschau KD, Birchall N, Episkopou V, Vassaux G, Evans MJ, Colledge WH, Cuthbert AW, et al. 1997. Complementation of null CF mice with a human CFTR YAC transgene. *EMBO J* **16:** 4238–4249.

Marino CR, Matovcik LM, Gorelick FA, Cohn JA. 1991. Localization of the cystic fibrosis transmembrane conductance regulator in pancreas. *J Clin Invest* **88:** 712–716.

Martinez-Anton A, Debolos C, Garrido M, Roca-Ferrer J, Barranco C, Alobid I, Xaubet A, Picado C, Mullol J. 2006. Mucin genes have different expression patterns in healthy and diseased upper airway mucosa. *Clin Exp Allergy* **36:** 448–457.

Matsui H, Grubb BR, Tarran R, Randell SH, Gatzy JT, Davis CW, Boucher RC. 1998. Evidence for periciliary liquid layer depletion, not abnormal ion composition, in the pathogenesis of cystic fibrosis airways disease. *Cell* **95:** 1005–1015.

Matsui H, Verghese MW, Kesimer M, Schwab UE, Randell SH, Sheehan JK, Grubb BR, Boucher RC. 2005. Reduced three-dimensional motility in dehydrated airway mucus prevents neutrophil capture and killing bacteria on airway epithelial surfaces. *J Immunol* **175:** 1090–1099.

McAllister F, Henry A, Kreindler JL, Dubin PJ, Ulrich L, Steele C, Finder JD, Pilewski JM, Carreno BM, Goldman SJ, et al. 2005. Role of IL-17A, IL-17F, and the IL-17 receptor in regulating growth-related oncogene-α and granulocyte colony-stimulating factor in bronchial epithelium: Implications for airway inflammation in cystic fibrosis. *J Immunol* **175:** 404–412.

Meyerholz DK, Stoltz DA, Namati E, Ramachandran S, Pezzulo AA, Smith AR, Rector MV, Suter MJ, Kao S, McLennan G, et al. 2010. Loss of cystic fibrosis transmembrane conductance regulator function produces abnormalities in tracheal development in neonatal pigs and young children. *Am J Respir Crit Care Med* **182:** 1251–1261.

Monzon ME, Forteza RM, Casalino-Matsuda SM. 2011. MCP-1/CCR2B-dependent loop upregulates MUC5AC and MUC5B in human airway epithelium. *Am J Physiol Lung Cell Mol Physiol* **300:** L204–L215.

Moskwa P, Lorentzen D, Excoffon KJ, Zabner J, McCray PB Jr, Nauseef MW, Dupuy C, Banfi B. 2007. A novel host defense system of airways is defective in cystic fibrosis. *Am J Respir Crit Care Med* **175:** 174–183.

Mulberg AE, Weyler RT, Altschuler SM, Hyde TM. 1998. Cystic fibrosis transmembrane conductance regulator expression in human hypothalamus. *Neuroreport* **9:** 141–144.

Park RW, Grand RJ. 1981. Gastrointestinal manifestations of cystic fibrosis: A review. *Gastroenterology* **81:** 1143–1161.

Park JA, He F, Martin LD, Li Y, Chorley BN, Adler KB. 2005. Human neutrophil elastase induces hypersecretion of mucin from well-differentiated human bronchial epithelial cells in vitro via a protein kinase Cδ-mediated mechanism. *Am J Pathol* **167:** 651–661.

Park SW, Zhen G, Verhaeghe C, Nakagami Y, Nguyenvu LT, Barczak AJ, Killeen N, Erle DJ. 2009. The protein disulfide isomerase AGR2 is essential for production of intestinal mucus. *Proc Natl Acad Sci* **106:** 6950–6955.

Parmley RR, Gendler SJ. 1998. Cystic fibrosis mice lacking Muc1 have reduced amounts of intestinal mucus. *J Clin Invest* **102:** 1798–1806.

Penque D, Mendes F, Beck S, Farinha C, Pacheco P, Nogueira P, Lavinha J, Malho R, Amaral MD. 2000. Cystic fibrosis F508del patients have apically localized CFTR in a reduced number of airway cells. *Lab Invest* **80:** 857–868.

Peterson M, Leblond CP. 1964. Synthesis of complex carbohydrates in the Golgi region, as shown by radioautography after injection of labeled glucose. *J Cell Biol* **21:** 143–148.

Pickles RJ. 2004. Physical and biological barriers to viral vector-mediated delivery of genes to the airway epithelium. *Proc Am Thorac Soc* **1:** 302–308.

Puchelle E, Gaillard D, Ploton D, Hinnrasky J, Fuchey C, Boutterin MC, Jacquot J, Dreyer D, Pavirani A, Dalemans W. 1992. Differential localization of the cystic fibrosis transmembrane conductance regulator in normal and cystic fibrosis airway epithelium. *Am J Respir Cell Mol Biol* **7:** 485–491.

Quinton PM. 1983. Chloride impermeability in cystic fibrosis. *Nature* **301:** 421–422.

Quinton PM. 1990. Cystic fibrosis: A disease in electrolyte transport. *FASEB J* **4:** 2709–2717.

Quinton PM. 2010. Role of epithelial HCO$_3$ transport in mucin secretion: Lessons from cystic fibrosis. *Am J Physiol Cell Physiol* **299:** C1222–C1233.

Rada B, Gardina P, Myers TG, Leto TL. 2011. Reactive oxygen species mediate inflammatory cytokine release and EGFR-dependent mucin secretion in airway epithelial cells exposed to *Pseudomonas pyocyanin*. *Mucosal Immunol* **4:** 158–171.

Reid CJ, Gould S, Harris A. 1997a. Developmental expression of mucin genes in the human respiratory tract. *Am J Respir Cell Mol Biol* **17:** 592–598.

Reid CJ, Hyde K, Ho SB, Harris A. 1997b. Cystic fibrosis of the pancreas: Involvement of MUC6 mucin in obstruction of pancreatic ducts. *Mol Med* **3:** 403–411.

Rhodes JM. 1989. Colonic mucus and mucosal glycoproteins: The key to colitis and cancer? *Gut* **30:** 1660–1666.

Ribeiro CM, Hurd H, Wu Y, Martino ME, Jones L, Brighton B, Boucher RC, O'Neal WK. 2009. Azithromycin treatment alters gene expression in inflammatory, lipid metabolism, and cell cycle pathways in well-differentiated human airway epithelia. *PLoS ONE* **4:** e5806.

Robert R, Thoreau V, Norez C, Cantereau A, Kitzis A, Mettey Y, Rogier C, Becq F. 2004. Regulation of the cystic fibrosis transmembrane conductance regulator channel by β-adrenergic agonists and vasoactive intestinal peptide in rat smooth muscle cells and its role in vasorelaxation. *J Biol Chem* **279:** 21160–21168.

Rochwerger L, Buchwald M. 1993. Stimulation of the cystic fibrosis transmembrane regulator expression by estrogen in vivo. *Endocrinology* **133:** 921–930.

Rogan MP, Taggart CC, Greene CM, Murphy PG, O'Neill SJ, McElvaney NG. 2004. Loss of microbicidal activity and increased formation of biofilm due to decreased lactoferrin activity in patients with cystic fibrosis. *J Infect Dis* **190:** 1245–1253.

Rogers CS, Stoltz DA, Meyerholz DK, Ostedgaard LS, Rokhlina T, Taft PF, Rogan MP, Pezzulo AA, Karp PH, Itani OA, et al. 2008. Disruption of the CFTR gene produces a model of cystic fibrosis in newborn pigs. *Science* **321:** 1837–1841.

Rose MC. 1988. Epithelial mucous glycoproteins and cystic fibrosis. *Horm Metab Res* **20:** 601–608.

Rose MC. 2006. Mucus. In *Encyclopedia of respiratory medicine* (ed. Laurent GJ, Shapiro SJ), pp. 62–66. Academic Press, San Diego.

Rose MC, Voynow JA. 2006. Respiratory tract mucin genes and mucin glycoproteins in health and disease. *Physiol Rev* **86:** 245–278.

Rose MC, Lynn WS, Kaufman B. 1979. Resolution of the major components of human lung mucosal gel and their capabilities for reaggregation and gel formation. *Biochemistry* **18:** 4030–4037.

Rose MC, Voter WA, Brown CF, Kaufman B. 1984. Structural features of human tracheobronchial mucus glycoprotein. *Biochem J* **222:** 371–377.

Rose MC, Brown CF, Jacoby JZ III, Lynn WS, Kaufman B. 1987. Biochemical properties of tracheobronchial mucins from cystic fibrosis and non-cystic fibrosis individuals. *Pediatr Res* **22:** 545–551.

Rousseau K, Kirkham S, Johnson L, Fitzpatrick B, Howard M, Adams EJ, Rogers DF, Knight D, Clegg P, Thornton DJ. 2008. Proteomic analysis of polymeric salivary mucins: No evidence for MUC19 in human saliva. *Biochem J* **413:** 545–552.

Roussel P, Lamblin G. 2003. The glycosylation of airway mucins in cystic fibrosis and its relationship with lung infection by *Pseudomonas aeruginosa*. *Adv Exp Med Biol* **535:** 17–32.

Roussel P, Lamblin G, Degand P. 1975. Heterogeneity of the carbohydrate chains of sulfated bronchial glycoproteins isolated from a patient suffering from cystic fibrosis. *J Biol Chem* **250:** 2114–2122.

Roy MG, Rahmani M, Hernandez JR, Alexander SN, Ehre C, Ho SB, Evans CM. 2011. Mucin production during prenatal and postnatal murine lung development. *Am J Respir Cell Mol Biol* **44:** 755–760.

Sasaki Y. 2003. New aspects of neurotransmitter release and exocytosis: Rho-kinase-dependent myristoylated alanine-rich C-kinase substrate phosphorylation and regulation of neurofilament structure in neuronal cells. *J Pharmacol Sci* **93:** 35–40.

Sasaki M, Ikeda H, Nakanuma Y. 2007. Expression profiles of MUC mucins and trefoil factor family (TFF) peptides in the intrahepatic biliary system: Physiological distribution and pathological significance. *Prog Histochem Cytochem* **42:** 61–110.

Schachter H, Dixon GH. 1965. A comparative study of the proteins in normal meconium and in meconium from meconium ileus patients. *Can J Biochem Physiol* **43:** 381–397.

Cite this article as *Cold Spring Harb Perspect Med* doi: 10.1101/cshperspect.a009589

Schulz BL, Sloane AJ, Robinson LJ, Prasad SS, Lindner RA, Robinson M, Bye PT, Nielson DW, Harry JL, Packer NH, et al. 2007. Glycosylation of sputum mucins is altered in cystic fibrosis patients. *Glycobiology* **17:** 698–712.

Sears PR, Davis CW, Chua M, Sheehan JK. 2011. Mucociliary interactions and mucus dynamics in ciliated human bronchial epithelial cell cultures. *Am J Physiol Lung Cell Mol Physiol* **301:** L181–L186.

Seksek O, Biwersi J, Verkman AS. 1996. Evidence against defective trans-Golgi acidification in cystic fibrosis. *J Biol Chem* **271:** 15542–15548.

Seminario-Vidal L, Kreda S, Jones L, O'Neal W, Trejo J, Boucher RC, Lazarowski ER. 2009. Thrombin promotes release of ATP from lung epithelial cells through coordinated activation of rho- and Ca^{2+}-dependent signaling pathways. *J Biol Chem* **284:** 20638–20648.

Sheehan JK, Kirkham S, Howard M, Woodman P, Kutay S, Brazeau C, Buckley J, Thornton DJ. 2004. Identification of molecular intermediates in the assembly pathway of the MUC5AC mucin. *J Biol Chem* **279:** 15698–15705.

Shen BQ, Finkbeiner WE, Wine JJ, Mrsny RJ, Widdicombe JH. 1994. Calu-3: A human airway epithelial cell line that shows cAMP-dependent Cl^- secretion. *Am J Physiol* **266:** L493–L501.

Singer M, Martin LD, Vargaftig BB, Park J, Gruber AD, Li Y, Adler LB. 2004. A MARCKS-related peptide blocks mucus hypersecretion in a mouse model of asthma. *Nat Med* **10:** 193–196.

Smith JJ, Travis SM, Greenberg EP, Welsh MJ. 1996. Cystic fibrosis airway epithelia fail to kill bacteria because of abnormal airway surface fluid. *Cell* **85:** 229–236.

Snouwaert JN, Brigman KK, Latour AM, Malouf NN, Boucher RC, Smithies O, Koller BH. 1992. An animal model for cystic fibrosis made by gene targeting. *Science* **257:** 1083–1088.

Song KS, Lee WJ, Chung KC, Koo JS, Yang EJ, Choi JY, Yoon JH. 2003. Interleukin-1β and tumor necrosis factor-α induce MUC5AC overexpression through a mechanism involving ERK/p38 mitogen-activated protein kinases-MSK1-CREB activation in human airway epithelial cells. *J Biol Chem* **278:** 23243–23250.

Song KS, Lee TJ, Kim K, Chung KC, Yoon JH. 2008. cAMP-responding element-binding protein and c-Ets1 interact in the regulation of ATP-dependent MUC5AC gene expression. *J Biol Chem* **283:** 26869–26878.

Spicer SS, Chakrin LW, Wardell JR Jr, Kendrick W. 1971. Histochemistry of mucosubstances in the canine and human respiratory tract. *Lab Invest* **25:** 483–490.

Spurr-Michaud S, Argueso P, Gipson I. 2007. Assay of mucins in human tear fluid. *Exp Eye Res* **84:** 939–950.

Strong TV, Boehm K, Collins FS. 1994. Localization of cystic fibrosis transmembrane conductance regulator mRNA in the human gastrointestinal tract by in situ hybridization. *J Clin Invest* **93:** 347–354.

Sturgess J. 1982. Morphological characteristics of the bronchial mucosa in cystic fibrosis. In *Fluid and electrolyte abnormalities in exocrine glands in cystic fibrosis* (ed. Quinton PM, et al.), pp. 254–270. San Francisco Press, San Francisco.

Sturgess J, Imrie J. 1982. Quantitative evaluation of the development of tracheal submucosal glands in infants with cystic fibrosis and control infants. *Am J Pathol* **106:** 303–311.

Sun X, Sui H, Fisher JT, Yan Z, Liu X, Cho HJ, Joo NS, Zhang Y, Zhou W, Yi Y, et al. 2010. Disease phenotype of a ferret CFTR-knockout model of cystic fibrosis. *J Clin Invest* **120:** 3149–3160.

Takeyama K, Dabbagh K, Lee HM, Agusti C, Lausier JA, Ueki IF, Grattan KM, Nadel JA. 1999. Epidermal growth factor system regulates mucin production in airways. *Proc Natl Acad Sci* **96:** 3081–3086.

Taylor C, Allen A, Dettmar PW, Pearson JP. 2004. Two rheologically different gastric mucus secretions with different putative functions. *Biochim Biophys Acta* **1674:** 131–138.

Ten Hagen KG, Fritz TA, Tabak LA. 2003. All in the family: The UDP-GalNAc:polypeptide *N*-acetylgalactosaminyltransferases. *Glycobiology* **13:** 1R–16R.

Thai P, Loukoianov A, Wachi S, Wu R. 2008. Regulation of airway mucin gene expression. *Annu Rev Physiol* **70:** 405–429.

Thornton DJ, Rousseau K, McGuckin MA. 2007. Structure and function of the polymeric mucins in airways mucus. *Annu Rev Physiol* **70:** 459–486.

Tirouvanziam R, Khazaal I, Peault B. 2002. Primary inflammation in human cystic fibrosis small airways. *Am J Physiol Lung Cell Mol Physiol* **283:** L445–L451.

Tizzano EF, Buchwald M. 1995. CFTR expression and organ damage in cystic fibrosis. *Ann Intern Med* **123:** 305–308.

Tizzano EF, Chitayat D, Buchwald M. 1993. Cell-specific localization of CFTR mRNA shows developmentally regulated expression in human fetal tissues. *Hum Mol Genet* **2:** 219–224.

Tizzano EF, O'Brodovich H, Chitayat D, Benichou JC, Buchwald M. 1994a. Regional expression of CFTR in developing human respiratory tissues. *Am J Respir Cell Mol Biol* **10:** 355–362.

Tizzano EF, Silver MM, Chitayat D, Benichou JC, Buchwald M. 1994b. Differential cellular expression of cystic fibrosis transmembrane regulator in human reproductive tissues. Clues for the infertility in patients with cystic fibrosis. *Am J Pathol* **144:** 906–914.

Trapnell BC, Chu CS, Paakko PK, Banks TC, Yoshimura K, Ferrans VJ, Chernick MS, Crystal RG. 1991. Expression of the cystic fibrosis transmembrane conductance regulator gene in the respiratory tract of normal individuals and individuals with cystic fibrosis. *Proc Natl Acad Sci* **88:** 6565–6569.

Trezise AE, Buchwald M. 1991. In vivo cell-specific expression of the cystic fibrosis transmembrane conductance regulator. *Nature* **353:** 434–437.

Trezise AE, Chambers JA, Wardle CJ, Gould S, Harris A. 1993. Expression of the cystic fibrosis gene in human foetal tissues. *Hum Mol Genet* **2:** 213–218.

Tuvim MJ, Mospan AR, Burns KA, Chua M, Mohler PJ, Melicoff E, Adachi R, Ammar-Aouchiche Z, Davis CW, Dickey BF. 2009. Synaptotagmin 2 couples mucin granule exocytosis to Ca^{2+} signaling from endoplasmic reticulum. *J Biol Chem* **284:** 9781–9787.

Verdugo P. 1991. Mucin exocytosis. *Am Rev Respir Dis* **144:** S33–S37.

Verdugo P. 2012. Supramolecular dynamics of mucus. *Cold Spring Harb Perspect Med* **2:** a009597.

Voynow JA, Rubin BK. 2009. Mucins, mucus, and sputum. *Chest* **135:** 505–512.

Voynow JA, Young LR, Wang Y, Horger T, Rose MC, Fischer BM. 1999. Neutrophil elastase increases MUC5AC mRNA and protein expression in respiratory epithelial cells. *Am J Physiol* **276:** L835–L843.

Voynow JA, Fischer BM, Malarkey DE, Burch LH, Wong T, Longphre M, Ho SB, Foster WM. 2004. Neutrophil elastase induces mucus cell metaplasia in mouse lung. *Am J Physiol Lung Cell Mol Physiol* **287:** L1293–L1302.

Voynow JA, Fischer BM, Roberts BC, Proia AD. 2005. Basal-like cells constitute the proliferating cell population in cystic fibrosis airways. *Am J Respir Crit Care Med* **172:** 1013–1018.

Welsh MJ, Tsui L-C, Boat TF, Beaudet AL. 1995. Cystic fibrosis. In *The metabolic basis of inherited disease*, 4th ed. (ed. Scriver CR, et al.), pp. 3799–3876. McGraw-Hill, New York.

Wine JJ, Joo NS. 2004. Submucosal glands and airway defense. *Proc Am Thorac Soc* **1:** 47–53.

Wine JJ, Joo NS, Choi JY, Cho HJ, Krouse ME, Wu JV, Khansaheb M, Irokawa T, Ianowski J, Hanrahan JW, et al. 2011. Measurement of fluid secretion from intact airway submucosal glands. *Methods Mol Biol* **742:** 93–112.

Wu R, Yankaskas J, Cheng E, Knowles MR, Boucher RC. 1985. Growth and differentiation of human nasal epithelial cells in culture: Serum-free, hormone-supplemented medium and proteoglycan synthesis. *Am Rev Respir Dis* **132:** 311–320.

Wu R, Martin WR, Robinson CB, St George JA, Plopper CG, Kurland G, Last JA, Cross CE, McDonald RJ, Boucher R. 1990. Expression of mucin synthesis and secretion in human tracheobronchial epithelial cells grown in culture. *Am J Respir Cell Mol Biol* **3:** 467–478.

Wu DY, Wu R, Chen Y, Tarasova N, Chang MM. 2007a. PMA stimulates MUC5B gene expression through an Sp1-based mechanism in airway epithelial cells. *Am J Respir Cell Mol Biol* **37:** 589–597.

Wu JV, Krouse ME, Wine JJ. 2007b. Acinar origin of CFTR-dependent airway submucosal gland fluid secretion. *Am J Physiol Lung Cell Mol Physiol* **292:** L304–L311.

Xia B, Royall JA, Damera G, Sachdev GP, Cummings RD. 2005. Altered *O*-glycosylation and sulfation of airway mucins associated with cystic fibrosis. *Glycobiology* **15:** 747–775.

Xu WM, Shi QX, Chen WY, Zhou CX, Ni Y, Rowlands DK, Yi LG, Zhu H, Ma ZG, Wang XF, et al. 2007. Cystic fibrosis transmembrane conductance regulator is vital to sperm fertilizing capacity and male fertility. *Proc Natl Acad Sci* **104:** 9816–9821.

Yang Y, Raper SE, Cohn JA, Engelhardt JF, Wilson JM. 1993. An approach for treating the hepatobiliary disease of cystic fibrosis by somatic gene transfer. *Proc Natl Acad Sci* **90:** 4601–4605.

Yankaskas JR, Haizlip JE, Conrad M, Koval D, Lazarowski E, Paradiso AM, Rinehart CA Jr, Sarkadi B, Schlegel R, Bou-cher RC. 1993. Papilloma virus immortalized tracheal epithelial cells retain a well-differentiated phenotype. *Am J Physiol* **264:** C1219–C1230.

Yoshimura K, Nakamura H, Trapnell BC, Chu CS, Dalemans W, Pavirani A, Lecocq JP, Crystal RG. 1991. Expression of the cystic fibrosis transmembrane conductance regulator gene in cells of non-epithelial origin. *Nucleic Acids Res* **19:** 5417–5423.

Yu H, Li Q, Kolosov VP, Perelman JM, Zhou X. 2011. Regulation of cigarette smoke-induced mucin expression by neuregulin1β/ErbB3 signalling in human airway epithelial cells. *Basic Clin Pharmacol Toxicol* **109:** 63–72.

Yuan-Chen WD, Wu R, Reddy SP, Lee YC, Chang MM. 2007. Distinctive epidermal growth factor receptor/extracellular regulated kinase-independent and -dependent signaling pathways in the induction of airway mucin 5B and mucin 5AC expression by Phorbol 12-Myristate 13-Acetate. *Am J Pathol* **170:** 20–32.

Zabner J, Smith JJ, Karp PH, Widdicombe JH, Welsh MJ. 1998. Loss of CFTR chloride channels alters salt absorption by cystic fibrosis airway epithelia in vitro. *Mol Cell* **2:** 397–403.

Zalewska A, Zwierz K, Zolkowski K, Gindzienski A. 2000. Structure and biosynthesis of human salivary mucins. *Acta Biochim Pol* **47:** 1067–1079.

Zeng W, Lee MG, Yan M, Diaz J, Benjamin I, Marino CR, Kopito R, Freedman S, Cotton C, Muallem S, et al. 1997. Immuno and functional characterization of CFTR in submandibular and pancreatic acinar and duct cells. *Am J Physiol* **273:** C442–C455.

Zhang Y, Doranz B, Yankaskas JR, Engelhardt JF. 1995. Genotypic analysis of respiratory mucous sulfation defects in cystic fibrosis. *J Clin Invest* **96:** 2997–3004.

Zhou Z, Treis D, Schubert SC, Harm M, Schatterny J, Hirtz S, Duerr J, Boucher RC, Mall MA. 2008. Preventive but not late amiloride therapy reduces morbidity and mortality of lung disease in βENaC-overexpressing mice. *Am J Respir Crit Care Med* **178:** 1245–1256.

Zhou Z, Duerr J, Johannesson B, Schubert SC, Treis D, Harm M, Graeber SY, Dalpke A, Schultz C, Mall MA. 2011. The ENaC-overexpressing mouse as a model of cystic fibrosis lung disease. *J Cyst Fibros* **10:** S172–S182.

Zhu Y, Ehre C, Abdullah LH, Sheehan JK, Roy M, Evans CM, Dickey BF, Davis CW. 2008. MUNC13-2$^{-/-}$ baseline secretion defect reveals source of oligomeric mucins in mouse airways. *J Physiol* **586:** 1977–1992.

Zhu L, Lee PK, Lee WM, Zhao Y, Yu D, Chen Y. 2009. Rhinovirus-induced major airway mucin production involves a novel TLR3-EGFR-dependent pathway. *Am J Respir Cell Mol Biol* **40:** 610–619.

Zuelzer WW, Newton WA Jr. 1949. The pathogenesis of fibrocystic disease of the pancreas; a study of 36 cases with special reference to the pulmonary lesions. *Pediatrics* **4:** 53–69.

Zuhdi Alimam M, Piazza FM, Selby DM, Letwin N, Huang L, Rose MC. 2000. Muc-5/5ac mucin messenger RNA and protein expression is a marker of goblet cell metaplasia in murine airways. *Am J Respir Cell Mol Biol* **22:** 253–260.

Supramolecular Dynamics of Mucus

Pedro Verdugo

Friday Harbor Laboratories, University of Washington, Friday Harbor, Washington 98250

Correspondence: verdugo@u.washington.edu

Our purpose here is not to address specific issues of mucus pathology, but to illustrate how polymer networks theory and its remarkable predictive power can be applied to study the supramolecular dynamics of mucus. Avoiding unnecessary mathematical formalization, in the light of available theory, we focus on the rather slow progress and the still large number of missing gaps in the complex topology and supramolecular dynamics of airway mucus. We start with the limited information on the polymer physics of respiratory mucins to then converge on the supramolecular organization and resulting physical properties of the mucus gel. In each section, we briefly discuss progress on the subject, the uncertainties associated with the established knowledge, and the many riddles that still remain.

A ROAD MAP

Mucus has a complex set of functions in the airway, including its central role in mucociliary clearance. Defective mucus flow rests at the base of some of the most critical pulmonary pathology, including chronic obstructive pulmonary disease (COPD), cystic fibrosis (CF), and asthma. Research on the intricate mechanisms that regulate mucus rheology, particularly in CF, has been largely focused on mucins and on the defective function of the cystic fibrosis transmembrane conductance regulator (CFTR). The biochemistry and molecular biology of mucins (Perez-Villar and Hill 1999; Chen et al. 2001; Thornton et al. 2008) and the biophysics of the CFTR (Riordan 2005; Sheppard and Welsh 2006) have all received a great deal of attention during this last decade. Even so, the pathophysiology of defective mucus remains largely phenomenological, and an understanding of the mechanisms that make airway mucus transportable is still limited. Mucus is a polymer gel, and this question rests by and large in the domains of the physics and physical chemistry of polymer gels. The matrix of polymer gels forms a supramolecular network with emerging properties that cannot be understood only on the basis of the properties of the molecules that make it. When polymers are constrained to lie close together by physical or chemical bonds, their behavior departs from those of free polymers. A new set of features emerges that is largely determined by polymer–polymer and polymer–solvent interactions, as well as the topological features of the gel's matrix, including the nature and density of interconnections (low/high energy bonds and physical tangles) and the conformational state and mechanical properties of the polymers that make it. Nonetheless, most of the emphasis on mucus research remains focused on the biochemistry and molecular biology of mucins, the giant polymers found in the mucus gel matrix.

One of the oldest controversies in mucus research questions the nature of interconnections among mucin chains: whether the mucus is a chemical gel with a matrix containing branched polymers interconnected by covalent S:S bonds, or a physical gel in which linear polymers are held together by tangles and low-energy electrostatic and hydrophobic bonds (Lee et al. 1977; Clamp et al. 1978; Verdugo et al. 1983). Because the rheological properties of chemical gels depend on the degree of covalent cross-links among polymer chains (Flory 1953), defective control of S:S bonding was thought to be the source of mucus pathology (Clamp et al. 1978). Conversely, the rheological properties of physical gels depend on the degree of swelling of the gel's matrix, that is, the number of tangles per unit of gel volume (Edwards and Grant 1973a). Mucus pathology in this case is likely to result from defective swelling (Verdugo 1984, 1990, 1991, 1998; Chen et al. 2010). The demonstration that mucus remains condensed while stored in the granule to then undergo massive swelling upon exocytosis (Verdugo 1984, 1991) validated the argument that swelling is a necessary and sufficient condition for mucus to reach its normal flow properties. It also rendered this controversy largely moot because, chemically or physically interconnected, the mucus gel must swell to become transportable because in either case it emerges from a condensed conformation. Mucus pathology in either case is likely to result primarily from defective swelling—with one caveat: if the mucus matrix is covalently bonded, swelling will reach a limit and eventually expand to a maximum volume determined by the degree of interchain (intermolecular) bonding (Flory 1953); the more the bonding, the less it can swell and the harder it should be to transport. However, there no limit to swelling if the matrix is tangled. The mucus matrix will freely relax to reach an equilibrium swelling that is controlled only by the amount of available water and particularly the composition of its solvent (e.g., the ASL). Because the rheological properties of tangled-matrix gels depend on tangle density (Edwards and Grant 1973a) and interchain tangles decrease exponentially as mucus volume grows (de Gennes and Leger 1982; Doi and Edwards 1984), small changes in swelling equilibrium result in large changes in mucus rheology, and deficient swelling could be a significant source of mucus pathology (Verdugo 1984, 1990, 1991). Defective mucus swelling became a central question in mucus pathology. Airway surface dehydration had been postulated to be the culprit (Matsui et al. 1998; Boucher 2007). However, regardless of the water supply, unless Ca is dislodged from the mucus matrix, mucus hydration will be drastically limited (Verdugo et al. 1987b; Aitken and Verdugo 1989; Verdugo 1998; Chen et al. 2010). The required physiological chelator responsible for withdrawing Ca from the mucus matrix turns out to be HCO_3, providing a powerful basis to understand CF mucus pathology (Garcia et al. 2009; Chen et al. 2010; Muchekehu and Quinton 2010; Quinton 2010).

Direct evidence from EM, DLS, and AFM portray mucins as linear chains (Sheehan et al. 1984; Marianne et al. 1987; Gupta et al. 1990; Matthews et al. 2002; Round et al. 2002; Bansil et al. 2005; Hong et al. 2005). However, biochemical evidence indicates that mucins in gastrointestinal mucus might be S:S-bonded (Ambort et al. 2011). These different matrix topologies may reveal two corresponding strategies of mucus function, which are particularly adapted to different local conditions of the guts and airways. In the gastrointestinal tract, a dense layer of protective mucus is required to defend the mucosa from being digested, and because mucus is exposed to large volumes of solvent, it seems fitting to limit mucus swelling to preserve a tight protective molecular mesh. In the airway, mucociliary flow requires delicate dynamic regulation of mucus rheology. This is probably achieved by strict control of mucus swelling equilibrium by managing water movement across the mucosa and delicate control of Na, pH, and particularly of Ca and HCO_3 concentrations in the ASL. An additional level of complexity is that in chronic airway pathology, the polymer tangled topology of airway mucus might become covalently cross-linked (Thornton et al. 1991; Sheehan et al. 1999). In this case, the normal mechanisms that control

Cite this article as *Cold Spring Harb Perspect Med* doi: 10.1101/cshperspect.a009597

mucus hydration would become ineffective in restoring normal mucus flow properties.

We first introduce a definition of polymer gels and give a few words regarding emergence in polymer networks. Gels are formed by a three-dimensional (3D) polymer matrix embedded in a solvent, which is water in the case of biological hydrogels. Solvent prevents the collapse of the network, and the network entraps the solvent, creating microenvironments that are in thermodynamic equilibrium with the surrounding media. "Gels" are a unique form of hierarchically structured supramolecular organization in which the polymers that form the gel matrix form a 3D network interconnected by chemical or physical cross-links that keep these chains in a statistically stable close neighborhood (Tanaka 1981). The chemical and physical characteristics of the individual polymer chains (including polyelectrolyte properties, hydrophobicity, size, linear or branched structure, etc.) and the nature of their interconnections determine the topology, chemical reactivity, and bulk physical properties of gel networks and how gels interact with solvents, smaller solutes, and microorganisms. Covalently cross-linked chemical gels have a defined limit swelling volume and cannot anneal. That is, polymers are tightly bound to each other and cannot migrate out of their network to interpenetrate neighboring gels and anneal, forming larger gels. Conversely, in physical gels, polymers can axially slide past each other and swell indefinitely. Axial mobility is an important feature of tangled networks—namely, it allows polymers from neighboring gels to interpenetrate and anneal to form larger gels (de Gennes and Leger 1982). In response to subtle physical or chemical shifts of their environment both cross-linked and entangled gels can undergo abrupt reversible phase transitions from a flexible and highly permeable swollen phase to a dense collapsed phase, and vice versa (Tanaka et al. 1980). As pointed out below, this mechanism allows volume-occupying mucus to collapse inside the granule as a densely compacted network.

"Emergence" refers to new properties that emerge from the collective behavior of multiple interacting components. For instance, only hydrogen and oxygen and no other elements can make water, and only a few features of these gases are required for water synthesis. However, once they combine to make water, a unique set of new properties emerges. In fact, the emergent properties of water have little to do with the two gases that make it, and little can be predicted about the emergent properties of water by speculating on the properties of these gases. Complex multiscale systems like polymer gels are characterized by a hierarchy of molecular order in which the impact of individual molecules in the bulk physical properties of the system is strictly conditioned by their mutual association with the supramolecular gel matrix. Just a few of a broad range of properties of the polymer components are important for gel assembly and bulk properties. Here again, the emergent properties of gels are not found in their component polymer chains, nor can they be unequivocally predicted from the chemical/physical properties of their individual polymer components.

Important advances in mucin biochemistry and molecular biology portray a remarkable complexity of mucin expression. There is a broad range of genetic variations that is producing an ever-growing molecular taxonomy of mucins, making it extremely intricate and arbitrary to specify structure–function assignments of these polymers to predict the complex features of a mucus gel explicitly. These studies make up a body of excellent science, but only a fraction of it can serve to reliably—and even then to only partially—forecast the properties of respiratory mucus. Most of the relevant physical information pertaining to macromolecular dynamics required to predict the role of mucin polymers in determining the bulk properties of mucus is still missing.

POLYMER PHYSICS OF AIRWAY MUCINS

Conformation of Mucin Chains

The molecular structure of mucins is reviewed in detail in Ambort et al. (2012b). Herewith we focus only on those features required to explain the assembly of mucus networks. MUC5AC and MUC5B, the main structural polymers in the matrix of airway mucus, are among the largest biopolymers known. They have a random-coil conformation in solution and are broadly poly-dispersed with a molecular mass ranging from 2×10^6 to 1×10^9 Da and a chain (contour) length varying between 500 nm and several micrometers (Matthews et al. 2002; Round et al. 2002; Thornton et al. 2008; Kesimer et al. 2010). They consist of a core apoprotein in which polyanionic glycosylated blocks containing sialic and sulfate ester terminals alternate periodically with hydrophobic "naked" cysteine-rich blocks. This pattern of alternating negatively charged hydrophilic blocks with globular hydrophobic blocks resembles the typical architecture of synthetic amphiphilic brush multiblock copolymers (Bates and Fredrickson 1990; Dobrynin and Rubinstein 2005). It gives mucins a broad potential for folding and forming ordered collapsed structures. Among the most critical characteristics of mucins for their role in respiratory mucus is their linear, unbranched conformation. Early work using dynamic laser scattering (DLS) revealed that mucin chains in cervical and respiratory mucus can diffuse along their axis (Lee et al. 1977; Verdugo et al. 1983). In chemically cross-linked gels, polymers can show only lateral local motion associated to the chain sections between crosslinks (Doi and Edwards 1984). However, in entangled gels, polymers can diffuse both locally and axially, the latter resembling the motion of a snake inside a randomly convoluted tube. These random movements are reported by two corresponding DSL sets of relaxation times—namely, short relaxation times reflecting limited local lateral polymer mobilities and long relaxation times associated with translational diffusion along the polymer axis that de Gennes and Leger (1982) called "reptation." DSL revealed that both sets of random diffusion mobilities are present in cow estrous cervical mucus (Lee et al. 1977) and in human bronchial mucus (Verdugo et al. 1983). Because reptation cannot take place in networks held together by covalent cross-links or made out of branched polymers (Edwards and Grant 1973b; de Gennes and

Leger 1982; Doi and Edwards 1984), these results provided the first objective indication that the polymer matrix of respiratory and cervical mucus is made of linear mucins that most likely are not chemically cross-linked but are tangled, forming a loosely woven network. Early EM images (Sheehan et al. 1984; Marianne et al. 1987) and more recent AFM imaging methods confirmed that respiratory MUC5B and MUC5AC show a characteristic linear conformation with contour lengths that can reach several micrometers (Gupta et al. 1990; Round et al. 2002; Bansil et al. 2005; Hong et al. 2005). Recent work using DLS and viscometry indicates that purified pig gastric mucins resemble an S:S-linked linear collection of dumbbell units forming daisy chain-like linear polymers (Yakubov et al. 2007). Similar results were reported by Bansil et al. (2005). DLS results in human mucins (Gupta et al. 1990) agree with the model that the molecular conformation of mucins consist of subunits connected by covalent (disulfide) bonds forming linear chains.

Mucins' linear conformation and large size are critical features for the airway mucus matrix. Swelling expands tangled networks requiring polymers to slide past each other, making linear conformation a paramount condition for axial diffusion and expansion. Regarding polymer size, the stability of tangled networks depends on the second power of the contour length of the polymers that form it (de Gennes and Leger 1982). This is because to move inside the network, polymers must diffuse axially, to "reptate" their way through multiple tangles, like a randomly wriggling snake. Because random walk times are proportional to the second power of the distance traveled (in this case, the contour length of the snake), small changes of mucin length result in drastic changes in mucus swelling kinetics. Upon exocytosis, mucus undergoes massive quick swelling, and mucin chain length can be a major hindrance to network relaxation (expansion). Thus, small changes of mucin length result in dramatic changes in swelling rate and swelling equilibrium. Conversely, mucin shortening by cleavage of S:S bonds of the apomucin backbone can readily disperse

Cite this article as *Cold Spring Harb Perspect Med* doi: 10.1101/cshperspect.a009597

the mucus matrix. Enzymes immobilized inside the condensed mucus granule could be activated following exocytosis and potentially play a role in peptide cleavage in mucus conditioning (J Sheehan, unpubl.). Unfortunately, the actual size of mucins in their native state remains uncertain. Biochemical separation procedures result in broad polydispersity due to extensive cleavage yielding molecular dimensions that are far from their original size and conformation (Sheehan and Thornton 2000; Round et al. 2002). Recent developments of methods to isolate secretory granules (Marszalek et al. 1997) allow working with uncontaminated material released from isolated electroporated granules. Mucin polymers released from isolated goblet cell granules spontaneously dispersed onto Ca-free, 140 mM Na, pH 6 medium. They show a narrow 0.1–0.3 polydispersity with mean square radius of gyration $R_g \approx 230$ nm (EY Chen and WC Chin, pers. comm.). These large dimensions agree with DSL results in salivary mucins (Kesimer et al. 2010) and with the multimicrometer contour length observed by atomic force microscopy in ocular and airway MUC5AC (Matthews et al. 2002; Round et al. 2007). In summary, despite their critical roles in mucus bulk properties, the contour length and solvent composition-dependent conformation of mucin chains in respiratory mucus remain uncertain.

Polyelectrolyte and Hydrophobic Features of Mucins

Sialic and sulfated sugar residues make mucins strongly polyanionic with an isoelectric pH point of ~ 2.5 (Bansil et al. 1995; Thornton et al. 2008). This feature has two important implications. First, it means that swelling of mucus must be driven by fixed charges and governed by a Donnan process (Katchalsky et al. 1951), an issue validated (Tam and Verdugo 1981) and addressed in detail in the next section on polymer physics of the mucus gel. Second, the elastic properties and equilibrium conformation of polyelectrolytes depend very strongly on the dielectric properties, counterion composition, and pH of the solvent. Depending on solvent

composition, polyelectrolytes can be flexible or rigid and brittle. These features can directly affect the rheological properties of the gels they form (Doi and Edwards 1984). Theoretically, the polyanionic charge of mucins makes the conformation, hydrodynamic diameter, and, especially, the elastic properties of these polymers highly sensitive to pH and to cations, particularly Ca and Na present in the ASL. Atomic force microscopy (AFM) reveals that conformation and folding of gastric and ocular mucin chains are critically dependent on pH (Bansil et al. 1995; Round et al. 2004). In summary, several lines of evidence confirm that airway mucins show a linear conformation and large molecular dimensions. However, the effects of Ca, Na, and pH on polymer elasticity, folding, and conformation of airway mucins have not been conducted. It is likely that mucins will undergo globule/coil transition upon exposure to solvent compositions that mimic those that induce macroscopic condensation of the mucus gel, namely, a low Na/Ca ratio and pH (Verdugo et al. 1992; Verdugo 1994). Direct evaluation of the effect of solvent composition on airway mucin conformation is an important still pending task, because change of polymer elasticity can directly affect the viscoelastic properties of mucus.

Direct measurements of mucin zeta (ζ) potential to assess changes in polyanionic charge density have not been performed in undegraded respiratory mucins at different Na/Ca ratios and pH. This information is crucial because variations of charge density of mucins could strongly affect the mucin electrostatic interactions and bulk properties of mucus, as well as bacterial binding to mucus (van Loosdrecht et al. 1990; Rayner and Wilson 1997).

Equally unexplored is the effect of small anions on the conformational dynamics of mucin polymers. Most of the few advances have been focused on effects of salt cations, whereas the anions of the supporting electrolyte have been largely ignored. Direct or indirect effects of anions on mucus, particularly by HCO_3, have only recently begun to be systematically investigated, revealing the crucial role of this anion on mucus swelling equilibrium (Garcia et al.

2009; Chen et al. 2010; Muchekehu and Quinton 2010; Quinton 2010).

Finally, the role of hydrophobic interactions in mucus–bacteria adhesion is well established (Rayner and Wilson 1997). Hydrophobic interactions play critical roles on gastric mucin conformation, folding, and interchain binding (Bansil et al. 1995). However, similar studies in respiratory mucins have not been conducted.

Mucin Polymer Folding

Hydrated mucins are volume-occupying polymers, well suited to form the matrix of mucus. However, while in storage the mucus gel remains in a condensed phase (Verdugo et al. 1992, 1994) with mucin chains highly compacted and constrained within a drastically limited intragranular volume. This state is hardly compatible with a simple collapsed random-coil conformation. Mucins, however, have the characteristic features of triblock copolymers with comb glycosylated hydrophilic rigid blocks flanked by flexible globular hydrophobic blocks (Di Cola et al. 2008; Waigh and Papagiannopoulos 2010). A characteristic feature of block copolymers is that they can fold to form ordered compact metastable liquid crystalline phases (Halperin 1991). Nematic crystalline ordered domains are, indeed, present in the mucus of the banana slug *Ariolimax columbianus* (Viney et al. 1993a,b). Mucus released from slug granules can swell at virtually explosive rates, going from 4 to 300 μm diameter in ~20 ms, with a corresponding volume increase close to 4.2×10^4 times their condensed size (Verdugo et al. 1992). At this rate of expansion, the internal frictional dissipations of polymers sliding with respect to one another would be enormous. Unfolding of ordered mucin blocks present between tangles can explain how the mucin matrix of these granules undergoes explosive swelling expansion with little reptational dissipation that otherwise would be required if mucins were to form a random network (Fig. 1). Thus, nematic ordered phases may account for both high-density folded packing and for reducing internal shear dissipation upon swelling via an unfolding of ordered mucin blocks. Important features of

block copolymers predicted by theory are starting to be verified in synthetic block copolymer networks (Bates and Frederickson 1990) and are critical to understand the complex polymer dynamics of mucin gels. Verification with regard to respiratory mucus remains urgently important unfinished work.

In summary, the biochemistry and molecular biology of respiratory mucins have produced a rich bounty of data. Unfortunately, however, most of this information otherwise rich in implications for many aspects of mucins functions is of limited significance to the network formation and supramolecular dynamics that are essential to predict mucus rheology in the airways. We are still missing fundamental pieces of the puzzle to build a precise multidimensional phase diagram of mucin-size dynamics and changes of conformation resulting from exposure to different Na/Ca ratios, pH, HCO_3, temperature, and the presence of amphiphilic and organic polycations normally found in the ASL. This information is critical to predict mucin interactions within the mucus matrix.

PHYSICS OF THE MUCUS GEL

Diffusive Mobilities of Mucins in the Mucus Matrix

We now turn our attention from the physics of mucin chains to the physics of the mucus matrix. As pointed out above, results from DLS studies in cow cervical mucus (Lee et al. 1977) and in human bronchial mucus sampled by bronchoscopy (Verdugo et al. 1983) showed the presence of reptational axial diffusion. These results ruled out the idea that these mucins are held together by interchain covalent bonds (de Gennes and Leger 1982) and provide strong support for the hypothesis that mucus is a physical gel held together by tangles and low-energy interactions. Because the rheological properties of gels containing a tangled matrix depend on tangle density, which decreases with the square of the gel volume (Edwards and Grant 1973b; de Gennes and Leger 1982), these observations implied that swelling should be the primary influence on mucus rheology, with a

     Cite this article as *Cold Spring Harb Perspect Med* doi: 10.1101/cshperspect.a009597

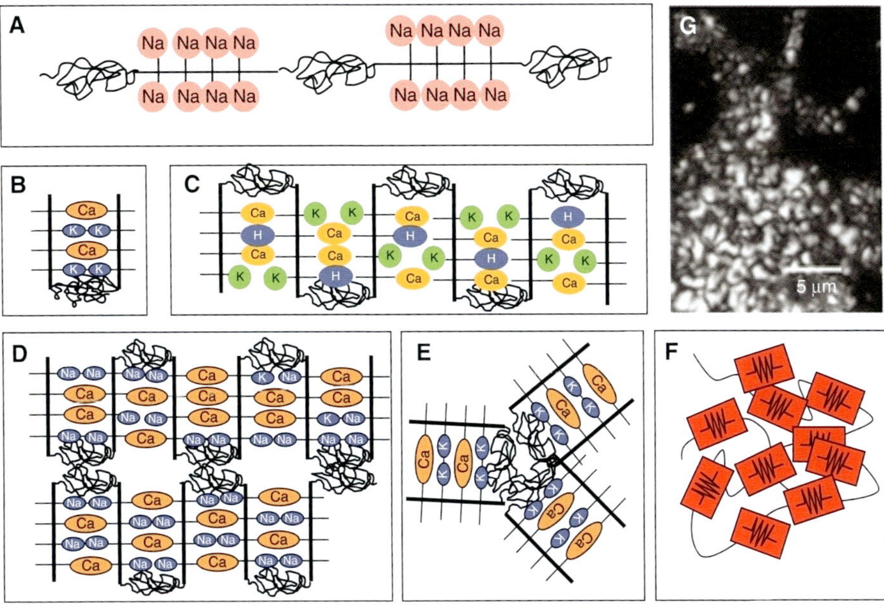

Figure 1. Block copolymer model for airway mucus. The typical periodic hydrophilic/hydrophobic domains and lineal conformation give airway mucins the characteristic profile of block copolymers. The elementary length structure in this model is a binary amphiphilic triblock copolymer consisting of alternating hydrophilic glycosylated polyanionic blocks—resembling densely brush-grafted blocks—flanked by globular peptide hydrophobic blocks (Waigh and Papagiannopoulos 2010). (*A*) In extended conformation—outside the granule—blocks are interconnected forming large linear chains with contour lengths reaching up to several micrometers. (*B*) While inside the granule, chains are condensed with polyanionic glycosylated blocks fully shielded by counterions and folded forming loops stabilized by H and Ca links. (*C*) These U-shaped blocks are interconnected, forming long densely collapsed lattices. (*E*) Lattices can associate laterally via hydrophobic bonds and assemble into large supramolecular nematic liquid crystalline ordered arrays as shown in *D* (planar view) that in side view may hold triplet or even quadruple interlinked collapsed chains. (*F*) Supramolecular arrays, like those illustrated in *E*, are interconnected, forming a tangled random network. In condensed phase, water inside the matrix is constrained to within the Debye field of the polymer charged sites, with a dielectric permittivity that keeps counterions bound. Upon formation of the secretory pore, release of protons and entry of water into the gels matrix result in a drastic change of dielectric property, changing long-range solvent-mediated dynamic interactions among blocks, triggering high-cooperativity counterion dissociation from binding sites. Donnan swelling—resulting from Na/Ca exchange—results in splitting Ca cross-links and repulsive polyanionic interactions that quickly unfold the mucin arrays (represented by boxes). The ordered topology of the network (expansion of the boxes) allows fast isotropic low-dissipative spreading out of the mucus chains without forming knots and loops that would otherwise be expected in a randomly tangled matrix expanding only on the basis of polymer reptation. (Modified from a napkin sketch drawing during our last sushi brainstorming session with my late friend Toyo Tanaka in Boston, early May 2000.) (*G*) An actual polarized microscopy micrograph of nematic liquid crystalline formation in mucus (Viney et al. 1993a).

quadratic dependence further implying that slight swelling deficits should result in thick, highly viscous non-transportable mucus (Verdugo 1990, 1991). They further suggested that mucus must start from an unswollen state, opening the chase for the physical–chemical control of mucus swelling.

Donnan Control of Mucus Swelling

In polymer gels, interchain bonds and tangles constrain the mobility of polymers and keep them inside the gel matrix. Swelling of gels is driven by forces generated by osmotic pressure; that is, if the chains forming the gel's matrix generate higher osmotic pressure than solutes in the bulk solvent outside the gel, the gel swells. In gels made of neutral polymers, the swelling rate is slow because, in this case, osmotic pressure results from limited hindered diffusion of the large polymers that form the gel's matrix. However, in gels made of polyionic polymers, their ionized charged sites are fixed and cannot move out of the matrix, because they are part of the polymers. As a result, electroneutrality holds small mobile counterions inside the matrix. A Donnan equilibrium is established (Katchalsky et al. 1951). In this case, higher concentrations of mobile ions inside the gel than in the bulk solvent outside the gel generate almost all of the osmotic pressure that drives quick swelling. Swelling equilibrium is reached when the elastic properties of the matrix balance the expanding forces generated by osmotic pressure. Katchalsky's findings (Katchalsky et al. 1951) that in ion exchange gels, swelling is driven by a Donnan potential, led the way to verify that mucus hydration is governed by a Donnan equilibrium process (Tam and Verdugo 1981). These observations gave the first objective physical–chemical foundation for understanding the role of solvent composition in the regulation of mucus swelling (Verdugo 1984, 1990, 1991). The significance of these observations became apparent following our discovery that the mucus gel is stored in a highly compacted form undergoing swelling following its release from the cell.

Mucus Polymer Gel Phase Transition

The experimental demonstration that mucus, like all polymer gels, can undergo reversible phase transition from a condensed to a solvated phase (Verdugo et al. 1992; Verdugo 1994) provided a strong indication that Dušek–Tanaka's theory (Dušek and Patterson 1968; Tanaka et al. 1980; Tanaka 1981) can, indeed, be applied to understand the mechanisms for mucus storage in and release from secretory granules (Verdugo et al. 1992; Verdugo 1994). Phase transition is a characteristic property of all polymer gels. Dušek–Tanaka's theory establishes that gels can exist in either a condensed or a solvated phase. Just as transitions of liquids to solids or

to gas are reversible and occur at precise critical temperatures and pressures, a polymer gel phase transition (PGPT) from a condensed to a solvated phase occurs within a narrow range of critical conditions, shows high cooperativity, and is fully reversible (Tanaka et al. 1980; Tanaka 1981). The polymer matrices of most secretory granules including mucins, chromogranins, heparin, SP-1, and the like have the typical features of polyionic polymer gels. During storage in secretory granules, the matrix remains in condensed phase, in an intragranular environment of high Ca and low pH (Izutsu et al. 1985; Verdugo et al. 1987a; Villalon et al. 1988). Condensation is not a simple process of dehydration. The matrix of isolated goblet and mast cell granules remains condensed in water for as long as the ionic composition of the medium mimics intragranular environment, that is, low (<500 μM) Na, high (>10 mM) Ca, and <6 pH. However, exposure to solutions containing 140 mM Na, 2 mM Ca at pH 7, which may mimic ASL, triggers PGPT and quick matrix swelling (Fernandez et al. 1991; Verdugo et al. 1992; Verdugo 1994). Moreover, swollen mucus can be readily recondensed by exposure to solutions that mimic the intragranular environment (Verdugo 1991, 1994; Verdugo et al. 1992). In the secretory cell, following the formation of a secretory pore, protons must quickly escape to the extracellular medium, and Ca inside the network must be rapidly exchanged by Na (Forstner and Forstner 1975; Izutsu et al. 1985; Villalon et al. 1988). Because negative charges of the mucins are fixed (cannot move out of the matrix), exchange of divalent Ca by monovalent Na causes the gel to swell. Donnan swelling in this case results from doubling the number of counterions, as required by electroneutrality (two Na atoms for each Ca). An increased number of ions inside the gel results in increased osmotic pressure and swelling. With one caveat, ASL Ca should remain low (Verdugo et al. 1987b; Verdugo 1998). The missing chelating agent that buffers Ca in the ASL was found to be HCO_3, providing a straightforward explanation for the mucus pathology in CF (Garcia et al. 2009; Chen et al. 2010; Muchekehu and Quinton 2010; Quinton 2010).

In airway goblet cells, the matrix of exocytosed mucus undergoes quick swelling, forming microspheres that increase their diameter from $\sim$1 μm to $\sim$4–5 μm with $\sim$6.25 folds or 625% volume expansion in 3–5 s (Verdugo 1984, 1998; Verdugo et al. 1987b). Fast swelling in a randomly tangled network should result in highly frictional dissipations and anisotropic expansion, resulting in tight knots and large loops. Instead isotropic mucus swelling strongly suggests the presence of ordered domains that can readily unfold in the mucus matrix. This feature was soon confirmed by the demonstration of nematic liquid crystalline order in mucus (Viney et al. 1993a,b), and set the basis for the model illustrated in Figure 1. Nematic anisotropies were observed by polarized light at visible wavelength and reveal that mucus ordered clusters are extremely large. Compacting of the matrix during storage inside the granule is likely to require a higher order. Based on EM images, Ambort et al. (2012a) have proposed a new model for mucin folding. However, inherent fixation artifacts weaken the significance of the proposed model, leaving mucin-folding topology still unresolved (Verdugo 2012). X-ray diffraction studies to verify ordered supramolecular mucin assembly in isolated mucus granules are unfortunately still missing.

A rich body of available theory—well verified in synthetic block copolymer networks—can be used to explain the characteristic ordered supramolecular arrays responsible for nematic features found in mucus (Halperin 1991). Ordered folding of mucins permits high-density packing and quick expansion of the network upon release from the cell. Ca is required to shield charges and cross-link the folded condensed phase of the mucus matrix. Upon release from the granule, Na/Ca exchange thrusts Donnan-driven swelling and releases Ca cross-links, unraveling the mucin network. Thus, an adequate supply of water (Matsui et al. 1998; Boucher 2007) and particularly the removal of Ca from the ASL are critical for normal mucus hydration (Verdugo et al. 1987b; Aitken and Verdugo 1989; Verdugo 1998; Garcia et al. 2009; Chen et al. 2010; Muchekehu and Quinton 2010; Quinton 2010).

Mucus Swelling Kinetics during Product Release in Exocytosis

Swelling is a necessary and sufficient condition for respiratory mucus to become transportable, and physical theories provide testable predictions for its regulation. However, experimental models to validate these ideas became available only after the introduction of primary tissue culturing methods of airway mucosa, allowing direct monitoring of ciliary activity and mucus exocytosis (Verdugo 1980, 1984). In agreement with Tanaka and Fillmore's (1979) theory of swelling of polymer gels, the volume expansion of exocytosed mucus follows a typical first-order kinetics (Fig. 2). It has the distinctive features of a diffusion process in which the characteristic relaxation time of expansion is proportional to the second power of the linear dimension, in this case, the increase of the radius of the exocytosed swelling mucus matrix (Verdugo 1984, 1998; Verdugo et al. 1987b; Aitken and Verdugo 1989; Fernandez et al. 1991; Dodd et al. 1998; Espinosa et al. 2002; Chin et al. 2006; Chen et al. 2010). The slope of this line has the dimensions of centimeters squared per second (cm^2/sec) and corresponds to the diffusivity (D) of the gel in the solvent, which in the case of airway mucus is the ASL.

Diffusivity (D) is an extremely sensitive parameter for assessing mucus swelling resulting from changes of composition of the ASL like pH, Na/Ca ratio, organic polycations, anions, and the like (Verdugo et al. 1987b; Aitken and Verdugo 1989; Verdugo 1998; Espinosa et al. 2002; Chen et al. 2010). Measurements of swelling kinetics during exocytosis in primary cultures of airway rabbit goblet cells rendered the first indication that pH changes in the direction of the mucin isoelectric point can—as predicted by theory—decrease the mucus-swelling rate. Changes of pH from 7.5 to 7, and 6.5 in a solvent containing 140 mM Na and 1.5 mM Ca, results in changes of D from 1.2×10^{-7} to 4×10^{-8} and 1×10^{-8} cm^2/sec, respectively, with a corresponding decrease of mucus swelling equilibrium that ultimately establishes the micro rheological load on cilia (Verdugo 1984). Collapsing the ASL-mucus gel Ca gradient—in this

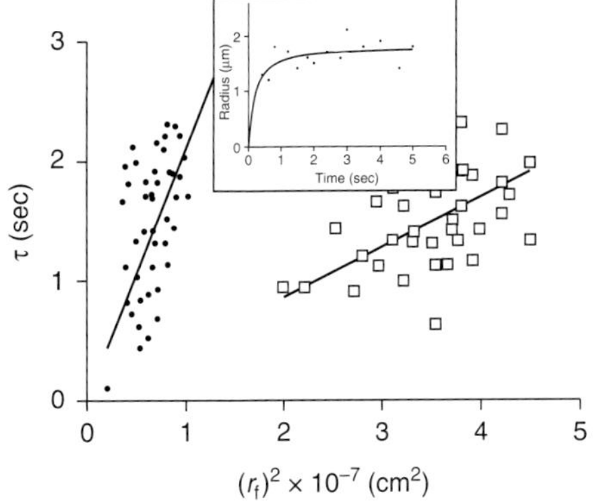

Figure 2. Illustrated here are data from studies published earlier (Verdugo 1998). Measurements from digitized video images show that the radial expansion of exocytosed goblet cell granules follows characteristic first-order kinetics, lending the process to be formalized in light of Tanaka's theory of swelling of polymer gels (Tanaka and Fillmore 1979). (*Inset*) A typical plot in which the continuous line is a nonlinear least-squares fitting of the data points to $r(t) = r_f - (r_f - r_i) \times e^{-t/\tau}$, where r_i and r_f are the initial and final radii of the granule and τ is the characteristic relaxation time of swelling, with r_i assumed to be 0.25 μm. Data points were collected as soon as the mucus micro gels became visible in phase contrast microscopy images. Notice that the square of the final radius is a linear function of τ. The slope of this line has the characteristic dimensions of diffusion (cm²/sec⁻¹), and, according to Tanaka's theory, it reflects the diffusion of the gel in the solvent. Mucus granules exocytosed by cultured goblet cells from biopsied nasal polyps of normal subjects and from CF patients swell when cell equilibrated in Hanks' solution containing 1 mM Ca (pH 7.2). D in normal mucus (open squares) reached $3.1 \pm 1.2 \times 10^{-7}$ cm²/sec. CF mucus granules swelled at almost an order of magnitude slower than normal (closed circles), yielding a $D = 7.4 \pm 2 \times 10^{-8}$ cm²/sec and also reaching a much smaller final swelling volume equilibrium. These differences become much more pronounced when cells were equilibrated in Hanks' solution containing 2.5 mM Ca. In this case, exocytosed mucus from cells of normal subjects reached $D = 2.5 \pm 0.9 \times 10^{-7}$ cm²/sec, whereas in CF mucus, D reached only $9.1 \pm 1.8 \times 10^{-9}$. D values correspond to the average $\pm$ SEM of 96 exocytosis-swelling events. These observations suggest that, in addition to the CF defective Ca-HCO₃ chelation, it is likely that mucin–Ca affinity might be increased, resulting in decreased Na/Ca exchange in CF. Alternatively, hydrophobic bonding, still largely unexplored, might be increased in CF.

case, by increasing Ca in the medium—must prevent removal of Ca cross-links among mucin polymers and inhibit mucus swelling. The experimental results showed a decrease of D from 2.5×10^{-7} to 2×10^{-8} cm²/sec when Ca in the bathing medium increased from 1 to 4 mM in 140 mM Na at neutral pH (Verdugo et al. 1987b). Recent experimental results confirm that increased extracellular Ca can drastically decrease both the swelling kinetics and swelling equilibrium in cervical mucus (Espinosa et al. 2002).

Measurements of D to investigate mucus pathology revealed that in goblet cells from CF patients, Ca inhibition of mucus swelling is about five times stronger than in goblet cells from normal individuals (Verdugo 1998). These observations provided the first objective demonstration that in human goblet cells, ASL Ca can inhibit mucus swelling, and that Na/Ca exchange in CF mucus might be defective. Considering that the payload of goblet cell granules includes large amounts of Ca that is released to the ASL during exocytosis (Izutsu et al. 1985; Villalon et al. 1988), chelation of ASL Ca is a critical requisite for normal mucus swelling (Verdugo et al. 1987b; Verdugo 1998). Recent observations show that HCO₃ can function as the actual chelator of Ca, suggesting that defective mucosal control of HCO₃ might be responsible for altered mucus rheology in CF (Garcia et al. 2009; Chen et al. 2010; Muchekehu and Quinton 2010; Quinton 2010). These results are consistent with early reports indicating that Ca is increased in the mucus of CF patients (Potter et al. 1967). The effect of polyions and particularly counterions on swelling kinetics increases with the second power of their valence (Ohmine and Tanaka 1982). In this regard, micromolar concentrations of organic polycations like cathelicidin, normally released by the airway epithelium (Felgentreff et al. 2006), or spermine or spermidine—the condensing agents of DNA—released with nucleic acids from dead cells or bacteria, could in theory cross-link the mucus matrix and drastically inhibit mucus swelling particularly in infected airways, and their role still awaits to be investigated.

Cite this article as *Cold Spring Harb Perspect Med* doi: 10.1101/cshperspect.a009597

Hydrophobic Interactions in Mucus

Electrostatic interactions are just one among several low-energy interactions that could cross-link the mucus matrix. Equally important but less explored is the potential contribution of hydrophobic interactions (Meyer et al. 2006). Hydrophobic interactions have been investigated in gastric mucus (Bansil and Turner 2006) but remain virtually unexplored in airway mucus. In polyelectrolyte systems in salt solutions—like the mucus gel—intermolecular interactions become independent of polymer charge and remain proportional to the presence of amphiphilic moieties over a large range of concentrations (Diamant and Andelman 1999). Strong interactions are known to take place in polymer–surfactant cosolutes even at very low concentrations (Kuhn et al. 1998). Thus, the amphiphilic nature of mucins presents the likely prospect that hydrophobic interactions are at work in respiratory mucus, between mucins, between mucins and lipids present in mucus (Widdicombe 1987), or between mucins and other amphiphilic molecules like surfactants released by Clara cells (Griese 1999) or exopolymer substances (EPSs) released by bacteria in infected airway (Rayner and Wilson 1997). These issues represent significant gaps still to be investigated in airway mucus.

CLOSING REMARKS

Much remains to explore to fully understand and predict the complex supramolecular dynamics of mucus. Significant effort has been invested in studying many features of mucins that, although highly significant to molecular biology or biochemistry, have so far served largely as the basis of speculation in attempting to explain what makes airway mucus a transportable gel. These studies portray a remarkable complexity of mucin expression with a broad range of genetic variations that is producing an ever-growing molecular taxonomy of mucins, making it extremely intricate and arbitrary to specify structure–function assignments of these polymers to explicitly explain the complex features of a mucus gel. We are still missing critical features specifically relevant to the role of mucins in mucus function. Mucins' ζ potential, the size, conformation, and elastic properties of native undegraded mucins, all remain to be investigated. Equally unexplored is the supramolecular organization of the mucus polymer network. In tangled networks, it is the mesh topology of interwoven mucins that determines the relaxation dynamics of the gel upon hydration. Transition from a condensed to a solvated phase requires programmed unfolding of large mucin chains. However, we know little regarding airway mucus supramolecular topology. The critical role of ions on mucus swelling has been shown, but much remains to be explored, and the significance of hydrophobic interactions in airway mucus remains greatly neglected.

ACKNOWLEDGMENTS

Over the last 30 plus years of NIH and later NSF funding, our research program gained strength from a remarkable crew of students and postdoctoral fellows who contributed with ideas, created a joyful collegial but critical laboratory atmosphere, and conducted the bulk of the work reported here. I am particularly grateful to Dr. Patrick Tam, Dr. Wiley Lee, Dr. Moira Aitken, Dr. Manuel Villalon, Dr. Monica Orellana, Dr. Ivan Quezada, Dr. Thien Nguyen, and Dr. Wei Chun Chin. Many of the ideas we explored in the past were born out of enlightening discussions with my colleagues Alex Silberberg, Sir Sam Edwards, Ralph Nossal, Ferenc Horkay, John Sheehan, Bill Davis, Adrian Parsegean, and Karel Dušek, and particularly my dear late friends Carol Basbaum and Toyo Tanaka. I am thankful to them all, for they pointed the compass to guide new exploration and exciting discoveries. Last but not least, this paper was kindly edited by Paul Quinton to whom I am indebted for his patience and many excellent suggestions.

REFERENCES

*Reference is also in this collection.

Aitken ML, Verdugo P. 1989. Donnan mechanism of mucin release and conditioning in goblet cells: The role of polyions. *J Exp Biol* **53**: 73–79.

Ambort D, Van der Post S, Johansson MEV, Mackenzie J, Thomsson E, Krengel U, Hansson GC. 2011. Function of the CysD domain of the gel-forming MUC2 mucin. *Biochem J* **436**: 461–470.

Ambort D, Johansson MEV, Gustafsson JK, Nilsson HE, Ermund A, Johansson BR, Koeck PJ, Hebert H, Hansson GC. 2012a. Calcium and pH-dependent packing and release of the gel-forming MUC2 mucin. *Proc Natl Acad Sci* **109**: 5645–5650.

* Ambort D, Johansson MEV, Gustafsson JK, Ermund A, Hansson GC. 2012b. Perspectives on mucus properties and formation—Lessons from the biochemical world. *Cold Spring Harb Perspect Med* **2**: a014159.

Bansil R, Turner BS. 2006. Mucin structure, aggregation, physiological functions and biomedical applications. *Curr Opin Coll Interf Sci* **11**: 164–170.

Bansil R, Stanley HE, Lamont JT. 1995. Mucin biophysics. *Ann Rev Physiol* **57**: 635–657.

Bansil R, Cao X, Bhaskar KR, LaMont JT. 2005. Dynamic light scattering study of gelation and aggregation of gastric mucin. *SPIE* **2982**: 116–125.

Bates FS, Fredrickson GH. 1990. Block copolymer thermodynamics: Theory and experiment. *Annu Rev Phys Chem* **41**: 525–557.

Boucher RC. 2007. Evidence for airway surface dehydration as the initiating event in CF airway disease. *J Intern Med* **261**: 5–16.

Chen Y, Zhao YH, Di YP, Wu R. 2001. Characterization of human mucin 5B gene expression in airway epithelium and the genomic clone of the amino-terminal and 5′-flanking region. *Am J Respir Cell Mol Biol* **25**: 542–553.

Chen EY, Yang N, Quinton PM, Chin WC. 2010. A new role for bicarbonate in mucus formation. *Am J Physiol Lung Cell Mol Physiol* **299**: L542–L549.

Chin WC, Quesada I, Steed J, Verdugo P. 2006. Modeling polyanion-Ca crosslinking in secretory networks: Assessment of charge density and bond affinity in polyanionic secretory networks. *Macromol Symp* **227**: 89–96.

Clamp JR, Allen A, Gibbons RA, Roberts GP. 1978. Chemical aspects of respiratory mucus. *Br Med Bull* **34**: 39–41.

de Gennes PG, Leger L. 1982. Dynamics of entangled polymer-chains. *Annu Rev Phys Chem* **33**: 49–61.

Diamant H, Andelman D. 1999. Onset of self-assembly in polymer surfactant systems. *Europhys Lett* **48**: 170–176.

Di Cola E, Yakubov G, Waigh TA. 2008. Double-globular structure of porcine stomach mucin: A small-angle X-ray scattering study. *Biomacromolecules* **9**: 3216–3222.

Dobrynin AV, Rubinstein M. 2005. Theory of polyelectrolytes in solutions and at surfaces. *Prog Polym Sci* **30**: 1049–1118.

Dodd S, Place GA, Hall RL, Harding SE. 1998. Hydrodynamic properties of mucins secreted by primary cultures of Guinea-pig tracheal epithelial cells: Determination of diffusion coefficients by analytical ultracentrifugation and kinetic analysis of mucus gel hydration and dissolution. *Eur Biophys J* **28**: 38–47.

Doi M, Edwards SF. 1984. *The theory of polymer dynamics*. Oxford University Press, Oxford.

Dušek K, Patterson D. 1968. Transition in swollen polymer networks induced by intramolecular condensation. *J Polym Sci B Polym Phys* **6**: 1209–1216.

Edwards SF, Grant JWV. 1973a. The effect of entanglements on the viscosity of a polymer melt. *J Phys A* **6**: 1171–1180.

Edwards SF, Grant JWV. 1973b. Effect of entanglements on diffusion in a polymer melt. *J Phys A* **6**: 1169–1185.

Espinosa M, Noé G, Troncoso C, Ho SB, Villalón M. 2002. Acidic pH and increasing Ca^{2+} reduce the swelling of mucins in primary cultures of human cervical cells. *Hum Reprod* **17**: 1964–1972.

Felgentreff K, Beisswenger C, Griese M, Gulder T, Bringmann G, Bals R. 2006. The antimicrobial peptide cathelicidin interacts with airway mucus. *Peptides* **27**: 3100–3106.

Fernandez JM, Villalon M, Verdugo P. 1991. Reversible recondensation of mast cell secretory products. *Biophys J* **59**: 1022–1027.

Flory P. 1953. *Principles of polymer chemistry.* Cornell University Press, Ithaca, NY.

Forstner JF, Forstner GG. 1975. Calcium binding to intestinal goblet cell mucin. *Biochim Biophys Acta* **386**: 283–292.

Garcia MA, Yang N, Quinton PM. 2009. Normal mouse intestinal mucus release requires cystic fibrosis transmembrane regulator-dependent bicarbonate secretion. *J Clin Invest* **119**: 2613–2622.

Griese M. 1999. Pulmonary surfactant in health and diseases: State of the art. *Eur Resp J* **13**: 1455–1476.

Gupta R, Jentoft N, Jamieson A M, Blackwell J. 1990. Structural analysis of purified human tracheobronchial mucins. *Biopolymers* **29**: 347–355.

Halperin A. 1991. On the collapse of multiblock copolymers. *Macromolecules* **24**: 1418–1419.

Hong Z, Chasan B, Bansil R, Turner BS, Bhaskar KR, Afdhal NH. 2005. Atomic force microscopy reveals aggregation of gastric mucin at low pH. *Biomacromolecules* **6**: 3458–3466.

Izutsu K, Johnson D, Schubert M, Wang E, Ramsey B, Tamarin A, Truelove E, Ensign W, Young M. 1985. Electron microprobe analysis of human labial gland secretory granules in cystic fibrosis. *J Clin Invest* **75**: 1951–1956.

Katchalsky A, Lifson S, Eisenberg HJ. 1951. Equation of swelling for polyelectrolyte gels. *J Polymer Sci* **7**: 571–574.

Kesimer M, Makhov AM, Griffith JD, Verdugo P, Sheehan JK. 2010. Unpacking a giant glycoprotein polymer from its storage granule: The mechanism of MUC5B mucin expansion after cellular release. *Am J Physiol Lung Cell Mol Physiol* **298**: L15–L22.

Kuhn PS, Levin Y, Barbosa MC. 1998. Complex formation between polyelectrolytes and ionic surfactants. *Chem Phys Lett* **298**: 51–56.

Lee WI, Verdugo P, Blandau RJ. 1977. The molecular structure of cervical mucus: A reevaluation by laser light scattering spectroscopy. *Gynec Invest* **8**: 254–266.

Marianne T, Perini JM, Lafitte JJ, Houdret N, Pruvot FR, Lamblin G, Slayter HS, Roussel P. 1987. Peptides of human bronchial mucus glycoproteins. Size determination by electron microscopy and by biosynthetic experiments. *Biochem J* **248**: 189–195.

Cite this article as *Cold Spring Harb Perspect Med* doi: 10.1101/cshperspect.a009597

Marszalek PE, Farrell B, Verdugo P, Fernandez JM. 1997. Kinetics of release of serotonin from isolated granules. (I) Amperometric detection of serotonin from electroporated granules. *Biophys J* **73**: 1160–1168.

Matsui H, Grubb BR, Tarran R, Randell SH, Gatzy JT, Davis CW, Boucher RC. 1998. Evidence for periciliary liquid layer depletion, not abnormal ion composition, in the pathogenesis of cystic fibrosis airways disease. *Cell* **95**: 1005–1015.

Matthews WG, Davis CW, Boucher RC. 2002. Atomic force microscopy studies of the structure of single mucin molecules and their viscoelastic properties. In *Biophysical Society 46th Annual Meeting (Abstract 814 poster), San Francisco, CA*. Biophysical Society, Bethesda, MD.

Meyer EE, Rosenberg KJ, Israelachvili J. 2006. Recent progress in understanding hydrophobic interactions. *Proc Natl Acad Sci* **43**: 15739–15746.

Muchekehu RW, Quinton PM. 2010. A new role for bicarbonate secretion in cervico-uterine mucus release. *J Physiol* **588**: 2329–2342.

Ohmine I, Tanaka T. 1982. Salt effects on the phase transitions of ionic gels. *J Chem Phys* **83**: 89–99.

Perez-Villar J, Hill RL. 1999. The structure and assembly of secreted mucins. *J Biol Chem* **274**: 31751–31754.

Potter JL, Mattheus LW, Spector W, Lemm J. 1967. Studies on pulmonary secretions. II: Osmolality, and ionic environment of pulmonary secretions from patients with cystic fibrosis, bronchiectasis, and laringectomy. *Am Rev Resp Dis* **96**: 83–97.

Quinton PM. 2010. Role of epithelial HCO_3^- transport in mucin secretion: Lessons from cystic fibrosis. *Am J Physiol Cell Physiol* **299**: C1222–C1233.

Rayner C, Wilson R. 1997. Mucus–bacteria interactions. In *Airway mucus: Basic mechanisms and clinical perspectives* (ed. Rogers DF, et al.), pp. 210–226. Birkhauser Verlag, Basel, Switzerland.

Riordan JR. 2005. Assembly of functional CFTR chloride channels. *Annu Rev Physiol* **67**: 701–718.

Round AN, Berry M, McMaster TJ, Stoll S, Gowers D, Corfield AP, Milles MJ. 2002. Heterogeneity and persistence length in human ocular mucins. *Biophys J* **83**: 1661–1670.

Round AN, Berry M, McMaster TJ, Corfield AP, Milesa MJ. 2004. Glycopolymer charge density determines conformation in human ocular mucin gene products: An atomic force microscope study. *J Struct Biol* **145**: 246–253.

Round AN, McMaster TJ, Miles MJ, Corfield AP, Berry M. 2007. The isolated *MUC5AC* gene product from human ocular mucin displays intramolecular conformational heterogeneity. *Glycobiology* **17**: 578–585.

Sheehan JK, Thornton DJ. 2000. Heterogeneity and size distribution of gel-forming mucins. *Methods Mol Biol* **125**: 87–96.

Sheehan JK, Oates K, Carlstedt I. 1984. Electron microscopy of cervical, gastric, and bronchial mucus glycoproteins. *Biochem J* **239**: 147–153.

Sheehan JK, Howard M, Richardson PS, Longwill T, Thornton DJ. 1999. Physical characterization of a low-charge glycoform of the MUC5B mucin comprising the gel-phase of an asthmatic respiratory mucous plug. *Biochem J* **338**: 507–513.

Sheppard DN, Welsh MJ. 2006. Structure and function of the CFTR chloride channel. *Physiol Rev* **79**: 23–45.

Tam PY, Verdugo P. 1981. Control of mucus hydration as a Donnan equilibrium process. *Nature* **292**: 340–342.

Tanaka T. 1981. Gels. *Sci Am* **244**: 124–138.

Tanaka T, Fillmore DJ. 1979. Kinetics of swelling of gels. *J Chem Phys* **70**: 1214–1218.

Tanaka T, Fillmore D, Sun S, Nishio I, Wislow GS, Shah A. 1980. Phase transitions in ionic gels. *Phys Rev Lett* **45**: 1636.

Thornton DJ, Sheehan JK, Carlstedt I. 1991. Heterogeneity of mucus glycoproteins from cystic fibrotic sputum: Are there different families of mucins? *Biochem J* **276**: 677–682.

Thornton DJ, Rousseau K, McGuckin A. 2008. Structure and function of the polymeric mucins in airways mucus. *Annu Rev Physiol* **70**: 459–486.

van Loosdrecht MC, Lyklema J, Norde W, Zehnder AJ. 1990. Influence of interfaces on microbial activity. *Microb Rev* **54**: 75–87.

Verdugo P. 1980. Ca^{2+}-dependent hormonal stimulation of ciliary activity. *Nature* **283**: 764–765.

Verdugo P. 1984. Hydration kinetics of exocytosed mucins in cultured secretory cells of the rabbit trachea: A new model. In *Mucus and mucosa (Ciba Foundation Symposium)* (ed. Nuget J, O'Connor M), Vol. 109, pp. 212–234. Pitman, London.

Verdugo P. 1990. Goblet cells and mucus secretion. *Annu Rev Physiol* **52**: 157–176.

Verdugo P. 1991. Mucin exocytosis. *Am Rev Resp Dis* **144**: S33–S37.

Verdugo P. 1994. Polymer gel phase transition in condensation–decondensation of secretory products. *Adv Poly Sci* **110**: 145–156.

Verdugo P. 1998. Polymer biophysics of mucus in cystic fibrosis. In *Cilia mucus and mucociliary interactions* (ed. Baum GL, et al.), pp. 167–190. Marcel Dekker, New York.

Verdugo P. Mucus supramolecular topology: An elusive riddle. *Proc Natl Acad Sci* (submitted).

Verdugo P, Tam PY, Butler J. 1983. Conformational structure of respiratory mucus studied by laser correlation spectroscopy. *Biorheology* **20**: 223–230.

Verdugo P, Deyrup-Olsen L, Aitken ML, Villalon MJ, Johnson D. 1987a. Molecular mechanism of mucin secretion: The role of intragranular charge shielding. *J Dent Res* **66**: 506–508.

Verdugo P, Aitken ML, Langley L, Villalon MJ. 1987b. Molecular mechanism of product storage and release in mucin secretion. II. The role of extracellular Ca^{2+}. *Biorheology* **24**: 625–633.

Verdugo P, Deyrup-Olsen I, Martin AW, Luchtel DL. 1992. Polymer gel phase transition: The molecular mechanism of product release in mucin secretion. In *Swelling of polymer networks, NATO ASI Series H* (ed. Karalis E), Vol. 64, pp. 671–681. Springer-Verlag, Heidelberg, Germany.

Villalon M, Basbaum CB, Johnson DE, Verdugo P. 1988. X-ray microanalysis of secretory granules from respiratory goblet cells. *FASEB J* **2:** A958.

Viney C, Huber AE, Verdugo P. 1993a. Liquid crystalline order in mucus. *Macromolecules* **26:** 852–257.

Viney C, Huber AE, Verdugo P. 1993b. Processing biological polymers in the liquid crystalline state. In *Biodegradable polymers and packaging* (ed. Ching C, et al.), pp. 209–224. Technomics, Lancaster, PA.

Waigh TA, Papagiannopoulos A. 2010. Biological and biomimetic comb polyelectrolytes. *Polymers* **2:** 57–70.

Widdicombe JG. 1987. Role of lipids in airway function. *Eur J Respir Dis* **153:** 197–204.

Yakubov GE, Papagiannopoulos A, Rat E, Easton RL, Waigh TA. 2007. Molecular structure and rheological properties of short-side-chain heavily glycosylated porcine stomach mucin. *Biomacromolecules* **8:** 3467–3477.

Cite this article as *Cold Spring Harb Perspect Med* doi: 10.1101/cshperspect.a009597

Perspectives on Mucus Properties and Formation—Lessons from the Biochemical World

Daniel Ambort, Malin E.V. Johansson, Jenny K. Gustafsson, Anna Ermund, and Gunnar C. Hansson

Department of Medical Biochemistry, University of Gothenburg, 405 30 Gothenburg, Sweden

Correspondence: gunnar.hansson@medkem.gu.se

Our model of the MUC2 mucin shows a well-organized netlike gel that is cross-linked by six different covalent and noncovalent bonds. When the MUC2 mucin is packed in the mucin granule it is organized by an amino-terminal concatenated ring platform formed at high calcium and low pH. This packing allows an ordered release and a normal mucin expansion when calcium is removed and pH increased by bicarbonate. This process is defective in the absence of cystic fibrosis transmembrane conductance regulator (CFTR)-dependent bicarbonate transport. The expanded secreted mucin is suggested to be self-organizing by properties inherited in the MUC2 mucin and by proteolytic processes.

Most of the organs affected in cystic fibrosis (CF) patients, except the sweat glands, can be attributed to the CF viscous mucus phenotype. Tubular organs, such as exocrine pancreas and epididymis, are often clogged by mucus already at birth. The small intestine can also be obstructed by the mucus-rich meconium to give meconium ileus at birth. CF adults often also experience intestinal pain and sometimes DIOS (distal ileum obstruction syndrome). However, most commonly and seriously, the slow and stagnant mucus of the lungs give persistent bacterial lung infections.

What is mucus then? Mucus is the gel-like material found on all mucosal surfaces. Its major component is one or several gel-forming mucins, discussed by Kreda et al. (2012). Mucus, depending on the organ, contains a number of specific components in addition to the mucins, such as, for example, exfoliated cells and other trapped materials. The function of mucus is to protect the underlying epithelium and to act like glue, which can trap and transport away debris and microorganisms found in the mucus gel. The CF pathologies can, in a simplified way, be understood as mucus that is trapped and stickier than normal.

Understanding mucus requires an understanding of mucins, which has, however, proven to be difficult as is reflected by this and Verdugo (2012). Verdugo (2012) and this article together give two almost completely contradictory models of how mucins are organized, largely because each group of investigators takes a different scientific starting point. Verdugo views mucus from "above" and explains its properties in terms of biophysical and macromolecular approaches generated in the polymer science of large synthetic polymers. The view in the present work, however, is based on recently acquired knowledge of mucin biochemistry and the characterizations of molecular details of gel-forming mucins. This views mucus from "below" and tries to assemble more detailed knowledge into models of mucus and mucus function. At first glance, these two models may seem totally incongruent. However, this is probably not the case because different mucus gels, formed by different mucins, show different organization and function. Before describing the biochemical model of mucins, one major feature must be discussed, the glycans.

THE DENSELY *O*-GLYCOSYLATED MUCIN DOMAINS

A high content of glycans, especially the *O* glycans (or as they are often incorrectly referred to, the mucin glycans) is typical of mucins. These glycans are densely clustered in the so-called mucin domains of the protein core to extend and make the mucin domains rodlike (Andersch-Björkman et al. 2007; Lang et al. 2007; Holmen Larsson et al. 2009). A bottle brush with its metal core in the center as the mucin protein backbone and the glycans forming the bristles are often used to explain this. The protein core is rich in the amino acids serine and threonine that anchor the *O* glycans. These amino acids are interspersed with prolines giving the peptide a random coil structure that allows the peptidyl-GalNAc transferases of the Golgi apparatus to attach *N*-acetylgalactosamine as the first *O*-linked sugar. The amino acid sequences of these domains have been poorly conserved during evolution, but can be identified as PTS sequences in the genome as long stretches of nucleotides that encode high frequencies of the three amino acids: proline, threonine, and serine (PTS) (Lang et al. 2004, 2007). These PTS sequences often contain more or less perfectly repeated sequences, but this is not always the case. These sequences are often called VNTR for variable number of tandem repeats, but this name is misleading as there are a number of mu-

cins with heavily degenerated repeats and no repeated sequences can be identified. Nonetheless, long stretches of >100 amino acids with a high frequency of the proline, threonine, and serine amino acids are characteristic of these domains.

The highly glycosylated mucin domains have, for mucins, the important biophysical property of binding water. This and their polymeric nature generate the gel-like properties typical for mucins. The numerous hydroxyl groups of the individual monosaccharides account for most of the high water-binding capacity. The presence of negatively charged groups in the form of carboxyl groups on the sialic acids and sulfate groups contributes further to water- and ion-binding capacities. As the glycosylation is encoded by specific transferases, the glycans on one and the same mucin protein core can vary enormously in different organs and even more so among the same organ in different species. These differences are, for example, determined by blood-group-type glycan epitopes (i.e., ABH, Lewis, etc., histo-blood-group systems). Such polymorphisms are beneficial for the host in constant battle against pathogenic organisms (Gagneux and Varki 1999). Although still debated, it is less likely that specific mucin glycans contribute very much to differences in mucus properties.

THE BIOCHEMICAL PERSPECTIVE ON MUCINS

Our view of mucins and how they form mucus starts with the MUC2 mucin (Johansson et al. 2011). This is the major mucin in the small and large intestine and as such can be expected to be exposed to the harshest conditions of all mucins (Gum et al. 1994). This will probably place the MUC2 mucin at one extreme and as such be one reason for its properties as compared with the linear polymer model traditionally given to mucins as discussed by Verdugo (2012). A major conclusion from our work on the MUC2 mucin is that mucus gels are very well-organized structures, much more than considered previously. This organization is based on the structure of its major building block, the MUC2 mucin monomer (Fig. 1).

     Cite this article as *Cold Spring Harb Perspect Med* doi: 10.1101/cshperspect.a014159

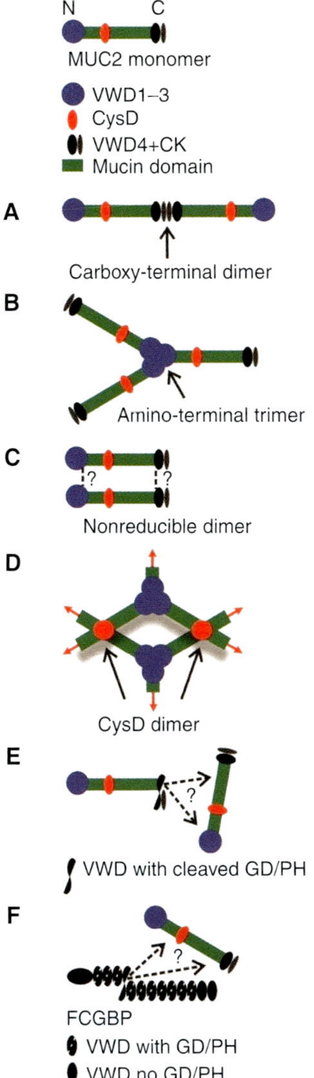

Figure 1. A schematic view of the MUC2 mucin (*top*). The amino-terminal von Willebrand D1, D2, and D3 (VWD1–3) domains comprising the first 1200 amino acids are given as a blue dot. The two mucin domains are green. The CysD domains are red (only one of the two is shown). The carboxy-terminal 980 amino acids, including the von Willebrand D4 (VWD4) and the cysteine-knot domains are given in black. (*A*) Disulfide-bonded dimeric carboxy terminus as formed in the endoplasmic reticulum. (*B*) Disulfide-bonded trimeric amino terminus as formed in the secretory vesicles. (*C*) Nonreducible covalent bond formed in the late secretory pathway. (*D*) CysD domains form noncovalent dimeric complexes. (*E*) The autocatalytic cleavage at the GD/PH bond (between aspartate and proline in the sequence

The MUC2 mucin builds the structure of the two-layered mucus of colon (Johansson et al. 2008). The inner mucus layer that is 50- to 100-μm thick (mouse) is anchored to the epithelium and impenetrable to bacteria. Thus, this layer keeps all the bacteria at a distance from the underlying epithelial cells. This inner layer is converted to an outer mucus layer that is less dense and forms the habitat of the commensal bacteria (Johansson et al. 2011). The inner layer has a stratified appearance and seems well organized. The organization of the MUC2 mucin in the mucus must be spontaneous owing to built-in properties and interactions.

As discussed by Verdugo (2012), the view of mucins as long linear polymers that give gel-like properties by entanglement has dominated the mucin field for decades. In contrast, we argue that the mucus gel is formed by a heavily cross-linked network with intrinsic biochemical interactions and properties. Today we know of six different types of covalent and noncovalent interactions within the MUC2 mucin that are all, except one, incompatible with a model of entangled linear polymers as a structural basis for mucins.

The MUC2 monomer is made of 5174 amino acids organized in an amino terminus with three complete and one partial von Willebrand D domains (VWD), two PTS sequences that give two mucin domains, and a carboxy-terminal end of 840 amino acids with one VWD and a cysteine-knot (CK) domain (Fig. 1, top) (Gum et al. 1994). The MUC2 mucin also contains two CysD domains in the central PTS-containing region (only the second one between the two mucin domains is shown in Fig. 1).

After folding in the endoplasmic reticulum, the primary translational MUC2 product is directly dimerized via its carboxy-terminal end

---

Figure 1. (*Continued*) glycine-aspartate-proline-histidine) in the carboxy-terminal von Willebrand D4 domain generates a new carboxy terminus with an internal anhydride that can cross-link MUC2. (*F*) FCGBP has 11 von Willebrand D domains (VWD) with a cleavage GD/PH site and two without. As for MUC2, the GD/PH bond is cleaved and will cross-link, for example, MUC2.

CK (Fig. 1A). Disulfide bonds between the CK domains form this dimer as in other dimeric proteins with CK domains (Asker et al. 1995, 1998; Lidell et al. 2003b). The dimers are then transported into the Golgi apparatus to be *O*-glycosylated. The mass of the dimer increases about fivefold owing to the glycans, to ~5 MDa.

In the *trans*-Golgi network, the mucin dimers are sorted to the regulated secretory pathway and the MUC2 amino termini form disulfide-bonded trimers (Fig. 1B) (Godl et al. 2002). The trimerization takes place in the VWD3 domain, a part of the molecule that is very dense and has the appearance of a trefoil when studied by molecular electron microscopy. There have been some discussions on the trimeric nature of the MUC2 amino termini, but these have been based solely on migration on SDS-polyacrylamide gels (Heazlewood et al. 2008). The reason for this discussion is that the nonreduced trimer is just a bit larger than two times the size of the reduced monomer. However, this is in fact proof of the nonreduced MUC2 being trimeric as the nonreduced MUC2 monomer is compact and migrates faster than the reduced form. Also, the porcine submaxillary mucin (PSM) forms amino-terminal trimers (Perez-Vilar and Hill 1998). Unfortunately, we do not have any details of the other human mucins, as these have not been expressed as truncated forms that would allow any final conclusions about their dimeric or trimeric amino-terminal nature. However, recent studies of the MUC6 amino termini also suggest a trimeric organization (Leir and Harris 2011). It is possible that some mucins still have amino-terminal dimeric forms as first assumed (Gum et al. 1994), parallel to the von Willebrand factor that is dimeric, and thus form linear polymers (Sadler 2009). Some biophysical and biochemical data may suggest that the MUC5B mucin and perhaps the MUC5AC are dimeric (Sheehan et al. 2004; Thornton et al. 2008).

During the later stages of the secretory pathway, probably after sorting to the regulated pathway and at the time of trimerization, the MUC2 mucin becomes insoluble in normal buffers (Axelsson et al. 1998). This occurs at the same time the nonreducible bond is formed as indicated by the appearance of nonreducible

dimers after reduction of the disulfide bonds analyzed by gel electrophoresis. The same type of bond was first observed in mature, secreted mucus from the intestine, where the MUC2 mucin is also insoluble in chaotropic salts like guanidinium chloride (Carlstedt et al. 1993; Herrmann et al. 1999). The nature and localization of the nonreducible covalent bond remains a mystery (Fig. 1C).

The MUC2, MUC5AC, and MUC5B mucins all contain two or more CysD domains. This domain is 100 amino acids long and contains 10 conserved Cys residues. After intensive studies on the second CysD of the MUC2 mucin we have concluded that these domains form noncovalent dimers that are held together by strong hydrophobic forces (Fig. 1D) (Ambort et al. 2011). That no disulfide bonds are involved in forming the dimers became evident from the observation that all five disulfide bonds could be attributed to intramolecular disulfide bonds. It was previously suggested that the CysD domains of MUC5AC do not interact (Perez-Vilar et al. 2004), but our present and unpublished results strongly indicate that like the MUC2 CysD domains, MUC5AC CysD domains also form noncovalent dimers.

During our studies of the MUC2 mucin, we have also found two additional types of MUC2 cross-links. An autocatalytic cleavage of the VWD4 domain at the GD/PH amino acid sequence was found to be enhanced at a pH close to 6 (Fig. 1E) (Lidell et al. 2003a). This type of cleavage also occurs in the VWD4 of the MUC5AC mucin (Lidell and Hansson 2006). The new carboxy-terminal aspartate forms an internal anhydride between its new carboxy-terminal carboxyl group and its side chain. Such anhydrides are very reactive, both with primary amines and hydroxyl groups. As such groups are numerous on mucins, especially hydroxyl groups, it is very likely that covalent bonds will be formed in mucus and thereby stabilize the mucin network just as fibrin is stabilized for mechanical strength. It should also be pointed out that the MUC2 and MUC5AC mucin polymers are still held together over the GD/PH cleavage by one or several disulfide bonds that bridge this sequence.

Recently, we showed that another protein, FCGBP, is covalently bound to MUC2 mucin in the large intestine (Fig. 1F) (Johansson et al. 2009). This protein was erroneously named immunoglobulin G Fc-binding protein. It contains 13 VWD domains that are closely related to the VWD4 domain of MUC2 as 11 of 13 of the VWD contain a GD/PH sequence that can be cleaved like the VWD4 of MUC2. The cleaved FCGBP parts remain associated by disulfide bonds. The amino-terminal portion of the FCGBP was found covalently attached to MUC2, which seems compatible with binding via the emergent carboxy-terminal aspartates. Also in this case, there is no detailed information on how and where FCGBP is bound, but its structure suggests that it can cross-link several MUC2 mucins to stabilize the mucus gel.

To summarize, our model of mature MUC2 mucin suggests a well-organized netlike gel that is cross-linked by five different covalent bonds and one noncovalent bond. It also suggests that the mucus is self-organized, as determined by properties inherent in the MUC2 mucin molecule.

PACKING OF MUCINS

The oligomeric status of the mucin amino termini is not only important for mucus, but also for how mucins are stored in the goblet cell granules. The von Willebrand factor multimers are packed as long spirals that make up the Weibel-Palade bodies (vesicles) in endothelial cells. The von Willebrand factor is unpacked by a slow unwinding of these spirals into a long polymeric "rope" that is pulled out by and follows the blood flow (Huang et al. 2008; Sadler 2009). This slow unwinding differs from the release of mucins in which mucin granules are released by a quick burst within seconds or milliseconds. To avoid severe mucin entanglement that would hinder the 1000-fold expansion in volume on release, the mucin release must be well organized. Thus mucins must be packed differently from the von Willebrand factor protein.

We recently showed that the MUC2 amino-terminal mucin is packed in goblet cell granules as condensed concatenated rings formed by one amino-terminal trimer in each corner of largely six-sided rings, but five- and seven-sided rings have also been observed (Ambort et al. 2012). The amino-terminal covalent trimers (Fig. 2, blue) interact by noncovalent forces generated after the VWD1 and VWD2 domains are folded back onto the VWD3 domain. This organization in amino-terminal rings is induced by high Ca^{2+}-ion concentration and low pH created along the secretory pathway. In our experimental system, rings were formed by 50 mM Ca^{2+} and a pH of 6.2 (Ambort et al. 2012). These rings concatenate into relatively flat surfaces from which the rest of the mucin extends vertically (or perpendicularly) as illustrated in Fig. 2. In this model, the MUC2 mucin is shown in a two-dimensional (2D) reconstruction, and the mucin domains as extended linear rectangles (rods in 3D). The dimeric carboxyl termini of the MUC2 mucins are joined at the other end (Fig. 2A). This model gives the expected, well-organized packing of mucins in the granules, in contrast to previous models of mucin packing (Verdugo et al. 1987), which was suggested to be solely because of shielding of negative charges. Nevertheless, in our model, mucins with negatively charged O glycans must have

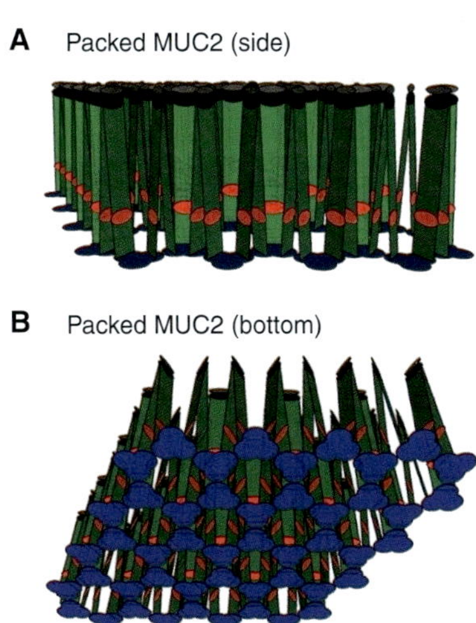

Figure 2. Schematic model of the packed assembled MUC2 polymer as found in the mucin granules of the goblet cells. (*A*) *Side* view, and (*B*) *bottom* view. The MUC2 mucin is shown in a 2D projection. Colors as in Figure 1.

D. Ambort et al.

these repulsive electrostatic forces minimized by counterions such as sodium and calcium ions. However, packed mucins cannot be highly organized by solely electrostatic divalent cross-links between mucin glycans, and are better explained by the model in Figure 2.

THE ORGANIZED RELEASE OF MUCINS TO FORM THE MUCUS

Mucin packing as described in Fig. 2 is compatible with rapid mucin expansion and quick release when the Ca^{2+} ions are removed and pH increased. When these cations are removed, the noncovalent interactions maintaining the concatenated ring structure is weakened and the trimeric amino termini start to separate, as illustrated in Fig. 3 (Ambort et al. 2012). When the amino termini start to separate, the dimeric MUC2 carboxy-terminal end "descends" and allows the trimeric amino terminals to spread apart until the net becomes flat. The unfolded mucin will, because of its dimeric carboxy termini and trimeric amino termini, spontaneously form a flat netlike structure with an inherent tendency to spontaneously form sheets (Fig. 4A). The organization of mucus found in colon is likely to be of this type. Studies using both electron microscopy and fluorescent microsco-

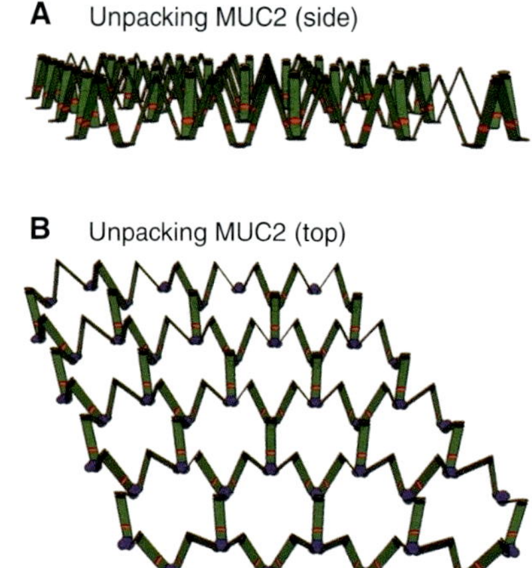

Figure 3. Schematic model of an early stage of unpacking of the assembled MUC2. (A) *Side* view, and (B) *top* view. The MUC2 mucin is shown in a 2D projection. Colors as in Figure 1.

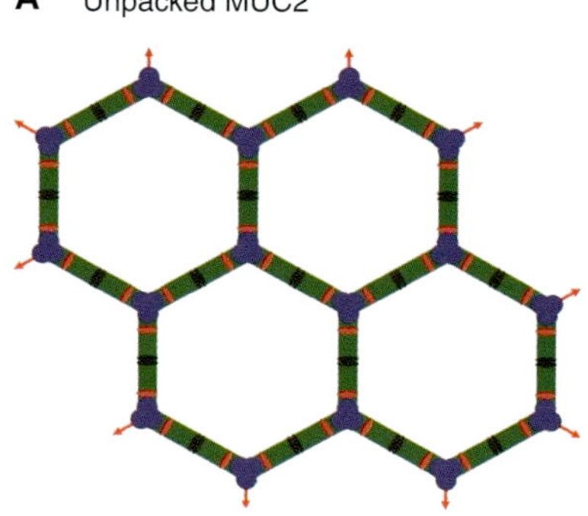

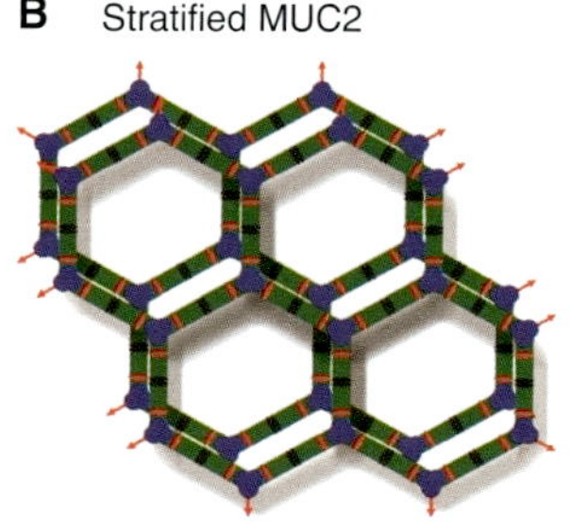

Figure 4. Model of expected organization of the stratified inner mucus layer of colon. Schematic drawing of the fully unfolded secreted MUC2 polymer netlike network (A) formed by the disulfide-bonded trimeric amino termini and dimeric carboxyl termini. (B) The dimerization of the CysD domains will interact maximally if each layer of MUC2 polymer rings is partly dislocated.

py support the case and show that the inner mucus layer is a laminated, stratified structure (Johansson et al. 2008; Ambort et al. 2012). Thus, we propose a model in which the mucins tend to spontaneously, upon secretion, reorganize into flat, netlike structures.

With these concepts in mind, the netlike structure of the MUC2 mucin forms stacked sheets, and with each layer of the net shifted slightly the CysD domains should fit together (Fig. 4B). Although we have no formal proof of this structure, it is tempting to use it to explain the molecular sieving property of the inner mucus layer of the colon. Typically, the CysD domain is located between mucin domains, and thus, the length of the extended mucin domains determines the pore size of the mucus. Interestingly, the MUC2 mucin that is found in intestines that must permeate nutrients have longer mucin domains, whereas MUC5AC and 5B mucins found in stomach and lungs, where only gases, water, and ions must permeate, have shorter mucin domains between the CysDs.

The model in Fig. 4 represents our current understanding of how MUC2 mucin is organized in the inner colon mucus layer. However, there is more to be discovered in how the actual mucus gel is organized. Proteolytic cleavage within the mucin polymers is an important additional event that will clearly have a large impact on the mucus structure. These processes will alter the mucus properties. Although we do not have any molecular details yet, it is clear that the conversion from the inner to the outer mucus layer of colon as well as the fourfold expansion in volume found in colon mucus is owing to proteolytic activities (Johansson et al. 2008, 2011). These proteolytic activities do not necessarily lead to polymer disruption as these can cleave the MUC2 in positions in which the intramolecular disulfide bonds can still maintain a covalent polymeric network, although the more penetrable mucus will also allow bacteria to enter (Johansson et al. 2011).

DIOS is a relatively common, poorly understood problem for CF patients. The mucus of the small intestine is, just as for colon, built around the MUC2 mucin. However, small intestinal mucus does not show an inner mucus layer like the colon (Atuma et al. 2001; Gustafsson et al. 2012). Instead, it has only one mucus layer that is penetrable to bacteria (Johansson and Hansson 2011). How MUC2 mucin generates such different mucus properties is still not fully understood, but most likely the differences between the large and small intestine are owing to specific proteolytic processing of MUC2 in the small intestine.

THE IMPORTANCE OF BICARBONATE

As discussed, the unpacking of mucins requires removal of Ca^{2+} and an increased pH. The natural agent that can accomplish this is HCO_3^-. The capacity of cystic fibrosis transmembrane conductance regulator (CFTR) to support HCO_3^- transport will thus be very important for mucin release as pointed out by Quinton (2008) and shown for both the cervix and small intestine (Garcia et al. 2009; Muchekehu and Quinton 2010). We have recently shown that the ileal mucus of CF mice appears anchored to the epithelium (Gustafsson et al. 2012), which is in contrast to the small intestine of wild-type animals where the mucus is easily aspirated and penetrable to beads the size of bacteria. In the CF mice, the mucus is two to three times more dense than normal (measured as MUC2 content per volume) and is impermeable to beads the size of bacteria. Remarkably, CF mucus properties appear to be reverted to normal when the mucins are excreted into 115 mM HCO_3^- or 20 mM EDTA (Gustafsson et al. 2012). The volume of the CF mucus also expands more normally when HCO_3^- is added to luminal fluid, which does not occur with wild-type mucus and indicates that the mucus is only partly expanded.

DISCUSSION

Although the above discussion takes its start in the biochemical nature and properties of the intestinal MUC2 mucin, there is no doubt that this knowledge can be applied also to the respiratory mucus. The surface mucin of the trachea is MUC5AC, the mucin with the most structural similarities to MUC2. We note that the MUC5AC mucin normally appears to form only one type of mucus in the lungs, whereas it too can form a two-layered system in the stomach, and thereby is similar to the different layers of the MUC2 mucin in the small and large intestine.

This review should be taken together with the views by Verdugo (2012). Our different views of mucus and mucins are obvious, and present in two different models: perfectly linear molecules or complex heavily cross-linked polymeric networks. In other aspects we are in agreement; for example, the importance of Ca^{2+} chelation for mucin release, although the explanations for the effects differ. Verdugo believes the major effect of Ca^{2+} is to shield the negative charges of the O glycans, whereas we believe the major effect is on the amino-terminal portion of the mucins as packed in the mucin granules. Despite our expertise in glycan chemistry, we believe that glycans are not that important for mucus formation, which contrasts directly with Verdugo. We do not claim that any of these models give a final answer, but that there are elements in both views that are likely correct. The highly cross-linked mucus is probably converted to less cross-linked forms by well-controlled proteolytic activities and thus mucus will become more linear, although it would never reach an ideal random linear thread. However, many biochemical details are still unknown. We have not yet studied the polymeric organization of any of the other gel-forming mucins or the packing organization in their storage granules. Mucins are complex and it will take quite some time until we are able to present a final and comprehensive model of normal mucus gel or of that found in CF patients. However, the importance of the CFTR channel for transporting bicarbonate for proper mucin release will provide novel therapeutic possibilities also, without a full understanding of the mucus gel.

ACKNOWLEDGMENTS

This work is supported by the Swedish Research Council (Nos. 7461, 21027), the Swedish Cancer Foundation, the Knut and Alice Wallenberg Foundation, IngaBritt and Arne Lundberg Foundation, Sahlgren's University Hospital (LUA-ALF), Wilhelm and Martina Lundgren's Foundation, Torsten och Ragnar Söderbergs Stiftelser, the Sahlgrenska Academy, National Institute of Allergy and Infectious Diseases (U01AI095473), the Swedish Foundation for Strategic Research— The Mucus-Bacteria-Colitis Center (MBC) of the Innate Immunity Program, the Swedish CF Foundation, the U.S. CF Foundation, Erica Lederhausen's Foundation, and Lederhausen's Center for CF Research at the University of Gothenburg.

REFERENCES

*Reference is also in this collection.

Ambort D, van der Post S, Johansson MEV, MacKenzie J, Thomsson E, Krengel U, Hansson GC. 2011. Function of the CysD domain of the gel-forming MUC2 mucin. *Biochem J* **436**: 61–70.

Ambort D, Johansson MEV, Gustafsson JK, Nilsson H, Ermund A, Johansson BR, Kock P, Hebert H, Hansson GC. 2012. Calcium and pH-dependent packing and release of the gel-forming MUC2 mucin. *Proc Natl Acad Sci* doi: 10.1073/pnas.1120269109.

Andersch-Björkman Y, Thomsson KA, Holmén Larsson JM, Ekerhovd E, Hansson GC. 2007. Large-scale identification of proteins, mucins and their O-glycosylation in the endocervical mucus during the menstrual cycle. *Mol Cell Proteomics* **6**: 708–716.

Asker N, Baeckstrom D, Axelsson MAB, Carlstedt I, Hansson GC. 1995. The human MUC2 mucin apoprotein appears to dimerize before O-glycosylation and shares epitopes with the "insoluble" mucin of rat small intestine. *Biochem J* **308**: 873–880.

Asker N, Axelsson MAB, Olofsson SO, Hansson GC. 1998. Human MUC5AC mucin dimerizes in the rough endoplasmic reticulum, similarly to the MUC2 mucin. *Biochem J* **335**: 381–387.

Atuma C, Strugula V, Allen A, Holm L. 2001. The adherent gastrointestinal mucus gel layer: Thickness and physical state in vivo. *Am J Physiol* **280**: G922–G929.

Axelsson MAB, Asker N, Hansson GC. 1998. O-glycosylated MUC2 monomer and dimer from LS 174T cells are water-soluble, whereas larger MUC2 species formed early during biosynthesis are insoluble and contain nonreducible intermolecular bonds. *J Biol Chem* **273**: 18864–18870.

Carlstedt I, Herrmann A, Karlsson H, Sheehan JK, Fransson L, Hansson GC. 1993. Characterization of two different glycosylated domains from the insoluble mucin complex of rat small intestine. *J Biol Chem* **268**: 18771–18781.

Gagneux P, Varki A. 1999. Evolutionary considerations in relating oligosaccharide diversity to biological function. *Glycobiology* **9**: 747–755.

Garcia MA, Yang N, Quinton PM. 2009. Normal mouse intestinal mucus release requires cystic fibrosis transmembrane regulator-dependent bicarbonate secretion. *J Clin Invest* **119**: 2613–2622.

Godl K, Johansson MEV, Karlsson H, Morgelin M, Lidell ME, Olson FJ, Gum JR, Kim YS, Hansson GC. 2002. The N-termini of the MUC2 mucin form trimers that are held together within a trypsin-resistant core fragment. *J Biol Chem* **277**: 47248–47256.

Gum JR, Hicks JW, Toribara NW, Siddiki B, Kim YS. 1994. Molecular cloning of human intestinal mucin (MUC2) cDNA. Identification of the amino terminus and overall sequence similarity to prepro-von Willebrand factor. *J Biol Chem* **269**: 2440–2446.

Gustafsson JK, Ermund A, Ambort D, Johansson MEV, Nilsson HE, Thorell K, Hebert H, Sjovall H, Hansson GC. 2012. Bicarbonate and functional CFTR channel *are* required for proper mucin secretion and link Cystic Fibrosis with its mucus phenotype. *J Exp Med* (in press).

Heazlewood CK, Cook MC, Eri R, Price GR, Tauro SB, Taupin D, Thornton DJ, Png CW, Crockford TL, Cornall RJ, et al. 2008. Aberrant mucin assembly in mice causes endoplasmic reticulum stress and spontaneous inflammation resembling ulcerative colitis. *PLoS Med* **5**: e54.

Herrmann A, Davies JR, Lindell G, Martensson S, Packer NH, Swallow DM, Carlstedt I. 1999. Studies on the "insoluble" glycoprotein complex from human colon. *J Biol Chem* **274**: 15828–15836.

Holmen Larsson JM, Karlsson H, Sjovall H, Hansson GC. 2009. A complex, but uniform *O*-glycosylation of the human MUC2 mucin from colonic biopsies analyzed by nanoLC/MS^n. *Glycobiology* **19**: 756–766.

Huang RH, Wang Y, Roth R, Yu X, Purvis AR, Heuser JE, Egelman EH, Sadler JE. 2008. Assembly of Weibel Palade body-like tubules from N-terminal domains of von Willebrand factor. *Proc Natl Acad Sci* **105**: 482–487.

Johansson MEV, Hansson GC. 2011. Keeping bacteria at a distance. *Science* **334**: 182–183.

Johansson MEV, Phillipson M, Petersson J, Holm L, Velcich A, Hansson GC. 2008. The inner of the two Muc2 mucin dependent mucus layers in colon is devoid of bacteria. *Proc Natl Acad Sci* **105**: 15064–15069.

Johansson MEV, Thomsson KA, Hansson GC. 2009. Proteomic analyses of the two mucus layers of the colon barrier reveal that their main component, the Muc2 mucin, is strongly bound to the Fcgbp protein. *J Proteome Res* **8**: 3549–3557.

Johansson MEV, Holmen Larsson JM, Hansson GC. 2011. The two mucus layers of colon are organized by the MUC2 mucin, whereas the outer layer is a legislator of host-microbial interactions. *Proc Natl Acad Sci* **108**: 4659–4665.

* Kreda SM, Davis CW, Rose MC. 2012. CFTR, mucins, and mucus obstruction in cystic fibrosis. *Cold Spring Harb Perspect Med* **2**: a009589.

Lang T, Alexandersson M, Hansson GC, Samuelsson T. 2004. Bioinformatic identification of polymerizing and transmembrane mucins in the puffer fish Fugu rubripes. *Glycobiology* **14**: 521–527.

Lang T, Hansson GC, Samuelsson T. 2007. Gel-forming mucins appeared early in metazoan evolution. *Proc Natl Acad Sci* **104**: 16209–16214.

Leir SH, Harris A. 2011. MUC6 mucin expression inhibits tumor cell invasion. *Exp Cell Res* **317**: 2408–2419.

Lidell ME, Hansson GC. 2006. Cleavage in the GDPH sequence of the C-terminal cysteine-rich part of the human MUC5AC mucin. *Biochem J* **399**: 121–129.

Lidell ME, Johansson MEV, Hansson GC. 2003a. An autocatalytic cleavage in the C-terminus of the human MUC2 mucin occurs at the low pH of the late secretory pathway. *J Biol Chem* **278**: 13944–13951.

Lidell ME, Johansson MEV, Mörgelin M, Asker N, Gum JR, Kim YS, Hansson GC. 2003b. The recombinant C-terminus of the human MUC2 mucin forms dimers in CHO cells and heterodimers with full-length MUC2 in LS 174T cells. *Biochem J* **372**: 335–345.

Muchekehu RW, Quinton PM. 2010. A new role for bicarbonate secretion in cervico-uterine mucus release. *J Physiol* **588**: 2329–2342.

Perez-Vilar J, Hill RL. 1998. Identification of the half-cystine residues in porchine submaxillary mucin critical for multimerization through the D-domains. *J Biol Chem* **273**: 34527–34534.

Perez-Vilar J, Randell SH, Boucher RC. 2004. C-mannosylation of MUC5AC and MUC5B Cys subdomains. *Glycobiology* **14**: 325–337.

Quinton PM. 2008. Cystic fibrosis: Impaired bicarbonate secretion and mucoviscidosis. *Lancet* **372**: 415–417.

Sadler JE. 2009. von Willebrand factor assembly and secretion. *J Thromb Haemost* **7**: 24–27.

Sheehan JK, Kirkham S, Howard M, Woodman P, Kutay S, Brazeau C, Buckley J, Thornton DJ. 2004. Identification of molecular intermediates in the assembly pathway of the MUC5AC mucin. *J Biol Chem* **279**: 15698–15705.

Thornton DJ, Rousseau K, McGuckin MG. 2008. Structure and function of the polymeric mucins in airways mucus. *Ann Rev Physiol* **70**: 459–486.

* Verdugo P. 2012. Supramolecular dynamics of mucus. *Cold Spring Harb Perspect Med* **2**: a009597.

Verdugo P, Deyrup-Olsen I, Aitken M, Villalon M, Johnson D. 1987. Molecular mechanism of mucin secretion: 1. The role of intragranular charge shielding. *J Dental Res* **66**: 506–508.

Structure and Function of the Mucus Clearance System of the Lung

Brenda M. Button[1] and Brian Button[2]

[1]Departments of AIRmed and Physiotherapy, The Alfred Hospital, Department of Medicine, Monash University, Melbourne, Australia

[2]Cystic Fibrosis Research and Treatment Center, University of North Carolina, Chapel Hill, North Carolina 27599

Correspondence: b.button@alfred.org.au; bbutton@med.unc.edu

In cystic fibrosis (CF), a defect in ion transport results in thick and dehydrated airway mucus, which is difficult to clear, making such patients prone to chronic inflammation and bacterial infections. Physiotherapy using a variety of airway clearance techniques (ACTs) represents a key treatment regime by helping clear the airways of thickened, adhered, mucus and, thus, reducing the impact of lung infections and improving lung function. This article aims to bridge the gap between our understanding of the physiological effects of mechanical stresses elicited by ACTs on airway epithelia and the reported effectiveness of ACTs in CF patients. In the first part of this review, the effects of mechanical stress on airway epithelia are discussed in relation to changes in ion transport and stimulation in airway surface layer hydration. The second half is devoted to detailing the most commonly used ACTs to stimulate the removal of mucus from the airways of patients with CF.

REGULATION OF AIRWAY SURFACE HYDRATION

The hydration state of the airway mucus layer is crucial for regulating the rate of mucociliary clearance (MCC) in the lung (Boucher 2007b). In the airways, the hydration state of the airway surface layer (ASL) fluid is controlled by the ion channels that regulate the mass of salt and water on the airway surfaces. Airway epithelia are highly water permeable because of their low transepithelial electrical resistances (Boucher 1994) and high expression of aquaporins (Kreda et al. 2001). The advantage of such epithelia is that they are capable of both absorbing fluid and secreting fluid, depending on the direction of the salt transport across airway epithelia. Because the concentration of NaCl within the ASL is essentially isotonic, the overall mass of NaCl in the ASL is critical for determining the total fluid volume of the ASL (Boucher 2007a). For example, if more salt is secreted into the ASL by the epithelium, water follows passively and the ASL volume will rapidly increase to maintain isotonicity. Conversely, if NaCl is absorbed, water also follows passively, resulting in a dehydration of the ASL. This ability to alternate between fluid absorption and secretion allows the airway epithelia to fine-tune the hydration state of the ASL and ensure efficient MCC.

Indeed, airway epithelia are capable of optimizing ASL hydration state by modifying the activities of apical ion channels to switch continuously between secretory and absorptive phenotypes (Button et al. 2007). In the airways, fluid absorption from the luminal airway surface to the submucosal compartment is regulated by net Na^+ absorption, mediated by the amiloride-sensitive epithelial sodium channel (ENaC) (Davis and Matalon 2007), whereas chloride (Cl^-) secretion across the apical membrane provides the driving force for net fluid secretion onto the airway surfaces. In the airways, Cl^- secretion is mediated by the cystic fibrosis transmembrane conductance regulator (CFTR), a member of the ATP-binding cassette (ABC) family of channels/transporters (Greger et al. 2001), and the calcium-activated chloride channel (or CaCC), which has been identified as TMEM16A, a member of the anoctamin family of anion channels (Rock et al. 2009).

DEFECTIVE ASL HYDRATION REGULATION IN CYSTIC FIBROSIS

In cystic fibrosis (CF), mutations of the gene coding CFTR are associated with a significant decrease in epithelial Cl^- secretion and excess Na^+ absorption (Boucher 2007a). The depletion of the ASL causes an excessive concentration of the mucus layer. When mucus concentrates, increases in its osmotic pressure results in drawing water osmotically from the PCL, collapsing the PCL and folding over the cilia. The sequence renders cilia ineffective at coordinated beating activity and thus unable to clear the adherent thickened mucus (Boucher 2007b; Button et al. 2012). This immobilization results in the apposition of the mucus layer to the cell-surface PCL, which has been hypothesized to lead to a physical interaction (adhesion) that is sufficiently strong to resist mucus clearance via ciliary beating or coughing. Stasis of lung secretions in CF eventually leads to mucus plugging, chronic bacterial infection and inflammation, and eventually, airway damage (bronchiectasis).

REGULATION OF ASL HYDRATION BY EXTRACELLULAR NUCLEOTIDES/ NUCLEOSIDES

Airway epithelia appear to be disconnected from the effects of systemic hormones that are known to regulate ion transport rates in other epithelia, including being resistant to the actions of mineralocorticoid and steroid hormones (Grubb and Boucher 1998). In contrast, airway epithelia rely on local extracellular signals in the ASL itself to maintain the ASL hydration state sufficient for basal mucus clearance, as well as report the stresses on airway surfaces that require an adjustment of ASL hydration and mucus clearance.

Key players in ASL hydration regulation that have been extensively characterized are the luminal concentrations of adenosine-based nucleotides (e.g., ATP or UTP) and nucleosides (e.g., adenosine [ADO]) (Schmid et al. 2011). These molecules act as major autocrine regulators of mucus clearance by their ability to modulate the activity of ion channels/transporters involved in ASL volume homeostasis. Specifically, studies have shown that exogenous application of ATP, UTP, or ADO produces coordinate regulatory effects on airway epithelia ion transport, including the inhibition of Na^+ absorption and activation of Cl^- secretion (Knowles et al. 1991; Benali et al. 1994), resulting in net secretion of fluid onto the airway surface. Indeed, inhaled UTP (and its derivatives) has been shown to stimulate both mucus clearance (Bennett et al. 1996) and sputum production (Johnson et al. 2002) in patients with obtrusive lung diseases.

The stimulation of net fluid secretion by nucleotides and nucleosides occurs via the activation of G-protein-coupled ATP-sensitive $P2Y_2$ and ADO-sensitive A_{2b} purinoceptors, located in the luminal membrane of superficial airway epithelial cells (Lazarowski and Boucher 2001, 2009). Stimulation of $P2Y_2$ receptors by extracellular ATP results in the activation of a heterotrimeric G protein (G_q) that increases the activity of both CaCC (TMEM16A), via an IP_3-dependent increase in intracellular calcium (Ca_i^{2+}) and CFTR, by a PKC-dependent mechanism (Fischer et al. 1998). In contrast, ENaC has been shown to be inactivated by $P2Y_2$ receptor activation (Stutts et al. 1995) via depletion of PIP_2 (Kunzelmann et al. 2005). The net effect is the stimulation of fluid secretion in the airways, facilitating mucus clearance. In addition

     Cite this article as *Cold Spring Harb Perspect Med* doi: 10.1101/cshperspect.a009720

to changes in ion transport by ATP, activation of P2Y$_2$ receptors can modulate the activity of other cellular processes, including stimulation in mucin secretion (Davis 2002) and increasing the rate of ciliary beating (Morse et al. 2001).

Adenosine, a product of ATP metabolism by extracellular enzymes (Picher et al. 2004), has also been shown to be a potent extracellular regulator of airway hydration state via the activation of G-protein (G$_s$)–coupled A$_{2b}$ purinoceptors on the luminal surface (Lazarowski et al. 1992). In airways, A$_{2b}$ stimulation results in cAMP-dependent activation of CFTR (Hanrahan et al. 1996) and inhibition of ENaC-mediated Na$^+$ absorption (Kunzelmann et al. 2001).

As with ATP, the net effect of increases in extracellular ADO is the stimulation of fluid secretion onto the airway surface.

Figure 1 presents a simplified schema depicting how these elements are integrated to participate in regulating the ATP/ADO levels, ASL hydration state, and, hence, the rate of mucus clearance from the lung. During conditions of ASL dehydration, the concentrations of ATP/ADO are elevated, resulting in a stimulation of fluid secretion. In contrast, the presence of "excess" fluid on the airways dilutes the extracellular ATP/ADO, resulting in net fluid absorption until the concentrations of ATP/ADO are sufficient to activate purinoceptors.

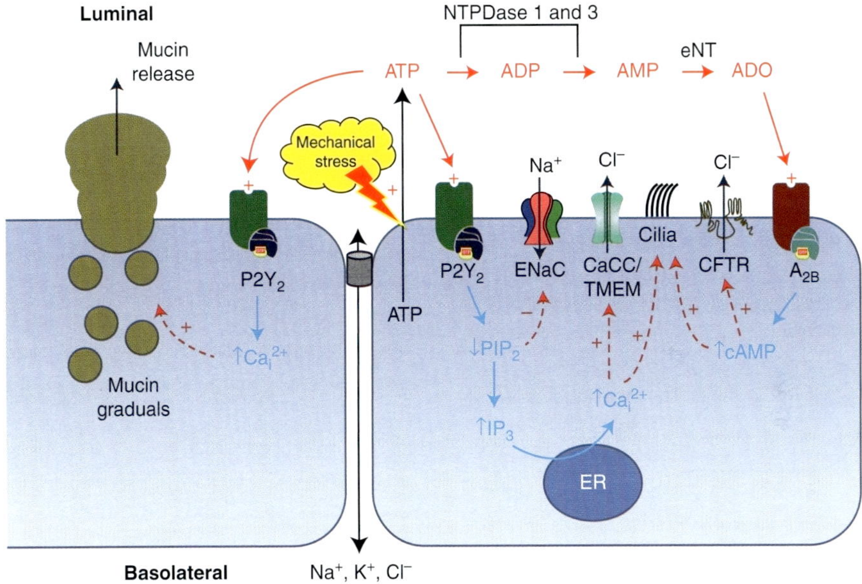

Figure 1. Model of purinergic signaling in airway epithelia. Mechanical stimulation of the epithelial surface by mechanical stress, irritants, or pathogens induces ATP release and elevates ASL (ATP). ATP is an agonist to a series of purinoceptors, including the G-protein-coupled P2Y$_2$ receptor located on the luminal surface of the airway lumen. Upon P2Y$_2$ receptor activation, the rate of PIP$_2$ hydrolysis increases, resulting in an increase in IP$_3$ and release of Ca^{2+} from stores. Increased levels of intracellular Ca^{2+} are responsible for activating CaCC (TMEM16a), resulting in a stimulation in Cl$^-$ secretion. Additionally, the reduction in PIP2 also results in a decrease in ENaC-mediated sodium absorption. The action of a series of ectonucleotidases converts the extracellular ATP to ADP, AMP, and eventually adenosine (ADO). Adenosine is a potent agonist of adenosine receptors, such as the A$_{2b}$ receptor, expressed on the airway surface. Activation of these receptors induces G-protein-coupled adenylate cyclase, followed by cAMP-dependent activation of protein kinase A. The result is stimulation of CFTR-mediated Cl$^-$ secretion. In addition to modulation of ion channel activity, stimulation of ATP release is a well-known agonist for stimulating mucin release (via P2Y$_2$ receptor activation) and cilia beat frequency (via both P2Y$_2$ and A$_{2b}$). Together, the action of mechanical stress is the stimulation of fluid secretion and acceleration of mucus transport rates.

STIMULATION OF ATP RELEASE BY MECHANICAL STRESSES

By virtue of their interfacial position in the lungs, the epithelial cells lining the airway are ideally situated to act as sensors of breathing-dependent changes in mechanical stress (such as airflow, stretching, and pressure). The role of mechanical forces in pulmonary health raises the possibility of relationships between the forces generated by tidal breathing, airway surface hydration, and the rates of mucociliary clearance.

As a mechanism to link these functions, substantial evidence suggests that the mass of nucleotides and nucleosides on airway surfaces is regulated by the rate of ATP release, triggered by breathing-induced mechanical stresses (Button and Boucher 2008). Consistent with this thesis are several studies that have reported that ATP is released from human airway epithelia subjected to physical forces. Examples of these ATP release-stimulating mechanical forces include forces generated by mechanical deformation (Grygorczyk and Hanrahan 1997), fluid shear stress (Tarran et al. 2005), compression/stretch (Button et al. 2007), or osmotic shock (Okada et al. 2006). It should be noted that although the ATP release pathway(s) are poorly understood, studies have suggested that it is the result of vesicular release (Kreda et al. 2007) and/or through conductive ion channels, such CFTR (Egan 2002) and Pannexins (Ransford et al. 2009).

MECHANICAL FORCES IN THE LUNG

The lung is a unique organ in that it is subjected to complex physical forces during normal tidal breathing, coughing, and vascular perfusion. Over the past few decades, it has been realized that these physical forces play an important role in regulating the structure, function, and metabolism of the lung—for example, the observation that normal breathing, interspersed with intermittent large volume breaths (sighs), is important for maintaining alveolar health and patency (Dietl et al. 2001). Furthermore, breathing movements observed as early as 10 wk of gestation in the human fetus suggest that these mechanical forces are important for cell differentiation and further lung development (Inanlou et al. 2005).

In the airways, two types of mechanical forces dominate. During normal breathing, airflow across the surface of the airway epithelium produces a wall shear stress. Because both airflow velocities and airway diameter decrease proportionally in the deeper lung, it has been estimated that the magnitude of the shear stress force varies little from the large to the small generations of the airways, centering on a value on the order of $\sim$0.5 dynes $\bullet$ cm^2 (Fredberg 1978; Tarran et al. 2005). The other mechanical stress that is relevant during respiration is the transairway pressure gradient. During normal expiration, a transmural pressure gradient of $\sim$8.5 cm H$_2$O is generated (Levitzky 1991), and can approach $\sim$20 cm H$_2$O in the proximal airways during exercise or forced expirations (Jones et al. 1975).

Abnormal physical forces exerted on lung tissues are known to contribute to pathological situations. For example, in extreme conditions, such as patients receiving positive-pressure mechanical ventilation or during severe bronchoconstriction associated with asthma, cells in the lung are compressed and/or stretched to supranormal levels, leading to an array of pathological outcomes, including inflammation, airway remodeling, and even cell death (Ressler et al. 2000). Although transient, mechanical stresses on airways are even more extreme during coughing (Leith 1985). The combination of a closed glottis with the rapid expiratory muscle contraction produces very rapid changes (600–1600 cm H$_2$O/sec) in transpulmonary pressures, reaching pressures as high as 200 cm H$_2$O (Basser et al. 1989). During the expiration phase of coughing, peak airflow rates of >500 L/min can be reached, producing wall surface shear stresses estimated to reach as high as 1700 dynes/cm^2 (Basser et al. 1989).

MECHANICAL FORCES AND STIMULATION IN MUCUS CLEARANCE

As detailed in the sections below, airway clearance therapy (ACT) represents an effective technique to facilitate the clearing of retained mucus secretions from the lungs of patients with chronic lung diseases like CF. Although the mechanism of action of such maneuvers on mucus clearance is still poorly understood, it is widely believed that the mechanical forces elicited by these techniques reduce the viscoelasticity of the bronchial mucus secretions. Such forces would be analogous to imparting the "shear-thinning" ketchup; that is, shake the bottle and the semisolid paste becomes a runny liquid. In addition, studies have also shown that ACT can act directly as a physical mucolytic (Sutton et al. 1983), by breaking down of high-molecular-weight DNA in CF sputum samples, to levels comparable to those obtained by exogenous rhDNAse treatment (App et al. 1998). Finally, the ability to generate fluid secretion by the action of mechanically stimulated ATP release during ACT would be expected to reduce the concentration of the mucus (Button et al. 2007). Combined, these effects will make the adhered mucus easier to clear from the lungs.

EXPECTORATION TECHNIQUES

Cough

Coughing is the body's natural backup mechanism for airway clearance. The cough is achieved by rapid airflow acceleration and extremely high flow rates and, when coupled with dynamic airway compression, is very effective in squeezing and clearing mucus from the airways (Cloutier 2007). Coughing is mostly reflexive and, thus, does not allow as much conscious control of starting lung volume or pressures developed. However, coughing can be directed in trained patients. During coughing, a deeper inspiration (often around one and one-half times the tidal volume) occurs, increasing elastic recoil. High intrapulmonary pressures are built up behind the glottis, and when the glottis opens, supramaximal, turbulent expiratory flows are generated. During the cough, the posterior membranous portion of the trachea is compressed, and the tracheal diameter is narrowed to about one-sixth of normal. With an increase in flow rate of sevenfold during a cough, the linear velocity of air increases 42-fold. Airflow at the site of compression is turbulent and makes the sound that we call a cough. As lung volume decreases and the elastic recoil pressure decreases, the equal pressure point moves upstream toward the alveoli (Cloutier 2007).

The equal pressure point plays a critical role in the effectiveness of coughing, because a substantial increase in airflow velocity occurs at points of narrowing (choke points). High linear airflow velocity provides the turbulent flow, high shearing forces along the airway walls, and high kinetic energies that move secretions cephalad. Because of the extremely high intraluminal and extraluminal pressures generated with a cough, there is more potential for significant dynamic airway collapse at the equal pressure point, especially in unstable airways. Cartilaginous support decreases from the trachea and larger bronchi to the smaller bronchi and is probably minimal within bronchioles. Although smooth muscles may aid in maintaining smaller airway patency, a cough may compress those airways too much to allow effective clearance (Lapin 2002). During a cough, alveolar, pleural, and subglottal pressures may rise as much as 200 cm H_2O (Frownfelter and Massery 2006). Coughing will clear to approximately the sixth or seventh generation of segmental bronchi. If a patient has retained secretions, distal airway clearance techniques must move the secretions to an area where the cough can be effective (Frownfelter and Massery 2006).

Movement of the Equal Pressure Point (EPP) during a Forced Expiration ("Huff")

"Huffing" is a forced expiratory maneuver, usually initiated from mid-to-low lung volumes via an open glottis, and is another way to produce supramaximal airflow and high linear velocities (Rossman et al. 1982). The equal pressure point (EPP) is defined as the point at which the intraluminal pressure and extraluminal pressure are equal. Beyond that point (downstream toward the mouth), the external pressure around the airway (extraluminal or pleural pressure) is greater than the pressure within it (intraluminal), and the airway is compressed, which limits flow. During a huff, the pressure in the airway (intraluminal pressure) decreases from the peripheral airways to the mouth, because of frictional pressure loss and convective acceleration pressure loss (Mead et al. 1967). The pressure also decreases because of the movement of air from the periphery with a cumulatively large cross-sectional area, centrally, with gradual cumulative cross-sectional area decreasing. During the exhalation, the extraluminal (pleural) pressure remains relatively constant, and the intraluminal pressure gradually decreases. Thus, there is a wave of equal pressure points moving deeper (more peripherally) into the airways as exhalation proceeds and intraluminal pressures fall.

The site of the equal pressure point is determined by the amount of expiratory force and the elastic recoil. A higher expiratory force shifts the equal pressure point peripherally (toward the alveoli). Likewise, if the forced exhalation is initiated at a lower lung volume, the pressure from the static elastic recoil will be less, as will the intraluminal pressure, and the resulting equal pressure point will be more peripheral.

Mucus clearance has been studied using a two-phase gas–liquid flow mechanism in an in vitro flow model. The study found that the peak expiratory flow rate should exceed the peak inspiratory flow rate by at least 10% for mucus to move proximally. Second, the peak expiratory flow rate should exceed 30–60 L/min to overcome the shear force to which the mucus is attached in order to move the mucus. This study also found that fluids with low viscosity and elasticity required higher expiratory flow rates than viscous fluid to move (Kim et al. 1987).

Asynchronous Ventilation (Differences in Lung Filling) and Interdependence

For mucus clearance to occur, air must get behind (distal to) the mucus. Even in normal lungs, time constants vary among lung regions, and asynchronous ventilation occurs secondary to regional and stratified *inhomogeneity* (Murray 1986). Mucus obstruction (partial or complete) in disease states may increase inhomogeneity, substantially impairing the ability to get air past mucus obstructions (Laube et al. 1989). Some airway clearance techniques include an

inspiratory pause to compensate for asynchronous ventilation (Lapin 2002).

Inspiration dilates airways as negative pleural pressure causes the lungs to inflate, causing a transient decrease in resistance to airflow that aids in getting air behind the mucus. Interdependence between adjacent parenchymal lung units occurs because of the elasticity of the surrounding interstitium and also between the lungs and chest wall (Mead et al. 1970). These mechanisms help to preserve uniform ventilation distribution. As lung volume decreases during expiration, expiratory flow limitation occurs.

Collateral Ventilation

The various lobes of the lung are subdivided by fissures. The division into lobes is incomplete, which allows for collateral ventilation. Collateral ventilation is an accessory pathway that connects air spaces supplied by other airways. The accessory pathways in the lung include (1) the channels of Lambert, which connect respiratory bronchioles and terminal bronchioles to air spaces supplied by other airways; (2) the pores of Kohn, which are openings in the alveolar walls that connect adjacent alveoli; and (3) the interbronchiolar channels of Martin. These accessory pathways help to prevent collapse of terminal respiratory units when their supplying airway becomes obstructed (atelectasis) (Cloutier 2007).

Airway Clearance Therapy in CF: Rationale and Practice

Airway clearance techniques (ACTs) are usually commenced as soon as the diagnosis of CF is made, often soon after birth. In infants, airway clearance techniques are performed by parents and/or caregivers. As children mature, they are taught techniques that can be performed independently of an assistant (Hodson 2000). Patients/caregivers are required to perform this treatment preventatively when well and more intensively during acute exacerbations (Williams 1994). A Consensus Conference Report on CF adult care describes airway clearance therapy as a "cornerstone" of treatment (Yanka-

     Cite this article as *Cold Spring Harb Perspect Med* doi: 10.1101/cshperspect.a009720

skas et al. 2004). Airway clearance techniques are required to have a sound basis in physiology and need to be efficient and effectively applied across the life span.

BREATHING TECHNIQUES TO PROMOTE MUCUS CLEARANCE

Autogenic Drainage

Autogenic drainage (AD) is an airway clearance technique underpinned by basic physics, fluid dynamics, pulmonary anatomy, and physiology, together with breathing mechanics (Mead et al. 1967; Dab and Alexander 1979; Chevaillier 1984; Schoni 1989; Lapin 2002; Van Ginderdeuren 2009). J. Chevaillier, the developer of this technique, states that mucus clearance is based on two different systems: the effects of ciliary clearance and increased airflow applying shearing forces to the mucus adhering to the airway wall, similar to the effects of erosion. "The higher the velocity of the medium, the stronger the erosive effect." To achieve the necessary shearing forces to remove secretions from airways, it is necessary to modulate the inspiratory and expiratory airflow.

AD aims to generate higher expiratory airflow in all generations of bronchi without causing dynamic collapse. This goal is achieved using repetitive controlled tidal volume breathing, starting at lower lung volumes in the vicinity of residual volume (RV) and gradually and progressively increasing lung volumes toward inspiratory reserve volume (IRV) as secretions are loosened, collected, and moved up toward the mouth for expectoration (Van Ginderdeuren 2009). This technique relies on the EPP to move into the more peripheral airways and significantly increase flow rate and linear velocity in these airways (Lapin 2002).

AD was found to be as effective as other airway clearance techniques when compared in shorter and longer randomized controlled trials (Pfleger et al. 1992; Giles et al. 1995; Miller et al. 1995; McIlwaine et al. 2010; Pryor et al. 2010). The process of airway clearance in AD is based on an active or passive assisted autogenic drainage (AAD) modulation of the airflow and

lung volume-based level of breathing. AAD is a technique that can be used with infants and very young children.

Active Cycle of Breathing Technique

The active cycle of breathing consists of (1) breathing control, (2) thoracic expansion exercises, and (3) a forced expiration technique. Breathing control (BC), which is gentle, relaxed breathing at the patient's own tidal volume and resting respiratory rate, is followed by thoracic expansion exercises (TEEs), which are deep, slow, relaxed inspirations to IRV with or without 2- to 3-sec breath-holds with quiet unforced expirations (elastic recoil). BC is interspersed throughout the cycle to allow recovery and prevent an increase in airflow obstruction. These TEEs help maximize ventilation via collateral channels and interdependence and ameliorate the problem of asynchronous ventilation and blocked airways. The forced expiration technique is a combination of one or more forced expirations (huffs) and periods of breathing control. Huffing to low lung volumes should assist in loosening and mobilizing excess secretions from smaller peripheral airways to larger central airways. When secretions have reached the larger airways, a huff or cough with high lung volume can be used to clear them (Lapin 2002; Pryor 2009). ACBT is either performed in upright sitting or recumbent positions (Cecins et al. 1999). During a forced expiratory maneuver, there is compression of the airway downstream (toward the mouth) from the equal pressure point. The squeezing action, which moves peripherally with decreasing lung volume, together with the increase in air speed as air flows through the narrowed segment, facilitates the movement of secretions along the airway (West 2004). ACBT is comparable to other airway clearance therapies in outcomes commonly used in clinical trials (Pryor et al. 2010; Robinson et al. 2010).

Positive Expiratory Pressure (PEP) Therapy to Enhance Mucus Clearance

Positive expiratory pressure (PEP) therapy, a flow-regulated technique, was developed in Denmark in the late 1970s using a mask, one-way valve, and selected expiratory orifice resistor. A pressure manometer can be inserted between the valve and the resistor (Falk et al. 1984; Tonnesen and Stovring 1984; Groth et al. 1985; Lannefors 2009). PEP is used to recruit closed or obstructed peripheral airways by mobilizing, transporting, and evacuating secretions in spontaneously breathing patients and is, by definition, combined with the forced expiration technique (FET) described above.

The aim of this flow-regulated technique is to obtain a temporary increase in functional residual capacity (FRC) that allows the tidal volume (TV) to be above the opening volume for otherwise closed or obstructed airways. The 12–15 consecutive TV breaths comprising each cycle should be only slightly active while achieving a mid-expiratory pressure of 10–20 cm H_2O. Patients should be taught how to recognize a temporary increase in FRC. The number of treatment cycles within a treatment session and the frequency of treatment are individually selected according to specific needs. Owing to the elastic recoil of the lung tissue at this temporarily increased FRC level, collateral ventilation should increase and open up these closed airways. The air in the recruited lung volumes can be used behind secretions and combined with either the forced expiration technique or the technique of autogenic drainage to mobilize, transport, and evacuate airway secretions. This technique can be applied in upright sitting or horizontal positions. PEP can be used in infants held on the arm of the caregiver with the mask applied over the nose and mouth. PEP was found to be superior to postural drainage in short- and longer-term studies (Lannefors and Wollmer 1992; McIlwaine et al. 1997). Contraindications to use of PEP are undrained pneumothorax and frank hemoptysis.

High-Pressure Positive Expiratory Pressure (Hi-PEP) Mucus Clearance Technique

High-pressure positive expiratory pressure (Hi-PEP) airway clearance physiotherapy was developed in Graz, Austria during the early 1980s. Hi-PEP uses forced expiratory maneuvers against

the PEP-mask's expiratory resistor for mobilizing and transporting intrabronchial secretions (Oberwaldner et al. 1986). The equipment used is the same as that described in the previous section.

The aim of Hi-PEP is to improve mucus clearance in a physiological way. During the "mobilization" phase, the aim is to increase collateral airflow to underventilated regions and to squeeze "Pendelluft" from hyperinflated into unobstructed and atelectatic lung units. This is achieved using a forced expiration against a marked resistive load. Furthermore, mobilization of mucous plugs is achieved by back pressure-effected dilation of airways. During the "transportation" phase, incorporation of the peripheral airways into the compressed downstream (toward the mouth) segment is a prerequisite for efficacy. An in-depth understanding of the downstream and upstream movement of the equal pressure point is required by therapists teaching patients this technique (Oberwaldner et al. 1991; Pfleger et al. 1992; Zach and Oberwaldner 1992; Oberwaldner 2009).

PEP breathing is undertaken through the mask, two one-way valves, and individually selected expiratory resistors for eight to 10 cycles using moderately increased tidal breathing. This maneuver is followed by inhalation to total lung capacity and one or more forced expiratory maneuvers against the expiratory resistor exhaling toward residual volume. The consequent mobilization of secretions usually results in coughing from low lung volumes into the PEP mask against the resistor, splinting the airways open and avoiding dynamic compression of unstable airways. The sustained expiratory pressures achieved usually range between 40 and 100 cm H_2O. The individually selected resistor and optimal expiratory pressure is determined by a spirometry-assisted method. Hi-PEP can be applied as a passive physiotherapy technique in infants and patients who are unable to cooperate actively. Contraindications to Hi-PEP include pneumothorax, cardiac disease, frank hemoptysis, lung surgery, and asthma. Hi-PEP is not recommended for patients who are exhausted and unable to meet the demands of this energy-consuming technique.

Oscillating Positive Expiratory Pressure (OscPEP) Therapy to Enhance Mucus Clearance

Oscillating positive expiratory pressure using the Flutter device was developed in Europe in the late 1980s (Althaus 1989) and approved by the U.S. FDA in 1994. This was followed by the development of the Acapella device in the United States in the early twenty-first century. The Flutter device uses an oscillating positive expiratory pressure that causes vibrations of the airway wall, which, in turn, loosen secretions, change the rheological properties of sputum to become less viscous, perhaps via promoting ATP release, and prevent premature closure of the bronchi during exhalations (App et al. 1998; Althaus 2009). The Flutter device consists of a molded plastic mouthpiece, a plastic cone, a steel ball, and a perforated cover. By tuning this frequency to the individual's own ventilatory abilities, the patient induces maximal vibrations of the bronchial walls, which promote mucus clearance. The Flutter generates PEP in the range 18–35 cm H_2O, and the angle at which the device is held determines the oscillation frequency, usually between 6 and 26 Hz. The patient's expiratory effort determines the pressure. The Acapella device uses a counterweighted plug and magnet to create airflow oscillation. The Acapella device has an adjustable PEP dial that is manipulated to provide more or less PEP during oscillations depending on individual requirements. The performance of these two devices is similar. However, unlike the Flutter, the performance of the Acapella is not dependent on device orientation and may be easier for some patients to use, particularly at low expiratory flows (Volsko et al. 2003). Following use of an oscillating PEP device, mobilized secretions should then be cleared after each cycle by using the forced expiration technique described previously.

Contraindications to oscillating PEP are frank hemoptysis and an undrained pneumothorax. Caution should be exercised in patients who are unable to tolerate the increased work of breathing and those with increased intracranial pressure; hemodynamic instability; recent facial, oral, or esophageal surgery; acute sinusitis; or middle ear pathology (Althaus 2009). Oscillating PEP therapy has been found to be as effective as other commonly used airway clearance techniques (Morrison and Agnew 2009; Pryor et al. 2010; Sontag et al. 2010).

High-Frequency Chest Wall Oscillation (HFCWO) Therapy

High-frequency chest wall oscillation (HFCWO) is a patient-delivered form of airway clearance therapy consisting of an inflatable vest and an air-pulse generator, also known as high-frequency chest compression (HFCC). This mechanical system delivers high-frequency chest wall oscillation capable of superimposing oscillatory airflow throughout a patient's inspiratory and expiratory cycle. This is performed by the vest being inflated to a nearly constant background pressure with a superimposed frequency of air pressure oscillations (Hansen and Warwick 1990). During HFCWO, pressure pulses within the vest oscillate the thoracic wall and generate transient increases in airflow within the airways. It has been proposed that HFCWO assists sputum removal by increasing airflow at low lung volumes, increasing expiratory flow bias (creating a peak expiratory flow rate that is 10% greater than peak inspiratory flow rate), resulting in an increased annular flow of mucus toward the mouth, and decreasing the viscoelasticity of mucus by reducing cross-linking (Fink and Mahlmeister 2002).

HFCWO is usually commenced at low pressures and frequencies and then increased to therapeutic optimum as the patient tolerates. Different devices allow a different range of oscillation and frequency settings. For example, "The Vest" has an adjustable frequency setting that provides delivery of HFCWO at frequencies from 5 to 20 Hz. The pressure delivered to "The Vest" is adjustable with a range of $\sim$3–50 cm H_2O and mean pressure of $\sim$25 cm H_2O. The HFCWO should be paused approximately every 5 min for huffing and coughing.

HFCWO is not recommended with unstable neck injury, intravenous access port being accessed under the vest, pulmonary embolism, lung contusion, current hemoptysis, hemodynamic instability, rib fractures, large pleural effusion, or emphysema. Precautions include end-stage disease (end expiratory volume may fall below closing capacity), port under the vest (not currently accessed), recent esophageal surgery, distended abdomen, bronchospasm, osteoporosis, and coagulopathy (AARC 1991). HFCWO therapy has been compared with other airway clearance techniques in CF and other disease groups with pulmonary problems (Arens et al. 1994; Braggion et al. 1995; Kluft et al. 1996; Scherer et al. 1998; Oermann et al. 2001; Phillips et al. 2004; Yuan et al. 2010; Fainardi et al. 2011). A simulation tool to study high-frequency compression energy transfer mechanisms and waveforms has been developed for applications in pulmonary disease (O'Clock et al. 2010). Recently, a large multicenter trial comparing HFCWO therapy with PEP was conducted in 12 centers across Canada in 107 children and adults over a 12-mo period. The HFCWO group experienced significantly more acute exacerbations requiring oral, inhaled, or IV antibiotics than the PEP group. Both groups maintained their lung function over the 1-yr period of the study, although the treatment burden was significantly higher in the HFCWO group. The results of this study favor PEP and do not support the use of HFCWO as the primary form of airway clearance in patients with CF (McIlwaine et al. 2013).

Intrapulmonary Percussive Vibration (IPV)

Intrapulmonary percussive ventilation (IPV) is a patient-activated form of physiotherapy that is administered via the mouth to the airways, enhances the mobilization and cephalad movement of pulmonary secretions, and increases resting lung volumes. IPV is used to treat and ventilate (short term, during treatment) patients with an obstructive and/or restrictive, acute or chronic respiratory disease.

IPV is an open breathing circuit combining a "Phasitron" with an aerosol converter. The Phasitron is a pressure flow converter that transforms small volumes of gas at high pressure and low flow into larger volumes of gas at low pressure and high flow. A high-output nebulizer provides a dense aerosol to deliver medications and hydrate secretions. During IPV, high-frequency minibursts of gas (at 100–300 cycles/min) are superimposed on the patient's own respiration at pressures of 5–35 cm H_2O. In practice, the driving pressure and frequency are individually titrated to patient comfort and thoracic movement.

IPV creates a global effect of internal percussion of the lungs, which could promote clearance of the peripheral bronchial tree. The high-frequency gas pulses are also proposed to expand the lungs, vibrate and enlarge the airways, and deliver gas into distal lung units, beyond accumulated mucus. Three forms of therapy are provided during IPV: percussive oscillatory vibrations to loosen retained secretions, high-density aerosol delivery to hydrate viscous mucous plugs, and positive expiratory pressure (PEP) to recruit alveolar lung units. IPV has been compared with other airway clearance therapies in CF and other disease groups (Natale 1994; Homnick 1995; Newhouse 1998; Scherer et al. 1998; Toussaint 2003).

CONSIDERATIONS IN POSITIONING TO OPTIMIZE MUCUS CLEARANCE THERAPY

Head-Down Tilted Positioning

A study of CF patients using radiolabeled sputum clearance compared postural drainage, PEP in upright sitting, and cycle exercise versus a control activity (Lannefors and Wollmer 1992). The most effective sputum clearance occurred using PEP in upright sitting. Surprisingly, when the majority of patients were in the left decubitus (head-down) position, there was greater sputum clearance from the dependent left lung compared with the "gravity-assisted" right lung. This study indicates that gravity results in increased ventilation in the dependent regions in positions commonly used during airway clearance, thus effecting increased sputum clearance from the dependent regions. Numerous side effects have been reported in head-down positions, including headaches, sinus pain (Cecins et al. 1999), desaturation of oxygen (Falk et al.

 Cite this article as *Cold Spring Harb Perspect Med* doi: 10.1101/cshperspect.a009720

1984), symptomatic and silent gastroesophageal reflux (Button et al. 1997, 1998, 2003, 2005; Orenstein 2003), and poor adherence (Myers and Horn 2006). Another relevant study compared the effects of body position on mean expiratory pressure (MEP) and peak expiratory flow rate (PEFR), surrogate markers of huff, and cough strength in adults with stable CF. Repeated measures of MEP and PEFR were performed across seven positions (standing, chair-sitting, sitting in bed with backrest vertical, sitting in bed with backrest at 45°, supine, side-lying, and side-lying with head-down tilt 20°) in random order. During testing, reflux sensation and oxygenation were monitored. MEP was significantly reduced in side-lying and in the head-down tilt position. PEFRs were significantly reduced in the three-quarters sitting, supine, side-lying, and head-down positions. Oxygenation and reflux scores were worst in the head-down position (Elkins et al. 2005). Patients with respiratory muscle fatigue may have increased respiratory distress in head-down positions due to the added resistive loading beneath the diaphragm because of the weight of the viscera. This may negatively impact airway clearance therapy.

Upright Sitting with Feet Dependent

Studies using radioactive inert gas with a radiation counter over the chest wall have shown that when radiolabeled gas is inhaled by an individual in the sitting position, radiation counts are lowest in the upper lung fields, intermediate in the mid-lung fields, and greatest in the lower lung fields. This effect of position is gravity dependent. The lower lung fields are preferentially ventilated compared with the upper lung fields. The causes of regional differences in ventilation can be explained in terms of the anatomy of the lung and the mechanics of breathing. An intrapleural pressure gradient exists down the lung. In the upright position, intrapleural pressure tends to be more negative at the top of the lung and becomes progressively less negative toward the bottom of the lung. This pressure is thought to reflect the weight of the suspended lung (Frownfelter and Dean 2006).

Supine

In the supine position, the apices and bases are ventilated comparably, and the lowermost lung fields are better ventilated than the uppermost. The supine position inherent in bed rest alters the configuration of the chest wall. The normal anteroposterior configuration becomes more transverse. The hemidiaphragms are displaced upward in a cephalad direction, which reduces functional residual capacity (FRC) in this position. If the supine position is adopted for long periods, pulmonary secretions tend to pool on the dependent sides of the airways. The upper side of the airway may dry out, exposing the patient to infection and obstruction (Frownfelter and Dean 2006).

Prone

The prone position shifts the mobile structures of the thoracic and abdominal cavities. The heart and great vessels are displaced anteriorly. The liver and spleen and kidneys shift caudally. The prone position increases arterial oxygen tension, tidal volume, and dynamic lung compliance. The pleural pressure gradient is homogenized; hence, the distribution of alveolar ventilation and alveolar inflation is augmented (Frownfelter and Dean 2006).

Side-Lying

In the side-lying position, there is greater gas exchange in the dependent lung. The lower lung fields are preferentially ventilated compared with the upper lung fields. The left lung is smaller than the right to accommodate the position of the heart. This accentuates anteroposterior expansion at the expense of transverse excursion of the dependent chest wall. In this position, the dependent hemidiaphragm is displaced in a cephalad direction because of the compression of the viscera. This results in greater excursion during respiration and to greater contribution to the ventilation to that lung and to gas exchange as a whole in adults (Frownfelter and Dean 2006). Positioning needs careful consideration in each individual to optimize airway clearance therapy on an ongoing basis.

Percussion and Vibration

Manual percussion and vibrations are sometimes used as an adjunct to positioning and the active cycle of breathing. Percussion is used to mobilize mucus and may stimulate coughing in infants and children. Percussion is performed using a cupped hand with a rhythmical flexion and extension action of the wrist. In adults, percussion can be performed with one or two hands. In infants, percussion is performed using two or three fingers of one hand. A literature review by Gallon (1992) suggests that percussion is only indicated in patients with excessive sputum production. Percussion has been shown to cause an increase in hypoxemia (Falk et al. 1984), but, when combined with thoracic expansion exercises, no fall was seen in oxygen saturation (Pryor et al. 1990). Rib fractures were reported in a neonate (Purohit et al. 1975). Vibration augments the expiratory flow and may help mobilize mucus (McCarren and Alison 2006). Percussion and vibration are contraindicated in patients with severe osteoporosis, frank hemoptysis, fractured ribs, and chest injuries. Caution should be exercised in patients with hyper-reactive airways and severe bronchospasm.

Physical Exercise as Airway Clearance

Whole body physical exercise increases minute ventilation and oxygen uptake via the cardiopulmonary system, uses shear forces to loosen mucus from the airway wall, increases resting lung volumes, and increases regional ventilation via gravitational effects. Exercising in different positions such as upright, sitting, supine, sidelying, and prone leads to the mobilization of pulmonary secretions and enhances airway clearance while concurrently increasing/maintaining cardiorespiratory fitness, general muscle strength, joint mobility, and postural alignment (Zach et al. 1982; Salh et al. 1989; Lannefors and Wollmer 1992; Hebestreit et al. 2001, 2010; Lannefors et al. 2004). Some individuals with mild lung disease and good lung function use exercise together with forced expiration (huffing), coughing, and expectoration as stand-alone airway clearance therapy. Others, with more extensive lung disease and larger volumes of sputum, use exercise as an adjunct to a formal airway clearance therapy regimen (Lannefors 2009; Kuys et al. 2011).

In a recent study of adults with CF, ease of expectoration improved following exercise (Dwyer et al. 2011). Ventilation and respiratory flow were significantly higher during treadmill and cycle exercise compared with control, and there was a significantly greater decrease in sputum mechanical impedance following treadmill walking compared with control but not with cycle ergometry. This last finding is thought to be because of trunk oscillations and increased ventilation and expiratory flow, resulting in increased propulsion of mucus and the creation of shear forces in the airways to augment ATP release airway surface liquid height, and mucus transport toward the oropharynx. Caution should be applied with patients with pulmonary hypertension, cor pulmonale, acute exacerbation with fever, and exercise-induced bronchospasm. Aerobic exercise is recommended as an adjunctive therapy for airway clearance and for its additional benefits to overall health (Flume et al. 2009).

Timing of Airway Clearance and Inhalational Therapies

Timing of therapies is important in terms of optimizing treatment outcomes. Bronchodilators, if prescribed, are generally inhaled before airway clearance therapy. The mucolytic agent of hypertonic saline is generally prescribed before or during airway clearance (Dentice et al. 2012; Elkins and Dentice 2012). The mucolytic agent dornase alfa is recommended either 30 min before or after airway clearance (Dentice and Elkins 2011).

CONCLUSIONS

Understanding of mucus clearance from the lung is still expanding and is the focus of a range of diverse investigations. Discoveries made in basic science research assist in further understanding the rationale that underpins the different airway clearance therapies. There are many different airway clearance techniques and regimens from which to choose for individual patients. Complete descriptions of the different techniques can be found on the Cystic Fibrosis Worldwide website (www.cfww.org): International Physiotherapy Group/CF, *Physiotherapy for people with CF throughout life* (4th ed., 2009). These techniques need to be continuously reevaluated and adapted according to changes in pathophysiology, developmental stages, lifestyle, or other considerations.

REFERENCES

AARC. 1991. AARC (American Association for Respiratory Care) clinical practice guideline. Postural drainage therapy. *Respir Care* **36:** 1418–1426.

Althaus P. 1989. The bronchial hygiene assisted by the flutter VRP1 (Module regulator of a positive pressure oscillation on expiration). *Eur Respir J* **2** (Suppl 8): 693.

Althaus P. 2009. *Oscillating PEP–Flutter therapy.* International Physiotherapy Group/Cystic Fibrosis, www.cfww.org.

App EM, Kieselmann R, Reinhardt D, Lindemann H, Dasgupta B, King M, Brand P. 1998. Sputum rheology changes in cystic fibrosis lung disease following two different types of physiotherapy: Flutter vs autogenic drainage. *Chest* **114:** 171–177.

Arens R, Gozal D, Omlin KJ, Vega J, Boyd KP, Keens TG, Woo MS. 1994. Comparison of high frequency chest compression and conventional chest physiotherapy in hospitalized patients with cystic fibrosis. *Am J Respir Crit Care Med* **150:** 1154–1157.

Basser PJ, McMahon TA, Griffith P. 1989. The mechanism of mucus clearance in cough. *J Biomech Eng* **111:** 288–298.

Benali R, Pierrot D, Zahm JM, de Bentzmann S, Puchelle E. 1994. Effect of extracellular ATP and UTP on fluid transport by human nasal epithelial cells in culture. *Am J Respir Cell Mol Biol* **10:** 363–368.

Bennett WD, Olivier KN, Zeman KL, Hohneker KW, Boucher RC, Knowles MR. 1996. Effect of uridine 5′-triphosphate plus amiloride on mucociliary clearance in adult cystic fibrosis. *Am J Respir Crit Care Med* **153:** 1796–1801.

Boucher RC. 1994. Human airway ion transport. Part one. *Am J Respir Crit Care Med* **150:** 271–281.

Boucher RC. 2007a. Airway surface dehydration in cystic fibrosis: Pathogenesis and therapy. *Annu Rev Med* **58:** 157–170.

Boucher RC. 2007b. Evidence for airway surface dehydration as the initiating event in CF airway disease. *J Intern Med* **261:** 5–16.

Braggion C, Cappelletti LM, Cornacchia M, Zanolla L, Mastella G. 1995. Short-term effects of three chest phys-

iotherapy regimens in patients hospitalized for pulmonary exacerbations of cystic fibrosis: A cross-over randomized study. *Pediatr Pulmonol* **19:** 16–22.

Button B, Boucher RC. 2008. Role of mechanical stress in regulating airway surface hydration and mucus clearance rates. *Respir Physiol Neurobiol* **163:** 189–201.

Button BM, Heine RG, Catto-Smith AG, Phelan PD, Olinsky A. 1997. Postural drainage and gastro-oesophageal reflux in infants with cystic fibrosis. *Arch Dis Child* **76:** 148–150.

Button BM, Heine RG, Catto-Smith AG, Phelan PD. 1998. Postural drainage in cystic fibrosis: Is there a link with gastro-oesophageal reflux? *J Paediatr Child Health* **34:** 330–334.

Button BM, Heine RG, Catto-Smith AG, Olinsky A, Phelan PD, Ditchfield MR, Story I. 2003. Chest physiotherapy in infants with cystic fibrosis: To tip or not? A five-year study. *Pediatr Pulmonol* **35:** 208–213.

Button BM, Roberts S, Kotsimbos TC, Levvey BJ, Williams TJ, Bailey M, Snell GI, Wilson JW. 2005. Gastroesophageal reflux (symptomatic and silent): A potentially significant problem in patients with cystic fibrosis before and after lung transplantation. *J Heart Lung Transplant* **24:** 1522–1529.

Button B, Picher M, Boucher RC. 2007. Differential effects of cyclic and constant stress on ATP release and mucociliary transport by human airway epithelia. *J Physiol* **580:** 577–592.

Button B, Cai L-H, Ehre C, Kesimer M, Hill DB, Sheehan JK, Boucher RC, Rubinstein M. 2012. A periciliary brush promotes the lung health by separating the mucus layer from airway epithelia. *Science* **337:** 937–941.

Cecins NM, Jenkins SC, Pengelley J, Ryan G. 1999. The active cycle of breathing techniques—To tip or not to tip? *Respir Med* **93:** 660–665.

Chevaillier J. 1984. Autogenic drainage. In *Cystic fibrosis horizons* (ed. Lawson D), p. 235. Wiley, New York.

Cloutier M. 2007. *Respiratory physiology.* Mosby Elsevier, Philadelphia.

Dab I, Alexander F. 1979. The mechanism of autogenic drainage studied with flow volume curves. *Monogr Paediatr* **10:** 50–53.

Davis CW. 2002. Regulation of mucin secretion from in vitro cellular models. *Novartis Found Symp* **248:** 113–131, 277–282.

Davis IC, Matalon S. 2007. Epithelial sodium channels in the adult lung—Important modulators of pulmonary health and disease. *Adv Exp Med Biol* **618:** 127–140.

Dentice R, Elkins M. 2011. Timing of dornase alfa inhalation for cystic fibrosis. *Cochrane Database Syst Rev* CD007923.

Dentice RL, Elkins MR, Bye PT. 2012. Adults with cystic fibrosis prefer hypertonic saline before or during airway clearance techniques: A randomised crossover trial. *J Physiother* **58:** 33–40.

Dietl P, Haller T, Mair N, Frick M. 2001. Mechanisms of surfactant exocytosis in alveolar type II cells in vitro and in vivo. *News Physiol Sci* **16:** 239–243.

Dwyer TJ, Alison JA, McKeough ZJ, Daviskas E, Bye PT. 2011. Effects of exercise on respiratory flow and sputum properties in patients with cystic fibrosis. *Chest* **139**: 870–877.

Egan ME. 2002. CFTR-associated ATP transport and release. *Methods Mol Med* **70**: 395–406.

Elkins M, Dentice R. 2012. Timing of hypertonic saline inhalation for cystic fibrosis. *Cochrane Database Syst Rev* **2**: CD008816.

Elkins MR, Alison JA, Bye PT. 2005. Effect of body position on maximal expiratory pressure and flow in adults with cystic fibrosis. *Pediatr Pulmonol* **40**: 385–391.

Fainardi V, Longo F, Faverzani S, Tripodi MC, Chetta A, Pisi G. 2011. Short-term effects of high-frequency chest compression and positive expiratory pressure in patients with cystic fibrosis. *J Clin Med Res* **3**: 279–284.

Falk M, Kelstrup M, Andersen JB, Kinoshita T, Falk P, Stovring S, Gothgen I. 1984. Improving the ketchup bottle method with positive expiratory pressure, PEP, in cystic fibrosis. *Eur J Respir Dis* **65**: 423–432.

Fink JB, Mahlmeister MJ. 2002. High-frequency oscillation of the airway and chest wall. *Respir Care* **47**: 797–807.

Fischer H, Illek B, Machen TE. 1998. Regulation of CFTR by protein phosphatase 2B and protein kinase C. *Pflugers Arch* **436**: 175–181.

Flume PA, Mogayzel PJ Jr, Robinson KA, Goss CH, Rosenblatt RL, Kuhn RJ, Marshall BC. 2009. Cystic fibrosis pulmonary guidelines: Treatment of pulmonary exacerbations. *Am J Respir Crit Care Med* **180**: 802–808.

Fredberg JJ. 1978. A modal perspective of lung response. *J Acoust Soc Am* **63**: 962–966.

Frownfelter D, Dean E. 2006. *Cardiovascular and pulmonary physical therapy: Evidence and practice.* Mosby & Elsevier Health Sciences, St. Louis, MO.

Frownfelter D, Massery M. 2006. Facilitating airway clearance with coughing techniques. In *Cardiovascular and pulmonary physical therapy: Evidence and practice* (ed. Frownfelter D, Dean E), pp. 363–376. Mosby & Elsevier Health Sciences, St. Louis, MO.

Gallon A. 1992. The use of percussion. *Physiotherapy* **78**: 85–89.

Giles DR, Wagener JS, Accurso FJ, Butler-Simon N. 1995. Short-term effects of postural drainage with clapping vs autogenic drainage on oxygen saturation and sputum recovery in patients with cystic fibrosis. *Chest* **108**: 952–954.

Greger R, Schreiber R, Mall M, Wissner A, Hopf A, Briel M, Bleich M, Warth R, Kunzelmann K. 2001. Cystic fibrosis and CFTR. *Pflugers Arch* **443**: S3–S7.

Groth S, Stafanger G, Dirksen H, Andersen JB, Falk M, Kelstrup M. 1985. Positive expiratory pressure (PEP-mask) physiotherapy improves ventilation and reduces volume of trapped gas in cystic fibrosis. *Bull Eur Physiopathol Respir* **21**: 339–343.

Grubb BR, Boucher RC. 1998. Effect of in vivo corticosteroids on Na^+ transport across airway epithelia. *Am J Physiol* **275**: C303–C308.

Grygorczyk R, Hanrahan JW. 1997. CFTR-independent ATP release from epithelial cells triggered by mechanical stimuli. *Am J Physiol* **272**: C1058–C1066.

Hanrahan JW, Mathews CJ, Grygorczyk R, Tabcharani JA, Grzelczak Z, Chang XB, Riordan JR. 1996. Regulation of the CFTR chloride channel from humans and sharks. *J Exp Zool* **275**: 283–291.

Hansen LG, Warwick WJ. 1990. High-frequency chest compression system to aid in clearance of mucus from the lung. *Biomed Instrum Technol* **24**: 289–294.

Hebestreit A, Kersting U, Basler B, Jeschke R, Hebestreit H. 2001. Exercise inhibits epithelial sodium channels in patients with cystic fibrosis. *Am J Respir Crit Care Med* **164**: 443–446.

Hebestreit H, Kieser S, Junge S, Ballmann M, Hebestreit A, Schindler C, Schenk T, Posselt HG, Kriemler S. 2010. Long-term effects of a partially supervised conditioning programme in cystic fibrosis. *Eur Respir J* **35**: 578–583.

Hodson ME. 2000. Treatment of cystic fibrosis in the adult. *Respiration* **67**: 595–607.

Homnick DN, Anderson K, Marks JH. 1998. Comparison of the flutter device to standard chest physiotherapy in hospitalized patients with cystic fibrosis: A pilot study. *Chest* **114**: 993–997.

Inanlou MR, Baguma-Nibasheka M, Kablar B. 2005. The role of fetal breathing-like movements in lung organogenesis. *Histol Histopathol* **20**: 1261–1266.

Johnson FL, Donohue JF, Shaffer CL. 2002. Improved sputum expectoration following a single dose of INS316 in patients with chronic bronchitis. *Chest* **122**: 2021–2029.

Jones DP, Ellam SV, Riddle H, Watson BW. 1975. The measurement of air flow in a forced expiration using a pressure-sensitive transistor. *Med Biol Eng* **13**: 71–77.

Kim CS, Iglesias AJ, Sackner MA. 1987. Mucus clearance by two-phase gas–liquid flow mechanism: Asymmetric periodic flow model. *J Appl Physiol* **62**: 959–971.

Kluft J, Beker L, Castagnino M, Gaiser J, Chaney H, Fink RJ. 1996. A comparison of bronchial drainage treatments in cystic fibrosis. *Pediatr Pulmonol* **22**: 271–274.

Knowles MR, Clarke LL, Boucher RC. 1991. Activation by extracellular nucleotides of chloride secretion in the airway epithelia of patients with cystic fibrosis. *N Engl J Med* **325**: 533–538.

Kreda SM, Gynn MC, Fenstermacher DA, Boucher RC, Gabriel SE. 2001. Expression and localization of epithelial aquaporins in the adult human lung. *Am J Respir Cell Mol Biol* **24**: 224–234.

Kreda SM, Okada SF, van Heusden CA, O'Neal W, Gabriel S, Abdullah L, Davis CW, Boucher RC, Lazarowski ER. 2007. Coordinated release of nucleotides and mucin from human airway epithelial Calu-3 cells. *J Physiol* **584**: 245–259.

Kunzelmann K, Schreiber R, Boucherot A. 2001. Mechanisms of the inhibition of epithelial Na^+ channels by CFTR and purinergic stimulation. *Kidney Int* **60**: 455–461.

Kunzelmann K, Bachhuber T, Regeer R, Markovich D, Sun J, Schreiber R. 2005. Purinergic inhibition of the epithelial Na^+ transport via hydrolysis of PIP2. *FASEB J* **19**: 142–143.

Kuys SS, Hall K, Peasey M, Wood M, Cobb R, Bell SC. 2011. Gaming console exercise and cycle or treadmill exercise provide similar cardiovascular demand in adults with cystic fibrosis: A randomised cross-over trial. *J Physiother* **57**: 35–40.

Lannefors L. 2009. *Positive expiratory pressure*. International Physiotherapy Group/Cystic Fibrosis, http://www.cfww.org.

Lannefors L, Wollmer P. 1992. Mucus clearance with three chest physiotherapy regimes in cystic fibrosis: A comparison between postural drainage, PEP and physical exercise. *Eur Respir J* **5:** 748–753.

Lannefors L, Button BM, McIlwaine M. 2004. Physiotherapy in infants and young children with cystic fibrosis: Current practice and future developments. *J R Soc Med* **97:** 8–25.

Lapin CD. 2002. Airway physiology, autogenic drainage, and active cycle of breathing. *Respir Care* **47:** 778–785.

Laube BL, Links JM, LaFrance ND, Wagner HN Jr, Rosenstein BJ. 1989. Homogeneity of bronchopulmonary distribution of 99mTc aerosol in normal subjects and in cystic fibrosis patients. *Chest* **95:** 822–830.

Lazarowski ER, Boucher RC. 2001. UTP as an extracellular signaling molecule. *News Physiol Sci* **16:** 1–5.

Lazarowski ER, Boucher RC. 2009. Purinergic receptors in airway epithelia. *Curr Opin Pharmacol* **9:** 262–267.

Lazarowski ER, Mason SJ, Clarke L, Harden TK, Boucher RC. 1992. Adenosine receptors on human airway epithelia and their relationship to chloride secretion. *Br J Pharmacol* **106:** 774–782.

Leith DE. 1985. The development of cough. *Am Rev Respir Dis* **131:** S39–S42.

Levitzky MG. 1991. *Pulmonary physiology*. McGraw-Hill, New York.

McCarren B, Alison JA. 2006. Physiological effects of vibration in subjects with cystic fibrosis. *Eur Respir J* **27:** 1204–1209.

McIlwaine PM, Wong LT, Peacock D, Davidson AG. 1997. Long-term comparative trial of conventional postural drainage and percussion versus positive expiratory pressure physiotherapy in the treatment of cystic fibrosis. *J Pediatr* **131:** 570–574.

McIlwaine M, Wong LT, Chilvers M, Davidson GF. 2010. Long-term comparative trial of two different physiotherapy techniques; postural drainage with percussion and autogenic drainage, in the treatment of cystic fibrosis. *Pediatr Pulmonol* **45:** 1064–1069.

McIlwaine MP, Alarie N, Davidson GF, Lands LC, Ratjen F, Milner R, Owen B, Agnew JL. 2013. Long-term multicentre randomised controlled study of high frequency chest wall oscillation versus positive expiratory pressure mask in cystic fibrosis. *Thorax* doi: 10.1136/thoraxjnl-2012-202915.

Mead J, Turner JM, Macklem PT, Little JB. 1967. Significance of the relationship between lung recoil and maximum expiratory flow. *J Appl Physiol* **22:** 95–108.

Mead J, Takishima T, Leith D. 1970. Stress distribution in lungs: A model of pulmonary elasticity. *J Appl Physiol* **28:** 596–608.

Miller S, Hall DO, Clayton CB, Nelson R. 1995. Chest physiotherapy in cystic fibrosis: A comparative study of autogenic drainage and the active cycle of breathing techniques with postural drainage. *Thorax* **50:** 165–169.

Morrison L, Agnew J. 2009. Oscillating devices for airway clearance in people with cystic fibrosis. *Cochrane Database Syst Rev* CD006842.

Morse DM, Smullen JL, Davis CW. 2001. Differential effects of UTP, ATP, and adenosine on ciliary activity of human nasal epithelial cells. *Am J Physiol Cell Physiol* **280:** C1485–C1497.

Murray JF. 1986. Ventilation. In *The normal lung (The basis for diagnosis and treatment of pulmonary disease)* (ed. Murray JF), pp. 114–117. W.B. Saunders, Philadelphia.

Myers LB, Horn SA. 2006. Adherence to chest physiotherapy in adults with cystic fibrosis. *J Health Psychol* **11:** 915–926.

Natale JE, Pfeifle J, Homnick DN. 1994. Comparison of intrapulmonary percussive ventilation and chest physiotherapy: A pilot study in patients with cystic fibrosis. *Chest* **105:** 1789–1793.

Newhouse PA, White F, Marks JH, Homnick DN. 1998. The intrapulmonary percussive ventilator and flutter device compared to standard chest physiotherapy in patients with cystic fibrosis. *Clin Pediatr* **37:** 427–432.

Oberwaldner B. 2009. Hi-PEP. International Physiotherapy Group/Cystic Fibrosis, www.cfww.org.

Oberwaldner B, Evans JC, Zach MS. 1986. Forced expirations against a variable resistance: A new chest physiotherapy method in cystic fibrosis. *Pediatr Pulmonol* **2:** 358–367.

Oberwaldner B, Theissl B, Rucker A, Zach MS. 1991. Chest physiotherapy in hospitalized patients with cystic fibrosis: A study of lung function effects and sputum production. *Eur Respir J* **4:** 152–158.

O'Clock GD, Lee YW, Lee J, Warwick WJ. 2010. A simulation tool to study high-frequency chest compression energy transfer mechanisms and waveforms for pulmonary disease applications. *IEEE Trans Biomed Eng* **57:** 1539–1546.

Oermann CM, Sockrider MM, Giles D, Sontag MK, Accurso FJ, Castile RG. 2001. Comparison of high-frequency chest wall oscillation and oscillating positive expiratory pressure in the home management of cystic fibrosis: A pilot study. *Pediatr Pulmonol* **32:** 372–377.

Okada SF, Nicholas RA, Kreda SM, Lazarowski ER, Boucher RC. 2006. Physiological regulation of ATP release at the apical surface of human airway epithelia. *J Biol Chem* **281:** 22992–23002.

Orenstein DM. 2003. Heads up! Clear those airways! *Pediatr Pulmonol* **35:** 160–161.

Pfleger A, Theissl B, Oberwaldner B, Zach MS. 1992. Self-administered chest physiotherapy in cystic fibrosis: A comparative study of high-pressure PEP and autogenic drainage. *Lung* **170:** 323–330.

Phillips GE, Pike SE, Jaffe A, Bush A. 2004. Comparison of active cycle of breathing and high-frequency oscillation jacket in children with cystic fibrosis. *Pediatr Pulmonol* **37:** 71–75.

Picher M, Burch LH, Boucher RC. 2004. Metabolism of P2 receptor agonists in human airways: Implications for mucociliary clearance and cystic fibrosis. *J Biol Chem* **279:** 20234–20241.

Pryor JA. 2009. *Active cycle of breathing techniques*. International Physiotherapy Group/Cystic Fibrosis, www.cfww.org.

Pryor JA, Webber BA, Hodson ME. 1990. Effect of chest physiotherapy on oxygen saturation in patients with cystic fibrosis. *Thorax* **45:** 77.

Pryor JA, Tannenbaum E, Scott SF, Burgess J, Cramer D, Gyi K, Hodson ME. 2010. Beyond postural drainage and percussion: Airway clearance in people with cystic fibrosis. *J Cyst Fibros* **9:** 187–192.

Purohit DM, Caldwell C, Levkoff AH. 1975. Letter: Multiple rib fractures due to physiotherapy in a neonate with hyaline membrane disease. *Am J Dis Child* **129:** 1103–1104.

Ransford GA, Fregien N, Qiu F, Dahl G, Conner GE, Salathe M. 2009. Pannexin 1 contributes to ATP release in airway epithelia. *Am J Respir Cell Mol Biol* **41:** 525–534.

Ressler B, Lee RT, Randell SH, Drazen JM, Kamm RD. 2000. Molecular responses of rat tracheal epithelial cells to transmembrane pressure. *Am J Physiol Lung Cell Mol Physiol* **278:** L1264–L1272.

Robinson KA, McKoy N, Saldanha I, Odelola OA. 2010. Active cycle of breathing technique for cystic fibrosis. *Cochrane Database Syst Rev* CD007862.

Rock JR, O'Neal WK, Gabriel SE, Randell SH, Harfe BD, Boucher RC, Grubb BR. 2009. Transmembrane protein 16A (TMEM16A) is a Ca^{2+}-regulated Cl^- secretory channel in mouse airways. *J Biol Chem* **284:** 14875–14880.

Rossman CM, Waldes R, Sampson D, Newhouse MT. 1982. Effect of chest physiotherapy on the removal of mucus in patients with cystic fibrosis. *Am Rev Respir Dis* **126:** 131–135.

Salh W, Bilton D, Dodd M, Webb AK. 1989. Effect of exercise and physiotherapy in aiding sputum expectoration in adults with cystic fibrosis. *Thorax* **44:** 1006–1008.

Scherer TA, Barandun J, Martinez E, Wanner A, Rubin EM. 1998. Effect of high-frequency oral airway and chest wall oscillation and conventional chest physical therapy on expectoration in patients with stable cystic fibrosis. *Chest* **113:** 1019–1027.

Schmid A, Clunes LA, Salathe M, Verdugo P, Dietl P, Davis CW, Tarran R. 2011. Nucleotide-mediated airway clearance. *Subcell Biochem* **55:** 95–138.

Schoni MH. 1989. Autogenic drainage: A modern approach to physiotherapy in cystic fibrosis. *J R Soc Med* **82:** 32–37.

Sontag MK, Quittner AL, Modi AC, Koenig JM, Giles D, Oermann CM, Konstan MW, Castile R, Accurso FJ. 2010. Lessons learned from a randomized trial of airway secretion clearance techniques in cystic fibrosis. *Pediatr Pulmonol* **45:** 291–300.

Stutts MJ, Lazarowski ER, Paradiso AM, Boucher RC. 1995. Activation of CFTR Cl^- conductance in polarized T84 cells by luminal extracellular ATP. *Am J Physiol* **268:** C425–C433.

Sutton PP, Lopez-Vidriero MT, Pavia D, Newman SP, Clarke SW. 1983. Effect of chest physiotherapy on the removal of mucus in patients with cystic fibrosis. *Am Rev Respir Dis* **127:** 390–391.

Tarran R, Button B, Picher M, Paradiso AM, Ribeiro CM, Lazarowski ER, Zhang L, Collins PL, Pickles RJ, Fredberg JJ, et al. 2005. Normal and cystic fibrosis airway surface liquid homeostasis. The effects of phasic shear stress and viral infections. *J Biol Chem* **280:** 35751–35759.

Tonnesen P, Stovring S. 1984. Positive expiratory pressure (PEP) as lung physiotherapy in cystic fibrosis: A pilot study. *Eur J Respir Dis* **65:** 419–422.

Toussaint M, de Win H, Steens M, Soudon P. 2003. Effects of intrapulmonary percussive ventilation on mucus clearance in Duchennes muscular dystrophy patients: A preliminary report. *Resp Care* **48:** 940–947.

Van Ginderdeuren F. 2009. *Autogenic drainage*. International Physiotherapy Group/Cystic Fibrosis, www.cfww.org.

Volsko TA, DiFiore J, Chatburn RL. 2003. Performance comparison of two oscillating positive expiratory pressure devices: Acapella versus Flutter. *Respir Care* **48:** 124–130.

West JB. 2004. *Respiratory physiology—The essentials*. Lippincott Williams & Wilkins, Baltimore.

Williams MT. 1994. Chest physiotherapy and cystic fibrosis. Why is the most effective form of treatment still unclear? *Chest* **106:** 1872–1882.

Yankaskas JR, Marshall BC, Sufian B, Simon RH, Rodman D. 2004. Cystic fibrosis adult care: Consensus conference report. *Chest* **125:** 1S–39S.

Yuan N, Kane P, Shelton K, Matel J, Becker BC, Moss RB. 2010. Safety, tolerability, and efficacy of high-frequency chest wall oscillation in pediatric patients with cerebral palsy and neuromuscular diseases: An exploratory randomized controlled trial. *J Child Neurol* **25:** 815–821.

Zach MS, Oberwaldner B. 1992. Effect of positive expiratory pressure breathing in patients with cystic fibrosis. *Thorax* **47:** 66–67.

Zach M, Oberwaldner B, Hausler F. 1982. Cystic fibrosis: Physical exercise versus chest physiotherapy. *Arch Dis Child* **57:** 587–589.

Cite this article as *Cold Spring Harb Perspect Med* doi: 10.1101/cshperspect.a009720

The Cystic Fibrosis Airway Microbiome

Susan V. Lynch[1] and Kenneth D. Bruce[2]

[1]Colitis and Crohn's Disease Microbiome Research Center, Division of Gastroenterology, Department of Medicine, University of California, San Francisco, San Francisco, California 94143

[2]Institute of Pharmaceutical Science, King's College, London, London SE1 9NH, United Kingdom

Repeated pulmonary exacerbation and progressive lung function decline characterize cystic fibrosis (CF) disease and represent the leading causes of mortality in this patient population. Recent studies have shown, using culture-independent assays, that multiple microbial species can be detected in airway samples from CF patients. Moreover, specific groups of bacteria within these bacterial communities or microbiota are highly associated with disease-associated factors such as antibiotic administration. This raises the possibility that, as in other human niches, pathogenic processes in the CF airways represent polymicrobial activities and that microbiome composition and perturbations to these communities define patient pulmonary health status. Airway samples are typically collected through the mouth, and are thus susceptible to contamination by upper airway secretions; hence, caution must be exercised in interpreting these data. Nonetheless, given the continuum of the upper and lower respiratory tract, understanding the contribution of these mixed-species assemblages to airway health is essential to improving CF patient care. This article aims to discuss recent advances in the field of CF airway microbiome research and interpret these findings in the context of CF pulmonary disease.

Cystic fibrosis (CF), an autosomal genetic disorder most prevalent in Caucasian populations, is attributable to mutations within the gene coding for cystic fibrosis transmembrane regulator (CFTR) protein. To date, more than 1600 mutations have been identified (http://www.genet.sickkids.on.ca/cftr/), but the functional implications of these mutations and their association with severity of disease symptoms have only been defined for a relatively small number of common CFTR genotypes. Functional consequences of CFTR mutation can be broadly divided into four classes: 1, absence of protein synthesis; II, inadequate processing; III, defective regulation; and IV, defective produc-

tion (Collins 1992). For example, the common ΔF508 mutation, which results in deletion of the phenylalanine amino acid residue at position 508, leads to protein misfolding and proteosomal degradation leading to apical epithelia lacking this crucial transporter. The majority of CF patients carry the ΔF508 mutation in at least one copy of the CFTR genes and those homozygous for this mutation show the most severe disease. CFTR plays an established role in chloride transport and more recently has been implicated in transport of thiocyanate (SCN^-), bicarbonate ($HCO3^-$), as well as proteins (Quinton 2001; Riordan 2008) across epithelial surfaces. Given the sheer prevalence of epithelial linings in the human host, it is no surprise that CF manifests as a systemic disease, characterized by pulmonary, gastrointestinal, and urogenital symptoms, including recurrent airway infection, aberrant nutrient absorption, and reproductive issues.

Disruption of chloride anion transport, one of the key underlying features of CF, leads to altered physiological conditions at epithelial surfaces. From a microbial viewpoint, however, the environment generated by CFTR mutation results in ideal conditions for colonization. For example, in the airways, a site in which microbial colonization is associated with recurrent infection, CFTR mutation results in reduced chloride secretion and increased sodium and water absorption at the epithelial surface, leading to a depleted airway surface liquid layer (ASL). ASL deficiency in turn leads to ciliary dyskinesis and impaired mucocilary clearance, a key component of the host innate immune response to microbes. As a result, mucosal surfaces of the airways show dehydrated mucus and are heavily colonized by microbes that contribute to progressive lung function decline over the lifetime of the patient.

Although it has long been acknowledged that the CF airway represents a permissive environment for microbial colonization, clinical laboratory testing for the presence of specific pathogenic species using culture-based approaches have led to a reductionist view of the microbiology in this niche. As a result, the dogma had been that the diversity of bacterial or fungal species present in CF airways was relatively restricted to a handful of "usual suspects" commonly detected by conventional laboratory culture. However, the last 10 years have witnessed a revolution in our appreciation of the complexity of microbes that exist within and on the human body. The advent of high-resolution molecular approaches, primarily developed in the field of environmental microbial ecology, has dramatically enhanced our ability to interrogate the true diversity of microbes present in a given niche. Application of these tools to human samples has led to the emergence of a new field of "human microbiome" research, focused on defining the types and abundance of microbial species at specific niches in the human body. The relatively recent establishment of human microbiome research initiatives in the United States, EU, and Asia, underscore the magnitude of global efforts aimed at defining the true microbial complexity resident in the human host and its role in maintenance of health or contribution to disease.

MICROBIOME PROFILING

As a backdrop for discussion of the CF microbiome, it is useful to briefly describe fundamentals of the culture-independent tools used to interrogate human microbiota. Molecular approaches to phylogenetically profile mixed-species members of a given microbial community are typically DNA based and commence with efficient extraction of nucleic acid from all members of the community in any given sample. Following successful extraction, profiling is primarily based on polymerase chain reaction (PCR) amplification of specific biomarker genes known to exist in all members of a specific taxonomic level (e.g., the 16S rRNA gene), ubiquitous to all bacteria or the fungal-specific internal transcribed spacer (ITS) region. Universal primers, specifically designed on conserved regions within a biomarker gene, are used in a PCR reaction to amplify a pool of, for example, 16S ribosomal RNA genes from the various genomes of community members. Although PCR is known to introduce certain biases in the profile (e.g., preferential amplification of specific targets), many researchers have implemented measures to minimize these issues. These include the use of degenerate primers to maximize the diversity-amplified, amplification across a gradient of annealing temperatures for each sample and pooling the amplified products before profiling, as well as minimizing the number of PCR cycles used to amplify the biomarker gene.

A large number of approaches boasting increasing resolution have emerged over the past 20 years to profile the mixed-species biomarker gene amplicons generated by culture-independent approaches. Among the more basic profiling approaches is separation of differential biomarker gene sequences by gel electrophoresis methods (e.g., terminal restriction fragment

Cite this article as *Cold Spring Harb Perspect Med* doi: 10.1101/cshperspect.a009738

length polymorphism [T-RFLP] or denaturing gradient gel electrophoresis [DGGE]), which provide a fingerprint of the most highly abundant members of the community and can be used to show shifts in the dominant community members across time or with treatment. More recently developed approaches include next-generation sequencing platforms, which generate enormous volumes of biomarker gene sequence data, permitting much deeper community coverage and information on the relative abundance of specific groups of microbes in a given community. Building on the recent expansion of microbial biomarker sequence data, groups have also developed phylogenetic microarrays, which, based on hybridization to oligonucleotides probes designed against discriminatory loci on specific biomarker genes, can profile thousands of organisms in a single parallel assay. These tools have more recently been applied to respiratory and other samples collected from CF patients and have, as for many other diseases, shed light on the complexity of the microbial communities present in the airways of patients with chronic inflammatory disease.

THE CF PULMONARY MICROBIOME

Although still in its infancy, some of the most seminal studies of the last decade pertaining to human health have emerged from the field of human microbiome research. For example, pioneering studies by Gordon and colleagues showed a wealth of bacterial diversity present in the gastrointestinal tract of humans (Ley et al. 2005; Turnbaugh et al. 2006). Moreover the investigators revealed that the composition of the bacterial consortia present in lean or obese adults was dramatically different and that characteristic shifts in community composition distinguished healthy and obese states (Ley et al. 2005). Other studies of the gastrointestinal microbiota have identified novel species, whose 16S rRNA gene has not before been sequenced (Eckburg et al. 2005), suggesting that our knowledge of the true diversity of bacteria resident in the human ecosystem is somewhat limited.

These important studies captured the imagination of researchers, garnered international interest, and certainly paved the way for investigations of other human host niches. However, earlier studies using gel-based separation approaches had already made their way into the public arena several years before the term "human microbiome" had even been coined. In their pioneering culture-independent study of CF patient samples, Rogers and colleagues showed for the first time using adult CF patient bronchoscopic and sputum samples, that multiple distinct 16S rRNA gene PCR products were present in each patient sample, indicating the presence of distinct bacterial phylotypes in this niche. Moreover, subsequent sequencing efforts identified, in addition to *Pseudomonas aeruginosa* and *Stenotrophomonas maltophilia*, multiple other species typically associated with the oral or gastrointestinal cavity including *Prevotella oris*, *Fusobacterium gonidiaformans*, and *Bacteroides fragilis* among others, which had not previously been associated with CF airways (Rogers et al. 2003). But the obvious question was whether the diversity of species identified with these culture-independent techniques was viable. The investigators quickly followed up their findings with a subsequent study of bacterial diversity in the airways based on RNA rather than DNA, to profile active members of the CF airway microbiota in adult patients. Total RNA extracted from 71 sputum samples was reverse transcribed and used as template for 16S rRNA gene PCR amplification. Amplicons generated by this approach represent community members who are actively transcribing their 16S rRNA gene and are hence viable members of the community. T-RFLP profiling of the amplicons identified 248 distinct bands, each originating from a viable bacterial member of the CF airway. Across patient samples, bacterial community richness (number of distinct phylotypes detected) was as high as 37 viable phylotypes in one patient, substantially greater than the typical one or two pulmonary pathogens reported through culture-based clinical laboratory testing to colonize CF airways. Follow-up cloning of T-RFLP bands from three patients in the study and sequencing 53 of these clones identified the presence of additional species not previously identified in CF airways including two *Abiotrophia* species, *Mycoplasma salivarium*, *Ralstonia taiwanensis*, *Rothia mucilaginosa*, *Treponema vincentii*, and *Veillonella atypical* (Rogers et al. 2004). These seminal studies showed that not only were bacterial species not typically associated with the CF airways present in these patients, but that they were evidently viable and thus possessed the potential to contribute to airway disease in this population, having significant implications for cystic fibrosis and other chronic pulmonary diseases.

Data generated by more recent culture-independent studies of CF patient samples, using progressively higher-resolution tools, have clearly showed the presence of even more complex and diverse microbiota in the airways of these patients (Harris et al. 2007; Cox et al. 2010; Klepac-Ceraj et al. 2010). These studies have further expanded the diversity of organisms detected in the airways of CF patients. However, moving beyond sheer description of microbiota and determining relationships between the composition of these assemblages and host health status is necessary to identify key features of the microbiome that contribute to patient health and its decline. In a cross-sectional study of respiratory samples collected from 45 clinically stable CF patients, aged between 9 months and 72 years, who had not received antibiotics for acute pulmonary exacerbation within 2 months of sample collection, Cox and colleagues examined changes in airway microbiota composition within this cohort. They first showed, as expected, a significant negative correlation ($-0.48; p < 0.0003$) existed between patient age and pulmonary function, confirming that, compared to the younger population, older CF patients in the cohort showed poorer airway health (Cox et al. 2010). To determine whether this decline in pulmonary health shown by older patients was associated with changes in the airway microbiota, the authors examined whether relationships existed between patient age and gross metrics of bacterial community composition. A significant negative correlation existed between patient age and community richness (number of types of bacteria present), evenness (relative distribution of community members), and diversity (index cal-culated based on richness and evenness metrics). In addition the communities present in the airways of older CF patients were comprised of phylogenetically related organisms belonging to the family Pseudomonadaceae (Cox et al. 2010). This indicated that substantial shifts in the airway microbiota composition and structure paralleled the decline in pulmonary health shown by the cohort providing the first evidence that microbiota composition was associated with airway function status.

To provide a clearer picture of precisely which members of the community were associated with younger and older CF patients, the authors examined relationships between relative abundance of all community members detected and patient age. Although substantial interpersonal variation has been shown to exist across healthy humans at other sites such as the GI tract (Wu et al. 2011), disease and associated treatments, particularly in the case of chronic illness, appears to act as a strong selective pressure on microbiome composition. Indeed, using simple linear regression, the abundance of >100 taxa were identified as either negatively ($n = 68$) or positively ($n = 45$) correlated with CF patient age. *Haemophilus influenzae* was among the negatively correlated, whereas *Pseudomonas aeruginosa* and *Stenotrophomonas maltophilia* showed strong and significant positive correlations with patient age, further supporting previous observations that these species are associated with early- or late-stage pathogenic processes, respectively. However, multiple other community members showed equally strong correlations with age, and although correlation does not automatically indicate causality and functional studies are now necessary to confirm their respective roles, at the very least, these species may modify microbiota function and pathogenic processes associated with younger and older CF patients. Further functional analyses of these communities to determine the microbial and host factors that contribute to airway health status are necessary and will further support DNA-based findings of diverse communities in the airways of CF patients. However, collectively, as for other host niches, these studies show that as we interrogate the human microbiome in

 Cite this article as *Cold Spring Harb Perspect Med* doi: 10.1101/cshperspect.a009738

distinct niches with more sophisticated tools providing progressively higher-resolution profiles of the species present, the diversity of organisms present increases. This fact was well shown recently by the European Human Microbiome consortium, who showed that when a sequence depth of ∼4 Gb was used to determine the number of strains common to two gastrointestinal samples, only 1% of strains were deemed common (135 strains). However, when the depth of sequencing was more than doubled to ∼11 Gb for each of these samples, the number of strains found in common increased by 25% (169 strains [Arumugam et al. 2011]), illustrating that conclusions must be interpreted in the context of the depth of profiling performed on a given sample.

A caveat with all lower airway studies is the need to sample through the oral cavity and upper airway. Goddard and colleagues recently showed, using explanted lungs from CF patients undergoing transplantation, that the diversity of bacteria in the lower airways was substantially less than that described in some recent culture-independent studies and that these communities were characterized by a handful of the usual CF pathogenic suspects (Goddard et al. 2012). By definition, patients undergoing lung transplant have severe airway disease. Examination of microbiome data from comparable adult CF patients with severe disease also shows broad concordance with these findings. The microbiota of these patients show a marked reduction in diversity and possess communities highly enriched for a small number of CF-associated pathogens such as *P. aeruginosa* and *Burkholderia* species (Cox et al. 2010). Indeed, the greatest bacterial diversity has been observed in pediatric airway samples from patients with milder disease and good pulmonary status. Whether this is a function of our inability to obtain clean lower airway samples in this population is a possibility. However, it is also conceivable, particularly in light of recent studies demonstrating a clear association between diminishing airway microbiome diversity in parallel with increased exposure to antibiotic treatments (Zhao et al. 2012), that greater bacterial diversity exists in the airways of younger CF patients compared to those with end-stage disease.

Although the microbiome studies described above have concentrated at the level of species, a further level of strain-based diversity and dynamics exists within these communities. Several studies have shown considerable strain diversity within populations of a given species, not least in *P. aeruginosa* strains isolated from cystic fibrosis airways (Mowat et al. 2011). Mowat and colleagues showed that major changes in the *P. aeruginosa* haplotype were evident in temporal samples from individual CF patients and that these populations were highly dynamic (Mowat et al. 2011). Such strain diversity presumably confers the capacity, within a species population, to house strains capable of withstanding a diversity of insults from both the host and other microbial species, likely contributing to species resilience within these diverse multispecies communities. Little work to date has interrogated the relationship between microbiome composition at the species and subspecies levels. However, with the recent advances in sequence-based capabilities, the ability to interrogate these "worlds within worlds" will improve substantially. Although the obvious need to develop bioinformatic skill sets to cope with such complex datasets is also clear.

ROLE OF MICROBIOME IN MODULATING IMMUNE RESPONSE

Despite the complexity of its microbiome, the human host has evolved a sophisticated immune system to recognize and discriminate commensal from pathogenic microbial species. Evidence from several gastrointestinal microbiota studies (Michail et al. 2011; Walker et al. 2011) has shown that, as in the CF airways, loss of microbiota diversity is a hallmark of chronic inflammatory disease. This suggests that immune homeostasis and colonization resistance (the ability to withstand invasion by pathogenic species) is predicated on appropriate mucosal colonization and that perturbations to the microbiota, particularly those that lead to loss of community diversity and increased abundance of specific immunogenic species, may drive these persistent inflammatory responses.

It is well established that specific pathogens, including key species involved in CF airway pathogenesis such as *P. aeruginosa*, express virulence factors enabling them to evade host immune responses (Kharazmi 1991). However, more recently it has been showed that through induction of host inflammatory responses, some bacterial species gain a significant competitive advantage over other members of the microbiota (Winter et al. 2010; Thiennimitr et al. 2011). For example, host inflammation induced in response to the gastrointestinal pathogen, *Salmonella enterica* serotype *Typhimurium* has been shown to promote the abundance of this species (Winter et al. 2010; Thiennimitr et al. 2011). This phenomenon has more recently been shown to be due, at least in part, to generation (via the interaction of inflammation-derived reactive oxygen species and luminal thiosulfate) of a novel electron acceptor, tetrathionate. *S. typhimurium*, unlike other species, is equipped with the ability to use this novel electron receptor for respiration, providing it with a significant growth advantage over fermentative microbiota in the intestine (Winter et al. 2010), and a mechanistic basis for its increased abundance under inflammatory conditions. Secondary to this growth advantage, host inflammation may also provide such species a competitive advantage by reducing colonization resistance of the host. It is conceivable that reactive oxygen species produced during the inflammatory response from, for example, neutrophils, may impact the viability of strict anaerobic species that colonize mucosal surfaces, particularly in the gastrointestinal tract. Loss of this protective ancillary barrier, which also appears to play a key role in immune homeostasis, may also contribute to pathogen outgrowth in these communities. Although these observations regarding inflammation and microbiome perturbation have, to date, been made in the gastrointestinal tract, an interesting observation in the Cox study supports this hypothesis. Younger CF patients who showed less airway inflammation and better pulmonary function possessed a greater diversity of airway bacterial species, including anaerobic community members, compared with older CF patients.

Factors that perturb the human microbiota have obvious implications for host health. A characteristic of CF patient health management is antibiotic administration. Human microbiome studies have shown dramatic and persistent effects of antibiotic administration on the gastrointestinal microbiota of pediatric subjects. Palmer and colleagues cataloged the fecal microbiota of 11 infants over the first year of life and showed that antimicrobial administration resulted in an acute loss of bacterial burden following treatment (Palmer et al. 2007). However the microbiota rebounded, within a matter of weeks, a community had reassembled that grossly resembled the pretreatment microbiota. Nonetheless, the authors pointed out that specific species present in appreciable abundance pretreatment, were neither detected in samples collected immediately following antimicrobial administration nor over the remainder of the study (in some cases up to 1 year). A recent study of the airway microbiome by Zhao and colleagues performed on samples collected over the course of a decade from CF patients showed a similar phenomenon (Zhao et al. 2012). Despite antimicrobial administration that caused a large perturbation to the assemblages, communities reassembled over time (Zhao et al. 2012). This study also echoed the findings of Cox et al. (2010), demonstrating that patients with progressive disease showed concomitant decreasing diversity. Although counterintuitive in the CF airways, the presence of a diversity of species appears to be associated with better pulmonary health. Given these data, it is tempting to speculate that recurrent antimicrobial administration for pulmonary function management of treatment of acute pulmonary infections, serves to serially select over time a more uneven and less diverse airway microbiota, a feature highly correlated with more severe inflammatory disease in multiple patient populations (Cox et al. 2010; Walker et al. 2011; Zhao et al. 2012).

Overall, this article has identified some of the important advances in microbiome research in relation to the CF airways. Microbiome studies will continue to evolve in parallel with more sophisticated technologies. These advances will clearly benefit expansion of cystic

 Cite this article as *Cold Spring Harb Perspect Med* doi: 10.1101/cshperspect.a009738

fibrosis studies on technical, practical, and conceptual levels. Increasingly, we envisage studies that integrate airway microbiome composition and function with therapy, the host immune response, and clinical outcomes. Such a systems-based approach will help shape our understanding of the lung microbiome and how this impacts on the well-being of patients with CF.

REFERENCES

Arumugam M, Raes J, Pelletier E, Le Paslier D, Yamada T, Mende DR, Fernandes GR, Tap J, Bruls T, Batto JM, et al. 2011. Enterotypes of the human gut microbiome. *Nature* **473:** 174–180.

Collins FS. 1992. Cystic fibrosis: Molecular biology and therapeutic implications. *Science* **256:** 774–779.

Cox MJ, Allgaier M, Taylor B, Baek MS, Huang YJ, Daly RA, Karaoz U, Andersen GL, Brown R, Fujimura KE, et al. 2010. Airway microbiota and pathogen abundance in age-stratified cystic fibrosis patients. *PLoS ONE* **5:** e11044.

Eckburg PB, Bik EM, Bernstein CN, Purdom E, Dethlefsen L, Sargent M, Gill SR, Nelson KE, Relman DA. 2005. Diversity of the human intestinal microbial flora. *Science* **308:** 1635–1638.

Goddard AF, Staudinger BJ, Dowd SE, Joshi-Datar A, Wolcott RD, Aitken ML, Fligner CL, Singh PK. 2012. Direct sampling of cystic fibrosis lungs indicates that DNA-based analyses of upper-airway specimens can misrepresent lung microbiota. *Proc Natl Acad Sci* **109:** 13769–13774.

Harris JK, De Groote MA, Sagel SD, Zemanick ET, Kapsner R, Penvari C, Kaess H, Deterding RR, Accurso FJ, Pace NR. 2007. Molecular identification of bacteria in bronchoalveolar lavage fluid from children with cystic fibrosis. *Proc Natl Acad Sci* **104:** 20529–20533.

Kharazmi A. 1991. Mechanisms involved in the evasion of the host defence by *Pseudomonas aeruginosa*. *Immunol Lett* **30:** 201–205.

Klepac-Ceraj V, Lemon KP, Martin TR, Allgaier M, Kembel SW, Knapp AA, Lory S, Brodie EL, Lynch SV, Bohannan BJ, et al. 2010. Relationship between cystic fibrosis respiratory tract bacterial communities and age, genotype, antibiotics and *Pseudomonas aeruginosa*. *Environ Microbiol* **12:** 1293–1303.

Ley RE, Backhed F, Turnbaugh P, Lozupone CA, Knight RD, Gordon JI. 2005. Obesity alters gut microbial ecology. *Proc Natl Acad Sci* **102:** 11070–11075.

Michail S, Durbin M, Turner D, Griffiths AM, Mack DR, Hyams J, Leleiko N, Kenche H, Stolfi A, Wine E. 2011. Alterations in the gut microbiome of children with severe ulcerative colitis. *Inflamm Bowel Dis* doi: 10.1002/ibd. 22860.

Mowat E, Paterson S, Fothergill JL, Wright EA, Ledson MJ, Walshaw MJ, Brockhurst MA, Winstanley C. 2011. *Pseudomonas aeruginosa* population diversity and turnover in cystic fibrosis chronic infections. *Am J Respir Crit Care Med* **183:** 1674–1679.

Palmer C, Bik EM, DiGiulio DB, Relman DA, Brown PO. 2007. Development of the human infant intestinal microbiota. *PLoS Biol* **5:** e177.

Quinton PM. 2001. The neglected ion: HCO3. *Nat Med* **7:** 292–293.

Riordan JR. 2008. CFTR function and prospects for therapy. *Annu Rev Biochem* **77:** 701–726.

Rogers GB, Hart CA, Mason JR, Hughes M, Walshaw MJ, Bruce KD. 2003. Bacterial diversity in cases of lung infection in cystic fibrosis patients: 16S ribosomal DNA (rDNA) length heterogeneity PCR and 16S rDNA terminal restriction fragment length polymorphism profiling. *J Clin Microbiol* **41:** 3548–3558.

Rogers GB, Carroll MP, Serisier DJ, Hockey PM, Jones G, Bruce KD. 2004. Characterization of bacterial community diversity in cystic fibrosis lung infections by use of 16s ribosomal DNA terminal restriction fragment length polymorphism profiling. *J Clin Microbiol* **42:** 5176–5183.

Thiennimitr P, Winter SE, Winter MG, Xavier MN, Tolstikov V, Huseby DL, Sterzenbach T, Tsolis RM, Roth JR, Baumler AJ. 2011. Intestinal inflammation allows *Salmonella* to use ethanolamine to compete with the microbiota. *Proc Natl Acad Sci* **108:** 17480–17485.

Turnbaugh PJ, Ley RE, Mahowald MA, Magrini V, Mardis ER, Gordon JI. 2006. An obesity-associated gut microbiome with increased capacity for energy harvest. *Nature* **444:** 1027–1031.

Walker AW, Sanderson JD, Churcher C, Parkes GC, Hudspith BN, Rayment N, Brostoff J, Parkhill J, Dougan G, Petrovska L. 2011. High-throughput clone library analysis of the mucosa-associated microbiota reveals dysbiosis and differences between inflamed and non-inflamed regions of the intestine in inflammatory bowel disease. *BMC Microbiol* **11:** 7.

Winter SE, Thiennimitr P, Winter MG, Butler BP, Huseby DL, Crawford RW, Russell JM, Bevins CL, Adams LG, Tsolis RM, et al. 2010. Gut inflammation provides a respiratory electron acceptor for *Salmonella*. *Nature* **467:** 426–429.

Wu GD, Chen J, Hoffmann C, Bittinger K, Chen YY, Keilbaugh SA, Bewtra M, Knights D, Walters WA, Knight R, et al. 2011. Linking long-term dietary patterns with gut microbial enterotypes. *Science* **334:** 105–108.

Zhao J, Schloss PD, Kalikin LM, Carmody LA, Foster BK, Petrosino JF, Cavalcoli JD, VanDevanter DR, Murray S, Li JZ, et al. 2012. Decade-long bacterial community dynamics in cystic fibrosis airways. *Proc Natl Acad Sci* **109:** 5809–5814.

The Cystic Fibrosis of Exocrine Pancreas

Michael Wilschanski[1] and Ivana Novak[2]

[1]Pediatric Gastroenterology, Hadassah University Hospital, Jerusalem 91240, Israel

[2]Molecular Integrative Physiology, Department of Biology, University of Copenhagen, DK 2100 Copenhagen Ø, Denmark

Correspondence: inovak@bio.ku.dk

The cystic fibrosis transmembrane conductance regulator (CFTR) protein is highly expressed in the pancreatic duct epithelia and permits anions and water to enter the ductal lumen. This results in an increased volume of alkaline fluid allowing the highly concentrated proteins secreted by the acinar cells to remain in a soluble state. This work will expound on the pathophysiology and pathology caused by the malfunctioning CFTR protein with special reference to ion transport and acid–base abnormalities both in humans and animal models. We will also discuss the relationship between cystic fibrosis (CF) and pancreatitis and outline present and potential therapeutic approaches in CF treatment relevant to the pancreas.

The pancreas is one of the organs earliest and most seriously affected by cystic fibrosis (CF). Although cystic fibrosis transmembrane conductance regulator (CFTR) is expressed only in a very small percentage of exocrine cells, its malfunction has catastrophic effects on the whole organ, resulting in its eventual destruction, leading to maldigestion and malnutrition. In recent years, new therapeutic approaches are being developed to improve anion/fluid balance, especially in the airways. Whether these will have any value for the pancreas requires a more detailed understanding of pancreatic function drawn from clinical and genetic studies and cell/organ studies of ion channels and transporters specific for pancreatic cells. In the present work, we try to raise some of the critical issues of the physiology and pathophysiology of the pancreas in CF.

EXOCRINE PANCREATIC ABNORMALITIES

Exocrine Pancreatic Function

The CFTR protein is highly expressed in pancreatic ductal epithelia and permits anions and fluid to enter the ductal lumen. There is evidence that CFTR is associated with bicarbonate transport directly or indirectly (see below). Indeed according to the Quinton hypothesis, it is the defect in bicarbonate transport that is the primary defect in CF leading to mucoviscidosis (Quinton 2008). The net result of ductal function is an increased volume of alkaline fluid, allowing the highly concentrated proteins secreted by the acinar cells to remain in a soluble state. Absent or reduced CFTR channel function impairs chloride and bicarbonate transport of the ducts, which results in reduced volume and hyperconcentration of macromolecules (Kopelman et al.

1985, 1988). The consequences of mutations in the *CFTR* gene have been shown by pancreatic function studies that indicate that CF patients have a low flow of secretions with a high protein concentration, which presumably will precipitate in the duct lumina causing obstruction and damage (Fig. 1).

These changes in the CF pancreas begin in utero and after delivery the process of small duct obstruction leading to large duct obstruction continues. At birth, and for several months afterward, there is a release into the blood stream of proteins originating in the pancreas. An example of this is immune reactive trypsinogen (IRT) that forms the basis for the neonatal screening test for CF. Interestingly, with this wholesale destruction of the exocrine pancreas occurring, the infant is asymptomatic. The rea-

son for this silent destruction is yet to be determined. Eventually, this process results in severe inflammation, obstruction of ducts by mucus and calcium containing debris, the destruction of acini, and generalized fibrosis. Contrary to popular belief that the pancreas is entirely nonfunctioning at birth, the high IRT does show that some exocrine pancreatic tissue is still present and this may have a bearing on possible small molecule therapies targeted at the remainder of the pancreas that may rescue enough tissue to preserve viability of the remaining pancreas.

One of the most remarkable observations is that genetic factors exquisitely influence the degree of pancreatic disease and its rate of progression. Large studies of CF patients resulted in their classification as pancreatic insufficient

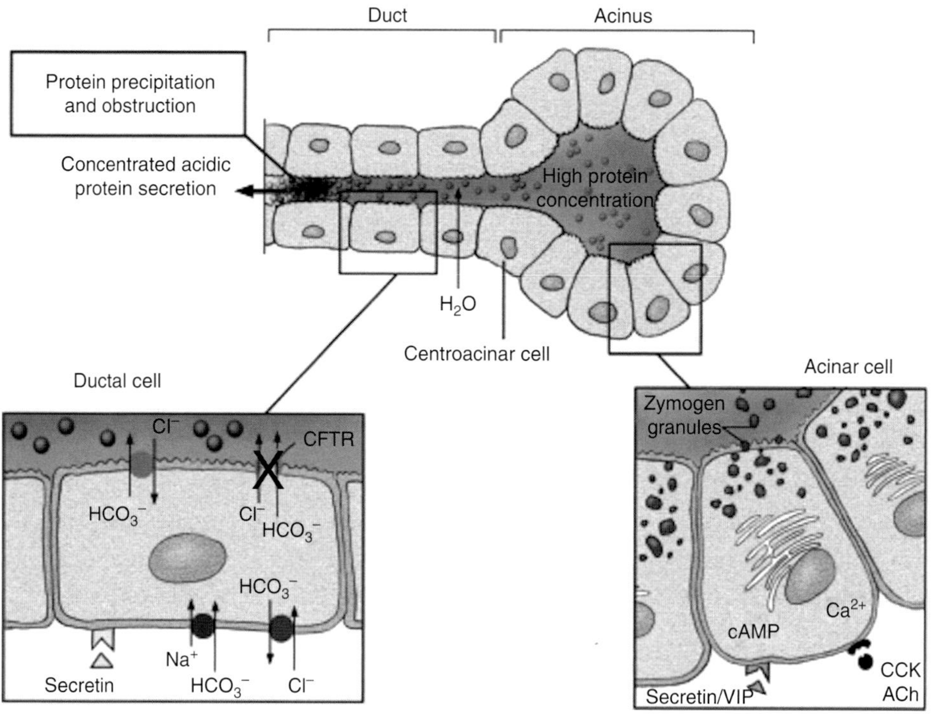

Figure 1. Pathogenesis of pancreatic disease in CF. Acinar cells secrete large quantities of protein, primarily in the form of digestive enzymes, into the acinar lumen. Under normal circumstances anions (Cl^- and HCO_3^-) are secreted into the ductal lumen (see detailed model in Fig. 3). This provides a driving force for the movement of fluid into the lumen of the duct and maintains the solubility of secreted proteins in a dilute, alkaline solution. In CF, impaired anion transport into the proximal ducts results in decreased secretion of more acidic fluid, which leads to precipitation of secreted proteins. Intraluminal obstruction of the ducts then causes progressive pancreatic damage and atrophy. (From Wilschanski and Durie 2007; reprinted, with permission.)

(PI) or pancreatic sufficient (PS). PI patients comprise 85% of all CF patients and have maldigestion as defined by evidence of steatorrhea following 72-hr fat balance studies. These PI patients require pancreatic enzyme replacement therapy with meals. In contrast, PS patients have evidence of pancreatic damage (these patients may be diagnosed by the high neonatal IRT test, which means that damage is occurring), but retain sufficient endogenous exocrine pancreatic function to sustain normal digestion.

Exocrine pancreatic status is directly linked to genotype (Kerem et al. 1990; Kristidis et al. 1992). Analysis of particular *CFTR* mutations in patients with these pancreatic phenotypes (PI vs. PS) revealed two categories of alleles: "severe" and "mild." Patients homozygous or compound heterozygous for severe alleles belonging to classes I, II, III, or VI confer pancreatic insufficiency, whereas a mild class IV or V allele sustains pancreatic function in a dominant fashion, even if the second mutation is severe. This observation appears plausible because all known mild alleles belong to class IV or V, all of which are (or predicted to be) associated with some residual chloride channel activity at the epithelial apical membranes. However, this classification system is not entirely consistent as there are some class I mutations with the stop codon at the end of the gene that are in fact PS. A small proportion (2%–3%) of patients carrying severe mutations on both alleles are PS at diagnosis, but most experience gradual transition from PS to PI. A few missense mutations (e.g., G85E) confer a variable pancreatic phenotype.

Although mild mutations confer sufficient CFTR function to prevent the pancreas from being completely destroyed, many PS patients have reduced exocrine pancreatic capacity and are associated with an increased risk of pancreatitis. Recurrent acute and chronic pancreatitis is a relatively infrequent complication of CF, first reported by Shwachman et al. (1975). In this retrospective study, only 0.5% of CF patients had pancreatitis. More recently, Durno et al. (2002) reported in a cohort of more than 1000 patients, followed over a period of 30 years, that the incidence was 1.7%. All the patients with pancreatitis were PS. In fact, this subgroup of PS patients appears to be highly susceptible to pancreatitis because almost one in five was affected by this complication. There have been suggestions in the literature that PI patients may also have pancreatitis, but most probably in these patients pancreatic function was not fully investigated (De Boeck et al. 2005). In the largest study to date of CF PS patients, Ooi et al. (2011), in a seminal paper, determined the association between severity of *CFTR* genotype and the risk of pancreatitis. They examined a large cohort of 277 PS patients from two CF centers of which 62 had well-documented pancreatitis. Using a novel pancreatic insufficiency prevalence score, the mutations were divided into three main groups: severe, moderate-severe, and mild. They found that the proportion of patients who developed pancreatitis was significantly greater for genotypes in the mild group than the moderate-severe group. Thus, the more mild mutations are associated with increased risk of pancreatitis.

Recurrent "Idiopathic" Pancreatitis

Several studies have shown that patients with idiopathic acute, recurrent, and chronic pancreatitis carry a significantly higher frequency of *CFTR* gene mutations than the general population (Cohn et al. 1998; Sharer et al. 1998). Bishop et al. (2005) prospectively examined 56 patients with idiopathic recurrent acute or chronic pancreatitis by performing extensive genotyping and transepithelial measures of ion channel function and comparing the findings with healthy controls, obligate CF heterozygotes, and patients with a prior diagnosis of CF-disease (PS and PI phenotypes). Genetic analysis revealed that 24 (40%) patients carried at least one *CFTR* mutation or variant, while six (10%) carried alterations on both alleles.

The sweat chloride and nasal potential difference (NPD) results in the patients with pancreatitis ranged from the values for healthy controls and obligate heterozygotes to the values for CF patients with PS and PI. Median sweat chloride and NPD results in patients with none or one mutation were clustered with values obtained in controls and obligate heterozygotes. In contrast, in patients with pancreatitis carrying *CFTR* mutations on both alleles, median ion transport values were intermediate between those of the controls and obligate heterozygotes and those of CF PS patients. Some individual values overlapped with the CF patients, and the diagnosis of CF could be confirmed in 21% of patients by abnormal ion channel measurements. Thus, CFTR-mediated ion channel abnormalities are influenced by the number or severity of the *CFTR* mutations and show a range of abnormalities similar to those in patients with mild or severe classic CF at one extreme, and controls and obligate CF heterozygotes on the other. This continuum of electrophysiological abnormalities is not surprising as PS patients have a 17% risk of developing pancreatitis and many of these presentations are in adulthood.

Similar observations have been made in individuals with other CF-like phenotypes, such as men with infertility caused by congenital bilateral absence of the vas deferens who are known to carry a high frequency of *CFTR*-gene mutations (Wilschanski et al. 2006). A relatively large population was examined, and similar to the patients with idiopathic pancreatitis, a wide range of electrophysiological abnormalities was observed. Abnormalities of CFTR function correlated closely with the number and severity of CFTR mutations.

In a recent publication, Schneider et al. (2011) investigated a large group of patients with "idiopathic" chronic pancreatitis and confirmed other studies that the combination of *CFTR* and serine protease inhibitor Kazal-type 1 (*SPINK1*) mutations markedly increase the risk of pancreatitis. However, a novel finding was that the variant R75Q of *CFTR* increases the risk of pancreatitis markedly. Patch-clamp studies on cells expressing this mutation showed that bicarbonate current is significantly impaired. This mutation is not associated with CF but this *CFTR* variant may be a new class of mutation that is specific to the pancreas, particularly correlating with the electrophysiological studies.

DUODENAL ACIDITY — PANCREAS AND OTHER ORGANS

One of the hallmarks of CF in the digestive system is hyperacidity in the duodenum. Hyperacidity reflects multiorgan contributions. The low duodenal pH has most severe consequences for the activity of pancreatic enzymes and, in the long run, for the energy balance of a patient. The malfunctioning pancreas may further contribute to this acid/base imbalance.

Problems with Pancreatic Enzymes but Not with Ulcers

The acidic duodenal condition contributes to inactivation of pancreatic enzymes, if still present in PS patients, precipitation of bile acids, and the development of meconium ileus (Freedman et al. 2001). In PI patients, duodenal acidity limits the action of replacement enzymes, especially that of lipase. Decreased lipase activity causes steatorrhea and fat malabsorption that are difficult to treat (Robinson et al. 1990). In addition, there are increased circulating and tissue levels of (n-6) fatty acids and inflammatory mediators (leukotrienes B4, IL, TNF-α), and there is oxidant stress and redox imbalance.

One would predict that the acidic duodenal environment would lead to ulcers. Paradoxically, it seems that CF patients do not have a higher incidence of duodenal ulcers (and even peptic ulcers may be diminished) (Kaunitz and Akiba 2006). Possibly, cellular pH buffering is elevated, as HCO_3^- is trapped because of dysfunctional CFTR and down-regulated Cl^-/HCO_3^- exchangers. For example, in $\Delta F508$ human or mouse models, pancreatic duct cells and enterocytes have higher resting intracellular pH (El-gavish 1991; Hirokawa et al. 2004). In addition, although HCO_3^- secretion is reduced, the duodenal acid/base barrier is only slightly impaired (Hirokawa et al. 2004).

How Does Hyperacidity Arise?

In CF patients, duodenal hyperacidity (below pH 4) is prominent in the postprandial period; however, resting gastric and duodenal pH values are normal (Robinson et al. 1990; Barraclough and Taylor 1996).

At first, duodenal hyperacidity was thought to be caused by gastric hypersecretion (Cox et al. 1982). Therefore, to lower the acid load to duodenum, gastric acid production is curbed by use of proton pump inhibitors and H_2 receptor antagonists. Indeed, omeprazole treatment seemed to improve fat digestion/absorption and improved patient weight (Barraclough and Taylor, 1996; Proesmans and De Boeck, 2003). Also in a recent successful model of CF, the ferret CFTR knockout, it was shown that omeprazole and ursodeoxycholic acid improved weight and survival of animals (Sun et al. 2010). Nevertheless, it seems that CFTR is also one of the Cl^- channels or transporters that is necessary for gastric acid secretion (Heitzmann and Warth, 2007; Kopic and Geibel 2010). For example, the $\Delta F508$ mutation in mice leads to decreased acid secretion (Sidani et al. 2007). As discussed in a review (Heitzmann and Warth 2007), it seems that the effect of CFTR on acid secretion may depend on the particular CFTR mutation and rescue by other Cl^- transporters/channels.

Normally, acid chyme is neutralized by HCO_3^- secreted by duodenal epithelia, pancreatic duct and bile duct secretions (Ainsworth et al. 1991). Therefore, duodenal hyperacidity in CF has been ascribed to loss-of-function in the CFTR transporter, especially in duodenal and pancreatic epithelia.

Pancreas—Lack of Bicarbonate Secretion and Other Effects

One may ask whether, in CF, lack of pancreatic HCO_3^- secretion contributes to duodenal acidification and whether acidity, in turn, has consequences for overall pancreatic function. Here, we will deal with pH-related effects on acini and whole pancreas function. H^+/HCO_3^- transport in pancreatic ducts will be discussed in the next section.

Secretion originating from healthy acini (e.g., stimulated with cholecystokinin) is neutral or even alkaline in pH, and contains enzymes (Sewell and Young 1975; You et al. 1983; Case and Argent 1993). In the pancreas with impaired duct function, secretion is not only low in volume and high in enzyme concentration, but it also has a relatively low pH. For example, $CFTR^{-/-}$ ($CFTR^{tm1UNC}$) mouse pancreatic-biliary juice after secretin stimulation was four-fold lower in volume and had pH 6.6 compared with pH 8.1 in wild-type mice (Freedman et al. 2001). In the $CFTR^{-/-}$ pig, pancreatic juice also had pH 5.7 compared with pH 8.4 in wild-type pigs (Uc et al. 2011). In addition, acidification of ductal and acinar lumens ($CFTR^{-/-}$ mice) can lead to secondary impairment of apical trafficking of zymogen granule membranes and solubilization of secretory (pro)enzymes (Freedman et al. 2001). Nevertheless, recent data on isolated acini of normal mice show that luminal/extracellular space is acidic, presumably owing to acidic secretory granules that contain the vacuolar type H^+ pump (Behrendorff et al. 2010). Taken together, it seems that optimal enzyme secretion processes relies on neutralizing fluid secretion of adjacent ducts and/or normal fluid secretion of acini.

In addition to the role of the pancreas in the duodenal pH/enzyme environment, duodenal hyperacidity has also indirect effects on pancreas. Increased release of gut hormones, such as secretin, results in increased signaling to the exocrine pancreas that would upregulate pancreatic HCO_3^- secretion. Indeed, in CF patients, increased plasma secretin levels are detected (Windstetter et al. 1997). In agreement, one study on $CFTR^{-/-}$ mice shows that mRNA for secretin (in duodenum) and vasoactive intestinal peptide (VIP) (in pancreas) were significantly increased, as well as pancreatic cAMP levels (De Lisle et al. 2001). Another study shows that such a situation may lead also to added stress, as indicated by increased expression of stress-/inflammation-related genes (Kaur et al. 2004). When the duodenal pH was experimentally corrected, expression of stress genes was also corrected.

Pancreatic dysfunction in CF involves the defective coupling of both ductal and acinar functions in the exocrine pancreas. Because CFTR is mainly expressed in ducts, the following section will consider pancreatic duct function on the cellular and integrated level.

CFTR AND OTHER TRANSPORTERS IN PANCREATIC DUCT SECRETION— A CELLULAR APPROACH

Pancreatic juice HCO_3^- concentrations vary with secretory rates, and they are inversed to changes in Cl^- concentrations. Interestingly, the relation between secretory rates and HCO_3^- concentrations in pancreatic juice collected from the pancreas of various species fall within the same pattern (Fig. 2). This indicates that the basic mechanism of HCO_3^- secretion/salvaging may be similar (Fig. 3), but perhaps the duct mass is different. Taking this as a starting point, ion transport models based on studies of cells and tissues from various animals are suitable models for basic secretion mechanisms and for CF models.

CFTR and Anion Exchanger

The fingerprinting of cellular mechanisms for pancreatic duct ion transport began when it was discovered that secretin-/cAMP- activated Cl^- channels were functionally located on the luminal membranes of isolated rat pancreatic ducts, and the first ion transport models were proposed

(Gray et al. 1988; Novak and Greger 1988b). Almost at the same time, CFTR was discovered (Kerem et al. 1989; Riordan et al. 1989), and soon thereafter CFTR was shown to have properties of a Cl^- channel, including in the pancreatic ducts (Gray et al. 1989, 1993; Tabcharani et al. 1991). Subsequently, CFTR was immunolocalized in rodent and human pancreas to intercalated and small intralobular ducts that also express aquaporins and carbonic ahydrases (Kumpulainen and Jalovaara 1981; Marino et al. 1991; Hyde et al. 1997; Burghardt et al. 2003).

The question of how CFTR Cl^- channels could underlie HCO_3^- secretion of pancreas, has been a challenging problem ever since. The first proposal was that by coupling of Cl^- channels to Cl^-/HCO_3^- exchange operating in parallel would result in efflux of HCO_3^- into the lumen. The anion exchangers belonging to the solute carrier families SLC26A6 and SLC26A3 were found expressed in pancreatic ducts, and the proposed transport ratio of $2HCO_3^-:1Cl^-$ for the first transporter would make it a candidate for secreting ducts (Lohi et al. 2000; Greeley et al. 2001; Dorwart et al. 2008). Until now, studies of SLC26A6 null mice showed

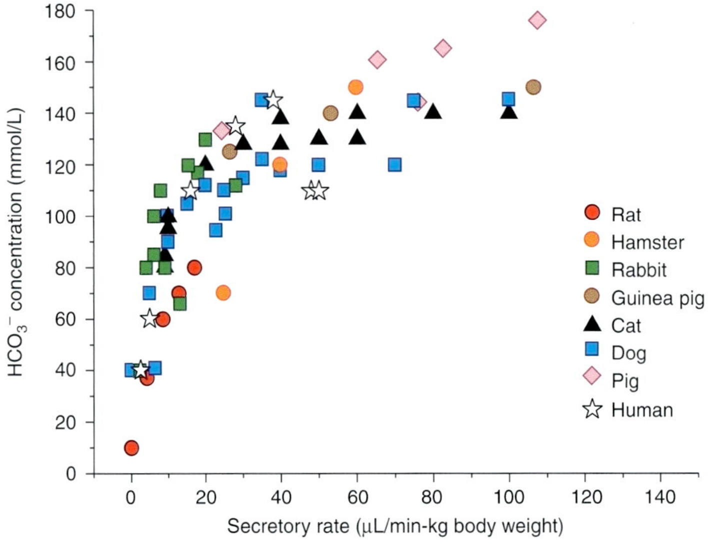

Figure 2. The relation between secretory rates and HCO_3^- concentrations in pancreatic juice of various species. Secretion was stimulated by secretin and secretory rate was collected for body weight. For details, see Novak et al. (2011).

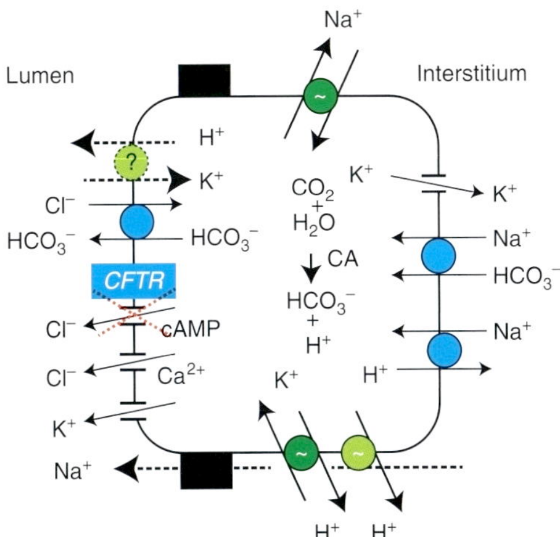

Figure 3. Cellular model for HCO_3^- secretion in pancreatic ducts. Primary active transporters (pumps) and several secondary active transporters and ion channels, including Ca^{2+}-activated K^+ and Cl^- channels, are involved in creating the chemical and electrical driving gradient necessary for production of $NaHCO_3$-rich pancreatic juice in healthy pancreas, as described in text. The CFTR has central role pancreatic ducts; in CF, pancreatic juice volume and pH are reduced.

different effects on duct/pancreas secretion (Wang et al. 2006; Ishiguro et al. 2007; Stewart et al. 2009; Yang et al. 2009). Nevertheless, it seems that there is a functional coupling between the R domain of CFTR and sulfate transporter anti-sigma (STAS) domains of this exchanger (Dorwart et al. 2008).

There are several studies that indicate other ways to achieve HCO_3^- secretion. For example, the CFTR channel is HCO_3^- permeable or CFTR regulates anion selectivity of another channel/transporter. Thus, guinea pig pancreatic ducts secrete fluid (and presumably HCO_3^-) in the absence of luminal Cl^-, in which Cl^-/HCO_3^- exchange would not favor HCO_3^- secretion (Ishiguro et al. 1998, 2009). Another study on HEK293 cells expression system showed that CFTR mutants associated with PI (e.g., I1489T) did not support HCO_3^- transport, whereas those associated with PS (e.g., R117H) show reduced transport (Choi et al. 2001). Indeed, many studies, also on pancreatic ducts, investigated whether CFTR conducts HCO_3^- (Becq

et al. 1993; Gray et al. 1993; O'Reilly et al. 2000). Until very recently, the consensus was that the permeability ratio $PHCO_3^-/PCl^-$ was >0.2–0.4 in conditions with intracellular Cl^- and pH values expected for most epithelia. A newer study also shows similar permeability ratio and it remains independent of ionic conditions (Tang et al. 2009).

Nevertheless, a recent study shows that human and rodent tissues express volume/Cl_i^- sensitive WNK1 kinase and two downstream kinases, SPAK and OSR1. Using patch-clamp recording, it was reported that $PHCO_3^-/PCl^-$ increased from 0.24 to 1.09 when the Cl^- concentration in the pipette was decreased from 150 to 4 mM, and investigators ascribed this permeability to CFTR. The investigators postulated that WNK1-OSR1/SPAK pathway is a molecular switch operating in distal ducts, which leads to increased concentration of incoming HCO_3^- from 80 to 150 mM (Park et al. 2010). Although attractive, the model does not yet explain how increased concentration of HCO_3^- would associate with fluid transport, and it cannot account for the fact that proximal small ducts are the richest sites of CFTR/AQP/carbonic anhydrase expression. Also, other studies show that WNKs inhibit CFTR (Yang et al. 2007, 2011). No doubt this exciting field of volume-sensitive kinases may not only be of relevance to kidney and blood pressure regulation, but also to salt and water transport in other epithelia.

NHE and NBC

In any case, HCO_3^- transport across the luminal membrane relies on provision of cellular HCO_3^-. This source could be achieved by carbonic-anhydrase catalyzed hydration of CO_2, and extrusion of resulting H^+ to the interstitium via a basolateral Na^+/H^+ exchanger, NHE1. Alternatively, or in addition, HCO_3^- could enter across the basolateral membrane by pancreatic Na^+-HCO_3^- cotransporter, pNBC, which transports $1Na^+$:$2HCO_3^-$ (also named NBCe1) (Zhao et al. 1994; Ishiguro et al. 1996; Abuladze et al. 1998). Recent studies show that IRBIT (inositol-3-phosphate receptor-binding protein) activates pNBC and CFTR, and thus could be a coordinating factor for transepithelial ion transport (Shirakabe et al. 2006; Yang et al., 2009, 2011). Thus, Ca^{2+}/IP_3 activating agonists could regulate HCO_3^- secretion, if the appropriate ion transporters on the luminal membrane are present (see below).

Proton Pumps

In addition to the secondary active transporters, NHE1 and pNBC, several earlier studies searched for primary active transporters (Fig. 3). One obvious candidate is the vacuolar H^+-pump, but evidence at the molecular level is missing, and the function in stimulated ducts is unclear (Zhao et al. 1994; Villanger et al. 1995; de Ondarza and Hootman 1997; Cheng et al. 1998). Recently, another study focused on other types of pumps, and it was shown that rodent pancreatic ducts express both gastric and nongastric H^+/K^+ pumps, which significantly contribute to secretin-stimulated duct secretion (Novak et al. 2011).

Other Roles of CFTR

CFTR can regulate other transporters including Cl^-/HCO_3^- exchange in pancreatic tissue (Lee et al. 1999). In other respiratory epithelia, CFTR and ENaC are inversely regulated (Donaldson and Boucher 2007). Freshly isolated rodent duct and in vivo pancreas studies show no evidence for functional ENaC (Novak and Hansen 2002), although on culture, interlobular murine ducts develop sensitivity to amiloride and in some situations can even absorb (Zeiher et al. 1995; Pascua et al. 2009).

One important feature of epithelial secretion of CF patients is the high mucus content. This phenomenon may not be only a consequence of lack of hydration or increased number of mucus secreting cells. It is proposed that HCO_3^- is a chaotropic anion important in mucus expansion and, therefore, lack of CFTR function will impair mucus hydration (Quinton 2008; De Lisle 2009; Garcia et al. 2009; Chong et al. 2013). This may be of relevance to distal pancreatic ducts that contain numerous mucus secreting cells.

A number of studies on various cells indicate that CFTR is involved in release of ATP from intracellular to extracellular spaces, and CFTR has been ascribed the role of an ATP transporter or regulator. Currently, a number of other ATP release mechanisms are more favored (Novak 2011).

Ca^{2+}-Activated Cl^- Channels

In addition to CFTR, which is mainly regulated by cAMP/PKA signaling, Ca^{2+}-activated Cl^- channels (CaCC) could potentially drive secretion (Fig. 3). In rodent and human pancreatic ducts, many studies have shown that a number of agonists increase cellular Ca^{2+}, increase Cl^- conductance (although transiently), change pH_i, and evoke fluid production in isolated ducts (Pahl and Novak 1993; Hug et al. 1994; Winpenny et al. 1998; Szalmay et al. 2001; Pascua et al. 2009). The question is whether this secretion is also HCO_3^- rich, that is, if machinery similar to that operated by CFTR could be recruited, and/or if such CaCC are also HCO_3^- permeable. If IRBIT stimulates pNBC (see above), HCO_3^- permeability on the luminal membrane would be required.

Interestingly, the identity of such CaCC has been elusive and a number of candidates were proposed earlier, including ClC-2 and bestrophins (Duran et al. 2010). Recently, three independent reports have shown that the TMEM16/Anoctamine families are good candidates for CaCC (Caputo et al. 2008; Schroeder et al. 2008; Yang et al. 2008). TMEM16A is expressed in rodent acinar cells (Yang et al. 2008) and knockout of TMEM16A in mice caused defects in CaCC in pancreatic acini (Ousingsawat et al. 2009). TMEM16A is also expressed in CFPAC-1 cells (Caputo et al. 2008) and in human pancreatic duct cell lines expressing normal CFTR, in which it determines transepithelial transport (Wang et al. 2013).

K^+ Channels

K^+ channels are not usually included in pancreatic duct cell models. Nevertheless, there are two indications that K^+ transport is also important in pancreatic ducts. First of all, pancreatic juice contains K^+ higher than in plasma. The K^+ conductance is increased with several agonists, and it is clear that K^+ channels are not only important for setting the resting membrane potential, but they also keep the driving force for anion secretion (Novak and Greger 1988a, 1991). We know the identity of some K^+ channels; this includes BK (maxi-K, SLO, KCNMA1), IK (KCNN4), TASK-2 (KCNK5), and others. Some are located on both the luminal and basolateral membranes, which may be operated by different regulatory systems (Gray et al. 1990; Fong et al. 2003; Hede et al. 2005; Jung et al. 2006; Hayashi et al. 2012).

Regulation of Pancreatic Duct Secretion

The classical bicarbonate-evoking secretagogue is secretin, although a number of other hormones and transmitters can also evoke and coregulate (HCO_3^-) secretion. Even cholinergic stimulation and CCK can evoke HCO_3^- secretion in some species, and they can potentiate the secretin effect on the volume of secretion (You et al. 1983; Holst 1993; Park et al. 1998; Chey and Chang 2001), although there are exceptions (Evans et al. 1996). Whether human pancreatic juice is also rich in HCO_3^- under these conditions is not clear.

One of the novel additions to regulation of pancreatic fluid secretion is the purinergic signaling, which coordinates acini-duct functions. This pathway relies on extracellular ATP and adenosine that, via purinergic and adenosine receptors, regulates specific epithelial transporters (Novak 2008, 2011). Pancreatic acini release ATP, some of which is stored in zymogen granules (Sørensen and Novak 2001; Haanes and Novak 2010). Pancreatic ducts express several types of P2 receptors, including P2Y2, P2Y4, P2Y11, P2X4, and P2X7 receptors, and adenosine A2A and A2B receptors. Luminal ATP can then upregulate anion and fluid secretion, and this activity involves regulation of CaCC and IK (Hug et al. 1994; Ishiguro et al. 1999; Hede et al. 2005; Jung et al. 2006; Novak et al. 2010; Hayashi et al. 2012; Wang et al. 2013). ATP also stimulates luminal anion exchange in duct epi-

thelia expressing functional CFTR (Namkung et al. 2003). In addition, ATP/uridine triphosphate (UTP) also potentiate cAMP-evoked mucin secretion (Jung et al. 2010). From the basolateral membrane, ATP may be released by nerves and/or distended epithelium; some purinergic receptors are inhibitory to secretion (e.g., P2Y2 receptors inhibit BK channels) whereas other receptors may have positive effects on secretion (Hede et al. 1999, 2005; Ishiguro et al. 1999; Wang et al. 2013).

ANIMAL MODELS FOR CF OF PANCREAS— INTEGRATED APPROACH

Although physiological studies on isolated ducts and human duct cell lines have taught us much about ion transport and regulation thereof, it is not enough to understand the impact of CFTR mutations on whole gland pathophysiology in CF. That is, we have to understand the link between dysfunction in HCO_3^- secretion, enzyme secretion, mucus plugging, and changes in pancreas morphology at an integrated level. Therefore, CF animal models have been invaluable, although challenging our understanding at times. The first mouse models of CF were developed shortly after the discovery of CFTR, and recently they were followed by very promising CF models in pigs and ferrets.

Mouse CF Models

There are a number of murine CF models developed, although a variety are gene-targeting strategies. These mice showed processing of mutated proteins, presence of other rescuing transporters, and last but not least, diversity in murine genetic background. These variables have led to heterogeneity of disease manifestations in mouse models (Ostedgaard et al. 2007). Generally, defects in airways and pancreas have been milder and more subtle to detect, whereas abnormalities in ion transport and morphology of intestinal tissues are marked and newborn mice suffer from intestinal obstruction. This is quite different from humans, in which only 10% of newborns have meconium ileus but already 80%–90% have PI. Thus, there is apparent discordance between intestinal and pancreatic CF phenotype in newborns. However, the rodent pancreas is different from the human pancreas in structure and development. The rodent pancreas is immature at birth and undergoes significant weight increase and maturation development postnatally, regarding enzyme and bicarbonate secretion (Githens 1994; Scaglia et al. 1997).

In the following section, we will focus on CF mouse models with respect to pancreas. The first CF mouse model was generated at the University of North Carolina and is denoted CFTR^{tm1UNC}. Only 5% of animals survived, exhibiting severe intestinal problems, but the pancreas morphology was not markedly affected at the point of the initial examination (Snouwaert et al. 1992). In a similar knockout model generated in Cambridge, CFTR^{tm1CAM}, about 40% of animals survived and about half of the mice showed some pancreatic pathology (i.e., blockage of pancreatic ducts), possibly because these animals lived longer (Ratcliff et al. 1993). In the Baylor College model (CFTR^{tm1BAY}), newborn knockout animals had normal pancreas, but with increasing age, some animals showed dilatation and inflammation of main ducts as well as some acinar atrophy (O'Neal et al. 1993). In the Edinburgh model (CFTR^{tm1HGU}), there remained 10% of wild-type CFTR mRNA, 95% mice survived, and the phenotype was milder (Dorin et al. 1992, 1994).

As the reports above indicate, development of CF in murine pancreas may depend on time. If animals were put on a complete supplemented liquid diet, they were able to survive longer, develop manifestations of CF in many organs, including pancreas, such as duct lumen dilatation and obstruction and acinar atrophy (Durie et al. 2004). Knockout animals still had lower body weight, lower pancreas mass and pancreatic enzyme content. Possibly, this lower weight was not only caused by the CF defect, but was partly attributable to malnutrition (Ip et al. 1996).

First, we summarize what is known about the acinar-related pathology. Acini of CFTR^{tm1UNC} homozygous mice had fewer zymogen granules and the major sulfated glycoprotein, gp300, normally present in ZG, was lining distended acinar lumens. There was also impaired apical membrane trafficking compared to wild-type mice (De Lisle 1995, 2001). In isolated acini from knockout animals, there was enhanced secretory protein response with dibutyryl-cAMP (dbcAMP) and carbachol, which may contribute to micro-precipitation and development of acinar dysfunction (Tang et al. 1999). Increased levels of mRNA for secretin and VIP and pancreatic cAMP levels indicated that pancreas was chronically stimulated (De Lisle et al. 2001), and there was impaired stress-gene expression (Kaur et al. 2004). Also knockout animals had mild PI, higher baseline proinflammatory states, and an antiapoptotic phenotype, which may sensitize them to develop more severe acute pancreatitis with a marked pancreatic inflammatory response (DiMagno et al. 2005).

Second, let us examine whether knockout mice also showed signs of pancreatic ductal disorders. CFTR^{tm1UNC} knockout mice produced lower volume and acidic pancreatic-biliary juice after secretin stimulation, and there was also reduced response to CCK (Freedman et al. 2001; DiMagno et al. 2005). In cultured duct monolayer from CFTR^{tm1UNC} knockout mice, forskolin stimulated small short-circuit currents (Isc) compared to the wild type, but ionomycin stimulated currents that were 3× higher in knockout than in wild-type preparation, indicating that CaCC was up-regulated (Clarke et al. 1994). Pancreatic ducts isolated from another CFTR-null mouse (CFTR^{tm1CAM}) secreted 60% less fluid when stimulated with forskolin, and apparently about 30% less with secretin, although the secretin response was much lower even in wild-type mice. In the same preparation, carbachol also induced secretion, but to about the same extent in both wild-type and CFTR-null ducts, indicating that CaCC was well expressed and not up-regulated in knockouts (Pascua et al. 2009). Similar conclusions were reached using whole-cell patch-clamp on duct cells from the same mutant type. Ducts of knockout mice did not have any cAMP-activated currents (i.e., CFTR), but ionomycin-stimulated currents that were about 10× higher than

 Cite this article as *Cold Spring Harb Perspect Med* doi: 10.1101/cshperspect.a009746

CFTR-currents, and these are about the same in both wild-type and knockout cells (Winpenny et al. 1995). In CFTR^{tm1HGU} model, there is residual CFTR activity (10% CFTR mRNA is expressed), but ionomycin-induced currents (i.e., CaCC) were larger and again similar in both wild-type and knockout cells (Gray et al. 1994). Thus, most studies show that CaCC was not up-regulated in CFTR$^{-/-}$.

Nevertheless, the duct studies mentioned above leave some issues. First, the ducts used were short- and long-term cultured, and some endogenous receptors were down-regulated. Therefore, experiments necessitate use of agents that stimulate "simple" signaling pathways than one might expect for receptors. Second, all mouse studies are performed on larger ducts that would not be primary secretors and can even absorb (Zeiher et al. 1995; Pascua et al. 2009).

In murine models of ΔF508 (CFTR^{tm2CAM}, CFTR^{tm2KTH}, CFTR^{tm1EUR}), there can be a gradient for severity of CFTR-ΔF508 processing defects that is more serious in humans than in mouse or pig (Ostedgaard et al. 2007). That is, ΔF508 CFTR is partially processed in mouse. Animals survive into adulthood and display several abnormalities seen in human ΔF508 patients. They also show little or no pancreatic histopathology (van Doorninck et al. 1995; Zeiher et al. 1995), although functional studies indicate some abnormalities. Cultured duct epithelium from CFTR^{tm2KTH} mice shows that a cAMP agonist can induce secretion and Isc changes in wild type but not in ΔF508. Thus, one could expect at least mild pancreas phenotype (Zeiher et al. 1995). Also, forskolin stimulated anion exchanger activity in luminal membranes of perfused main ducts was lower in mutants compared to wild type (Lee et al. 1999). In contrast, similar ducts express less NHE3, and CFTR interaction with NHE3 postulated in HCO$_3^-$ salvage mechanisms (Ahn et al. 2001) could affect anionic composition of juice. In humans, *CFTR* gene mutations associate with recurrent acute pancreatitis in patients with PS. Both CFTR$^{-/-}$ and ΔF508 mice develop caerulein-induced pancreatitis (DiMagno et al. 2005, 2010).

G551D and G480C mutants would be of interest for pancreas, as these mutations are associated with PI. However, mice with this mutation have milder symptoms compared to null mice (e.g., reduced risk of fatal intestinal blockage), and they have residual ($\sim$4%) mutant activity, and no pancreas and lung pathology was detected or studied (Delaney et al. 1996; Dickinson et al. 2002).

Pig and Ferret CF Models

Relatively mild changes in airway and pancreatic epithelia of mouse models spurred development of CF models in other animals. In contrast to mice, pigs have many physiological and anatomical features similar to humans. Recently, a pig model of CF was made by targeted disruption of both CFTR alleles. Newborn CFTR$^{-/-}$ pigs exhibit many CF features, including exocrine pancreatic pathology (dilated and obstructed ducts with mucus incretion, residual acini, interstitial zymogen material, limited inflammation, and fibrosis (Rogers et al. 2008; Meyerholz et al. 2010). There were also intestinal lesions (meconium ileus and microcolon), and if surgically treated, animals reached developing lung abnormalities similar to CF patients. Also, degeneration of the pancreas continued, becoming more fatty, with markedly decreased exocrine cells, and obstructed ducts. Pig "cystic fibrosis of the pancreas" closely resembled Dorothy Anderson's original descriptions (Stoltz et al. 2010).

The ferret CFTR$^{-/-}$ model also has many characteristics of human CF disease including defective airway Cl$^-$ transport and submucosal gland secretion, pancreatic, liver, and vas deferens disease, and variable intestinal disorder (Sun et al. 2010). In the ferret, the pancreas of newborn animals showed dilated acini and ducts with inspissated eosinophilic zymogen secretions, but there was significantly less destruction as seen in newborn CF pigs.

Taken together, animal models are invaluable tools for studying basic and CF-related mechanisms of integrated pancreatic function, and the effects of therapeutics. Although pig and ferret models may develop CF more in line with humans, mouse models still offer a lot of potential especially in our understanding of regulation of cAMP versus Ca^{2+}-driven anion secretion.

THERAPEUTIC POTENTIALS FOR CF OF PANCREAS

Therapeutic approaches for pancreatic treatments are very challenging, as most CF patients are born with PI or get it soon after, and the pancreatic mass is destroyed. In worst cases, when the liver is also affected, pancreas–liver transplantations are performed in a few centers. Another approach would be to reconstitute the pancreas mass using stem cell approaches. This would involve both acini and duct regeneration. At the moment, this task is not high on the research priority list in the CF field, perhaps because very good advances have been made with symptomatic strategies. The focus on treating PI patients is to provide sufficient nutrition, vitamins, and pancreatic enzyme replacement therapy to PI patients to improve nutritional status of the patient. Nevertheless, more progress is still needed, especially for lipase preparations, as enzymes do not work optimally in acidic duodenal environments. Therefore, there are special efforts to design improved lipase preparations by molecular design and recombinant technology (Colin et al. 2010). For example, recombinant acid-resistant lipase of plant origin Merispase (Meristem Therapeutics) is being evaluated in phase II trials. Another product in clinical trials contains microbial lipase, and also protease and amylase preparation (Lipromatase, Lilly). These types of products may be essential as an alternative to porcine-based products.

In patients that still have some pancreas function (PS), the strategy must be to prevent duct/acini degradation, inflammation, and fibrosis, that is, development of pancreatitis. The problem is that PS patients have increased risk of pancreatitis. A number of anti-inflammatory and antioxidant agents are currently in use, including gluthathione, sildenafil, simavastatin, etc. However, further research and novel strategies are needed to improve the clinical care of

CF patients (Innis and Davidson 2008; Jones and Helm 2009).

Nevertheless, because the culprit is CFTR expressed in ducts, the main therapeutic strategies should be to modulate the basic defect here, which could possibly improve duct function in PI and PS patients and allow the pancreas to regenerate and develop. For CFTR protein rescue strategies, see Rowe and Verkman (2013).

Another approach is to bypass defective CFTR and activate CaCC (e.g., TMEM16A), although it is not clear whether the optimal target is the channel, intracellular Ca^{2+} levels, or a receptor regulating either or both (Cuthbert 2011). Activation of CaCC by the lantibiotic duramycin (Moli1901, Lancovutide, Lantibio) is being tested as a strategy and is in phase II trial in Europe. (The effect may be via raising intracellular calcium levels.) The outcome for pancreas is unclear, as tests on cell lines Panc-1 and CFPAC-1 indicate that the effects are nonspecific (Oliynyk et al. 2010). Also, studies on CF mouse models and human pancreatic cell lines indicate that CaCC does not take over from CFTR (see above). Another channel that may be relevant for some epithelia is ClC-2 that is activated by lubiprostone, although the effect is via the prostanoid receptor EP4 (Cuthbert 2011), and ClC-2 is only expressed in pancreatic acini.

An extension of this strategy would be to enhance natural regulator of CaCC, such as P2 receptors (e.g., P2Y2 receptors), as has been envisaged for airways (Deterding et al. 2007; Lazarowski and Boucher 2009). Although the basic research and strategy was well planned, Denufasol tetrasodium (INS37217, Inspire) was taken out of phase III trials recently. For pancreas, one would need to target the receptors of the luminal membrane because the basolateral receptors can, in fact, inhibit secretion rather than stimulate it (see above). For obvious reasons, delivery of intraluminal UTP derivative is unrealistic.

CONCLUDING REMARKS

Translational research is identifying a number of compounds that offer promising pharmacotherapy for CF. However, regarding the pancreas, therapeutic approaches are more complicated than for airways as they require systemic administration, and testing of pancreatic parameters in PS patients would need to be implemented. The possibility of increasing the function of a mutation that ordinarily confers PI to one that confers PS via a potentiator or corrector is appearing. However, this may result in increasing the risk of pancreatitis in a CF patient who did not suffer from this before. Before we move in this direction, we clearly need more drug studies on pancreatic function in good animal models.

ACKNOWLEDGMENTS

Work cited (I.N.) is supported by the Danish Natural Science Council and the Lundbeck Foundation.

REFERENCES

*Reference is also in this collection.

Abuladze N, Lee I, Newman D, Hwang J, Boorer K, Pushkin A, Kurtz I. 1998. Molecular cloning, chromosomal localization, tissue distribution, and functional expression of the human pancreatic sodium bicarbonate cotransporter. *J Biol Chem* **273:** 17689–17695.

Ahn W, Kim KH, Lee JA, Kim JY, Choi JY, Moe OW, Milgram SL, Muallem S, Lee MG. 2001. Regulatory interaction between the cystic fibrosis transmembrane conductance regulator and. *J Biol Chem* **276:** 17236–17243.

Ainsworth MA, Ladegaard L, Svendsen P, Cantor P, Olsen O, Schaffalitzky de Muckadell OB. 1991. Pancreatic, hepatic, and duodenal mucosal bicarbonate secretion during infusion of secretin and cholecystokinin. Evidence of the importance of hepatic bicarbonate in the neutralization of acid in the duodenum of anaesthetized pigs. *Scand J Gastroenterol* **26:** 1035–1041.

Barraclough M, Taylor CJ. 1996. Twenty-four hour ambulatory gastric and duodenal pH profiles in cystic fibrosis: Effect of duodenal hyperacidity on pancreatic enzyme function and fat absorption. *J Ped Gastroenterol Nutr* **23:** 45–50.

Becq F, Hollande E, Gola M. 1993. Phosphorylation-regulated low-conductance Cl^- channels in a human pancreatic duct cell line. *Pflügers Arch* **425:** 1–8.

Behrendorff N, Floetenmeyer M, Schwiening C, Thorn P. 2010. Protons released during pancreatic acinar cell secretion acidify the lumen and contribute to pancreatitis in mice. *Gastroenterology* **139:** 1711–1720.

Bishop MD, Freedman SD, Zielenski J, Ahmed N, Dupuis A, Martin S, Ellis L, Shea J, Hopper I, Corey M, et al. 2005. The cystic fibrosis transmembrane conductance regulator gene and ion channel function in patients with idiopathic pancreatitis. *Hum Genet* **118:** 372–381.

Burghardt B, Elkaer ML, Kwon TH, Racz GZ, Varga G, Steward MC, Nielsen S. 2003. Distribution of aquaporin water channels AQP1 and AQP5 in the ductal system of the human pancreas. *Gut* **52:** 1008–1016.

Caputo A, Caci E, Ferrera L, Pedemonte N, Barsanti C, Sondo E, Pfeffer U, Ravazzolo R, Zegarra-Moran O, Galietta LJ. 2008. TMEM16A, a membrane protein associated with calcium-dependent chloride channel activity. *Science* **322:** 590–594.

Case RM, Argent BE. 1993. Pancreatic duct cell secretion. Control and mechanism of transport. In *The pancreas biology, pathobiology, and disease* (ed. Go VLW, et al.), pp. 301–350. Raven, New York.

Cheng HS, Leung PY, Cheng Chew SB, Leung PS, Lam SY, Wong WS, Wang ZD, Chan HC. 1998. Concurrent and independent HCO_3^- and Cl^- secretion in a human pancreatic duct cell line (CAPAN-1). *J Membr Biol* **164:** 155–167.

Chey WY, Chang T. 2001. Neural hormonal regulation of exocrine pancreatic secretion. *Pancreatology* **1:** 320–335.

Choi JY, Muallem D, Kiselyov K, Lee MG, Thomas PJ, Muallem S. 2001. Aberrant CFTR-dependent HCO_3^- transport in mutations associated with cystic fibrosis. *Nature* **410:** 94–97.

*Chong PA, Kota P, Dokholyan NV, Forman-Kay JD. 2013. Dynamics intrinsic to cystic fibrosis transmembrane conductance regulator function and stability. *Cold Spring Harb Perspect Med.* **3:** a009522.

Clarke LL, Grubb BR, Yankaskas JR, Cotton CU, McKenzie A, Boucher RC. 1994. Relationship of a non-cystic fibrosis transmembrane conductance regulator-mediated chloride conductance to organ-level disease in Cftr$^{-/-}$ mice. *Proc Natl Acad Sci* **91:** 479–483.

Cohn JA, Friedman KJ, Noone PG, Knowles MR, Silverman LM, Jowell PS. 1998. Relation between mutations of the cystic fibrosis gene and idiopathic pancreatitis. *N Engl J Med* **339:** 653–658.

Colin DY, Prez-Beauclair P, Silva N, Infantes L, Kerfelec B. 2010. Modification of pancreatic lipase properties by directed molecular evolution. *Protein Eng Des Sel* **23:** 365–373.

Cox KL, Isenberg JN, Ament ME. 1982. Gastric acid hypersecretion in cystic fibrosis. *J Pediatr Gastroenterol Nutr* **1:** 559–565.

Cuthbert A. 2011. New horizons in the treatment of cystic fibrosis. *Br J Pharmacol* **163:** 173–183.

De Boeck K, Weren M, Proesmans M, Kerem E. 2005. Pancreatitis among patients with cystic fibrosis: Correlation with pancreatic status and genotype. *Pediatrics* **115:** e463–e469.

Delaney SJ, Alton EW, Smith SN, Lunn DP, Farley R, Lovelock PK, Thomson SA, Hume DA, Lamb D, Porteous DJ, et al. 1996. Cystic fibrosis mice carrying the missense mutation G551D replicate human genotype-phenotype correlations. *EMBO J* **15:** 955–963.

De Lisle RC. 1995. Increased expression of sulfated gp300 and acinar tissue pathology in pancreas of CFTR$^{-/-}$ mice. *Am J Physiol* **268:** G717–G723.

De Lisle RC. 2009. Pass the bicarb: The importance of HCO_3^- for mucin release. *J Clin Invest* **119:** 2535–2537.

De Lisle RC, Isom KS, Ziemer D, Cotton CU. 2001. Changes in the exocrine pancreas secondary to altered small intestinal function in the CF mouse. *Am J Physiol Gastrointest Liver Physiol* **281:** G899–G906.

de Ondarza J, Hootman SR. 1997. Confocal microscopic analysis of intracellular pH regulation in isolated guinea pig pancreatic ducts. *Am J Physiol* **272:** G124–G134.

Deterding RR, LaVange LM, Engels JM, Mathews DW, Coquillette SJ, Brody AS, Millard SP, Ramsey BW. 2007. Phase 2 randomized safety and efficacy trial of nebulized denufosol tetrasodium in cystic fibrosis. *Am J Respir Crit Care Med* **176:** 362–369.

Dickinson P, Smith SN, Webb S, Kilanowski FM, Campbell IJ, Taylor MS, Porteous DJ, Willemsen R, De Jonge HR, Farley R, et al. 2002. The severe G480C cystic fibrosis mutation, when replicated in the mouse, demonstrates mistrafficking, normal survival and organ-specific bioelectrics. *Hum Mol Genet* **11:** 243–251.

DiMagno MJ, Lee SH, Hao Y, Zhou SY, McKenna BJ, Owyang C. 2005. A proinflammatory, antiapoptotic phenotype underlies the susceptibility to acute pancreatitis in cystic fibrosis transmembrane regulator $(-/-)$ mice. *Gastroenterology* **129:** 665–681.

DiMagno MJ, Lee SH, Owyang C, Zhou SY. 2010. Inhibition of acinar apoptosis occurs during acute pancreatitis in the human homologue DeltaF508 cystic fibrosis mouse. *Am J Physiol Gastrointest Liver Physiol* **299:** G400–G412.

Donaldson SH, Boucher RC. 2007. Sodium channels and cystic fibrosis. *Chest* **132:** 1631–1636.

Dorin JR, Dickinson P, Emslie E, Clarke AR, Dobbie L, Hooper ML, Halford S, Wainwright BJ, Porteous DJ. 1992. Successful targeting of the mouse cystic fibrosis transmembrane conductance regulator gene in embryonal stem cells. *Transgenic Res* **1:** 101–105.

Dorin JR, Stevenson BJ, Fleming S, Alton EW, Dickinson P, Porteous DJ. 1994. Long-term survival of the exon 10 insertional cystic fibrosis mutant mouse is a consequence of low level residual wild-type CFTR gene expression. *Mamm Genome* **5:** 465–472.

Dorwart MR, Shcheynikov N, Yang D, Muallem S. 2008. The solute carrier 26 family of proteins in epithelial ion transport. *Physiology* **23:** 104–114.

Duran C, Thompson CH, Xiao Q, Hartzell HC. 2010. Chloride channels: Often enigmatic, rarely predictable. *Annu Rev Physiol* **72:** 95–121.

Durie PR, Kent G, Phillips MJ, Ackerley CA. 2004. Characteristic multiorgan pathology of cystic fibrosis in a long-living cystic fibrosis transmembrane regulator knockout murine model. *Am J Pathol* **164:** 1481–1493.

Durno C, Corey M, Zielenski J, Tullis E, Tsui LC, Durie P. 2002. Genotype and phenotype correlations in patients with cystic fibrosis and pancreatitis. *Gastroenterology* **123:** 1857–1864.

Elgavish A. 1991. High intracellular pH in CFPAC: A pancreas cell line from a patient with cystic fibrosis is lowered by retrovirus-mediated CFTR gene transfer. *Biochem Biophys Res Commun* **180:** 342–348.

Evans RL, Ashton N, Elliott AC, Green R, Argent BE. 1996. Interaction between secretin and acetylcholine in the regulation of fluid secretion by isolated rat pancreatic ducts. *J Physiol* **461:** 265–273.

Fong P, Argent BE, Guggino WB, Gray MA. 2003. Characterization of vectorial chloride transport pathways in the human pancreatic duct adenocarcinoma cell line, HPAF. *Am J Physiol Cell Physiol* **285:** C433–C445.

Freedman SD, Kern HF, Scheele G. 2001. Pancreatic acinar cell dysfunction in $CFTR^{-/-}$ mice is associated with impairments in luminal pH and endocytosis. *Gastroenterology* **121:** 950–957.

Garcia MA, Yang N, Quinton PM. 2009. Normal mouse intestinal mucus release requires cystic fibrosis transmembrane regulator-dependent bicarbonate secretion. *J Clin Invest* **119:** 2613–2622.

Githens S. 1994. Differentiation and development of the pancreas in animals. In *The exocrine pancreas biology, pathology and diseases* (ed. Go VLW, et al.), pp. 21–56. Raven, New York.

Gray MA, Greenwell JR, Argent BE. 1988. Secretin-regulated chloride channel on the apical plasma membrane of pancreatic duct cells. *J Membr Biol* **105:** 131–142.

Gray MA, Harris A, Coleman L, Greenwell JR, Argent BE. 1989. Two types of chloride channel on duct cells cultured from human fetal pancreas. *Am J Physiol* **257:** C240–C251.

Gray MA, Greenwell JR, Garton AJ, Argent BE. 1990. Regulation of maxi-K^+ channels on pancreatic duct cells by cyclic AMP-dependent phosphorylation. *J Membr Biol* **115:** 203–215.

Gray MA, Plant S, Argent BE. 1993. cAMP-regulated whole cell chloride currents in pancreatic duct cells. *Am J Physiol Cell Physiol* **264:** C591–C602.

Gray MA, Winpenny JP, Porteous DJ, Dorin JR, Argent BE. 1994. CFTR and calcium-activated chloride currents in pancreatic duct cells of a transgenic CF mouse. *Am J Physiol* **266:** C213–C221.

Greeley T, Shumaker H, Wang Z, Schweinfest CW, Soleimani M. 2001. Downregulated in adenoma and putative anion transporter are regulated by CFTR in cultured pancreatic duct cells. *Am J Physiol Gastrointest Liver Physiol* **281:** G1301–G1308.

Haanes KA, Novak I. 2010. ATP storage and uptake by isolated pancreatic zymogen granules. *Biochem J* **429:** 303–311.

Hayashi M, Wang J, Hede SE, Novak I. 2012. An intermediate-conductance Ca^{2+}-activated K^+ channel is important for secretion in pancreatic duct cells. *Am J Physiol Cell Physiol* **303:** C151–C159.

Hede SE, Amstrup J, Christoffersen BC, Novak I. 1999. Purinoceptors evoke different electrophysiological responses in pancreatic ducts. P2Y inhibits K^+ conductance, and P2X stimulates cation conductance. *J Biol Chem* **274:** 31784–31791.

Hede SE, Amstrup J, Klaerke DA, Novak I. 2005. $P2Y_2$ and $P2Y_4$ receptors regulate pancreatic Ca^{2+} activated K^+ channels differently. *Pflügers Arch* **450:** 429–436.

Heitzmann D, Warth R. 2007. No potassium, no acid: K^+ channels and gastric acid secretion. *Physiology* **22:** 335–341.

Hirokawa M, Takeuchi T, Chu S, Akiba Y, Wu V, Guth PH, Engel E, Montrose MH, Kaunitz JD. 2004. Cystic fibrosis gene mutation reduces epithelial cell acidification and injury in acid-perfused mouse duodenum. *Gastroenterology* **127**: 1162–1173.

Holst JJ. 1993. Neural regulation of pancreatic exocrine function. In *The pancreas biology, pathobiology, and disease* (ed. Go VLW, et al.), pp. 381–402. Raven, New York.

Hug M, Pahl C, Novak I. 1994. Effect of ATP, carbachol and other agonists on intracellular calcium activity and membrane voltage of pancreatic ducts. *Pflügers Arch* **426**: 412–418.

Hyde K, Reid CJ, Tebbutt SJ, Weide L, Hollingsworth MA, Harris A. 1997. The cystic fibrosis transmembrane conductance regulator as a marker of human pancreatic duct development. *Gastroenterology* **113**: 914–919.

Innis SM, Davidson AG. 2008. Cystic fibrosis and nutrition: Linking phospholipids and essential fatty acids with thiol metabolism. *Annu Rev Nutr* **28**: 55–72.

Ip WF, Bronsveld I, Kent G, Corey M, Durie PR. 1996. Exocrine pancreatic alterations in long-lived surviving cystic fibrosis mice. *Pediatr Res* **40**: 242–249.

Ishiguro H, Steward MC, Lindsay ARG, Case RM. 1996. Accumulation of intracellular HCO_3^- by Na^+-HCO_3^- cotransport in interlobular ducts from guinea-pig pancreas. *J Physiol* **495**: 169–178.

Ishiguro H, Naruse S, Steward MC, Kitagawa M, Ko SB, Hayakawa T, Case RM. 1998. Fluid secretion in interlobular ducts isolated from guinea-pig pancreas. *J Physiol* **511**: 407–422.

Ishiguro H, Naruse S, Kitagawa M, Hayakawa T, Case RM, Steward MC. 1999. Luminal ATP stimulates fluid and HCO_3^- secretion in guinea-pig pancreatic duct. *J Physiol* **519**: 551–558.

Ishiguro H, Namkung W, Yamamoto A, Wang Z, Worrell RT, Xu J, Lee MG, Soleimani M. 2007. Effect of Slc26a6 deletion on apical Cl^-/HCO_3^- exchanger activity and cAMP-stimulated bicarbonate secretion in pancreatic duct. *Am J Physiol Gastrointest Liver Physiol* **292**: G447–G455.

Ishiguro H, Steward MC, Naruse S, Ko SB, Goto H, Case RM, Kondo T, Yamamoto A. 2009. CFTR functions as a bicarbonate channel in pancreatic duct cells. *J Gen Physiol* **133**: 315–326.

Jones AM, Helm JM. 2009. Emerging treatments in cystic fibrosis. *Drugs* **69**: 1903–1910.

Jung SR, Kim K, Hille B, Nguyen TD, Koh DS. 2006. Pattern of Ca^{2+} increase determines the type of secretory mechanism activated in dog pancreatic duct epithelial cells. *J Physiol* **576**: 163–178.

Jung SR, Hille B, Nguyen TD, Koh DS. 2010. Cyclic AMP potentiates Ca^{2+}-dependent exocytosis in pancreatic duct epithelial cells. *J Gen Physiol* **135**: 527–543.

Kaunitz JD, Akiba Y. 2006. Review article: Duodenal bicarbonate–mucosal protection, luminal chemosensing and acid-base balance. *Aliment Pharmacol Ther* **24**: 169–176.

Kaur S, Norkina O, Ziemer D, Samuelson LC, De Lisle RC. 2004. Acidic duodenal pH alters gene expression in the cystic fibrosis mouse pancreas. *Am J Physiol Gastrointest Liver Physiol* **287**: G480–G490.

Kerem B, Rommens JM, Buchanan JA, Markiewicz D, Cox TK, Chakravarti A, Buchwald M, Tsui LC. 1989. Identification of the cystic fibrosis gene: Genetic analysis. *Science* **245**: 1073–1080.

Kerem E, Corey M, Kerem BS, Rommens J, Markiewicz D, Levison H, Tsui LC, Durie P. 1990. The relation between genotype and phenotype in cystic fibrosis—Analysis of the most common mutation (delta F508). *N Engl J Med* **323**: 1517–1522.

Kopelman H, Durie P, Gaskin K, Weizman Z, Forstner G. 1985. Pancreatic fluid secretion and protein hyperconcentration in cystic fibrosis. *N Engl J Med* **312**: 329–334.

Kopelman H, Corey M, Gaskin K, Durie P, Weizman Z, Forstner G. 1988. Impaired chloride secretion, as well as bicarbonate secretion, underlies the fluid secretory defect in the cystic fibrosis pancreas. *Gastroenterology* **95**: 349–355.

Kopic S, Geibel J. 2010. Update on the mechanisms of gastric acid secretion. *Curr Gastroenterol Rep* **12**: 458–464.

Kristidis P, Bozon D, Corey M, Markiewicz D, Rommens J, Tsui LC, Durie P. 1992. Genetic determination of exocrine pancreatic function in cystic fibrosis. *Am J Hum Genet* **50**: 1178–1184.

Kumpulainen T, Jalovaara P. 1981. Immunohistochemical localization of carbonic anhydrase isoenzymes in the human pancreas. *Gastroenterology* **80**: 796–799.

Lazarowski ER, Boucher RC. 2009. Purinergic receptors in airway epithelia. *Curr Opin Pharmacol* **9**: 262–267.

Lee MG, Choi JY, Luo X, Strickland E, Thomas PJ, Muallem S. 1999. Cystic fibrosis transmembrane conductance regulator regulates luminal Cl^-/HCO_3^- exchange in mouse submandibular and pancreatic ducts. *J Biol Chem* **274**: 14670–14677.

Lohi H, Kujala M, Kerkela E, Saarialho-Kere U, Kestila M, Kere J. 2000. Mapping of five new putative anion transporter genes in human and characterization of SLC26A6, a candidate gene for pancreatic anion exchanger. *Genomics* **70**: 102–112.

Marino CR, Matovcik LM, Gorelick FS, Cohn JA. 1991. Localization of the cystic fibrosis transmembrane conductance regulator in pancreas. *J Clin Invest* **88**: 712–716.

Meyerholz DK, Stoltz DA, Pezzulo AA, Welsh MJ. 2010. Pathology of gastrointestinal organs in a porcine model of cystic fibrosis. *Am J Pathol* **176**: 1377–1389.

Namkung W, Lee JA, Ahn W, Han W, Kwon SW, Ahn DS, Kim KH, Lee MG. 2003. Ca^{2+} activates cystic fibrosis transmembrane conductance regulator- and Cl^--dependent HCO_3 transport in pancreatic duct cells. *J Biol Chem* **278**: 200–207.

Novak I. 2008. Purinergic receptors in the endocrine and exocrine pancreas. *Purinergic Signal* **4**: 237–253.

Novak I. 2011. Purinergic signalling in epithelial ion transport-regulation of secretion and absorption. *Acta Physiol (Oxf)* **202**: 501–522.

Novak I, Greger R. 1988a. Electrophysiological study of transport systems in isolated perfused pancreatic ducts: Properties of the basolateral membrane. *Pflügers Arch* **411**: 58–68.

Novak I, Greger R. 1988b. Properties of the luminal membrane of isolated perfused rat pancreatic ducts: Effect of cyclic AMP and blockers of chloride transport. *Pflügers Arch* **411**: 546–553.

Novak I, Greger R. 1991. Effect of bicarbonate on potassium conductance of isolated perfused rat pancreatic ducts. *Pflügers Arch* **419**: 76–83.

Novak I, Hansen MR. 2002. Where have all the Na^+ channels gone? In search of functional ENaC in exocrine pancreas. *Biochim Biophys Acta* **1566:** 162–168.

Novak I, Jans IM, Wohlfahrt L. 2010. Effect of P2X7 receptor knockout on exocrine secretion of pancreas, salivary glands and lacrimal glands. *J Physiol* **588:** 3615–3627.

Novak I, Wang J, Henriksen KL, Haanes KA, Krabbe S, Nitschke R, Hede SE. 2011. Pancreatic bicarbonate secretion involves two proton pumps. *J Biol Chem* **286:** 280–289.

Oliynyk I, Varelogianni G, Roomans GM, Johannesson M. 2010. Effect of duramycin on chloride transport and intracellular calcium concentration in cystic fibrosis and non-cystic fibrosis epithelia. *APMIS* **118:** 982–990.

O'Neal WK, Hasty P, McCray PB Jr, Casey B, Rivera-Perez J, Welsh MJ, Beaudet AL, Bradley A. 1993. A severe phenotype in mice with a duplication of exon 3 in the cystic fibrosis locus. *Hum Mol Genet* **2:** 1561–1569.

Ooi CY, Dorfman R, Cipolli M, Gonska T, Castellani C, Keenan K, Freedman SD, Zielenski J, Berthiaume Y, Corey M, et al. 2011. Type of CFTR mutation determines risk of pancreatitis in patients with cystic fibrosis. *Gastroenterology* **140:** 153–161.

O'Reilly CM, Winpenny JP, Argent BE, Gray MA. 2000. Cystic fibrosis transmembrane conductance regulator currents in guinea pig pancreatic duct cells: Inhibition by bicarbonate ions. *Gastroenterology* **118:** 1187–1196.

Ostedgaard LS, Rogers CS, Dong Q, Randak CO, Vermeer DW, Rokhlina T, Karp PH, Welsh MJ. 2007. Processing and function of CFTR-DeltaF508 are species-dependent. *Proc Natl Acad Sci* **104:** 15370–15375.

Ousingsawat J, Martins JR, Schreiber R, Rock JR, Harfe BD, Kunzelmann K. 2009. Loss of TMEM16A causes a defect in epithelial Ca^{2+}-dependent chloride transport. *J Biol Chem* **284:** 28698–28703.

Pahl C, Novak I. 1993. Effect of vasoactive intestinal peptide, carbachol and other agonists on cell membrane voltage of pancreatic duct cells. *Pflügers Arch* **424:** 315–320.

Park HS, Lee YL, Kwon HY, Chey WY, Park HJ. 1998. Significant cholinergic role in secretin-stimulated exocrine secretion in isolated rat pancreas. *Am J Physiol* **274:** G413–G418.

Park HW, Nam JH, Kim JY, Namkung W, Yoon JS, Lee JS, Kim KS, Venglovecz V, Gray MA, Kim KH, et al. 2010. Dynamic regulation of CFTR bicarbonate permeability by Cl_i^- and its role in pancreatic bicarbonate secretion. *Gastroenterology* **139:** 620–631.

Pascua P, Garcia M, Fernandez-Salazar MP, Hernandez-Lorenzo MP, Calvo JJ, Colledge WH, Case RM, Steward MC, San Roman JI. 2009. Ducts isolated from the pancreas of CFTR-null mice secrete fluid. *Pflugers Arch* **459:** 203–214.

Proesmans M, De Boeck K. 2003. Omeprazole, a proton pump inhibitor, improves residual steatorrhoea in cystic fibrosis patients treated with high dose pancreatic enzymes. *Eur J Pediatr* **162:** 760–763.

Quinton PM. 2008. Cystic fibrosis: Impaired bicarbonate secretion and mucoviscidosis. *Lancet* **372:** 415–417.

Ratcliff R, Evans MJ, Cuthbert AW, MacVinish LJ, Foster D, Anderson JR, Colledge WH. 1993. Production of a severe cystic fibrosis mutation in mice by gene targeting. *Nat Genet* **4:** 35–41.

Riordan JR, Rommens JM, Kerem B, Alon N, Rozmahel R, Grzelczak Z, Zielenski J, Lok S, Plavsic N, Chou JL. 1989. Identification of the cystic fibrosis gene: Cloning and characterization of complementary DNA. *Science* **245:** 1066–1073.

Robinson PJ, Smith AL, Sly PD. 1990. Duodenal pH in cystic fibrosis and its relationship to fat malabsorption. *Dig Dis Sci* **35:** 1299–1304.

Rogers CS, Stoltz DA, Meyerholz DK, Ostedgaard LS, Rokhlina T, Taft PJ, Rogan MP, Pezzulo AA, Karp PH, Itani OA, et al. 2008. Disruption of the CFTR gene produces a model of cystic fibrosis in newborn pigs. *Science* **321:** 1837–1841.

* Rowe SM, Verkman AS. 2013. Cystic fibrosis transmembrane regulator correctors and potentiators. *Cold Spring Harb Perspect Med* doi: 10.1101/cshperspect.a009761.

Scaglia L, Cahill CJ, Finegood DT, Bonner-Weir S. 1997. Apoptosis participates in the remodeling of the endocrine pancreas in the neonatal rat. *Endocrinology* **138:** 1736–1741.

Schneider A, Larusch J, Sun X, Aloe A, Lamb J, Hawes R, Cotton P, Brand RE, Anderson MA, Money ME, et al. 2011. Combined bicarbonate conductance-impairing variants in CFTR and SPINK1 variants are associated with chronic pancreatitis in patients without cystic fibrosis. *Gastroenterology* **140:** 162–171.

Schroeder BC, Cheng T, Jan YN, Jan LY. 2008. Expression cloning of TMEM16A as a calcium-activated chloride channel subunit. *Cell* **134:** 1019–1029.

Sewell WA, Young JA. 1975. Secretion of electrolytes by the pancreas of the anaesthetized rat. *J Physiol* **252:** 379–396.

Sharer N, Schwarz M, Malone G, Howarth A, Painter J, Super M, Braganza J. 1998. Mutations of the cystic fibrosis gene in patients with chronic pancreatitis. *N Engl J Med* **339:** 645–652.

Shirakabe K, Priori G, Yamada H, Ando H, Horita S, Fujita T, Fujimoto I, Mizutani A, Seki G, Mikoshiba K. 2006. IRBIT, an inositol 1,4,5-trisphosphate receptor-binding protein, specifically binds to and activates pancreas-type Na^+/HCO_3^- cotransporter 1 (pNBC1). *Proc Natl Acad Sci* **103:** 9542–9547.

Shwachman H, Lebenthal E, Khaw KT. 1975. Recurrent acute pancreatitis in patients with cystic fibrosis with normal pancreatic enzymes. *Pediatrics* **55:** 86–95.

Sidani SM, Kirchhoff P, Socrates T, Stelter L, Ferreira E, Caputo C, Roberts KE, Bell RL, Egan ME, Geibel JP. 2007. Delta F508 mutation results in impaired gastric acid secretion. *J Biol Chem* **282:** 6068–6074.

Snouwaert JN, Brigman KK, Latour AM, Malouf NN, Boucher RC, Smithies O, Koller BH. 1992. An animal model for cystic fibrosis made by gene targeting. *Science* **257:** 1083–1088.

Sørensen CE, Novak I. 2001. Visualization of ATP release in pancreatic acini in response to cholinergic stimulus. Use of fluorescent probes and confocal microscopy. *J Biol Chem* **276:** 32925–32932.

Stewart AK, Yamamoto A, Nakakuki M, Kondo T, Alper SL, Ishiguro H. 2009. Functional coupling of apical Cl^-/HCO_3^- exchange with CFTR in stimulated HCO_3^-

secretion by guinea pig interlobular pancreatic duct. *Am J Physiol Gastrointest Liver Physiol* **296:** G1307–G1317.

Stoltz DA, Meyerholz DK, Pezzulo AA, Ramachandran S, Rogan MP, Davis GJ, Hanfland RA, Wohlford-Lenane C, Dohrn CL, Bartlett JA, et al. 2010. Cystic fibrosis pigs develop lung disease and exhibit defective bacterial eradication at birth. *Sci Transl Med* **2:** 29ra31.

Sun X, Sui H, Fisher JT, Yan Z, Liu X, Cho HJ, Joo NS, Zhang Y, Zhou W, Yi Y, et al. 2010. Disease phenotype of a ferret CFTR-knockout model of cystic fibrosis. *J Clin Invest* **120:** 3149–3160.

Szalmay G, Varga G, Kajiyama F, Yang XS, Lang TF, Case RM, Steward MC. 2001. Bicarbonate and fluid secretion evoked by cholecystokinin, bombesin and acetylcholine in isolated guinea-pig pancreatic ducts. *J Physiol* **535:** 795–807.

Tabcharani JA, Chang X-B, Riordan JR, Hanrahan JW. 1991. Phosphorylation-regulated Cl^- channel in CHO cells stably expressing the cystic fibrosis gene. *Nature* **352:** 628–631.

Tang S, Beharry S, Kent G, Durie PR. 1999. Synergistic effects of cAMP- and calcium-mediated amylase secretion in isolated pancreatic acini from cystic fibrosis mice. *Pediatr Res* **45:** 482–488.

Tang L, Fatehi M, Linsdell P. 2009. Mechanism of direct bicarbonate transport by the CFTR anion channel. *J Cyst Fibros* **8:** 115–121.

Uc A, Stoltz DA, Ludwig P, Pezzulo A, Griffin M, bu-El-Haija M, bu-El-Haija M, Meyerholz DK, Taft P, Welsh MJ. 2011. Pancreatic and biliary secretion differ in cystic fibrosis and wild-type pigs. *J Cyst Fibros* **10:** S69.

van Doorninck JH, French PJ, Verbeek E, Peters RH, Morreau H, Bijman J, Scholte BJ. 1995. A mouse model for the cystic fibrosis delta F508 mutation. *EMBO J* **14:** 4403–4411.

Villanger O, Veel T, Ræder MG. 1995. Secretin causes H^+/HCO_3^- secretion from pig pancreatic ductules by vacuolar-type H^+-adenosine triphosphatase. *Gastroenterology* **108:** 850–859.

Wang Y, Soyombo AA, Shcheynikov N, Zeng W, Dorwart M, Marino CR, Thomas PJ, Muallem S. 2006. Slc26a6 regulates CFTR activity in vivo to determine pancreatic duct HCO_3^- secretion: Relevance to cystic fibrosis. *EMBO J* **25:** 5049–5057.

Wang J, Haanes KA, Novak I. 2013. Purinergic regulation of CFTR and Ca^{2+}-activated Cl^- channels and K^+ channels in human pancreatic duct epithelium. *Am J Physiol Cell Physiol* **304:** C673–C684.

Wilschanski M, Durie PR. 2007. Patterns of GI disease in adulthood associated with mutations in the CFTR gene. *Gut* **56:** 1153–1163.

Wilschanski M, Dupuis A, Ellis L, Jarvi K, Zielenski J, Tullis E, Martin S, Corey M, Tsui LC, Durie P. 2006. Mutations in the cystic fibrosis transmembrane regulator gene and in vivo transepithelial potentials. *Am J Respir Crit Care Med* **174:** 787–794.

Windstetter D, Schaefer F, Scharer K, Reiter K, Eife R, Harms HK, Bertele-Harms R, Fiedler F, Tsui LC, Reitmeir P, et al. 1997. Renal function and renotropic effects of secretin in cystic fibrosis. *Eur J Med Res* **2:** 431–436.

Winpenny JP, Verdon B, McAlroy HL, Colledge WH, Ratcliff R, Evans MJ, Gray MA, Argent BE. 1995. Calcium-activated chloride conductance is not increased in pancreatic duct cells of CF mice. *Pflügers Arch* **430:** 26–33.

Winpenny JP, Harris A, Hollingsworth MA, Argent BE, Gray MA. 1998. Calcium-activated chloride conductance in a pancreatic adenocarcinoma cell line of ductal origin (HPAF) and in freshly isolated human pancreatic duct cells. *Pflugers Arch* **435:** 796–803.

Yang CL, Liu X, Paliege A, Zhu X, Bachmann S, Dawson DC, Ellison DH. 2007. WNK1 and WNK4 modulate CFTR activity. *Biochem Biophys Res Commun* **353:** 535–540.

Yang YD, Cho H, Koo JY, Tak MH, Cho Y, Shim WS, Park SP, Lee J, Lee B, Kim BM, et al. 2008. TMEM16A confers receptor-activated calcium-dependent chloride conductance. *Nature* **455:** 1210–1215.

Yang D, Shcheynikov N, Zeng W, Ohana E, So I, Ando H, Mizutani A, Mikoshiba K, Muallem S. 2009. IRBIT coordinates epithelial fluid and HCO_3^- secretion by stimulating the transporters pNBC1 and CFTR in the murine pancreatic duct. *J Clin Invest* **119:** 193–202.

Yang D, Li Q, So I, Huang CL, Ando H, Mizutani A, Seki G, Mikoshiba K, Thomas PJ, Muallem S. 2011. IRBIT governs epithelial secretion in mice by antagonizing the WNK/SPAK kinase pathway. *J Clin Invest* **121:** 956–965.

You CH, Rominger JM, Chey WY. 1983. Potentiation effect of cholecystokinin-octapeptide on pancreatic bicarbonate secretion stimulated by a physiologic dose of secretin in humans. *Gastroenterology* **85:** 40–45.

Zeiher BG, Eichwald E, Zabner J, Smith JJ, Puga AP, McCray PB Jr, Capecchi MR, Welsh MJ, Thomas KR. 1995. A mouse model for the ΔF508 allele of cystic fibrosis. *J Clin Invest* **96:** 2051–2064.

Zhao H, Star RA, Muallem S. 1994. Membrane localization of H^+ and HCO_3^- transporters in the rat pancreatic ducts. *J Gen Physiol* **104:** 57–85.

The Cystic Fibrosis Intestine

Robert C. De Lisle[1] and Drucy Borowitz[2]

[1]Anatomy and Cell Biology, University of Kansas School of Medicine, Kansas City, Kansas 66160
[2]State University of New York at Buffalo School of Medicine and Biomedical Sciences, Women and Children's Hospital of Buffalo, Buffalo, New York 14222

Correspondence: rdelisle@kumc.edu

The clinical manifestations of cystic fibrosis (CF) result from dysfunction of the cystic fibrosis transmembrane regulator protein (CFTR). The majority of people with CF have a limited life span as a consequence of CFTR dysfunction in the respiratory tract. However, CFTR dysfunction in the gastrointestinal (GI) tract occurs earlier in ontogeny and is present in all patients, regardless of genotype. The same pathophysiologic triad of obstruction, infection, and inflammation that causes disease in the airways also causes disease in the intestines. This article describes the effects of CFTR dysfunction on the intestinal tissues and the intraluminal environment. Mouse models of CF have greatly advanced our understanding of the GI manifestations of CF, which can be directly applied to understanding CF disease in humans.

The gene responsible for CF is *CFTR*, a cyclic adenosine monophosphate (cAMP)-regulated anion channel expressed at high levels in various epithelia and at much lower levels in many other cell types. As in most other epithelial organs, CFTR in the intestine mediates secretion of chloride, bicarbonate, and fluid. Mutations in the CFTR gene may result in partial or total loss of function (see Ferec and Cutting 2012), resulting in secreted fluid with abnormally low volume and aberrant electrolyte composition. The altered milieu on the epithelial surface is believed to be the major cause of pathogenesis in affected organs, including the intestine. This review will focus on pathologies of the intestine, including how luminal functions (digestion, host–microbial interactions) are affected, and therapeutic interventions based on our understanding of CF pathogenesis in the intestine.

CFTR DISTRIBUTION IN THE INTESTINE

The *CFTR* gene is weakly expressed in the stomach, but is more strongly expressed all along the intestinal tract (Strong et al. 1994). There is a cephalad-caudad gradient with *CFTR* messenger RNA (mRNA) levels highest in the duodenum, including high expression in the mucus secreting Brunner's glands, and levels decrease distally along the small intestine to the ileum. There is moderate *CFTR* expression in the large intestine. There is also a gradient of expression along the crypt–villus axis with greatest expression in the crypts in the small intestine and near the base of the crypts in the large intestine. Even though *CFTR* mRNA levels show a strong crypt-to-surface gradient, this is not necessarily indicative of protein levels. Immunolocalization data for the small intestine show CFTR protein levels in the villi that exceed predictions based on very low mRNA levels in these cells (Jakab et al. 2011). It has been suggested that because enterocytes have a short life span on the intestinal villi (3–5 d), CFTR protein can be synthesized in crypt cells where its mRNA is more abundant, and the protein product may last for the time that it takes the cells to migrate from the crypt up to the villus tip where they turn over.

The localization of *CFTR* expression in the intestine reflects the need for bicarbonate and fluid secretion. Bicarbonate secretion is highest in the proximal intestine that receives a high acid load from the stomach. This acidity is normally neutralized by bicarbonate to support optimal activity of digestive enzymes from the exocrine pancreas and solubility of bile salts from the biliary tract (Carey 1984). When CFTR function is deficient, gastric acid is not properly neutralized in the intestine and this contributes to poor digestive function in the intestine. Although the intestine is often thought of as an absorbing tissue, it also secretes significant volumes of fluid, largely from the crypts that have high CFTR levels, where it is thought to flush out this cul-de-sac to maintain a sterile environment to protect the stem cells at the bottom of the crypt. In addition, the intestine receives a large volume of bicarbonate-rich fluid from the pancreas (see Wilschanski and Novak 2013).

CLINICAL CONSEQUENCES OF CF IN THE INTESTINE

The gastrointestinal (GI) system is among the earliest parts of the body affected in CF, with a significant proportion of neonates showing damage. About 60% of patients with severe CFTR mutations (see article in this collection by Ferec and Cutting 2012) are born with exocrine pancreatic insufficiency (EPI) (Cipolli et al. 2007), which progresses with age to include 85%–90% of patients (see Wilschanski and Novak 2013). Loss of bicarbonate-rich pancreatic fluids in patients with EPI alters the intestinal intraluminal milieu. The lack of this major acid neutralizing power, the lack of fluid for hydration of the intestinal contents (Kopelman et al. 1985), and reduced enzymatic capacity to digest ingested food contributes in a multifactorial

way to GI symptoms and signs. Of note, patients with CF who are pancreatic sufficient have decreased bicarbonate and volume secretion (Kopelman et al. 1988) but produce sufficient flow to allow adequate enzyme secretion for digestion.

The liver and biliary tract are also affected in CF but more variably so. CF-related liver disease occurs in 5%–15% of patients with CF (Lindblad et al. 1999; Colombo et al. 2006), usually in the first decade of life. Clinically significant CF liver disease rarely develops de novo in patients over age 18 (Bhardwaj et al. 2009). Even in CF patients without clinical liver disease there is fecal loss of bile acids (O'Brien et al. 1993), which can lead to poor micelle formation, contributing to fat and vitamin malabsorption. However, in CF mice fecal loss of bile salts does not appear to affect fat absorption (reviewed in Wouthuyzen-Bakker et al. 2011).

Obstruction

Most or perhaps all of the wide range of effects of CF on the gut are secondary to the decrease in anion and fluid transport caused by loss of CFTR function. Loss of CFTR function results in a relatively dehydrated luminal environment, and deficiency in this bicarbonate secretion results in prolonged postprandial acidity in the proximal small intestine (Dalzell and Heaf 1990; Robinson et al. 1990; Barraclough and Taylor 1996; Gelfond et al. 2012). A major consequence of the altered luminal environment is the accumulation of mucus in the CF intestine. Along with adequate fluid volume, bicarbonate appears to be crucial to the normal expansion and solubility of intestinal mucus (Garcia et al. 2009; Verdugo 2012). It can be argued that excessive mucus accumulation is the most important pathological event in CF disease, as this contributes to most of the other sequelae of this disease.

The most serious acute complication of the intestine in CF is obstruction of the terminal ileum or proximal large intestine, which if untreated can result in rupture and sepsis. Such obstruction in CF neonates is called meconium ileus (MI). Ileus means diminished or absent peristalsis, in this case associated with failure

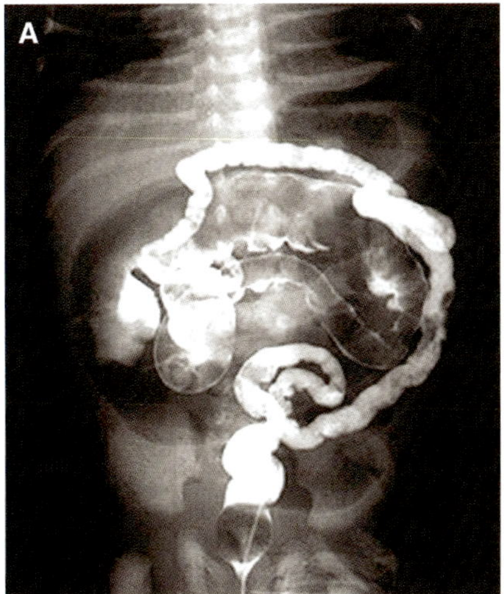

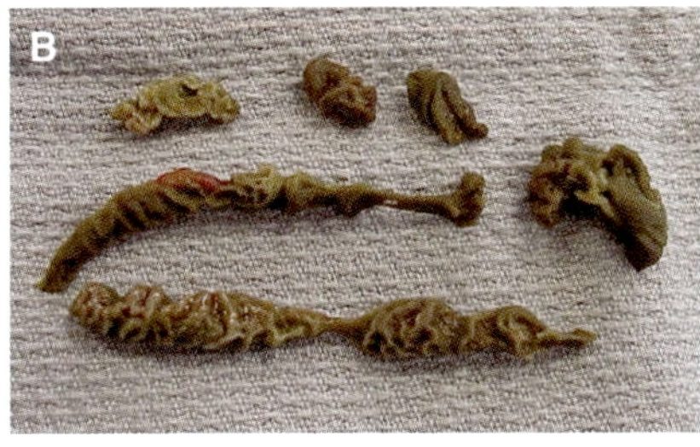

Figure 1. Obstruction of the intestines in CF. (*A*) Microcolon of disuse with prestenotic dilatation, and (*B*) meconium casts taken from an infant with cystic fibrosis and meconium ileus.

to pass meconium, the in utero intestinal contents. Normal meconium is already thick, being composed of shed intestinal epithelial cells, mucus, bile, amniotic fluid, and ingested fetal hair (lanugo). In CF, the meconium is more viscid in consistency and will even form casts of the intestine (Fig. 1). MI occurs in ~20% of CF neonates (CF Foundation Patient Registry 2010, www.CFF.org) and is associated with severe CFTR mutations (Feingold and Guilloud-Bataille 1999). MI often can be treated by radiologically guided contrast enema with a solution containing a nonabsorbable osmolyte (e.g., Gastrograffin, 3,5-diacetamido-2,4,6-triiodobenzoic acid), which draws fluid into the intestine allowing the meconium to be flushed out. In some cases, surgical removal of the blockage with or without intestinal resection is required.

Obstruction can also occur in older CF patients and is called distal intestinal obstructive syndrome (DIOS) (Houwen et al. 2010). The material forming the obstruction in DIOS has been described as being "mucofeculant" to indicate its composition of mucus and fecal-type matter (undigested materials and a high content of bacteria). This material is very thick with a putty-like consistency that adheres strongly to the mucosal surface. A bubbly granular mass can be seen in the right lower quadrant on abdominal radiographs (Fig. 2). Similar to MI, DIOS is often successfully treated using osmotic agents (Cleghorn et al. 1986). Although MI and DIOS are similar they are likely distinct entities because less than half of patients that develop DIOS had MI as infants (Houwen et al. 2010). Also, genetic twin studies of CF patients showed the risk for development of DIOS is caused primarily by nongenetic (environmental) factors, whereas there is strong evidence for effects of modifier genes for the occurrence of MI (see Blackman et al. 2006; Knowles and Drumm 2012).

Chronic, low-grade obstruction is also present in patients with CF as constipation or ob-

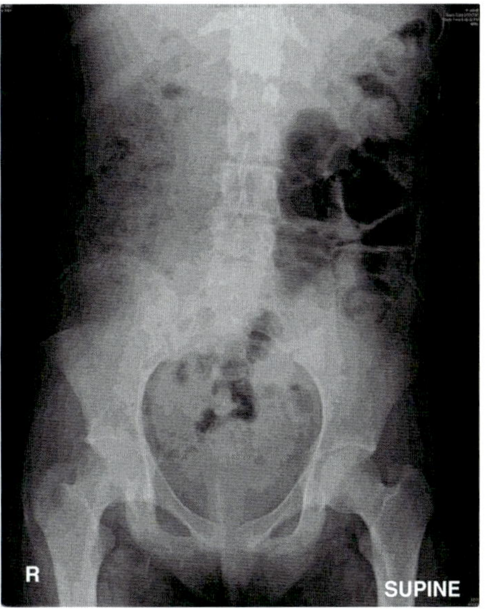

Figure 2. Bubbly granular mass in the right lower quadrant of a patient with distal intestinal obstruction syndrome (DIOS).

stipation. The prevalence has been reported to be between 26% and 46% of patients with CF (Rubinstein et al. 1986; van der Doef et al. 2010). This condition exists even in patients having bowel movements each day. A diagnostic schema has recently been developed that is based on symptoms of abdominal pain or a change in frequency or consistency of stools, plus relief of symptoms by use of an osmotic laxative such as polyethylene glycol (Houwen et al. 2010). A potential new treatment is lubiprostone, which was approved for chronic idiopathic constipation in the general population in 2006. Lubiprostone, a synthetic eicosanoid, was originally shown to activate the CLC2 chloride channel (Cuppoletti et al. 2004), which enhances fluid secretion into the gut thus increasing bowel movements. The fact that lubiprostone targeted a non-CFTR mechanism makes it an attractive candidate for CF therapy. However, lubiprostone may act via prostaglandin receptors to activate CFTR (Bijvelds et al. 2009; Wood 2010). Despite this uncertainty, there are some positive data using human CF tissues (Joo et al. 2009). Recently, a case series was published on the use of lubiprostone as a treatment for chronic constipation in adults with CF (O'Brien et al. 2010). These definitions and the efficacy of treatments that focus on increasing intraluminal fluid suggest that chronic constipation or obstipation likely results primarily from poor hydration of the intestinal lumen in individuals with CF.

Microbial Dysbiosis

One of the hallmarks of CF lung disease is the colonization of the usually nearly sterile airways with bacteria such as *Staphylococcus aureus*, *Pseudomonas aeruginosa*, and a range of other microorganisms. In contrast to the airways, the GI tract in healthy individuals is home to a wide variety of bacteria, the majority of which are localized to the cecum and colon, where they contribute to the bulk of fecal material. In patients with CF, the GI microbiota is altered in diversity, location, and cell density. In health, the summated actions of gastric acid, peristalsis, intestinal fluid, antibacterial proteins, and the ileocecal valve keep the quantity of bacteria in

the small intestine at low levels. Failure of any of these mechanisms can lead to small intestinal bacterial overgrowth (SIBO), a syndrome that can result in diarrhea, abdominal distension, flatulence, steatorrhea, macrocytic anemia, and weight loss (Singh and Toskes 2003). Weight loss may result from bacterial competition for ingested nutrients, intestinal inflammation from enterotoxic metabolites, and the ability of bacteria to deconjugate bile acids, thus reducing their ability to emulsify fat (Hofmann and Mysels 1992).

The most commonly used diagnostic for SIBO is breath testing after ingestion of a test sugar (Braden 2009). This approach measures exhaled hydrogen or methane, which reports the time at which the test sugars encounter bacteria in the gut. Hydrogen and methane cannot be produced by host-cell metabolism but are the products of microbial fermentation; the magnitude of the gas levels is proportional to the concentration of bacteria. Using breath testing, SIBO has been found in about one-half of CF patients (Lewindon et al. 1998; Fridge et al. 2007; Lisowska et al. 2009, 2010). Also using breath testing, several studies have reported that orocecal transit is frequently slower in CF (Bali et al. 1983; Dalzell et al. 1990; Escobar et al. 1992; Murphy et al. 1992; Lewindon et al. 1998). Although breath testing is easy to perform, there are limitations that complicate interpretation and affect the sensitivity and specificity of this approach (for a review, see Braden 2009).

Mucus accumulation in the intestinal lumen in CF creates a niche for abnormal microbial colonization, which can lead to SIBO, a form of microbial dysbiosis. Detailed characterization of the bacteria that overgrow the human CF small intestine requires direct sampling by the invasive technique of endoscopy. There is also the issue of what region to sample (duodenal vs. ileal) and what type of sample (luminal content vs. mucosa) should be obtained to study bacteria in the intestine (Bhat et al. 1980). In CF, the bacteria are likely to be predominantly in the mucus as they are in the CF mouse (Norkina et al. 2004), which is poorly soluble and adheres to the gut wall. A full understanding of the microbiota of the CF small intestine may therefore require, in addition to luminal samples, biopsy of the intestinal tissue and the adherent mucus to have the best chance of obtaining appropriate samples. At present, only the most proximal small intestine can be biopsied nonoperatively, excluding the jejunum and proximal ileum from sampling. Although enteroscopy can be used to obtain both tissue and aspirates from these parts of the small bowel, it is not part of routine practice. Altered small intestinal microbiota is associated with several clinical conditions (Cotter 2011), but this has not been studied yet in CF.

Because of the complexity of the gut microbiota and the fact that the majority of these bacteria have never been cultured, molecular techniques are much more appropriate for such investigations than traditional culture-based approaches. This involves use of PCR amplification and either analysis by electrophoresis or by sequencing of the bacterial 16S ribosomal gene, which has enough variability to identify bacteria often to the species level (Gill et al. 2006) and can be readily applied to the human fecal microbiota (Qin et al. 2010a,b). Analysis of the fecal microbiota is much easier than small intestine, and has begun to be applied to stool samples from patients with CF. One such study showed decreased species richness and less temporal stability of the CF fecal microbiota as compared with healthy controls (Duytschaever et al. 2012). Other work investigated early life development of the gut and airway microbiota in CF (Madan et al. 2012). The investigators suggested "that nutritional factors and gut colonization patterns are determinants of the microbial development of respiratory tract microbiota in infants with CF and present opportunities for early intervention in CF with altered dietary or probiotic strategies." Clearly, these are exciting times for novel insights into the roles of our microbiota in health and in disease.

In addition to the unique niche formed by static mucus in the CF intestine, there are other factors that contribute to microbial dysbiosis in CF, such as the need for periodic acute courses of antimicrobials as well as chronic antibacterial therapy to combat airway infections. Even when administered by inhalation, significant amounts of antibiotics will be ingested and thus also affect the gastrointestinal tract. Because large numbers of bacteria are normal residents of the healthy gut, antibiotic use is a two-edged sword (Sekirov et al. 2010). Antibiotic use suppresses normal commensal strains and can foster selective antibiotic-resistant bacterial overgrowth that will exploit the empty niche left by the eradicated organisms.

An important commensal organism commonly lost in CF owing to antibiotic use is *Oxalobacter formigenes* (Sidhu et al. 1998). These bacteria metabolize oxalate. Patients with CF are at increased risk of hyperoxaluria and formation of calcium-oxalate kidney stones. Conversely, an example of a pathogen acquired in CF is *Clostridium difficile*, which can be detected in fecal samples of >50% of CF patients (Yahav et al. 2006).

C. difficile infection is a common occurrence with antibiotic use even in non-CF patients. Interestingly, most CF patients with *C. difficile* infections are asymptomatic, although a small percentage can develop colitis. A major effect of this pathogen on the gut is induction of secretory diarrhea mediated by a toxin that activates CFTR-dependent Cl secretion. Deficiency of functional CFTR in CF may explain the lack of symptoms in patients infected with *C. difficile*.

Besides antibiotic use, the altered luminal environment and physiology of the CF small intestine is expected to contribute to microbial dysbiosis. The healthy intestine has low levels of bacteria proximally and greater numbers distally. There are especially high levels in the colon where trillions of bacteria reside. A number of factors act to limit bacterial load in the proximal intestine (Husebye 2005). The first is acidity in the stomach, which is strong enough to kill most ingested microbes. This barrier is especially important in patients with CF because of swallowed sputum. Many patients are on acid blockers to aid digestion. Others use acid blockers because of gastroesophageal reflux. Use of these medications allows more ingested bacteria to survive passage through the stomach into the small intestine.

Another major mechanism to maintain low bacterial load in the proximal small intestine is the combined activity of gel-forming soluble mucins secreted from goblet cells and intestinal motility (Nieuwenhuijs et al. 1998). Mucus has complex rheological properties in that it can be adherent and at the same time acts as a lubricant. In terms of controlling bacterial load in the proximal small intestine, mucus can be thought of as acting in a manner analogous to floor sweeping compound: As dirt and debris stick to floor sweeping compound, so do bacteria bind to the complex oligosaccharides on the mucin molecules. And, as a broom easily cleans up floor sweeping compound with its bound dirt, so does intestinal motility act as the broom that sweeps the mucus–bacteria complex distally toward the large intestine. The migrating motor complex in the small intestine creates strong, caudally directed waves of smooth muscle contraction during the interdigestive period between meals. In the CF intestine the mucus is excessively viscid and intestinal motility is abnormally slow. Under these conditions, bacteria behave opportunistically by binding to the static mucus resulting in abnormal colonization and overgrowth of the small intestine.

The microbiota of the gut has important effects on the whole body's immune system especially during postnatal development (Edelman and Kasper 2008; Hansen et al. 2010; Takahashi 2010; Quigley 2011). Thus, there are important implications of microbial dysbiosis in the gut on distant organs, including the airways. Although there has been much speculation regarding the potential therapeutic manipulation of the gut microbiota in CF using probiotics, "beneficial" bacteria, there have been very few actual studies (Bruzzese et al. 2004, 2007; Weiss et al. 2010). These limited results are promising but much work remains to be done to provide convincing evidence of benefits and to define which probiotic organisms should be administered and at what concentration and dosing interval.

Inflammation

The presence of intestinal inflammation in patients with CF has received little recognition until recently. Investigations into this aspect of CF GI disease came in the wake of the epidemic of

fibrosing colonopathy, a complication related to exposure to excessive amounts of pancreatic replacement therapy (PERT) seen in the early 1990s (Schwarzenberg et al. 1995). Endoscopic lavage of the CF small intestine showed increased levels of inflammatory markers in the lumen (Smyth et al. 2000). A conventional endoscopic study showed that the CF duodenum was morphologically normal but there were increased levels of several inflammatory markers in biopsied tissue (Raia et al. 2000). More recent work using video-equipped capsule endoscopy showed that morphological abnormalities including edema, erythema, mucosal breaks, and ulcerations occur in the jejunum and ileum in >60% of CF patients (Werlin et al. 2010). That study also reported significant elevations of fecal calprotectin (a neutrophil secretory product) in many CF patients, consistent with intestinal inflammation. However, whether fecal calprotectin indicates colonic or small intestinal inflammation is not clear. Although calprotectin is well correlated with colonic inflammation, it appears less reliable for inflammation in the proximal GI tract (Summerton et al. 2002; Montalto et al. 2007, 2010; Garcia-Sanchez et al. 2010).

Crohn's disease is a multigenic inflammatory bowel disease that occurs with increased frequency in CF. Crohn's susceptibility includes genetic (MacDonald and Monteleone 2005) as well as environmental factors, such as aberrant immunological interactions with gut microflora, pathogenic bacteria, or exposure to bacterial antigens, factors that are present in patients with CF. There have been reports of an increased prevalence of Crohn's disease, up to 12.5-fold greater in patients with CF than in the general population (Lloyd-Still 1994). Incident cases must be confirmed by biopsy, because serum markers for inflammatory bowel disease may be falsely positive or negative in patients with CF (Condino et al. 2005).

Celiac disease (CD) is a destructive autoimmune condition of the small intestinal mucosa that causes malnutrition and is also more prevalent in CF (Fluge et al. 2009; Walkowiak et al. 2010). The immune response in CD is elicited by wheat gliadin and prolamine protein present in grains. The occurrence of CD varies from country to country, and this is also reflected in patients with CF. In one study, no CD patients with CF were found among subjects from Denmark, a country with a very low incidence of CD, whereas the majority of CF with CD were Swedish, from an area of Sweden with a high incidence of CD (Fluge et al. 2009). Tissue transglutaminase immunoglobulin A (TTG-IgA), a serologic marker used to screen for CD, can be elevated in patients with CF in the absence of elevated endomysial antibodies, another serologic marker, and in the absence of histological evidence of active celiac disease (Fluge et al. 2009; Hasosah et al. 2009); these patients may have falsely positive TTG-IgA or may have latent CD. Histologic diagnosis depends on finding villus atrophy and elevated intraepithelial lymphocytes. It is possible that CF predisposes to development of CD owing to impaired protein digestion, which normally degrades the gliadin antigen, as a result of pancreatic insufficiency. Alternatively, CD susceptibility in CF could be a consequence of an already inflamed intestine. In any case, because celiac disease seriously affects nutrition and it is treatable by removal of grains from the diet, it is important to screen for its occurrence in patients with CF who are growing poorly, which requires small intestinal biopsies to verify the diagnosis of CD if it is suspected on the basis of serologic testing.

The Role of Acidity in the Intestinal Lumen

Control of acidity in the intestinal lumen is important in CF for the effectiveness of PERT. Because lipase and protease are irreversibly inactivated in acid, PERT is formulated as microbeads that are enteric coated so that the enzymes can pass unharmed through the acid environment of the stomach and then dissolve in the less acidic pH of the duodenum. In people with EPI gastric acid is not neutralized in the proximal small intestine and PERT capsules may dissolve more distally in the small intestine than is desired. The proximal absorptive surface of the upper small bowel will be bypassed, and the mixing of bile salts with ingested triglycerides and the digested fatty acids will be aberrant. This will result in loss of normal bile salt-mediated

micelle formation needed for efficient uptake of digested lipids by enterocytes. Additionally, the acidic pH of the proximal small intestine in CF can cause precipitation of bile salts, further impairing their ability to participate in lipid digestion and assimilation (Freudenberg et al. 2008).

Endoscopy has been used to measure pH in the intestinal tract of CF patients (Dalzell and Heaf 1990; Robinson et al. 1990; Barraclough and Taylor 1996). These studies showed that the proximal small intestine is more acidic and for a longer time in CF as compared with control patients. This abnormality was recently confirmed using capsule endoscopy (SmartPill), which showed that the CF small intestine was abnormally acidic for the first 30 min after gastric emptying (Gelfond et al. 2012), a critical time for PERT tablet dissolution and mixed micelle formation of bile salts with lipid digestion products.

Malnutrition and Failure to Thrive

A common issue in CF is maldigestion and/or malabsorption of lipids (steatorrhea) and fat-soluble vitamins (Rovner et al. 2007). The longevity of CF patients has improved dramatically over the past few decades, in good part owing to increased attention to nutritional support. Poor nutritional status in CF is highly correlated with deterioration of lung function and is a strong independent predictor of mortality as the disease progresses (Sharma et al. 2001; Steinkamp and Wiedemann 2002; Milla 2004; Courtney et al. 2007; Stallings et al. 2008). The cause and effect relationship between nutrition and airway function is not well understood (Borowitz 1996) but the correlation of the two is very strong (Stallings et al. 2008). Exocrine pancreatic insufficiency is the major cause of maldigestion, which is treated by PERT. However, even with optimal PERT, nutrition and growth is often not fully corrected (Baker et al. 2005; Borowitz et al. 2005). Furthermore, a significant proportion of pancreatic-sufficient patients have nutritional deficits as measured by low body mass index (Kumar et al. 2010). This indicates that pancreatic insufficiency cannot account for all nutritional issues in CF, which points to a potential problem in the small intestine.

Why fat digestion and assimilation of the digestive products is impaired in CF is only partly understood (Kalnins et al. 2007; Wouthuyzen-Bakker et al. 2011). Normal fat digestion and assimilation involves a complex series of biochemical events in the gut lumen and in the absorptive enterocyte. Investigation of fat assimilation in CF patients, using stable isotope-labeled fatty acids, revealed there was impaired postlipolytic solubilization and/or uptake of long chain fatty acids (Kalivianakis et al. 1999), indicating that intestinal mucosal abnormalities contribute to fat malabsorption. In other work, using biopsies of human tissue, it was shown that reesterification of absorbed fatty acids and their release from enterocytes was slower in CF (Peretti et al. 2006). Thus, there are defects at several points in the process of fat digestion and assimilation in the CF intestine, only some of which can be readily explained by the altered luminal environment. A further influence on malnutrition may be the altered microbiota of the CF intestine. As noted above, bacteria in the intestine compete with the host for ingested nutrients, can deconjugate bile salts reducing their effectiveness at emulsifying fats for digestion and absorption (Hofmann and Mysels 1992), and the inflammation they cause can affect mucosal digestive functions (Peuhkuri et al. 2010).

WHAT WE HAVE LEARNED FROM MOUSE MODELS OF CF

Use of transgenic mouse models of CF has contributed greatly to our understanding of normal electrolyte transport mechanisms in the intestine (Seidler et al. 2009; see also Bridges 2012; Frizzell and Hanrahan 2012; Park and Lee 2012). In addition they have proved excellent models for investigation of the CF intestinal phenotype. There are many mouse models of CF, which include conventional knockouts of the *Cftr* gene as well as targeted mutations of *Cftr* that are known to be disease causing in humans (for a review, see Grubb and Gabriel 1997; Guilbault et al. 2007). Unlike human CF, CF mice are pancreatic sufficient and have only mild airway involvement

(Grubb and Gabriel 1997). Thus, these models have helped elucidate the effects of intestinal abnormalities in CF without the confounding influence of EPI.

The major phenotypes in CF mouse models are intestinal obstruction, which can be lethal, and poor growth ("failure to thrive"), recapitulating symptoms and signs seen in humans with CF. The severity of the intestinal phenotype varies widely and depends on the degree of loss of functional Cftr. By comparing these different CF mouse lines it has been estimated that 10%–15% of normal Cftr function is sufficient to avoid lethal intestinal problems (Guilbault et al. 2007).

A working hypothesis for the pathophysiology of CF in the intestine is the following sequence of events: (1) loss of functional CFTR results in deficient anion and fluid transport; (2) the altered luminal environment impairs turnover and clearance of mucus; (3) static mucus allow abnormal bacterial colonization (microbial dysbiosis); (4) microbial dysbiosis alters immune system behavior; and (5) immune responses further stimulate mucus production, etc. Studies using CF mice support this model, as reviewed here.

Obstruction

The first CF mouse, which was a conventional knockout mouse generated shortly after discovery of the CF gene, confirmed the essential role of CFTR in Cl^- secretion in the intestine (Snouwaert et al. 1992). It was also recognized from the first that excessive mucus accumulation in the gut caused intestinal obstruction, the major phenotype of these mice. This is accompanied by increased density of goblet cells in the intestinal epithelium (Beharry et al. 2007) even though expression of the major intestinal mucin genes (*Muc2*, *Muc3*) is not elevated (Parmley and Gendler 1998; De Lisle et al. 2006).

The most severe model is knockout of *Cftr* resulting in total loss of expression (the *Cftr*[tm1UNC] mouse) (Snouwaert et al. 1992). However, even in this line of CF mice the intestinal phenotype varies depending on the genetic background of the mice (Gyömörey et al. 2000;

Norkina and De Lisle 2005), indicating the presence of modifier genes, similar to MI in human CF (see Knowles and Drumm 2012). On the C57BL/6 background, the phenotype is most severe and >95% of such $Cftr^{-/-}$ mice die of intestinal obstruction, about one-half before weaning and the rest succumbing within a week of weaning (Snouwaert et al. 1992). The majority that survive to weaning can be rescued by use of an elemental liquid diet (Peptamen) that is easily assimilated and lacks the high levels of dietary fiber found in regular solid chow (Eckman et al. 1995). An alternative is to replace the animals' drinking water with polyethylene glycol-based osmotic laxative solution, which allows the CF mice to survive on standard solid chow (Clarke et al. 1996).

Accumulation of mucus in the CF mouse intestine can be reduced by altering electrolyte transport pathways to compensate for loss of CFTR. In so-called "gut corrected" CF mice, in which human CFTR is expressed in villus cells driven by the fatty acid-binding protein promoter, the intestinal phenotype is reversed and these mice do not show intestinal mucus accumulation or obstruction (Zhou et al. 1994). There has been one study of the effects of the putative Clc2 chloride channel activator lubiprostone on the intestinal phenotype of CF mice, showing some improvement in immune activity, but unexpectedly it also increased mucus accumulation (De Lisle et al. 2010a). These data are more consistent with activation of prostaglandin receptors than activation of the Clc2 Cl^- channel.

CF mice have also been used to test the roles of candidate modifier genes. In one study it was shown that CF mice that are also transgenic for increased expression of *Clca3* (also known as *Gob5*) had less intestinal obstruction and improved survival (Young et al. 2007). When discovered, it was thought Clca3 was a calcium-regulated chloride channel, but subsequent work showed that this protein cannot be a channel because it is a soluble protein expressed in goblet cell mucin granules and released when the mucins are exocytosed (Mundhenk et al. 2006). The function of Clca3 is unknown at this time, so how it ameliorates the CF intestinal phenotype

remains to be discovered. Another gene, *Muc1*, which encodes a membrane-associated mucin that has signal transduction capabilities, was shown to affect the severity of the intestinal phenotype in CF mice. Knockout of *Muc1* in CF mice resulted in reduced mucus accumulation in the large intestine (Malmberg et al. 2006), but the mechanism by which this occurred is not known. Muc1 is not quantitatively important as a mucin, so its signaling properties may affect either mucus production or solubility of gel-forming secreted mucins in the CF intestine.

Infection

Studies in CF mice support the hypothesis that static mucus allows microbial dysbiosis with increased numbers as well as loss of species diversity (Table 1) (Norkina et al. 2004). CF mice given a diet of standard chow and an osmotic laxative are more susceptible to colonization with pathogenic bacteria (Clarke et al. 2004). This study showed that mucus accumulation

Table 1. Classification of bacteria in wild-type and CF small intestines by sequencing of cloned 16S polymerase chain reaction (PCR) products

	Number of clones	
Classification	Wild-type mouse	CF mouse
Uncultured bacteria	37	2
Klebsiella (oxytoca, pneumoniae, and *granulomatis)*	20	0
Enterobacter (aerogenes or *asburiae)*	13	0
Other Enterobacteriaceae	5	95
Lactococcus lactis	9	0
Acinetobacter sp.	5	0
Pantoea (sp. and *agglomerans)*	4	0
Unclassified β proteobacter	2	0
Helicobacter typhlonicus	1	1
Bacteroides acidofaciens	1	0
Streptococcus (gordonii or *parasanguinis)*	1	0
Clostridium perfringens	0	6

From Norkina et al. 2004; modified and reproduced, with permission, from the authors.

in the crypt lumen creates a physical blockage that prevents access of the Paneth cell antibacterial products to the intestinal lumen proper. The interpretation was that this mucus blockage impairs innate defense mechanisms of the Paneth cell thus resulting in microbial dysbiosis. In the same strain of *Cftr* knockout mice maintained on a liquid diet instead of laxative, there is overt SIBO (Norkina et al. 2004) with even greater mucus accumulation and crypt lumen blockage as compared with laxative-treated mice (De Lisle et al. 2007). Importantly, SIBO occurs within 4 days of birth in CF mice (Canale-Zambrano et al. 2010) and is largely reversed by laxative treatment given at weaning (De Lisle et al. 2007; Canale-Zambrano et al. 2010).

As discussed above, slow small intestinal transit can contribute to SIBO and is common in CF patients as well as in *Cftr* knockout mice (De Lisle 2007). Investigation in CF mice showed that slow intestinal transit could be corrected using laxative treatment (De Lisle et al. 2007). As laxative also results in eradication of SIBO, this shows an association of dysmotility to SIBO. However, direct eradication of SIBO with broad spectrum antibiotics did not correct transit in CF mice, and in fact resulted in slowed transit in wild-type mice (De Lisle 2007). This result highlights the complex relationship between the intestinal microbiota and physiological functioning of the intestines. It is known that the composition of the intestinal microbiota, not just their numbers, is important. Germ-free animals have slowed intestinal transit and colonization with specific bacteria can either maintain slow transit (e.g., *Escherichia coli*) or enhance transit to usual rates (e.g., primitive anaerobic fermenters) (Caenepeel et al. 1989; Husebye et al. 2001).

To investigate the mechanisms of slowed intestinal transit in the CF mouse, ex vivo studies of intestinal tissue were performed. It was found that the circular smooth muscle from CF mouse small intestine had abnormal spontaneous activity and did not respond to cholinergic stimulation or even direct depolarization with KCl (De Lisle et al. 2010b). This impaired smooth muscle activity was found to be associated with elevated PGE_2 levels. PGE_2 has a dominant relaxant effect on the circular smooth muscle of the intestine, resulting in inhibition of normal motility programs. Interestingly, broad spectrum antibiotics that perturb the microbiota result in similar changes with increased PGE_2 levels and decreased circular smooth muscle activity (De Lisle et al. 2010b). These data reinforce the important role of the microbiota in normal smooth muscle activity.

Inflammation

In addition to mucus accumulation and SIBO, CF mice have increased innate immune activity in the intestine, with increased levels of neutrophils and mast cells and up-regulation of various marker genes (Norkina et al. 2004). At this time, it is difficult to prove a direct cause and effect relationship between microbial dysbiosis and altered immune function in the CF intestine. However, there are associations between SIBO, altered immune system activity, and increased mucus production in CF mice. Laxative-treated CF mice have fewer neutrophils and mast cells and reduction of inflammation marker gene expression (De Lisle et al. 2007). If SIBO is directly eradicated with broad spectrum antibiotics, there is reduced mucus accumulation (De Lisle et al. 2006) as well as significantly less inflammation in the CF mouse intestine (Norkina et al. 2004). These results support the model that SIBO stimulates an immune response, a consequence of which is greater mucus release from goblet cells. This is not surprising, as mucus is normally protective of epithelia like that lining the intestine, and an appropriate response to bacterial overgrowth is increased mucus production. However, in the environment of the CF intestine increased mucus is detrimental because mucins do not expand properly and the aberrant mucus does not turn over and fails to carry away bacteria.

As SIBO develops, it likely increases mucus secretion via inflammatory pathways. An example of this is the elevated level of PGE_2 observed in the CF mouse intestine (De Lisle et al. 2008). Prostaglandins are increased in an unusual manner in the CF mouse intestine and in the intestine of wild-type mice treated with broad spectrum antibiotics. In both cases there is decreased expression of the major prostaglandin degradative enzymes, prostaglandin dehydrogenase (Hpgd) and prostaglandin reductase 1 (Ptgr1) (De Lisle et al. 2008, 2010b). Interestingly, it has been shown that *HPGD* is a tumor suppressor gene whose expression is often decreased in GI cancer (Myung et al. 2006). It is also known that there is a greatly elevated risk of cancers of the large and small intestines in CF (Maisonneuve et al. 2003) and decreased *HPGD* expression is a potential candidate for this increased cancer risk in CF.

Work crossing CF mice with other transgenic lines has shown the importance of inflammation and mucus in intestinal obstruction. In one study, CF mice were crossed with mice deficient in the lipopolysaccharide receptor (toll-like receptor 4, Tlr4) (Canale-Zambrano et al. 2010). It was found that CF mice with a single wild-type copy of *Tlr4* had increased survival (less mucus obstruction) but that CF mice with no functional *Tlr4* had even more rapid death in the postnatal period as compared with CF mice with two wild-type copies of *Tlr4*. This study illustrates the complex relationship between the immune system and CF pathologies.

Electrolyte Secretion by the Intestinal Mucosa

Studies in CF mice have shown that duodenal bicarbonate secretion by the epithelium involves two pathways: electroneutral secretion via a CFTR-assisted Cl^-/HCO_3^-) exchange process and an electrogenic secretion of HCO_3^- via a CFTR conductance pathway (Clarke and Harline 1998). During cAMP stimulation of the CF duodenum, a small net increase in base secretion can be measured owing to cAMP inhibition of the sodium–hydrogen exchanger (NHE) activity rather than increased HCO_3^- secretion (Clarke et al. 2001). Use of talniflumate (LoMucin), a pharmacological agent that inhibits Cl^-/HCO_3^- exchange involved in intestinal NaCl absorption, improved survival of CF mice and tended to reduce mucus accumulation (Walker et al. 2005). Similarly, loss of NaCl absorption mediated by NHE in the intestine reduced the incidence of intestinal obstruction in CF mice (Bradford et al. 2009).

Malnutrition and Failure to Thrive

CF mice are an excellent model to separate the influence of EPI from other GI factors on malnutrition and failure to thrive. CF mice on the liquid diet are only ~70% the body weight of wild-type mice on the same diet (Norkina et al. 2004) and weight gain can be improved by eradicating SIBO, either with broad spectrum antibiotics (Norkina et al. 2004) or less so using laxative (De Lisle et al. 2007). Interestingly, the CF mice consume 20%–25% less food, and pair-feeding this smaller amount of food to wild-type mice results in them being proportionately smaller than ad libitum fed mice. These observations implicate appetite and satiety control problems in the CF failure to thrive phenotype. There are some data in human CF showing alterations in appetite controlling hormones leptin and ghrelin (Stylianou et al. 2007) but the significance of such changes is not yet understood. Similar to human CF, CF mice have impaired lipolysis as well as impaired postlipolytic uptake of fatty acids, and the former was improved by inhibition of gastric acid (Bijvelds et al. 2005). A complete understanding of fat maldigestion in CF remains an important goal.

CONCLUDING REMARKS

We have gained much insight into CF pathogenesis in the intestine, in large part using CF mouse models to confirm and expand the limited data from CF patients. A strength of animal models is they allow more in-depth investigation and testing of potential therapeutic approaches. What we have learned also holds promise for therapeutics to improve quality of life in CF by reducing gut issues that cause "nonspecific" symptoms such as gas and bloating. For example, the many benefits of continuous osmotic laxative use in the CF mouse detailed above suggest that such use may be beneficial to prevent/alleviate GI symptoms in CF patients. There is a need for clinical trials to evaluate these agents as routine prophylaxis to prevent development of DIOS and potentially improve overall gut function, which may have benefits for nutrition, overall health, and quality of life for CF patients.

In addition to improving understanding of gut issues in CF, what we have learned from investigation of the CF mouse intestine is informative about the CF airways. Even though the CF mouse does not have airway disease, all the proposed steps in CF pathogenesis in the intestine are applicable to CF airway disease. Loss of CFTR-mediated anion and fluid transport results in mucus accumulation in the CF airway. Static mucus allows infection of the airways in CF. Unlike the gut, the airways normally maintain a pseudosterile environment, and when infection occurs in the CF airway there is a more pronounced immune response than in the gut. But, similar to the CF gut response to microbial dysbiosis, airway infection also increases mucus production, which exacerbates the situation. And, similar to use of laxatives for intestinal issues, inhalation of the osmotic agents hypertonic saline (Wark and McDonald 2009) and mannitol (Jaques et al. 2008; Bilton et al. 2011) have been shown to stabilize or improve airway function, probably by helping hydration of mucus and restoring the periciliary fluid layer required for normal cilia beating.

An important issue that needs more investigation is the intestinal microbiota in CF. CF patients receive frequent antibiotics to combat airway infection, but there are no data on the impact of these treatments on the CF gut. There is a need to define the effects of such antibiotics on the gut microbiota, the functional consequences of the altered GI microbiota on intestinal function, and the influence of such changes on the whole body. For example, recent work has drawn a link between the gut microbiota and susceptibility to development of the airway disease asthma (Couzin-Frankel 2011). CF gut microbial dysbiosis likely occurs shortly after birth because the intestine is already affected at this time. Thus, the immune system in CF may develop abnormally because of microbial dysbiosis in the gut. Through altered immune system development, there is a great potential that the gut microbiota has important consequences for the CF airway. Although this concept has been considered for other diseases it is just beginning to be investigated in CF using the *Cftr* knockout mouse (De Lisle et al. 2011). We are in the infancy of understanding how the gut and its microbiota interact to affect health and disease states. Nevertheless, targeting the gut microbiota with prebiotics and probiotics is an up-and-coming approach because microbial dysbiosis is involved in a wide range of immune disorders.

Recently, two new animal models of CF have been developed: transgenic CF pigs (Rogers et al. 2008; Stoltz et al. 2010) and ferrets (Sun et al. 2010). Importantly, these animals better reproduce airway and pancreatic pathologies characteristic of CF. They also have a higher prevalence of severe intestinal obstruction than human CF patients, more similar to CF mice. Although the new pig and ferret models more closely reproduce human CF disease, mouse models will continue to be useful for investigation of CF intestinal issues because they lack the complication of pancreatic insufficiency, they have a more rapid generation time, and are less expensive to maintain. What we learn investigating the small intestine can be informative of CF airway disease. The use of animal models, especially mouse models, to test potential new approaches for treatment of gastrointestinal disease may also result in benefits for therapy of CF airway disease.

ACKNOWLEDGMENTS

We appreciate the thoughtful comments of Drs. Daniel Gelfond and Daniel Leung. Work in R.C.D.'s laboratory is supported by the Cystic Fibrosis Foundation and the National Institutes of Health (AI083479).

REFERENCES

*Reference is also in this collection.

Baker SS, Borowitz D, Duffy L, Fitzpatrick L, Gyamfi J, Baker RD. 2005. Pancreatic enzyme therapy and clinical outcomes in patients with cystic fibrosis. *J Pediatr* **146:** 189–193.

Bali A, Stableforth DE, Asquith P. 1983. Prolonged small-intestinal transit time in cystic fibrosis. *Br Med J (Clin Res Ed)* **287:** 1011–1013.

Barraclough M, Taylor CJ. 1996. Twenty-four hour ambulatory gastric and duodenal pH profiles in cystic fibrosis: Effect of duodenal hyperacidity on pancreatic enzyme function and fat absorption. *J Pediatr Gastroenterol Nutr* **23:** 45–50.

Beharry S, Ackerley C, Corey M, Kent G, Heng YM, Christensen H, Luk C, Yantiss RK, Nasser IA, Zaman M, et al. 2007. Long-term docosahexaenoic acid therapy in a congenic murine model of cystic fibrosis. *Am J Physiol Gastrointest Liver Physiol* **292:** G839–G848.

Bhardwaj S, Canlas K, Kahi C, Temkit M, Molleston J, Ober M, Howenstine M, Kwo PY. 2009. Hepatobiliary abnormalities and disease in cystic fibrosis: Epidemiology and outcomes through adulthood. *J Clin Gastroenterol* **43:** 858–864.

Bhat P, Albert MJ, Rajan D, Ponniah J, Mathan VI, Baker SJ. 1980. Bacterial flora of the jejunum: A comparison of luminal aspirate and mucosal biopsy. *J Med Microbiol* **13:** 247–256.

Bijvelds MJ, Bronsveld I, Havinga R, Sinaasappel M, De Jonge HR, Verkade HJ. 2005. Fat absorption in cystic fibrosis mice is impeded by defective lipolysis and post-lipolytic events. *Am J Physiol Gastrointest Liver Physiol* **288:** G646–G653.

Bijvelds MJ, Bot AG, Escher JC, De Jonge HR. 2009. Activation of intestinal Cl- secretion by lubiprostone requires the cystic fibrosis transmembrane conductance regulator. *Gastroenterology* **137:** 976–985.

Bilton D, Robinson P, Cooper P, Gallagher CG, Kolbe J, Fox H, Jaques A, Charlton B. 2011. Inhaled dry powder mannitol in cystic fibrosis: An efficacy and safety study. *Eur Respir J* **38:** 1071–1080.

Blackman SM, Deering-Brose R, McWilliams R, Naughton K, Coleman B, Lai T, Algire M, Beck S, Hoover-Fong J, Hamosh A, et al. 2006. Relative contribution of genetic and nongenetic modifiers to intestinal obstruction in cystic fibrosis. *Gastroenterology* **131:** 1030–1039.

Borowitz D. 1996. The interrelationship of nutrition and pulmonary function in patients with cystic fibrosis. *Curr Opin Pulm Med* **2:** 457–461.

Borowitz D, Durie PR, Clarke LL, Werlin SL, Taylor CJ, Semler J, De Lisle RC, Lewindon PJ, Lichtman SM, Sinaasappel M, et al. 2005. Gastrointestinal outcomes and confounders in cystic fibrosis. *J Pediatr Gastroenterol Nutr* **41:** 273–285.

Braden B. 2009. Methods and functions: Breath tests. *Best Pract Res Clin Gastroenterol* **23:** 337–352.

Bradford EM, Sartor MA, Gawenis LR, Clarke LL, Shull GE. 2009. Reduced NHE3-mediated Na$^+$ absorption increases survival and decreases the incidence of intestinal obstructions in cystic fibrosis mice. *Am J Physiol Gastrointest Liver Physiol* **296:** G886–G898.

* Bridges RJ. 2012. Mechanisms of bicarbonate secretion: Lessons from the airways. *Cold Spring Harb Perspect Med* **2:** a015016.

Bruzzese E, Raia V, Gaudiello G, Polito G, Buccigrossi V, Formicola V, Guarino A. 2004. Intestinal inflammation is a frequent feature of cystic fibrosis and is reduced by probiotic administration. *Aliment Pharmacol Ther* **20:** 813–819.

Bruzzese E, Raia V, Spagnuolo MI, Volpicelli M, De MG, Maiuri L, Guarino A. 2007. Effect of Lactobacillus GG supplementation on pulmonary exacerbations in patients with cystic fibrosis: A pilot study. *Clin Nutr* **26:** 322–328.

Caenepeel P, Janssens J, Vantrappen G, Eyssen H, Coremans G. 1989. Interdigestive myoelectric complex in germ-free rats. *Dig Dis Sci* **34:** 1180–1184.

Canale-Zambrano JC, Auger ML, Haston CK. 2010. Toll-like receptor-4 genotype influences the survival of cystic fibrosis mice. *Am J Physiol Gastrointest Liver Physiol* **299:** G381–G390.

Carey MC. 1984. Bile acids and bile salts: Ionization and solubility properties. *Hepatology* **4:** 66S–71S.

Cipolli M, Castellani C, Wilcken B, Massie J, McKay K, Gruca M, Tamanini A, Assael MB, Gaskin K. 2007. Pancreatic phenotype in infants with cystic fibrosis identified by mutation screening. *Arch Dis Child* **92:** 842–846.

Clarke LL, Harline MC. 1998. Dual role of CFTR in cAMP-stimulated HCO$_3^-$ secretion across murine duodenum. *Am J Physiol Gastrointest Liver Physiol* **274:** G718–G726.

Clarke LL, Gawenis LR, Franklin CL, Harline MC. 1996. Increased survival of CFTR knockout mice with an oral osmotic laxative. *Lab Anim Sci* **46:** 612–618.

Clarke LL, Stien X, Walker NM. 2001. Intestinal bicarbonate secretion in cystic fibrosis mice. *JOP* **2:** 263–267.

Clarke LL, Gawenis LR, Bradford EM, Judd LM, Boyle KT, Simpson JE, Shull GE, Tanabe H, Ouellette AJ, Franklin CL, et al. 2004. Abnormal Paneth cell granule dissolution and compromised resistance to bacterial colonization in the intestine of CF mice. *Am J Physiol Gastrointest Liver Physiol* **286:** G1050–G1058.

Cleghorn GJ, Stringer DA, Forstner GG, Durie PR. 1986. Treatment of distal intestinal obstruction syndrome in cystic fibrosis with a balanced intestinal lavage solution. *Lancet* **1:** 8–11.

Colombo C, Russo MC, Zazzeron L, Romano G. 2006. Liver disease in cystic fibrosis. *J Pediatr Gastroenterol Nutr* **43:** S49–S55.

Condino AA, Hoffenberg EJ, Accurso F, Penvari C, Anthony M, Gralla J, O'Connor JA. 2005. Frequency of ASCA seropositivity in children with cystic fibrosis. *J Pediatr Gastroenterol Nutr* **41:** 23–26.

Cotter PD. 2011. Small intestine and microbiota. *Curr Opin Gastroenterol* **27:** 99–105.

Courtney JM, Bradley J, McCaughan J, O'Connor TM, Shortt C, Bredin CP, Bradbury I, Elborn JS. 2007. Predictors of mortality in adults with cystic fibrosis. *Pediatr Pulmonol* **42:** 525–532.

Couzin-Frankel J. 2011. Bacteria and asthma: Untangling the links. *Science* **330:** 1168–1169.

Cuppoletti J, Malinowska DH, Tewari KP, Li QJ, Sherry AM, Patchen ML, Ueno R. 2004. SPI-0211 activates T84 cell chloride transport and recombinant human ClC-2 chloride currents. *Am J Physiol Cell Physiol* **287:** C1173–C1183.

Dalzell AM, Heaf DP. 1990. Oro-caecal transit time and intra-luminal pH in cystic fibrosis patients with distal intestinal obstruction syndrome. *Acta Univ Carol [Med] (Praha)* **36:** 159–160.

Dalzell AM, Freestone NS, Billington D, Heaf DP. 1990. Small intestinal permeability and orocaecal transit time in cystic fibrosis. *Arch Dis Child* **65:** 585–588.

De Lisle RC. 2007. Altered transit and bacterial overgrowth in the cystic fibrosis mouse small intestine. *Am J Physiol Gastrointest Liver Physiol* **293:** G104–G111.

De Lisle RC, Roach EA, Norkina O. 2006. Eradication of small intestinal bacterial overgrowth in the cystic fibrosis mouse reduces mucus accumulation. *J Pediatr Gastroenterol Nutr* **42:** 46–52.

De Lisle RC, Roach E, Jansson K. 2007. Effects of laxative and *N*-acetylcysteine on mucus accumulation, bacterial load, transit, and inflammation in the cystic fibrosis mouse small intestine. *Am J Physiol Gastrointest Liver Physiol* **293:** G577–G584.

De Lisle RC, Meldi L, Flynn M, Jansson K. 2008. Altered eicosanoid metabolism in the cystic fibrosis mouse small intestine. *J Pediatr Gastroenterol Nutr* **47:** 406–416.

De Lisle RC, Mueller R, Roach E. 2010a. Lubiprostone ameliorates the cystic fibrosis mouse intestinal phenotype. *BMC Gastroenterol* **10:** 107.

De Lisle RC, Sewell R, Meldi L. 2010b. Enteric circular muscle dysfunction in the cystic fibrosis mouse small intestine. *Neurogastroenterol Motil* **22:** 341–e87.

De Lisle RC, Mueller R, Boyd M. 2011. Impaired mucosal barrier function in the small intestine of the cystic fibrosis mouse. *J Pediatr Gastroenterol Nutr* **53:** 371–379.

Duytschaever G, Huys G, Bekaert M, Boulanger L, De BK, Vandamme P. 2012. Dysbiosis of bifidobacteria and Clostridium cluster XIVa in the cystic fibrosis fecal microbiota. *J Cyst Fibros* doi: 10.1016/j.jcf.2012.10.003.

Eckman EA, Cotton CU, Kube DM, Davis PB. 1995. Dietary changes improve survival of CFTR S489X homozygous mutant mouse. *Am J Physiol Lung Cell Mol Physiol* **269:** L625–L630.

Edelman SM, Kasper DL. 2008. Symbiotic commensal bacteria direct maturation of the host immune system. *Curr Opin Gastroenterol* **24:** 720–724.

Escobar H, Perdomo M, Vasconez F, Camarero C, del Olmo MT, Suarez L. 1992. Intestinal permeability to 51Cr-EDTA and orocecal transit time in cystic fibrosis. *J Pediatr Gastroenterol Nutr* **14:** 204–207.

Feingold J, Guilloud-Bataille M. 1999. Genetic comparisons of patients with cystic fibrosis with or without meconium ileus. Clinical Centers of the French CF Registry. *Ann Genet* **42:** 147–150.

* Ferec C, Cutting GR. 2012. Assessing the disease-liability of mutations in CFTR. *Cold Spring Harb Perspect Med* **2:** a009480.

Fluge G, Olesen HV, Gilljam M, Meyer P, Pressler T, Storrosten OT, Karpati F, Hjelte L. 2009. Co-morbidity of cystic fibrosis and celiac disease in Scandinavian cystic fibrosis patients. *J Cyst Fibros* **8:** 198–202.

Freudenberg F, Broderick AL, Yu BB, Leonard MR, Glickman JN, Carey MC. 2008. Pathophysiological basis of liver disease in cystic fibrosis employing a ΔF508 mouse model. *Am J Physiol Gastrointest Liver Physiol* **294:** G1411–G1420.

Fridge JL, Conrad C, Gerson L, Castillo RO, Cox K. 2007. Risk factors for small bowel bacterial overgrowth in cystic fibrosis. *J Pediatr Gastroenterol Nutr* **44:** 212–218.

* Frizzell RA, Hanrahan JW. 2012. Physiology of epithelial chloride and fluid secretion. *Cold Spring Harb Perspect Med* **2:** a009563.

Garcia MAS, Yang N, Quinton PM. 2009. Normal mouse intestinal mucus release requires cystic fibrosis transmembrane regulator-dependent bicarbonate secretion. *J Clin Invest* **119:** 2613–2622.

Garcia-Sanchez V, Iglesias-Flores E, Gonzalez R, Gisbert JP, Gallardo-Valverde JM, Gonzalez-Galilea A, Naranjo-Rodriguez A, de Dios-Vega JF, Muntane J, Gomez-Camacho F. 2010. Does fecal calprotectin predict relapse in patients with Crohn's disease and ulcerative colitis? *J Crohns Colitis* **4:** 144–152.

Gelfond D, Ma C, Semler J, Borowitz D. 2012. Intestinal pH and gastrointestinal transit profiles in cystic fibrosis patients measured by wireless motility capsule. *Dig Dis Sci* doi: 10.1007/s10620-012-2209-1.

Gill SR, Pop M, Deboy RT, Eckburg PB, Turnbaugh PJ, Samuel BS, Gordon JI, Relman DA, Fraser-Liggett CM, Nelson KE. 2006. Metagenomic analysis of the human distal gut microbiome. *Science* **312:** 1355–1359.

Grubb BR, Gabriel SE. 1997. Intestinal physiology and pathology in gene-targeted mouse models of cystic fibrosis. *Am J Physiol Gastrointest Liver Physiol* **273:** G258–G266.

Guilbault C, Saeed Z, Downey GP, Radzioch D. 2007. Cystic fibrosis mouse models. *Am J Respir Cell Mol Biol* **36:** 1–7.

Gyömörey K, Rozmahel R, Bear CE. 2000. Amelioration of intestinal disease severity in cystic fibrosis mice is associated with improved chloride secretory capacity. *Pediatr Res* **48:** 731–734.

Hansen J, Gulati A, Sartor RB. 2010. The role of mucosal immunity and host genetics in defining intestinal commensal bacteria. *Curr Opin Gastroenterol* **26:** 564–571.

Hasosah M, Davidson G, Jacobson K. 2009. Persistent elevated tissue-transglutaminase in cystic fibrosis. *J Paediatr Child Health* **45:** 172–173.

Hofmann AF, Mysels KJ. 1992. Bile acid solubility and precipitation in vitro and in vivo: The role of conjugation, pH, and Ca^{2+} ions. *J Lipid Res* **33:** 617–626.

Houwen RH, van der Doef HP, Sermet I, Munck A, Hauser B, Walkowiak J, Robberecht E, Colombo C, Sinaasappel M, Wilschanski M. 2010. Defining DIOS and constipation in cystic fibrosis with a multicentre study on the incidence, characteristics, and treatment of DIOS. *J Pediatr Gastroenterol Nutr* **50:** 38–42.

Husebye E. 2005. The pathogenesis of gastrointestinal bacterial overgrowth. *Chemotherapy* **51:** 1–22.

Husebye E, Hellstrom PM, Sundler F, Chen J, Midtvedt T. 2001. Influence of microbial species on small intestinal myoelectric activity and transit in germ-free rats. *Am J Physiol Gastrointest Liver Physiol* **280:** G368–G380.

Jakab RL, Collaco AM, Ameen NA. 2011. Physiologic relevance of cell-specific distribution patterns of CFTR, NKCC1, NBCe1, and NHE3 along the crypt-villus axis in the intestine. *Am J Physiol Gastrointest Liver Physiol* **300:** G82–G98.

Jaques A, Daviskas E, Turton JA, Turton JA, McKay K, Cooper P, Stirling RG, Robertson CF, Bye PT, Lesouef PN, et al. 2008. Inhaled mannitol improves lung function in cystic fibrosis. *Chest* **133**: 1388–1396.

Joo NS, Wine JJ, Cuthbert AW. 2009. Lubiprostone stimulates secretion from tracheal submucosal glands of sheep, pigs, and humans. *Am J Physiol Lung Cell Mol Physiol* **296**: L811–L824.

Kalivianakis M, Minich DM, Bijleveld CM, Van Aalderen WM, Stellaard F, Laseur M, Vonk RJ, Verkade HJ. 1999. Fat malabsorption in cystic fibrosis patients receiving enzyme replacement therapy is due to impaired intestinal uptake of long-chain fatty acids. *Am J Clin Nutr* **69**: 127–134.

Kalnins D, Durie PR, Pencharz P. 2007. Nutritional management of cystic fibrosis patients. *Curr Opin Clin Nutr Metab Care* **10**: 348–354.

* Knowles MR, Drumm M. 2012. The influence of genetics on cystic fibrosis phenotypes. *Cold Spring Harb Perspect Med* **2**: a009548.

Kopelman H, Durie P, Gaskin K, Weizman Z, Forstner G. 1985. Pancreatic fluid secretion and protein hyperconcentration in cystic fibrosis. *N Engl J Med* **312**: 329–334.

Kopelman H, Corey M, Gaskin K, Durie P, Weizman Z, Forstner G. 1988. Impaired chloride secretion, as well as bicarbonate secretion, underlies the fluid secretory defect in the cystic fibrosis pancreas. *Gastroenterology* **95**: 349–355.

Kumar M, Potter E, Berschback N, McColley SA. 2010. Nutritional status in children with pancreatic sufficient cystic fibrosis. *Pediatr Pulmonol* **45**(Suppl 33): 412 (abstract).

Lewindon PJ, Robb TA, Moore DJ, Davidson GP, Martin AJ. 1998. Bowel dysfunction in cystic fibrosis: Importance of breath testing. *J Paediatr Child Health* **34**: 79–82.

Lindblad A, Glaumann H, Strandvik B. 1999. Natural history of liver disease in cystic fibrosis. *Hepatology* **30**: 1151–1158.

Lisowska A, Wojtowicz J, Walkowiak J. 2009. Small intestine bacterial overgrowth is frequent in cystic fibrosis: Combined hydrogen and methane measurements are required for its detection. *Acta Biochim Pol* **56**: 631–634.

Lisowska A, Madry E, Pogorzelski A, Szydlowski J, Radzikowski A, Walkowiak J. 2010. Small intestine bacterial overgrowth does not correspond to intestinal inflammation in cystic fibrosis. *Scand J Clin Lab Invest* **70**: 322–326.

Lloyd-Still JD. 1994. Crohn's disease and cystic fibrosis. *Dig Dis Sci* **39**: 880–885.

MacDonald TT, Monteleone G. 2005. Immunity, inflammation, and allergy in the gut. *Science* **307**: 1920–1925.

Madan JC, Koestler DC, Stanton BA, Davidson L, Moulton LA, Housman ML, Moore JH, Guill MF, Morrison HG, Sogin ML, et al. 2012. Serial analysis of the gut and respiratory microbiome in cystic fibrosis in infancy: Interaction between intestinal and respiratory tracts and impact of nutritional exposures. *MBio* **3**: e0025112-12.

Maisonneuve P, FitzSimmons SC, Neglia JP, Campbell PW III, Lowenfels AB. 2003. Cancer risk in nontransplanted and transplanted cystic fibrosis patients: A 10-year study. *J Natl Cancer Inst* **95**: 381–387.

Malmberg EK, Noaksson KA, Phillipson M, Johansson ME, Hinojosa-Kurtzberg M, Holm L, Gendler SJ, Hansson GC. 2006. Increased levels of mucins in the cystic fibrosis mouse small intestine and modulator effects of the Muc1 mucin expression. *Am J Physiol Gastrointest Liver Physiol* **291**: G203–G210.

Milla CE. 2004. Association of nutritional status and pulmonary function in children with cystic fibrosis. *Curr Opin Pulm Med* **10**: 505–509.

Montalto M, Santoro L, Curigliano V, D'Onofrio F, Cammarota G, Panunzi S, Ricci R, Gallo A, Grieco A, Gasbarrini A, et al. 2007. Faecal calprotectin concentrations in untreated coeliac patients. *Scand J Gastroenterol* **42**: 957–961.

Montalto M, Gallo A, Ianiro G, Santoro L, D'Onofrio F, Ricci R, Cammarota G, Covino M, Vastola M, Gasbarrini A, et al. 2010. Can chronic gastritis cause an increase in fecal calprotectin concentrations? *World J Gastroenterol* **16**: 3406–3410.

Mundhenk L, Alfalah M, Elble RC, Pauli BU, Naim HY, Gruber AD. 2006. Both cleavage products of the mCLCA3 protein are secreted soluble proteins. *J Biol Chem* **281**: 30072–30080.

Murphy MS, Brunetto AL, Pearson AD, Ghatei MA, Nelson R, Eastham EJ, Bloom SR, Green AA. 1992. Gut hormones and gastrointestinal motility in children with cystic fibrosis. *Dig Dis Sci* **37**: 187–192.

Myung SJ, Rerko RM, Yan M, Platzer P, Guda K, Dotson A, Lawrence E, Dannenberg AJ, Lovgren AK, Luo G, et al. 2006. 15-Hydroxyprostaglandin dehydrogenase is an in vivo suppressor of colon tumorigenesis. *Proc Natl Acad Sci* **103**: 12098–12102.

Nieuwenhuijs VB, Verheem A, Duijvenbode-Beumer H, Visser MR, Verhoef J, Gooszen HG, Akkermans LM. 1998. The role of interdigestive small bowel motility in the regulation of gut microflora, bacterial overgrowth, and bacterial translocation in rats. *Ann Surg* **228**: 188–193.

Norkina O, De Lisle RC. 2005. Potential genetic modifiers of the cystic fibrosis intestinal inflammatory phenotype on mouse chromosomes 1, 9, and 10. *BMC Genet* **6**: 29.

Norkina O, Burnett TG, De Lisle RC. 2004. Bacterial overgrowth in the cystic fibrosis transmembrane conductance regulator null mouse small intestine. *Infect Immun* **72**: 6040–6049.

O'Brien S, Mulcahy H, Fenlon H, O'Broin A, Casey M, Burke A, FitzGerald MX, Hegarty JE. 1993. Intestinal bile acid malabsorption in cystic fibrosis. *Gut* **34**: 1137–1141.

O'Brien CE, Anderson PJ, Stowe CD. 2010. Use of the chloride channel activator lubiprostone for constipation in adults with cystic fibrosis: A case series (March). *Ann Pharmacother* **44**: 577–581.

* Park HW, Lee MG. 2012. Transepithelial bicarbonate secretion: Lessons from the pancreas. *Cold Spring Harb Perspect Med* **2**: a009571.

Parmley RR, Gendler SJ. 1998. Cystic fibrosis mice lacking muc1 have reduced amounts of intestinal mucus. *J Clin Invest* **102**: 1798–1806.

Peretti N, Roy CC, Drouin E, Seidman E, Brochu P, Casimir G, Levy E. 2006. Abnormal intracellular lipid processing contributes to fat malabsorption in cystic fibrosis patients. *Am J Physiol Gastrointest Liver Physiol* **290**: G609–G615.

Peuhkuri K, Vapaatalo H, Korpela R. 2010. Even low-grade inflammation impacts on small intestinal function. *World J Gastroenterol* **16**: 1057–1062.

Qin J, Li R, Raes J, Arumugam M, Burgdorf KS, Manichanh C, Nielsen T, Pons N, Levenez F, Yamada T, et al. 2010a. A human gut microbial gene catalogue established by metagenomic sequencing. *Nature* **464**: 59–65.

Qin X, Sheth SU, Sharpe SM, Dong W, Lu Q, Xu D, Deitch EA. 2010b. The mucus layer is critical in protecting against ischemia/reperfusion-mediated gut injury and in the restitution of gut barrier function. *Shock* **35**: 275–281.

Quigley EM. 2011. Microflora modulation of motility. *J Neurogastroenterol Motil* **17**: 140–147.

Raia V, Maiuri L, De Ritis G, De Vizia B, Vacca L, Conte R, Auricchio S, Londei M. 2000. Evidence of chronic inflammation in morphologically normal small intestine of cystic fibrosis patients. *Pediatr Res* **47**: 344–350.

Robinson PJ, Smith AL, Sly PD. 1990. Duodenal pH in cystic fibrosis and its relationship to fat malabsorption. *Dig Dis Sci* **35**: 1299–1304.

Rogers CS, Stoltz DA, Meyerholz DK, Ostedgaard LS, Rokhlina T, Taft PJ, Rogan MP, Pezzulo AA, Karp PH, Itani OA, et al. 2008. Disruption of the CFTR gene produces a model of cystic fibrosis in newborn pigs. *Science* **321**: 1837–1841.

Rovner AJ, Stallings VA, Schall JI, Leonard MB, Zemel BS. 2007. Vitamin D insufficiency in children, adolescents, and young adults with cystic fibrosis despite routine oral supplementation. *Am J Clin Nutr* **86**: 1694–1699.

Rubinstein S, Moss R, Lewiston N. 1986. Constipation and meconium ileus equivalent in patients with cystic fibrosis. *Pediatrics* **78**: 473–479.

Schwarzenberg SJ, Wielinski CL, Shamieh I, Carpenter BL, Jessurun J, Weisdorf SA, Warwick WJ, Sharp HL. 1995. Cystic fibrosis-associated colitis and fibrosing colonopathy. *J Pediatr* **127**: 565–570.

Seidler U, Singh A, Chen M, Cinar A, Bachmann O, Zheng W, Wang J, Yeruva S, Riederer B. 2009. Knockout mouse models for intestinal electrolyte transporters and regulatory PDZ adaptors: New insights into cystic fibrosis, secretory diarrhoea and fructose-induced hypertension. *Exp Physiol* **94**: 175–179.

Sekirov I, Russell SL, Antunes LC, Finlay BB. 2010. Gut microbiota in health and disease. *Physiol Rev* **90**: 859–904.

Sharma R, Florea VG, Bolger AP, Doehner W, Florea ND, Coats AJ, Hodson ME, Anker SD, Henein MY. 2001. Wasting as an independent predictor of mortality in patients with cystic fibrosis. *Thorax* **56**: 746–750.

Sidhu H, Hoppe B, Hesse A, Tenbrock K, Bromme S, Rietschel E, Peck AB. 1998. Absence of Oxalobacter formigenes in cystic fibrosis patients: A risk factor for hyperoxaluria. *Lancet* **352**: 1026–1029.

Singh VV, Toskes PP. 2003. Small bowel bacterial overgrowth: Presentation, diagnosis, and treatment. *Curr Gastroenterol Rep* **5**: 365–372.

Smyth RL, Croft NM, O'Hea U, Marshall TG, Ferguson A. 2000. Intestinal inflammation in cystic fibrosis. *Arch Dis Child* **82**: 394–399.

Snouwaert JN, Brigman KK, Latour AM, Malouf NN, Boucher RC, Smithies O, Koller BH. 1992. An animal model for cystic fibrosis made by gene targeting. *Science* **257**: 1083–1088.

Stallings VA, Stark LJ, Robinson KA, Feranchak AP, Quinton H. 2008. Evidence-based practice recommendations for nutrition-related management of children and adults with cystic fibrosis and pancreatic insufficiency: Results of a systematic review. *J Am Diet Assoc* **108**: 832–839.

Steinkamp G, Wiedemann B. 2002. Relationship between nutritional status and lung function in cystic fibrosis: Cross sectional and longitudinal analyses from the German CF quality assurance (CFQA) project. *Thorax* **57**: 596–601.

Stoltz DA, Meyerholz DK, Pezzulo AA, Ramachandran S, Rogan MP, Davis GJ, Hanfland RA, Wohlford-Lenane C, Dohrn CL, Bartlett JA, et al. 2010. Cystic fibrosis pigs develop lung disease and exhibit defective bacterial eradication at birth. *Sci Transl Med* **2**: 29ra31.

Strong TV, Boehm K, Collins FS. 1994. Localization of cystic fibrosis transmembrane conductance regulator mRNA in the human gastrointestinal tract by in situ hybridization. *J Clin Invest* **93**: 347–354.

Stylianou C, Galli-Tsinopoulou A, Koliakos G, Fotoulaki M, Nousia-Arvanitakis S. 2007. Ghrelin and leptin levels in young adults with cystic fibrosis: Relationship with body fat. *J Cyst Fibros* **6**: 293–296.

Summerton CB, Longlands MG, Wiener K, Shreeve DR. 2002. Faecal calprotectin: A marker of inflammation throughout the intestinal tract. *Eur J Gastroenterol Hepatol* **14**: 841–845.

Sun X, Sui H, Fisher JT, Yan Z, Liu X, Cho HJ, Joo NS, Zhang Y, Zhou W, Yi Y, et al. 2010. Disease phenotype of a ferret CFTR-knockout model of cystic fibrosis. *J Clin Invest* **120**: 3149–3160.

Takahashi K. 2010. Interaction between the intestinal immune system and commensal bacteria and its effect on the regulation of allergic reactions. *Biosci Biotechnol Biochem* **74**: 691–695.

van der Doef HP, Kokke FT, Beek FJ, Woestenenk JW, Froeling SP, Houwen RH. 2010. Constipation in pediatric cystic fibrosis patients: An underestimated medical condition. *J Cyst Fibros* **9**: 59–63.

* Verdugo P. 2012. Supramolecular dynamics of mucus. *Cold Spring Harb Perspect Med* **2**: a009597.

Walker NM, Simpson JE, Levitt RC, Boyle KT, Clarke LL. 2005. Talniflumate increases survival in a cystic fibrosis mouse model of distal intestinal obstructive syndrome (DIOS). *J Pharmacol Exp Ther* **317**: 275–283.

Walkowiak J, Blask-Osipa A, Lisowska A, Oralewska B, Pogorzelski A, Cichy W, Sapiejka E, Kowalska M, Korzon M, Szaflarska-Poplawska A. 2010. Cystic fibrosis is a risk factor for celiac disease. *Acta Biochim Pol* **57**: 115–118.

Wark P, McDonald VM. 2009. Nebulised hypertonic saline for cystic fibrosis. *Cochrane Database Syst Rev* CD001506.

Weiss B, Bujanover Y, Yahav Y, Vilozni D, Fireman E, Efrati O. 2010. Probiotic supplementation affects pulmonary exacerbations in patients with cystic fibrosis: A pilot study. *Pediatr Pulmonol* **45**: 536–540.

Werlin SL, Benuri-Silbiger I, Kerem E, Adler SN, Goldin E, Zimmerman J, Malka N, Cohen L, Armoni S, Yatzkan-Israelit Y, et al. 2010. Evidence of intestinal inflammation in patients with cystic fibrosis. *J Pediatr Gastroenterol Nutr* **51:** 304–308.

* Wilschanski M, Novak I. 2013. The cystic fibrosis of exocrine pancreas. *Cold Spring Harb Perspect Med* **3:** a009746.

Wood JD. 2010. Enteric nervous system: Sensory physiology, diarrhea and constipation. *Curr Opin Gastroenterol* **26:** 102–108.

Wouthuyzen-Bakker M, Bodewes FA, Verkade HJ. 2011. Persistent fat malabsorption in cystic fibrosis; lessons from patients and mice. *J Cyst Fibros* **10:** 150–158.

Yahav J, Samra Z, Blau H, Dinari G, Chodick G, Shmuely H. 2006. *Helicobacter pylori* and *Clostridium difficile* in cystic fibrosis patients. *Dig Dis Sci* **51:** 2274–2279.

Young FD, Newbigging S, Choi C, Keet M, Kent G, Rozmahel RF. 2007. Amelioration of cystic fibrosis intestinal mucous disease in mice by restoration of mCLCA3. *Gastroenterology* **133:** 1928–1937.

Zhou L, Dey CR, Wert SE, DuVall MD, Frizzell RA, Whitsett JA. 1994. Correction of lethal intestinal defect in a mouse model of cystic fibrosis by human CFTR. *Science* **266:** 1705–1708.

Cystic Fibrosis Transmembrane Regulator Correctors and Potentiators

Steven M. Rowe[1] and Alan S. Verkman[2]

[1]Departments of Medicine; Pediatrics; Cellular, Developmental, and Integrative Biology; and the Gregory Fleming James Cystic Fibrosis Research Center, University of Alabama at Birmingham, Birmingham, Alabama 35294

[2]Departments of Medicine and Physiology, University of California, San Francisco, California 94143-0521

Correspondence: verkman@itsa.ucsf.edu

Cystic fibrosis (CF) is caused by loss-of-function mutations in the CF transmembrane conductance regulator (CFTR) protein, a cAMP-regulated anion channel expressed primarily at the apical plasma membrane of secretory epithelia. Nearly 2000 mutations in the CFTR gene have been identified that cause disease by impairing its translation, cellular processing, and/or chloride channel gating. The fundamental premise of CFTR corrector and potentiator therapy for CF is that addressing the underlying defects in the cellular processing and chloride channel function of CF-causing mutant CFTR alleles will result in clinical benefit by addressing the basic defect underlying CF. Correctors are principally targeted at F508del cellular misprocessing, whereas potentiators are intended to restore cAMP-dependent chloride channel activity to mutant CFTRs at the cell surface. This article reviews the discovery of CFTR potentiators and correctors, what is known regarding their mechanistic basis, and encouraging results achieved in clinical testing.

Cystic fibrosis (CF) is caused by loss-of-function mutations in the CF transmembrane conductance regulator (CFTR) protein, a cAMP-regulated chloride channel expressed primarily at the apical plasma membrane of secretory epithelia in the airways, pancreas, intestine, and other tissues. CFTR is a large, multidomain glycoprotein consisting of two membrane-spanning domains, two nucleotide-binding domains (NBD1 and NBD2) that bind and hydrolyze ATP, and a regulatory (R) domain that gates the channel by phosphorylation. There is limited crystal structure data on isolated cytoplasmic domains of CFTR, and low-resolution electron crystallographic data and homology modeling of full-length CFTR. Nearly 2000 mutations in the CFTR gene have been identified that produce the loss-of-function phenotype by impairing its translation, cellular processing, and/or chloride channel gating. The F508del mutation, which is present in at least one allele in ~90% of CF patients, impairs CFTR folding, stability at the endoplasmic reticulum and plasma membrane, and chloride channel gating (Dalemans et al. 1991; Denning et al. 1992; Lukacs et al. 1993; Du et al. 2005a). Other mutations primarily alter channel gating (e.g., G551D), conductance (e.g., R117H), or translation (e.g., G542X) (Welsh and Smith 1993). The fundamental premise of CFTR corrector and potentiator therapy for CF is that correction of the underlying defects in the cellular processing and chloride channel function of CF-causing mutant CFTR alleles will be of clinical benefit. Correctors are principally targeted at F508del cellular misprocessing, whereas potentiators are intended to restore cAMP-dependent chloride channel activity to mutant CFTRs at the cell surface. In contrast to current therapies, such as antibiotics, anti-inflammatory agents, mucolytics, nebulized hypertonic saline, and pancreatic enzyme replacement, which treat CF disease manifestations, correctors and potentiators correct the underlying CFTR anion channel defect.

This work is focused on corrector and potentiator therapy for CF. The goal is to normalize defective folding, plasma membrane targeting, surface stability, and channel function in cells expressing disease-causing CFTR mutants. An ideal therapy would be a single drug without off-target effects that normalizes mutant CFTR folding, processing, and function to resemble that of wild-type CFTR. Although remarkable progress has been made in the past decade in small-molecule correctors and potentiators in their discovery and rapid advancement to clinical trials, much work remains, in particular, for F508del correctors, in the identification of compounds with high efficacy and potency, understanding their mechanism of action, and establishing long-term clinical benefit.

STRATEGIES FOR CFTR DRUG DISCOVERY

Because of the paucity of structural information on full-length wild-type and F508del-CFTR, as well as the complexity of the defects caused by the F508del mutation, CFTR drug discovery has largely used phenotype assays based on CFTR chloride channel function. Genetically encoded, halide-sensing fluorescent proteins have been very useful in this regard. Yellow fluorescent protein (YFP) mutants have been identified whose fluorescence is strongly quenched (reduced) by iodide (Jayaraman et al. 2000), a halide that is efficiently transported by CFTR. The YFP mutant YFP-H148Q/I521L is brightly fluorescent and 50% quenched by 2–3 mM iodide (Galietta et al. 2001a). The general screening strategy for modulators of CFTR function is the generation of cells coexpressing CFTR (wild-type or mutant) together with the YFP iodide sensor. Test compounds can be added before assay of iodide influx, which involves measurement of the time course of cell fluorescence in response to iodide addition to the extracellular solution. Fisher rat thyroid (FRT) cells were found to be particularly useful for chloride channel drug discovery because of their epithelial origin and formation of tight junctions, rapid growth on uncoated plastic, strong adherence in multiwell plate format, ease of stable expression following transfection, and low basal halide permeability (Galietta et al. 2001b). An alternative screening approach has been the use of membrane potential-sensitive fluorescent dyes to measure CFTR-dependent membrane depolarization following chloride addition to the extracellular solution (Van Goor et al. 2006). An alternative assay approach for corrector testing is the appearance of mutant CFTR at the cell plasma membrane measured using an externally epitope-tagged CFTR with ELISA-based readout involving secondary antibodies (Carlile et al. 2007). A new generation of assays is under development that probes specific molecular interactions involved in F508del folding, such as domain–domain interactions or nucleotide-binding domain stability.

Utilizing cell-based high-throughput screening with YFP fluorescence readout, potent thiazolidinone (e.g., CFTR$_{inh}$-172) (Ma et al. 2002a) and glycine hydrazide (e.g., GlyH-101) (Muanprasat et al. 2004) inhibitors of wild-type CFTR have been identified (Fig. 1A), which are used widely as CF research tools. The glycine hydrazides target the external CFTR pore, which has allowed the development of nonabsorbable macromolecular conjugates that inhibit CFTR-dependent fluid secretion in the intestine (Sonawane et al. 2008). Recently, PPQ and related BPO classes of CFTR inhibitors, with IC$_{50}$ down to ~5 nM, have been discovered (Tradtrantip et al. 2009a; Snyder et al. 2011). Small-molecule CFTR inhibitors are in preclinical development for therapy of enterotoxin-mediated secretory diarrheas and polycystic kidney disease. High-throughput screening has also yielded small-molecule activators of wild-type CFTR that tar-

 Cite this article as *Cold Spring Harb Perspect Med* doi: 10.1101/cshperspect.a009761

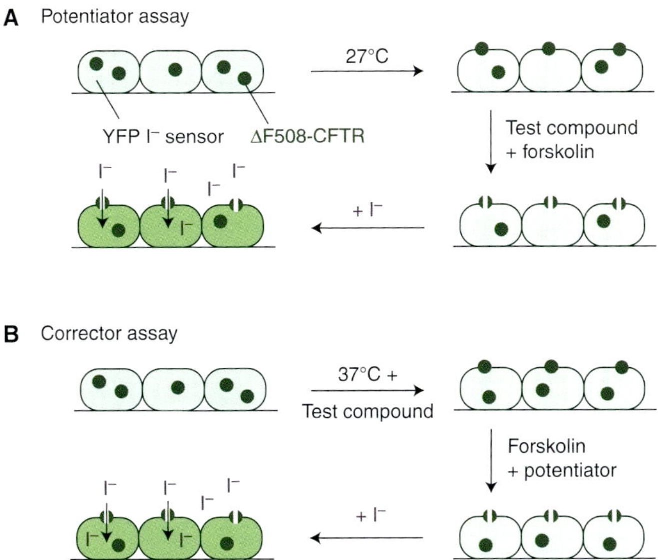

Figure 1. Cell-based screening assays for high-throughput identification of F508del-CFTR potentiators and correctors. (*A*) Potentiator assay: Fisher rat thyroid (FRT) cells coexpressing human F508del-CFTR and a halide-sensing yellow fluorescent protein (YFP) are incubated at reduced temperature (27°C) for 18–24 h before assay to target F508del-CFTR to the plasma membrane. Test compounds are added for 10 min in the presence of a cAMP agonist (forskolin) before iodide addition. F508del-CFTR function is assayed in a plate reader from the kinetics of YFP fluorescence quenching following iodide addition. (*B*) Corrector assay: Cells are incubated with test compounds at 37°C for 24 h. F508del-CFTR function is assayed by iodide addition in the presence of forskolin and a potentiator such as genistein.

get the channel directly (Ma et al. 2002b), as well as modulators of phosphodiesterase activity (Tradtrantip et al. 2009b).

Figure 2 diagrams high-throughput screening assays that have been used to identify small-molecule potentiators and correctors of F508del-CFTR. In these screens, potentiator activity is the ability of a compound to normalize mutant CFTR chloride channel gating when added just before assay of transport and generally in the presence of a cAMP agonist. Potentiator activity is assayed in F508del-CFTR-transfected epithelial cells following low-temperature incubation in which F508del-CFTR is targeted to the plasma membrane, with test compound (together with cAMP agonist) added just before assay (Yang et al. 2003). Potentiator-dependent restoration of F508del-CFTR channel function is assayed from the kinetics of decreasing YFP fluorescence following iodide addition. Corrector activity is the ability of a compound to promote cell-surface expression of F508del-CFTR

generally following prolonged incubation. Corrector activity is assayed in F508del-CFTR-expressing cells by >12 h incubation with test compound at 37°C, followed by washout and addition of a potentiator such as the flavone genistein (together with cAMP agonist) just before measurement of iodide influx (Pedemonte et al. 2005a). Similar assays have been used to identify inhibitors and activators of calcium-activated chloride channels (Namkung et al. 2011a,b). The latter have potential therapeutic efficacy in CF by restoring chloride transport independent of CFTR rescue.

POTENTIATORS

The practical drug discovery strategy used to date involves separate functional assays to screen for F508del-CFTR potentiators and correctors. We point out, however, that because channel gating and cellular processing are interrelated processes that each depend on F508del-CFTR

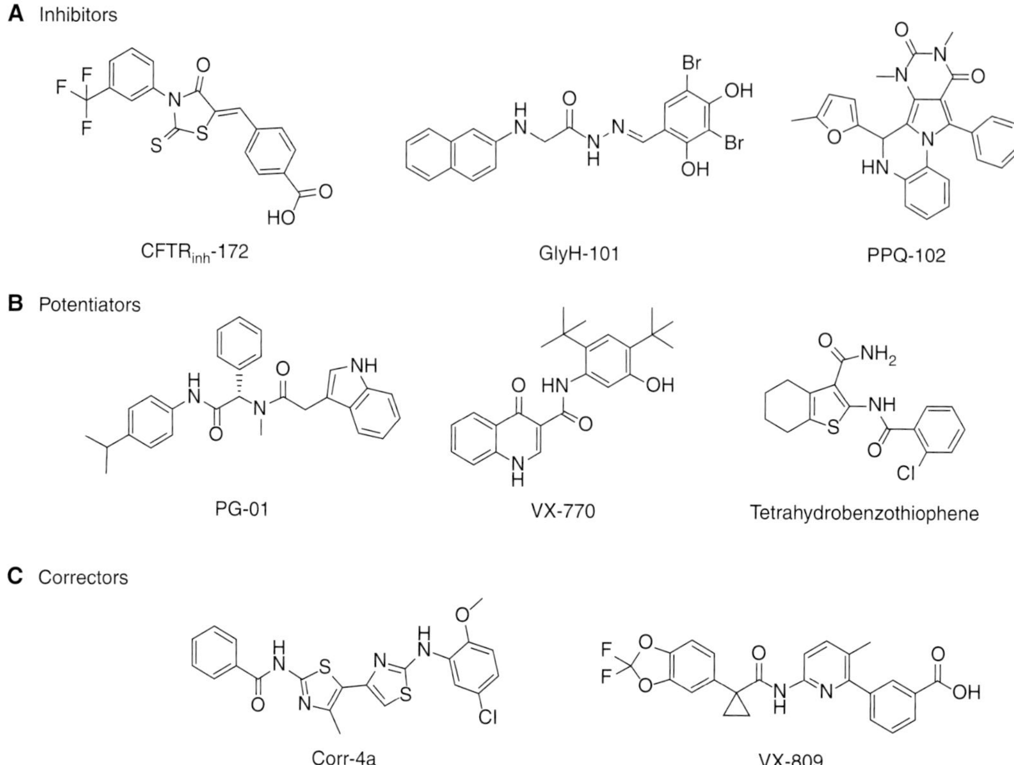

Figure 2. Chemical structures of small-molecule inhibitors of wild-type CFTR (*A*), potentiators (*B*), and correctors (*C*) of F508del-CFTR.

folding, the conventional, sharp distinction between potentiators and correctors is somewhat artificial and could hamper CF drug discovery, for example, in the identification of efficacious compounds for single-drug therapy of CF caused by the F508del mutation. Nevertheless, screening efforts have identified multiple chemical classes of potentiators and correctors, and, recently, a class of compounds with dual potentiator and corrector activities.

Before HTS, several chemical classes of efficacious F508del-CFTR potentiators were known, including flavones, xanthines, and benzimidazoles (Fig. 1B). A small screen of approved drugs and follow-on medicinal chemistry have identified dihydropyridine blockers of slow calcium channels as potentiators (Budriesi et al. 2011), although their development potential is unclear because of off-target cardiac and other effects. High-throughput screening efforts

by the Verkman laboratory (Yang et al. 2003; Pedemonte et al. 2005b) and by Vertex Pharmaceuticals (Van Goor et al. 2006) yielded several additional classes of small-molecule potentiators. Interestingly, a relatively high "hit rate" was found for identification of potentiators, with many potential development candidates; however, these potentiators do not have significant corrector activity. The phenylglycine PG-01 (Fig. 1B), when added together with a cAMP agonist, activates F508del-CFTR chloride conductance with low nanomolar potency, with maximum efficacy comparable to that of the flavone genistein, restoring F508del-CFTR open probability to approximately that of wild-type CFTR. Ivacaftor (VX-770), an investigational agent identified by Vertex Pharmaceuticals, has similar efficacy. PG-01 and ivacaftor also correct defective channel gating in the CF-causing mutant G551D-CFTR, a CFTR "gating" mutant

Cite this article as *Cold Spring Harb Perspect Med* doi: 10.1101/cshperspect.a009761

whose primary defect is reduced channel open probability after maximal cAMP stimulation. Ivacaftor has undergone development for human use, as described further below. Although potentiators may be clinically useful for CF patients having G551D and possibly other gating/conductance mutations, a potentiator alone is unlikely to have clinical benefit in CF caused by the F508del mutation because of the small amount of F508del-CFTR targeted to the cell plasma membrane, a finding substantiated by the clinical evaluation of ivacaftor in F508del homozygotes (see below). Another concern with some available potentiators is that superphysiological concentrations of cAMP agonists are needed for potentiator efficacy following low-temperature or corrector rescue, probably because the rescued F508del-CFTR is in a nonnative conformation. Wild-type CFTR, in contrast, is activated strongly at low concentrations of cAMP without the need for a potentiator.

There is limited information about the mechanism of action of potentiators, but they are likely to involve a diverse range of cellular mechanisms. The Rowe laboratory has investigated potentiator mechanisms with a focus on R-domain phosphorylation (Pyle et al. 2011). Results suggest that potentiation of cAMP-dependent CFTR activation occurs independent of R-domain phosphorylation, whereas agents that activate CFTR itself without altering the response to cAMP resulted in robust R-domain phosphorylation. Other results have suggested that VRT-532, a CFTR potentiator that enhances the cAMP responsiveness of CFTR, potentiates CFTR gating by directly interacting with G551D or F508del CFTR to restore its defective ATPase activity (Pasyk et al. 2009; Wellhauser et al. 2009). Agents that potentiate CFTR by altering its cAMP dependence likely interact with CFTR itself, perhaps facilitating the conformation change involved in CFTR gating, although disruption of protein–protein interactions may also be involved in the action of some potentiators (Eckford and Bear 2011). Other indirect mechanisms of CFTR activation, including activation of cAMP through activation of phosphodiesterases or inhibition of phosphatases, or even alteration of electrochemical gradients for chloride transport by action on basolateral K^+ channels, the NKCC cotransporter or the Na^+/K^+ ATPase are also possible, and may underlie the effects of some putative "potentiators" that may be better classified as nonspecific activators (Pyle et al. 2011).

CORRECTORS

The identification and development of correctors of F508del-CFTR cellular misprocessing presents a much greater challenge than that for potentiators because of the involvement of multiple components of the cellular quality control machinery that are expressed in a cell-specific manner, and because of the multiplicity of defects conferred by the F508del mutation (Du et al. 2005b; He et al. 2010; Thibodeau et al. 2010). Correctors could act as "pharmacological chaperones" by interacting with F508del-CFTR itself, facilitating its folding and cellular processing, or as "proteostasis regulators" by modulating the cellular quality-control machinery to alter F508del-CFTR recognition and processing (Mu et al. 2008; Powers et al. 2009). Although pharmacological chaperones have the potential for greater target selectivity and there is a precedent for their utility in other protein folding diseases, proteostasis regulators may produce greater efficacy, particularly in combination with agents that directly enhance CFTR folding.

Neutraceuticals and drugs approved for other indications have been reported to have F508del-CFTR corrector activity, including curcumin (Egan et al. 2004), miglustat (Noel et al. 2008; Norez et al. 2009), and sildenafil (Lubamba et al. 2008; Robert et al. 2008). However, follow-up studies, and in the case of curcumin, a small clinical trial (see below), have failed to confirm bona fide or robust corrector action of these compounds. The compound 4-phenylbutryate (Buphenyl), which is approved for an inherited disorder of urea metabolism, increases F508del-CFTR plasma membrane expression in cell culture models, probably acting as a proteostatis regulator. However, in a clinical trial involving nasal potential difference measurements, F508del CF patients showed minimal benefit (Zeitlin et al. 2002). Down-regulation of Aha1, an Hsp90 cochaperone (Wang et al. 2006), and inhibition of HDAC7 (Hutt et al. 2010), increase plasma membrane F508del-CFTR in cell culture models by facilitating F508del-CFTR folding and stability, acting by a proteostasis regulator mechanism.

Several classes of small-molecule F508del-CFTR correctors have been identified by high-throughput screening (Fig. 1C). The original study identified four classes of compounds, including the bithiazole corr-4a, which increased F508del-CFTR cell-surface expression after 12–24 h incubation at 37°C, resulting in increased chloride conductance in transfected cell models and human F508del-CFTR bronchial cell cultures (with addition of a potentiator and cAMP agonist) (Pedemonte et al. 2005b). Follow-up medicinal chemistry yielded bithiazole analogs with improved potency and pharmacological properties (Yu et al. 2008), with EC_{50} ∼300 nM for the most potent bithiazole. However, in F508del bronchial cell culture models, available correctors normalize chloride conductance to only 10%–15% of that in wild-type bronchial cell cultures. Screening by Vertex Pharmaceuticals has yielded other classes of correctors (Van Goor et al. 2006, 2010), with the most promising compound, VX-809, in phase II clinical trials as discussed below. Other small-scale screening efforts have yielded additional candidate correctors, including the approved drug glafanine (Robert et al. 2010), the phenyl-hydrazone RDR1 (Sampson et al. 2011), and a few compounds from computational screening (Kalid et al. 2010), although these candidate correctors have quite low efficacy.

The molecular mechanisms of corrector action are poorly understood. Initial studies on corr-4a supported a mechanism of direct interaction with F508del-CFTR, as corr-4a improved F508del-CFTR folding at the endoplasmic reticulum and did not correct other misfolded proteins, including some CFTR mutants (Pedemonte et al. 2005b). Cys-cross-linking data supported the idea that corr-4a interacts with F508del-CFTR in the endoplasmic reticulum to promote its folding (Loo et al. 2008), although indirect effects involving proteostasis regulation cannot be excluded. One study suggested that corr-4a

targets and stabilizes NBD2 (Grove et al. 2009), whereas another study suggested that corr-4a action may also involve interference with ubiquitination (Jurkuvenaite et al. 2010). There is also evidence for direct interaction of corrector VRT-325 with F508del-CFTR, as it alters CFTR ATPase activity (Kim Chiaw et al. 2010) and protease susceptibility (Yu et al. 2011). Notwithstanding these and other lines of suggestive evidence, compelling data remains lacking about the mechanism of action of available correctors.

Further mechanistic studies remain a high priority, as they may offer clues to understand the limited efficacy of available correctors. One commonality among F508del-CFTR correctors has been their limited efficacy of CFTR rescue following treatment by a single compound (or pathway, in the case of RNA interference). It has been suggested that single agents may have limited efficacy (therapeutic "ceiling") because of the complex, multiple defects in F508del-CFTR folding (Sloane and Rowe 2010). It remains to be determined whether the limited efficacy of available correctors can be overcome with improved understanding of the folding pathway and molecular partners of F508del-CFTR. As a more immediate approach to enhance efficacy, it has been shown that combinations of F508del correctors can have additive or even synergistic effects (Bridges 2010; Lin et al. 2010), supporting the idea that more than one cellular mechanism is relevant to the cellular recognition and degradation of F508del-CFTR. The use of CFTR potentiators in combination with CFTR correctors represents another option to overcome deficits in efficient functional rescue for the protein. For example, VX-770 approximately doubles the effect of VX-809 in F508del HBE cells (Van Goor et al. 2010).

There has been interest in the identification of single compounds with dual corrector and potentiator activities, with the rationale being the practicality of single versus double drug therapy. Of the available correctors, the bithiazoles and aminoarylthiazoles appear to at least partially normalize F508del-CFTR folding, as their prolonged incubation results in greater chloride conductance in response to a cAMP agonist (Pedemonte et al. 2011). In a proof-

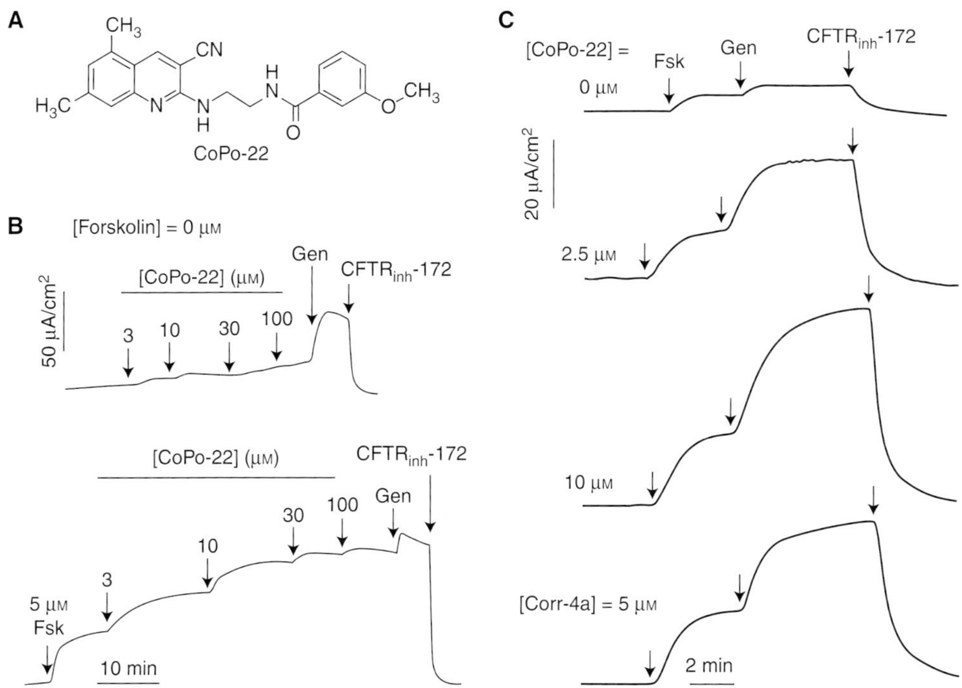

Figure 3. Dual potentiator and corrector activities of cyanoquinoline CoPo-22. (*A*) CoPo-22 structure. (*B*) Potentiator activity assayed by measurement of short-circuit current in FRT cells expressing F508del-CFTR. Assays performed at 0 and 5 μM forskolin as indicated. Effects of reference potentiator genistein shown, and CFTR inhibitor CFTR$_{inh}$-172. (*C*) Corrector activity measured following 18 h incubation with indicated concentrations of CoPo-22. Data for bithiazole corr-4a shown for comparison. Chloride conductance measured following addition of high concentrations of forskolin and genistein. (From Phuan et al. 2011; adapted, with permission, from the author.)

of-concept study, a hybrid bithiazole-phenylglycine corrector potentiator was synthesized, which, when cleaved by intestinal enzymes, yields an active bithiazole corrector and phenylglycine potentiator (Mills et al. 2010). Recently, the Verkman laboratory identified a cyanoquinoline class of F508del-CFTR correctors with independent potentiator activity (Phuan et al. 2011). Corrector-potentiator-22 (CoPo-22, Fig. 3A) rapidly increased chloride current when added to low-temperature rescued cells (potentiator activity, Fig. 3B), as well as increased cell-surface expression and chloride current after prolonged incubation (corrector activity, Fig. 3C). Although the corrector and potentiator efficacies of CoPo-22 are comparable to those of corr-4a and genistein, respectively, their EC$_{50}$ are in the low micromolar rather than nanomo-

lar range. Whether improved cyanoquinoline analogs can be development candidates for CF therapy is unclear, although these studies offer proof of concept for the possibility of dual-action corrector-potentiator compounds.

CLINICAL DEVELOPMENT OF CFTR POTENTIATORS

Progress in the clinical testing of potentiators and correctors of CFTR function in CF patients is among the most exciting developments in CF therapeutics. Although considerable challenges remain to bring therapeutic options to CF patients with different CFTR mutations, recent data have established proof of concept that rescue of CFTR-mediated anion transport can result in clinical benefit, and have opened a new era of CF treatment options that address the fundamental defect in the disease (Sloane and Rowe 2010). Table 1 summarizes clinical trial data.

Just as the discovery of CFTR correctors has been more challenging than that for potentiators, clinical progress toward developing a safe and efficacious CFTR modulator has moved more swiftly for CFTR potentiators, led by studies of ivacaftor. Following phase 1 testing in normal subjects and CF volunteers, ivacaftor was evaluated in a phase 2 randomized, double-blind, ascending dose trial that used both a crossover component and a confirmatory parallel group design. The trial focused on CF subjects with the G551D-CFTR allele, a gating mutation with low open probability at baseline, but relatively sensitive to ivacaftor in vitro (Van Goor et al. 2009). Ivacaftor was safe and well tolerated following administration for 2 and 4 weeks. Clinically significant and dose-dependent improvements in CFTR activity were observed in three ivacaftor dose groups (75 mg, 150 mg, and 250 mg, each bid) as measured by nasal potential difference and sweat chloride (Accurso et al. 2010). Improved pulmonary function was also observed, including an $\sim$8.7% increase in lung function as measured by FEV1% in the 150-mg bid group. The median reduction in sweat chloride of $\sim$59 mEq at the maximally effective dose resulted in a mean sweat chloride of $\sim$55 mEq, a value below the traditional diagnostic threshold of CF. CFTR-dependent anion transport measured by nasal potential difference (NPD) also improved a degree postulated to confer clinical benefit as predicted by genotype–phenotype correlations (Wilschanski et al. 2006). A summary of the ivacaftor data following 14-day administration is shown in Figure 4, which provided the first example of a systemic drug that restored CFTR activity and conferred meaningful clinical improvement in lung function (Accurso et al. 2010).

Phase 3 trials in G551D CF patients include two long-term randomized placebo-controlled clinical trials. In a trial in older children and adults (age 12 and above) (Ramsey et al. 2011), the primary end point was achieved, establish-ing an $\sim$10.5% improvement in FEV1% at 24 weeks, which was sustained at 48 weeks. In addition, secondary clinical end points showed significant improvement, including a 55% reduction in the probability of experiencing a pulmonary exacerbation during the course of the study, a 3.1-kg weight gain (compared with 0.9 kg in the placebo group), and an improvement in respiratory symptoms as assessed by the CFQ-R, a patient reported quality-of-life index. Similar to the phase 2 study, sweat chloride improved to mean level below the diagnostic threshold of CF (60 mEq/L). Similar results were reported in a smaller study that enrolled pediatric G551D CF patients age 6–12. The mean improvement in FEV1 was 12.5% following 24 weeks of treatment, with similar improvement in sweat chloride. In both clinical studies ivacaftor was safe and well tolerated, and potential mechanistic-based toxicities such as secretory diarrhea were not observed.

These results have established the proof-of-concept that targeting the restoration of CFTR activity using a small-molecule potentiator is safe and feasible, and could confer clinical benefit. The degree of improvement in spirometry among participants of the phase 3 trial of ivacaftor compares favorably to that of commonly used therapies for chronic CF care, including inhaled recombinant human DNase (Fuchs et al. 1994), inhaled tobramycin (Ramsey et al. 1999), azithromycin (Saiman et al. 2003), and hypertonic saline (Elkins et al. 2006). These results provide a rationale for continued evaluation of CFTR potentiators and correctors, and for their application to CF caused by other CFTR mutations including F508del.

Although principally designed as a safety study, ivacaftor has also been tested in CF patients homozygous for F508del-CFTR. Commensurate with predictions regarding its in vitro efficacy, a 4-month placebo-controlled trial revealed no statistically significant change in FEV1 following ivacaftor treatment (Flume et al. 2011). Interestingly, sweat chloride did improve by a small amount in the ivacaftor treatment group (3 mEq reduction compared with placebo), suggesting that low levels of F508del-CFTR at the cell surface can respond to a CFTR

Table 1. Summary of selected clinical trials evaluating CFTR potentiators and correctors in CF subjects

Study	Compound	Mutation	Enrollment age	N	Duration	Δ FEV1% predicted	Δ In CF exacerbation
CFTR potentiator trials							
Accurso et al. 2010	Ivacaftor (VX-770)	G551D	18 yr and older	39	14–28 d	10.8% Relative improvement[a]	N/A
Ramsey et al. 2011	Ivacaftor (VX-770)	G551D	6 yr and older	161	48 wk	10.6% Absolute improvement; 16.7% relative improvement	55% Reduction
Flume et al. 2012	Ivacaftor (VX-770)	F508del homozygous	18 yr and older	140	16 wk	No significant improvement	
CFTR corrector trials							
McCarty et al. 2002	CPX	F508del homozygous	18–38 yr	37	24 h	No significant change	N/A
Rubenstein and Zeitlin 1998	4-Phenyl butyrate	F508del homozygous	14 yr and older	18	1 wk	N/A	N/A
Zeitlin et al. 2002	4-Phenyl butyrate	F508del homozygous	18 yr and older	19	1 wk	No significant change	N/A
Goss et al. 2006	Curcuminoids	F508del homozygous	18 yr and older	9	14 d	No significant change	N/A
Clancy et al. 2011	VX-809	F508del homozygous	18 yr and older	89	28 d	No significant change	N/A

[a]Pooled data at 14 days from 14-d (part 1) and 28-d (part 2) study; all changes with the 150-mg dose group.
[b]20-g Dose group.
[c]−1.8 mV Change at day 4, but not sustained at day 7; 20-g dose group.

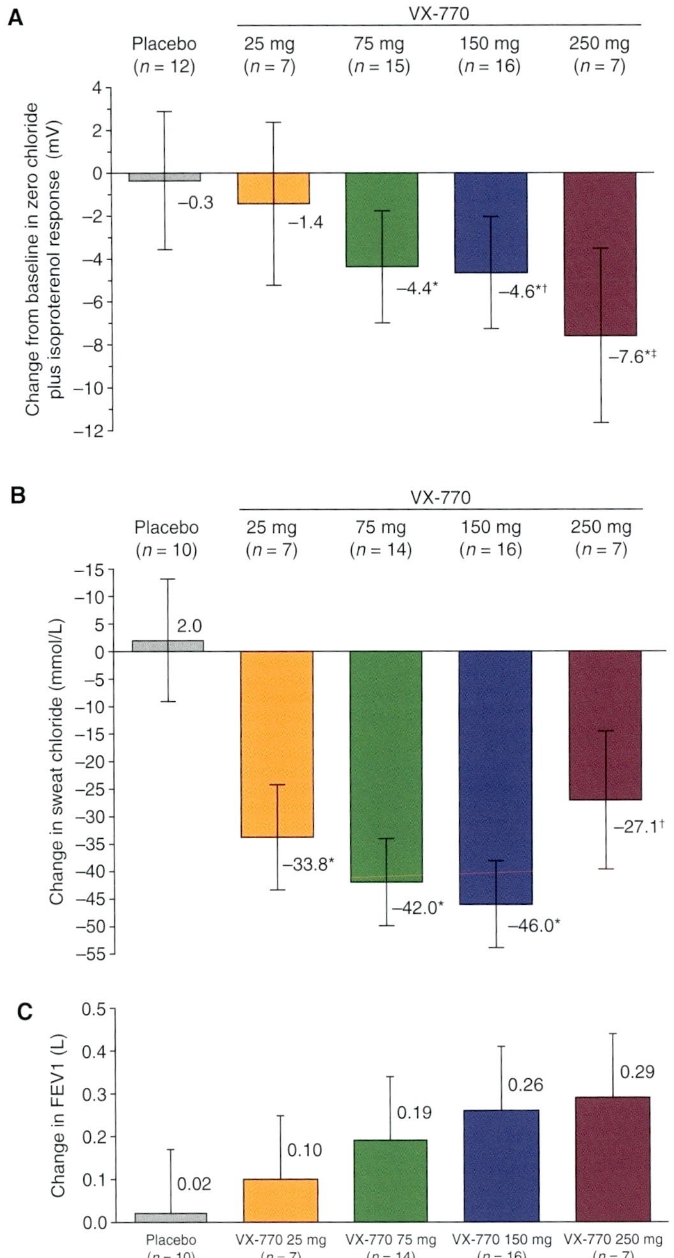

Figure 4. Results of a phase 2 clinical trial evaluating ivacaftor in G551D CF patients. Results show the relationship between CFTR activity measured by NPD: (A) *, $P < 0.01$ within subject; †, $P < 0.05$ vs. placebo; ‡, $P < 0.01$ vs. placebo, and sweat chloride; (B) *, $P < 0.001$ within subject and vs. placebo; †, $P < 0.001$ within subject and $P < 0.01$ vs. placebo with pulmonary function assessed by change in FEV1% predicted; (C) *, $P < 0.01$ within subject; †, $P < 0.05$ vs. placebo; ‡, $P < 0.01$ vs. placebo. (Data used in creation of figures from Accurso et al. 2010.)

potentiator, but at levels insufficient to confer clinical benefit. CFTR correctors will thus likely be necessary to restore clinically significant CFTR activity in vivo.

CLINICAL DEVELOPMENT OF CFTR CORRECTORS

A number of initial studies to establish the efficacy of correctors of F508del processing in CF patients were unsuccessful, but have advanced the clinical evaluation of CFTR modulators (Table 1). Among the first to be tested was the adenosine A1 receptor antagonist 8-cyclopentyl-1,3-dipropylxanthine (CPX), a corrector identified by Eidelman et al. (1992) (Guay-Broder et al. 1995). Although no efficacy was observed in a randomized multicenter trial that tested sweat chloride and nasal potential difference following 14-d administration, the study indicated the importance of standardization of biomarkers of CFTR activity, which was in its early stages at the time (McCarty et al. 2002). Subsequent efforts to characterize the activity of CFTR modulators would benefit from improvements, including standardization of sweat chloride testing using pilocarpine iontophoresis and the Macroduct collection device (Rowe et al. 2007). Similarly, nasal potential difference testing has benefited from a more rigorously standardized protocol that incorporates methodological improvements that enhance signal-to-noise ratio and allow electronic data capture for centralized and blinded interpretation (Solomon et al. 2010).

Although significant enthusiasm surrounded the initial discovery of curcumin as a putative CFTR corrector, curcuminoid derivatives did not improve CFTR activity in a two-center study that evaluated changes in nasal potential difference (Goss et al. 2006). Minimal levels of free curcumin were detected by mass spectroscopy, suggesting poor absorption could have contributed to results in the clinic, in addition to questions regarding lack of efficacy as a F508del processing corrector suggested by in vitro studies (Song et al. 2004; Grubb et al. 2006). As curcumin also shows robust activity as a CFTR potentiator (Grubb et al. 2006; Wang et al. 2007), its

activity as a direct CFTR activator may have confounded initial studies that relied heavily on a murine model that expresses detectable levels of F508del-CFTR at the cell surface (Egan et al. 2004; Ostedgaard et al. 2007).

The investigational agent VX-809 represents the first corrector molecule discovered by high-throughput screening to be tested in CF patients. Its efficacy as a corrector was validated in a panel of primary human bronchial epithelial cells derived from F508del homozygotes before human testing (Van Goor et al. 2010). Although VX-809 is less active in comparison to ivacaftor in G551D HBE cells (which restored CFTR-dependent short-circuit current to $\sim$35%–50% of wild-type levels), activity was sufficient to increase short-circuit current to $\sim$16% of wild-type levels. In F508del homozygous subjects, VX-809 induced a dose-dependent reduction in sweat chloride, with a maximal reduction of 8 mEq/L compared with placebo (Clancy et al. 2011). However, no significant change in CFTR-dependent PD or improvement in lung function were found. Thus, VX-809 was not sufficient to confer clinical improvement in a short-term (28-d) study when used as a single agent.

Because of limitations in the efficacy of VX-809 as a monotherapy in vitro and in vivo, and the effect of ivacaftor on VX-809-treated F508del HBE cells, the use of ivacaftor in combination with VX-809 represents a natural progression to enhance restoration of CFTR activity in the clinical setting, and is currently being tested in ongoing clinical studies.

RELATIONSHIP OF CLINICAL FINDINGS TO PRECLINICAL EFFICACY MODELS

Clinical studies testing potentiators and correctors provide insight on the use and relative performance of preclinical models. Although potentiators and correctors work through different cellular mechanisms, the degree of activity detected by changes in sweat chloride generally correlates across dose groups (and pharmacologic agents) to the degree of correction observed in primary human bronchial epithelial cells. For example, when the activity of various doses of ivacaftor was tested in G551D CF subjects, NPD and sweat chloride improved to ∼30% of the activity observed in normal (non-CF) patients (Rowe et al. 2010), a level compatible with less severe pulmonary dysfunction and a nonclassic CF phenotype based on known genotype–phenotype correlations in the disease (Wilschanski et al. 2006). This also matched the level of correction observed following in vitro treatment of primary human bronchial epithelial cells derived from individuals with at least one copy of the G551D mutation. In total, these data provide reasonable guidance on the efficacy limits necessary to observe a change in clinical outcome associated with "conversion" to a more mild CF phenotype (e.g., a change in lung function) (Rowe et al. 2010), and support the utility of electrophysiological measurements on primary human bronchial epithelial cell cultures as a useful preclinical surrogate of clinical efficacy. Whether this compares favorably with testing in new large animal models of CF that recapitulate many aspects of human disease remains to be determined (Rogers et al. 2008; Sun et al. 2010).

FUTURE CHALLENGES

The rapid progress over the past decade in small-molecule corrector and potentiator therapy for CF is perhaps the most exciting recent advance in the field, as it offers the possibility of a treatment that corrects the underlying CFTR defect in the appropriate target cells. However, the limited efficacy of first-generation correctors, and the lack of understanding of their mechanism of action, mandates the need for further basic and clinical research. Additional screening efforts are indicated to identify safe and potent correctors with high efficacy. However, because of the complex, multifaceted folding and thermodynamic defects imparted by the F508del mutation, the feasibility of identifying single correctors with high efficacy in normalizing F508del-CFTR cellular processing remains unproven. New screening paradigms, perhaps targeted to specific, well-characterized defects in F508del-CFTR folding and structure, may be needed, as will consideration of combination corrector therapy, such as compounds targeted to the F508del-CFTR protein combined with agents that alter the cellular proteostasis machinery. There also remains a need for better in vitro preclinical surrogate assays of compound action, and a need to better understand the determinants of clinical efficacy and the thresholds of CFTR activation required to produce sustained clinical benefit.

ACKNOWLEDGMENT

The authors thank Heather Hawthorne for assistance in preparing this manuscript.

REFERENCES

Accurso FJ, Rowe SM, Clancy JP, Boyle MP, Dunitz JM, Durie PR, Sagel SD, Hornick DB, Konstan MW, Donaldson SH, et al. 2010. Effect of VX-770 in persons with cystic fibrosis and the G551D-CFTR mutation. *N Engl J Med* **363:** 1991–2003.

Bridges RJ. 2010. Capitalizing on corrector mechanistic differences to achieve synergy in F508del CFTR expression. *Ped Pulmonol Suppl* **45:** 119–120.

Budriesi R, Ioan P, Leoni A, Pedemonte N, Locatelli A, Micucci M, Chiarini A, Galietta LJ. 2011. Cystic fibrosis: A new target for 4-Imidazo[2,1-b]thiazole-1,4-dihydropyridines. *J Med Chem* **54:** 3885–3894.

Carlile GW, Robert R, Zhang D, Teske KA, Luo Y, Hanrahan JW, Thomas DY. 2007. Correctors of protein trafficking defects identified by a novel high-throughput screening assay. *Chembiochem* **8:** 1012–1020.

Clancy JP, Rowe SM, Accurso FJ, Aitken ML, Amin RS, Ashlock MA, Ballmann M, Boyle MP, Bronsveld I, Campbell PW, et al. 2011. Results of a phase IIa study of VX-809, an investigational CFTR corrector compound, in subjects with cystic fibrosis homozygous for the F508del-CFTR mutation. *Thorax* **67:** 12–18.

Dalemans W, Barbry P, Champigny G, Jallat S, Dott K, Dreyer D, Crystal RG, Pavirani A, Lecocq JP, Lazdunski M. 1991. Altered chloride ion channel kinetics associated with the ΔF508 cystic fibrosis mutation. *Nature* **354:** 526–528.

Denning GM, Anderson MP, Amara JF, Marshall J, Smith AE, Welsh MJ. 1992. Processing of mutant cystic fibrosis transmembrane conductance regulator is temperature-sensitive. *Nature* **358:** 761–764.

Du K, Sharma M, Lukacs GL. 2005a. The ΔF508 cystic fibrosis mutation impairs domain-domain interactions and arrests post-translational folding of CFTR. *Nat Struct Mol Biol* **12:** 17–25.

Du K, Sharma M, Lukacs GL. 2005b. The ΔF508 cystic fibrosis mutation impairs domain-domain interactions and arrests post-translational folding of CFTR. *Nat Struct Mol Biol* **12:** 17–25.

Eckford PD, Bear CE. 2011. Targeting the regulation of CFTR channels. *Biochem J* **435:** e1–e4.

Egan ME, Pearson M, Weiner SA, Rajendran V, Rubin D, Glockner-Pagel J, Canny S, Du K, Lukacs GL, Caplan MJ.

2004. Curcumin, a major constituent of turmeric, corrects cystic fibrosis defects. *Science* **304:** 600–602.

Eidelman O, Guay-Broder C, van Galen PJ, Jacobson KA, Fox C, Turner RJ, Cabantchik ZI, Pollard HB. 1992. A1 adenosine-receptor antagonists activate chloride efflux from cystic fibrosis cells. *Proc Natl Acad Sci* **89:** 5562–5566.

Elkins MR, Robinson M, Rose BR, Harbour C, Moriarty CP, Marks GB, Belousova EG, Xuan W, Bye PT. 2006. A controlled trial of long-term inhaled hypertonic saline in patients with cystic fibrosis. *N Engl J Med* **354:** 229–240.

Flume PA, Borowitz DS, Liou TG, Li K, Yen K, Ordonez CL, Geller DE. 2011. VX-770 in subjects with cystic fibrosis who are homozygous for the *F508del-CFTR* mutation. *J Cyst Fibros* **10:** S16.

Flume PA, Liou TG, Borowitz DS, Li H, Yen K, Ordonez CL, Geller DE. 2012. Ivacaftor in subjects with cystic fibrosis who are homozygous for the *F508del-CFTR* mutation. *Chest* **142:** 718–724.

Fuchs HJ, Borowitz DS, Christiansen DH, Morris EM, Nash ML, Ramsey BW, Rosenstein BJ, Smith AL, Wohl ME. 1994. Effect of aerosolized recombinant human DNase on exacerbations of respiratory symptoms and on pulmonary function in patients with cystic fibrosis. *N Engl J Med* **331:** 637–642.

Galietta LJ, Haggie PM, Verkman AS. 2001a. Green fluorescent protein-based halide indicators with improved chloride and iodide affinities. *FEBS Lett* **499:** 220–224.

Galietta LV, Jayaraman S, Verkman AS. 2001b. Cell-based assay for high-throughput quantitative screening of CFTR chloride transport agonists. *Am J Physiol Cell Physiol* **281:** C1734–C1742.

Goss CH, Genatossio A, Rowbotham RK, Hamblett N, McNamara S, Knowles M, Brass-Ernst L, Aitken ML, Zeitlin PL, Boyle MP. 2006. A phase I safety and dose finding study of orally administered curcuminoids in adult subjects with cystic fibrosis who are homozygous for ΔF508 cystic fibrosis transmembrane conductance regulator. *Ped Pulmonol Suppl* **41:** 247.

Grove DE, Rosser MF, Ren HY, Naren AP, Cyr DM. 2009. Mechanisms for rescue of correctable folding defects in CFTRΔF508. *Mol Biol Cell* **20:** 4059–4069.

Grubb BR, Gabriel SE, Mengos A, Gentzsch M, Randell SH, Van Heeckeren AM, Knowles MR, Drumm ML, Riordan JR, Boucher RC. 2006. SERCA pump inhibitors do not correct biosynthetic arrest of ΔF508 CFTR in cystic fibrosis. *Am J Respir Cell Mol Biol* **34:** 355–363.

Guay-Broder C, Jacobson KA, Barnoy S, Cabantchik ZI, Guggino WB, Zeitlin PL, Turner RJ, Vergara L, Eidelman O, Pollard HB. 1995. A1 receptor antagonist 8-cyclopentyl-1,3-dipropylxanthine selectively activates chloride efflux from human epithelial and mouse fibroblast cell lines expressing the cystic fibrosis transmembrane regulator ΔF508 mutation. *Biochemistry* **34:** 9079–9087.

He L, Aleksandrov LA, Cui L, Jensen TJ, Nesbitt KL, Riordan JR. 2010. Restoration of domain folding and interdomain assembly by second-site suppressors of the ΔF508 mutation in CFTR. *FASEB J* **24:** 3103–3112.

Hutt DM, Herman D, Rodrigues AP, Noel S, Pilewski JM, Matteson J, Hoch B, Kellner W, Kelly JW, Schmidt A, et al. 2010. Reduced histone deacetylase 7 activity restores

function to misfolded CFTR in cystic fibrosis. *Nat Chem Biol* **6:** 25–33.

Jayaraman S, Haggie P, Wachter RM, Remington SJ, Verkman AS. 2000. Mechanism and cellular applications of a green fluorescent protein-based halide sensor. *J Biol Chem* **275:** 6047–6050.

Jurkuvenaite A, Chen L, Bartoszewski R, Goldstein R, Bebok Z, Matalon S, Collawn JF. 2010. Functional stability of rescued ΔF508 cystic fibrosis transmembrane conductance regulator in airway epithelial cells. *Am J Respir Cell Mol Biol* **42:** 363–372.

Kalid O, Mense M, Fischman S, Shitrit A, Bihler H, Ben-Zeev E, Schutz N, Pedemonte N, Thomas PJ, Bridges RJ, et al. 2010. Small molecule correctors of F508del-CFTR discovered by structure-based virtual screening. *J Comput Aided Mol Des* **24:** 971–991.

Kim Chiaw P, Wellhauser L, Huan LJ, Ramjeesingh M, Bear CE. 2010. A chemical corrector modifies the channel function of F508del-CFTR. *Mol Pharmacol* **78:** 411–418.

Lin S, Sui J, Cotard S, Fung B, Andersen J, Zhu P, El Messadi N, Lehar J, Lee M, Staunton J. 2010. Identification of synergistic combinations of F508del cystic fibrosis transmembrane conductance regulator (CFTR) modulators. *Assay Drug Dev Technol* **8:** 669–684.

Loo TW, Bartlett MC, Clarke DM. 2008. Correctors promote folding of the CFTR in the endoplasmic reticulum. *Biochem J* **413:** 29–36.

Lubamba B, Lecourt H, Lebacq J, Lebecque P, De Jonge H, Wallemacq P, Leal T. 2008. Preclinical evidence that sildenafil and vardenafil activate chloride transport in cystic fibrosis. *Am J Respir Crit Care Med* **177:** 506–515.

Lukacs GL, Chang XB, Bear C, Kartner N, Mohamed A, Riordan JR, Grinstein S. 1993. The ΔF508 mutation decreases the stability of cystic fibrosis transmembrane conductance regulator in the plasma membrane. Determination of functional half-lives on transfected cells. *J Biol Chem* **268:** 21592–21598.

Ma T, Thiagarajah JR, Yang H, Sonawane ND, Folli C, Galietta LJ, Verkman AS. 2002a. Thiazolidinone CFTR inhibitor identified by high-throughput screening blocks cholera toxin-induced intestinal fluid secretion. *J Clin Invest* **110:** 1651–1658.

Ma T, Vetrivel L, Yang H, Pedemonte N, Zegarra-Moran O, Galietta LJ, Verkman AS. 2002b. High-affinity activators of cystic fibrosis transmembrane conductance regulator (CFTR) chloride conductance identified by high-throughput screening. *J Biol Chem* **277:** 37235–37241.

McCarty NA, Standaert TA, Teresi M, Tuthill C, Launspach J, Kelley TJ, Milgram LJ, Hilliard KA, Regelmann WE, Weatherly MR, et al. 2002. A phase I randomized, multicenter trial of CPX in adult subjects with mild cystic fibrosis. *Pediatr Pulmonol* **33:** 90–98.

Mills AD, Yoo C, Butler JD, Yang B, Verkman AS, Kurth MJ. 2010. Design and synthesis of a hybrid potentiator-corrector agonist of the cystic fibrosis mutant protein ΔF508-CFTR. *Bioorg Med Chem Lett* **20:** 87–91.

Mu TW, Ong DS, Wang YJ, Balch WE, Yates JR 3rd, Segatori L, Kelly JW. 2008. Chemical and biological approaches synergize to ameliorate protein-folding diseases. *Cell* **134:** 769–781.

Muanprasat C, Sonawane ND, Salinas D, Taddei A, Galietta LJ, Verkman AS. 2004. Discovery of glycine hydrazide pore-occluding CFTR inhibitors: Mechanism, structure-activity analysis, and in vivo efficacy. *J Gen Physiol* **124:** 125–137.

Namkung W, Phuan PW, Verkman AS. 2011a. TMEM16A inhibitors reveal TMEM16A as a minor component of calcium-activated chloride channel conductance in airway and intestinal epithelial cells. *J Biol Chem* **286:** 2365–2374.

Namkung W, Yao Z, Finkbeiner WE, Verkman AS. 2011b. Small-molecule activators of TMEM16A, a calcium-activated chloride channel, stimulate epithelial chloride secretion and intestinal contraction. *FASEB J* **25:** 4048–4062.

Noel S, Wilke M, Bot AG, De Jonge HR, Becq F. 2008. Parallel improvement of sodium and chloride transport defects by miglustat (*n*-butyldeoxynojyrimicin) in cystic fibrosis epithelial cells. *J Pharmacol Exp Ther* **325:** 1016–1023.

Norez C, Antigny F, Noel S, Vandebrouck C, Becq F. 2009. A cystic fibrosis respiratory epithelial cell chronically treated by miglustat acquires a non-cystic fibrosis-like phenotype. *Am J Respir Cell Mol Biol* **41:** 217–225.

Ostedgaard LS, Rogers CS, Dong Q, Randak CO, Vermeer DW, Rokhlina T, Karp PH, Welsh MJ. 2007. Processing and function of CFTR-ΔF508 are species-dependent. *Proc Natl Acad Sci* **104:** 15370–15375.

Pasyk S, Li C, Ramjeesingh M, Bear CE. 2009. Direct interaction of a small-molecule modulator with G551D-CFTR, a cystic fibrosis-causing mutation associated with severe disease. *Biochem J* **418:** 185–190.

Pedemonte N, Lukacs GL, Du K, Caci E, Zegarra-Moran O, Galietta LJ, Verkman AS. 2005a. Small-molecule correctors of defective ΔF508-CFTR cellular processing identified by high-throughput screening. *J Clin Invest* **115:** 2564–2571.

Pedemonte N, Sonawane ND, Taddei A, Hu J, Zegarra-Moran O, Suen YF, Robins LI, Dicus CW, Willenbring D, Nantz MH, et al. 2005b. Phenylglycine and sulfonamide correctors of defective ΔF508 and G551D cystic fibrosis transmembrane conductance regulator chloride-channel gating. *Mol Pharmacol* **67:** 1797–1807.

Pedemonte N, Tomati V, Sondo E, Caci E, Millo E, Armirotti A, Damonte G, Zegarra-Moran O, Galietta LJ. 2011. Dual activity of aminoarylthiazoles on the trafficking and gating defects of the cystic fibrosis transmembrane conductance regulator chloride channel caused by cystic fibrosis mutations. *J Biol Chem* **286:** 15215–15226.

Phuan PW, Yang B, Knapp J, Wood A, Lukacs GL, Kurth MJ, Verkman AS. 2011. Cyanoquinolines with independent corrector and potentiator activities restore ΔF508-CFTR chloride channel function in cystic fibrosis. *Mol Pharmacol* 80: 683–693.

Powers ET, Morimoto RI, Dillin A, Kelly JW, Balch WE. 2009. Biological and chemical approaches to diseases of proteostasis deficiency. *Annu Rev Biochem* **78:** 959–991.

Pyle LC, Ehrhardt A, Mitchell LH, Fan L, Ren A, Naren AP, Li Y, Clancy JP, Bolger GB, Sorscher EJ, et al. 2011. Regulatory domain phosphorylation to distinguish the mechanistic basis underlying acute CFTR modulators. *Am J Physiol Lung Cell Mol Physiol* **301:** L587–L597.

Ramsey BW, Pepe MS, Quan JM, Otto KL, Montgomery AB, Williams-Warren J, Vasiljev KM, Borowitz D, Bowman CM, Marshall BC, et al. 1999. Intermittent administration of inhaled tobramycin in patients with cystic fibrosis. *N Engl J Med* **340:** 23–30.

Ramsey BW, Davies J, McElvaney NG, Tullis E, Bell SC, Drevinek P, et al. 2011. A CFTR potentiator in patients with cystic fibrosis and the G551D mutation. *New Engl J Med* **365:** 1663–1672.

Robert R, Carlile GW, Pavel C, Liu N, Anjos SM, Liao J, Luo Y, Zhang D, Thomas DY, Hanrahan JW. 2008. Structural analog of sildenafil identified as a novel corrector of the F508del-CFTR trafficking defect. *Mol Pharmacol* **73:** 478–489.

Robert R, Carlile GW, Liao J, Balghi H, Lesimple P, Liu N, Kus B, Rotin D, Wilke M, de Jonge HR, et al. 2010. Correction of the Δ Phe508 cystic fibrosis transmembrane conductance regulator trafficking defect by the bioavailable compound glafenine. *Mol Pharmacol* **77:** 922–930.

Rogers CS, Stoltz DA, Meyerholz DK, Ostedgaard LS, Rokhlina T, Taft PJ, Rogan MP, Pezzulo AA, Karp PH, Itani OA, et al. 2008. Disruption of the CFTR gene produces a model of cystic fibrosis in newborn pigs. *Science* **321:** 1837–1841.

Rowe SM, Accurso F, Clancy JP. 2007. Detection of cystic fibrosis transmembrane conductance regulator activity in early-phase clinical trials. *Proc Am Thorac Soc* **4:** 387–398.

Rowe SM, Clancy JP, Boyle M, Van Goor F, Ordonez C, Dong Q, Campbell P, Ashlock M, Accurso F. 2010. Parallel effects of VX-770 on transepithelial potential difference in vitro and in vivo. *J Cystic Fibrosis* **9** (Suppl): S20.

Rubenstein RC, Zeitlin PL. 1998. A pilot clinical trial of oral sodium 4-phenylbutyrate (Buphenyl) in ΔF508-homozygous cystic fibrosis patients: Partial restoration of nasal epithelial CFTR function. *Am J Respir Crit Care Med* **157:** 484–490.

Saiman L, Marshall BC, Mayer-Hamblett N, Burns JL, Quittner AL, Cibene DA, Coquillette S, Fieberg AY, Accurso FJ, Campbell PW III, et al. 2003. Azithromycin in patients with cystic fibrosis chronically infected with *Pseudomonas aeruginosa*: A randomized controlled trial. *JAMA* **290:** 1749–1756.

Sampson HM, Robert R, Liao J, Matthes E, Carlile GW, Hanrahan JW, Thomas DY. 2011. Identification of a NBD1-binding pharmacological chaperone that corrects the trafficking defect of F508del-CFTR. *Chem Biol* **18:** 231–242.

Sloane PA, Rowe SM. 2010. Cystic fibrosis transmembrane conductance regulator protein repair as a therapeutic strategy in cystic fibrosis. *Curr Opin Pulm Med* **16:** 591–597.

Snyder DS, Tradtrantip L, Yao C, Kurth MJ, Verkman AS. 2011. Potent, metabolically stable benzopyrimido-pyrrolo-oxazine-dione (BPO) CFTR inhibitors for polycystic kidney disease. *J Med Chem* **54:** 5468–5477.

Solomon GM, Konstan MW, Wilschanski M, Billings J, Sermet-Gaudelus I, Accurso F, Vermeulen F, Levin E, Hathorne H, Reeves G, et al. 2010. An international randomized multicenter comparison of nasal potential difference techniques. *Chest* **138:** 919–928.

Sonawane ND, Zhao D, Zegarra-Moran O, Galietta LJ, Verkman AS. 2008. Nanomolar CFTR inhibition by pore-occluding divalent polyethylene glycol-malonic acid hydrazides. *Chem Biol* **15:** 718–728.

Song Y, Sonawane ND, Salinas D, Qian L, Pedemonte N, Galietta LJ, Verkman AS. 2004. Evidence against the rescue of defective ΔF508-CFTR cellular processing by curcumin in cell culture and mouse models. *J Biol Chem* **279:** 40629–40633.

Sun X, Sui H, Fisher JT, Yan Z, Liu X, Cho HJ, Joo NS, Zhang Y, Zhou W, Yi Y, et al. 2010. Disease phenotype of a ferret CFTR-knockout model of cystic fibrosis. *J Clin Invest* **120:** 3149–3160.

Thibodeau PH, Richardson JM III, Wang W, Millen L, Watson J, Mendoza JL, Du K, Fischman S, Senderowitz H, Lukacs GL, et al. 2010. The cystic fibrosis-causing mutation ΔF508 affects multiple steps in cystic fibrosis transmembrane conductance regulator biogenesis. *J Biol Chem* **285:** 35825–35835.

Tradtrantip L, Sonawane ND, Namkung W, Verkman AS. 2009a. Nanomolar potency pyrimido-pyrrolo-quinoxalinedione CFTR inhibitor reduces cyst size in a polycystic kidney disease model. *J Med Chem* **52:** 6447–6455.

Tradtrantip L, Yangthara B, Padmawar P, Morrison C, Verkman AS. 2009b. Thiophenecarboxylate suppressor of cyclic nucleotides discovered in a small-molecule screen blocks toxin-induced intestinal fluid secretion. *Mol Pharmacol* **75:** 134–142.

Van Goor F, Straley KS, Cao D, Gonzalez J, Hadida S, Hazlewood A, Joubran J, Knapp T, Makings LR, Miller M, et al. 2006. Rescue of ΔF508-CFTR trafficking and gating in human cystic fibrosis airway primary cultures by small molecules. *Am J Physiol Lung Cell Mol Physiol* **290:** L1117–L1130.

Van Goor F, Hadida S, Grootenhuis PD, Burton B, Cao D, Neuberger T, Turnbull A, Singh A, Joubran J, Hazlewood A, et al. 2009. Rescue of CF airway epithelial cell function in vitro by a CFTR potentiator, VX-770. *Proc Natl Acad Sci* **106:** 18825–18830.

Van Goor F, Hadida S, Grootenhuis PD, Stack JH, Burton B, Olson E, Wine J, Frizzell RA, Ashlock M, Negulescu P. 2010. Rescue of the protein folding defect in cystic fibrosis in vitro by the investigational small molecule, VX-809. *J Cyst Fibros* **9:** S14.

Wang X, Venable J, LaPointe P, Hutt DM, Koulov AV, Coppinger J, Gurkan C, Kellner W, Matteson J, Plutner H, et al. 2006. Hsp90 cochaperone Aha1 downregulation rescues misfolding of CFTR in cystic fibrosis. *Cell* **127:** 803–815.

Wang W, Bernard K, Li G, Kirk KL. 2007. Curcumin opens cystic fibrosis transmembrane conductance regulator channels by a novel mechanism that requires neither ATP binding nor dimerization of the nucleotide-binding domains. *J Biol Chem* **282:** 4533–4544.

Wellhauser L, Kim Chiaw P, Pasyk S, Li C, Ramjeesingh M, Bear CE. 2009. A small-molecule modulator interacts directly with ΔPhe508-CFTR to modify its ATPase activity and conformational stability. *Mol Pharmacol* **75:** 1430–1438.

Welsh MJ, Smith AE. 1993. Molecular mechanisms of *CFTR* chloride channel dysfunction in cystic fibrosis. *Cell* **73:** 1251–1254.

Wilschanski M, Dupuis A, Ellis L, Jarvi K, Zielenski J, Tullis E, Martin S, Corey M, Tsui LC, Durie P. 2006. Mutations in the cystic fibrosis transmembrane regulator gene and in vivo transepithelial potentials. *Am J Respir Crit Care Med* **174:** 787–794.

Yang H, Shelat AA, Guy RK, Gopinath VS, Ma T, Du K, Lukacs GL, Taddei A, Folli C, Pedemonte N, et al. 2003. Nanomolar affinity small molecule correctors of defective ΔF508-CFTR chloride channel gating. *J Biol Chem* **278:** 35079–35085.

Ye L, Knapp JM, Sangwung P, Fettinger JC, Verkman AS, Kurth MJ. 2010. Pyrazolylthiazole as ΔF508-cystic fibrosis transmembrane conductance regulator correctors with improved hydrophilicity compared to bithiazoles. *J Med Chem* **53:** 3772–3781.

Yu GJ, Yoo CL, Yang B, Lodewyk MW, Meng L, El-Idreesy TT, Fettinger JC, Tantillo DJ, Verkman AS, Kurth MJ. 2008. Potent s-*cis*-locked bithiazole correctors of ΔF508 cystic fibrosis transmembrane conductance regulator cellular processing for cystic fibrosis therapy. *J Med Chem* **51:** 6044–6054.

Yu W, Kim Chiaw P, Bear CE. 2011. Probing conformational rescue induced by a chemical corrector of F508del-cystic fibrosis transmembrane conductance regulator (CFTR) mutant. *J Biol Chem* **286:** 24714–24725.

Zeitlin PL, Diener-West M, Rubenstein RC, Boyle MP, Lee CK, Brass-Ernst L. 2002. Evidence of CFTR function in cystic fibrosis after systemic administration of 4-phenylbutyrate. *Mol Ther* **6:** 119–126.

Antibiotic and Anti-Inflammatory Therapies for Cystic Fibrosis

James F. Chmiel[1], Michael W. Konstan[1], and J. Stuart Elborn[2]

[1]Department of Pediatrics, Case Western Reserve University School of Medicine, Rainbow Babies and Children's Hospital, Cleveland, Ohio 44106

[2]Medicine and Surgery, Queens University Belfast, Belfast City Hospital, Belfast BT9 7AB, Northern Ireland, United Kingdom

Correspondence: michael.konstan@case.edu

Cystic fibrosis (CF) lung disease is characterized by chronic bacterial infection and an unremitting inflammatory response, which are responsible for most of CF morbidity and mortality. The median expected survival has increased from <6 mo in 1940 to >38 yr now. This dramatic improvement, although not great enough, is due to the development of therapies directed at secondary disease pathologies, especially antibiotics. The importance of developing treatments directed against the vigorous inflammatory response was realized in the 1990s. New therapies directed toward the basic defect are now visible on the horizon. However, the impact of these drugs on downstream pathological consequences is unknown. It is likely that antibiotics and anti-inflammatory drugs will remain an important part of the maintenance regimen for CF in the foreseeable future. Current and future antibiotic and anti-inflammatory therapies for CF are reviewed.

Although cystic fibrosis (CF) impacts many organ systems, lung disease accounts for most of the morbidity and mortality (Davis et al. 1996). Abnormal or insufficient cystic fibrosis transmembrane conductance regulator (CFTR) leads from the defective gene and protein to an abnormal surface environment to a vicious cycle of obstruction, chronic infection, and inflammation (Chmiel et al. 2002a). Relieving obstruction with mucolytics and airway clearance, controlling infection with antibiotics, and reducing inflammation with anti-inflammatory drugs have been the cornerstones of a comprehensive pulmonary treatment program in CF.

INFECTION AND INFLAMMATION IN THE CF AIRWAY

CF airways are most susceptible to chronic infection with *Staphylococcus aureus*, *Hemophilus influenzae*, *Pseudomonas aeruginosa*, *Burkholderia cepacia* complex organisms, *Stenotrophomonas maltophilia*, and *Achromobacter xylosoxidans*. Most of these microbes form biofilms, thus serving as persistent inflammatory stimuli (Chmiel and Davis 2003). When local host defense mechanisms are challenged by intercurrent viral or bacterial infections, massive numbers of neutrophils are recruited into the airway. Although inflammation is meant to eradicate infection, this ultimately fails, and the exaggerated inflammatory response that ensues is responsible for much of the lung's pathology.

The CF inflammatory response begins early in life, becomes persistent, and is often excessive relative to the burden of infection. Bronchoalveolar lavage (BAL) fluid from CF patients, including infants and patients with mild disease, contains large concentrations of inflammatory mediators and cells, particularly neutrophils (Konstan et al. 1993, 1994; Birrer et al. 1994; Armstrong et al. 1995, 1997; Balough et al. 1995; Bonfield et al. 1995a; Khan et al. 1995; Kirchner et al. 1996). The presence of large concentrations of inflammatory mediators in the absence of detectable pathogens suggests either that the inflammatory response operates independently of infection or that there is failure to terminate the inflammatory response once the inciting stimulus has been removed (Khan et al. 1995). The inflammatory response could be triggered by a transient viral or bacterial infection that then cannot be stopped. In addition, BAL studies show that infected CF infants have more inflammation than do similarly infected non-CF infants (Noah et al. 1997; Muhlebach et al. 1999).

Neutrophils, present in massive quantities, release actin, DNA, proinflammatory cytokines and chemokines, oxidants, and proteases. B cells and T cells, particularly TH-17 cells, also contribute to CF lung disease (Dubin et al. 2007). The continued presence of bacteria triggers an unrelenting inflammatory response that drives the persistent generation of proinflammatory mediators including neutrophil chemoattractants IL-8 and LTB$_4$, which recruit more neutrophils into the airways, fueling the vicious cycle of inflammation that leads to lung destruction (Konstan and Berger 1997; Chmiel et al. 2002a). Because the neutrophil plays a central role in CF airway pathophysiology, any anti-inflammatory drug developed for CF must, either directly or indirectly, address the neutrophil and its products (Table 1).

Airway surface fluid from CF patients contains large concentrations of inflammatory mediators including TNF-α, IL-1β, IL-6, IL-8, IL-17, and GM-CSF (Bonfield et al. 1995a; McAllister et al. 2005). The synthesis of these mediators is promoted by a few transcription factors including AP-1, NF-κB, and MAPK. In addition to a heightened proinflammatory arm, there appears to be inappropriately decreased counter-regulatory pathways, particularly those involving IL-10 and nitric oxide (NO) (Bonfield et al. 1995b, 1999; Balfour-Lynn et al. 1996; Grasemann et al. 1997). When counter-regulatory controls are abnormal, an imbalance occurs, resulting in prolonged and excessive inflammatory mediator production. It is possible that a combination of these mechanisms fuels the destructive inflammatory cascade. It is unlikely that a single defect in one pathway accounts for the entirety of the exaggerated inflammatory response. Multiple pathways are probably up-regulated (i.e., pathways that activate the inflammatory response) or down-regulated (i.e., pathways that terminate the inflammatory response) in response to the cell's attempt to correct the underlying physiological abnormality due to abnormal CFTR. The end result is that the delicate balance between the pro- and anti-inflammatory arms is disrupted, and pathways promoting the activation and perpetuation of the inflammatory response are favored. A schematic of the CF airway is seen in Figure 1.

Table 1. Neutrophil chemoattractants and products targeted by anti-inflammatory drugs

I. Neutrophil chemoattractants
A. IL-8
B. LTB$_4$
C. Complement components: C5a, C5a-des-Arg
D. Bacterial products: N-formyl-Met-Leu-Phe

II. Neutrophil products
A. Proteases (elastase)
B. DNA
C. Oxidants (H_2O_2 and O_2^-)
D. IL-8
E. LTB$_4$

THERAPIES DIRECTED AT BACTERIAL INFECTION

The range of microorganisms, particularly bacteria that have been identified in the CF airway, is much wider than previously appreciated

Cite this article as *Cold Spring Harb Perspect Med* doi: 10.1101/cshperspect.a009779

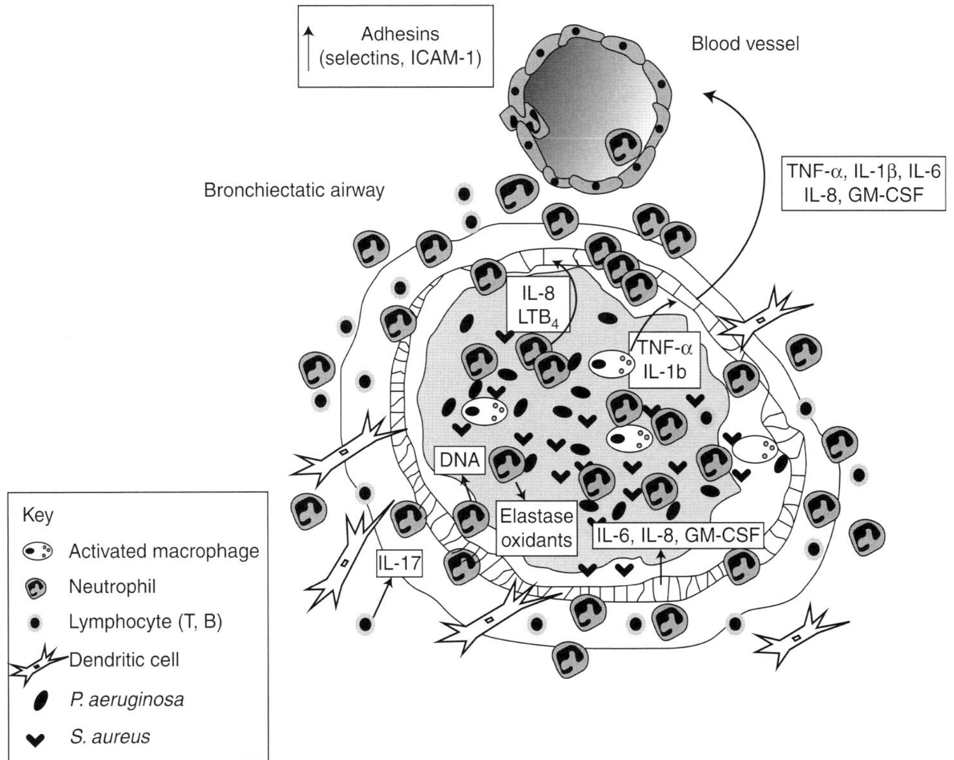

Figure 1. Infection and inflammation in the CF lung. CF lung disease is essentially an endobronchial/peribronchial process. The alveoli do not become involved until late in the disease course. In CF, the gene defect leads to an abnormal airway surface environment and then to airway obstruction with mucus (represented as an amorphous gray area in the center of the airway), chronic bacterial infection, and persistent inflammation. Although there are several different types of bacteria that infect the CF airway, *S. aureus* predominates early in life but typically yields to *P. aeruginosa*. However, most patients have polymicrobic infections, thus making selection of appropriate antibiotics for treatment of a pulmonary exacerbation difficult. The bacterial infection is associated with a vigorous inflammatory response that is characterized by a large neutrophilic infiltration. Other inflammatory cells are also involved in the inflammatory response including epithelial cells, macrophages, dendritic cells, B lymphocytes, and T lymphocytes. These cells release many inflammatory cytokines and mediators that further promote the inflammatory response. The inflammatory response is ineffectual in eliminating bacteria from the airway, but the excess inflammatory mediators released into the endobronchial and peribronchial spaces damage the airway wall architecture and ultimately lead to bronchiectasis. This figure focuses on the proinflammatory aspects of CF airway inflammation. The abnormalities in the counter-regulatory mechanisms in CF are not depicted.

(Tunney et al. 2008, 2011). Using molecular techniques, organisms that were previously unculturable, particularly microaerophilic and anaerobic bacteria, have been identified (Lipuma 2012). The virulence of these organisms and their complex interaction as an airway microbiome has yet to be fully established. However, there are some studies to suggest that the diversity or the lack of diversity in sputum and BAL from people with CF is associated with more severe and chronic disease, and this may in part be driven by antibiotic therapy (Tunney et al. 2011; Zhao et al. 2012)

Antimicrobial therapy in CF has a long history, and effective treatment of infection is thought to be one of the key advances that have resulted in improved outcomes over the past 50 years for people with CF (Döring et al. 2012).

S. aureus was identified in 1949 as the predominant pathogen in cultures taken from young children with CF. Subsequently, *P. aeruginosa* was identified as an organism associated with bronchiectasis and chronicity. From the late 1950s, *P. aeruginosa* was reported with increasing frequency from children with CF, and the mucoid phenotype for *P. aeruginosa* was first described in 1966 (J Littlewood; www.cfmedicine.com). Specific antimicrobial therapy in CF has, therefore, been directed against these organisms. Over the subsequent decades, a range of other dominant organisms has presented challenges for antibiotic therapy. These include members of the *B. cepacia* complex, *S. maltophilia*, *Achromobacter* species, and other Gram-negative infections. Nontuberculous mycobacterial infection is also emerging as a new challenge for antibiotic treatment for people with CF.

ANTIBIOTIC PROPHYLAXIS

The presence of *S. aureus* in airway secretions from people with CF prompted early clinicians to consider the use of prophylactic antibiotics to prevent and control Staphylococcal infection in the early years (Smyth and Walters 2012). This type of treatment has been and, in some countries, remains controversial. In some small studies, prophylactic treatment from diagnosis with flucloxacillin has been associated with a reduction in the frequency of positive *Staphylococcal* airway sample cultures and a reduction in admission to hospital, but no long-term improvements have been shown in lung function. These observations were followed up with a clinical trial of cephalexin, a more broad-spectrum antibiotic, which did not show any efficacy. But in patients treated with long-term prophylaxis, there was an increase in the frequency of new infections with *P. aeruginosa*.

These observations have been supported by several registry studies. In a recent updated *Cochrane Review*, a meta-analysis of four studies including 401 patients under the age of seven was reported (Smyth and Walters 2012). In this analysis, no significant increase in the number of isolates of *P. aeruginosa* was shown between treated and untreated groups from reported studies, although there was a trend toward a lower cumulative isolation rate of *P. aeruginosa* in the patients treated with long-term antibiotics at 2–3 yr and a higher isolation rate from 4 to 6 yr. These studies used a range of antibiotics including flucloxacillin, co-trimoxazole, cefadroxil, and cephalexin. It is unclear what the clinical benefits of simply reducing the frequency of culture of *S. aureus* might be.

Antibiotic Therapy for *S. aureus*

S. aureus is a very common infecting organism in people with CF. This is most commonly methicillin-sensitive *S. aureus* (MSSA), but methicillin-resistant *S. aureus* (MRSA) is becoming increasingly common in the sputum of people with CF. The increased incidence of MRSA infection in people with CF is particularly concerning because it is associated with a more rapid rate of decline in lung function and worse survival (Dasenbrook et al. 2008, 2010). When either MSSA or MRSA is identified in sputum, it makes sense to attempt eradication. For MSSA, cephalexin, flucloxacillin, co-trimoxazole, and, additionally, in adults, tetracyclines can be used to do this. MRSA similarly may be eradicated with a combination of drugs used by clinicians including vancomycin, linezolid, rifampin, rifampicin, fusidic acid, tetracycline, and co-trimoxazole-based regimes. Several studies are currently trying to determine the best regime to treat this organism. Similarly, antibiotic approaches can be used to treat exacerbations in individuals who are chronically infected with this organism.

ANTIBIOTIC THERAPY FOR INFECTION WITH *P. aeruginosa* ERADICATION THERAPY

In contrast to the debate around antibiotic prophylaxis for *S. aureus*, there is currently no debate for antibiotic prophylaxis against *P. aeruginosa* in people with CF (Döring et al. 2012). There is currently a clinical trial in Europe using avian immunoglobulin Y antibodies, which are gargled, to explore if this novel intervention of treating the oropharynx might prevent *P. aeruginosa* acquisition in the lower airways (EUDRACT-2011-000801-39).

There is, however, a compelling rationale for antibiotic eradication therapy for *P. aeruginosa* in CF. This approach was championed by the Copenhagen Centre, who in early studies of this intervention using a historical comparative cohort, showed that there was a strong indication that the combination of inhaled colistin and oral ciprofloxacin resulted in approximately an 80% success in *P. aeruginosa* eradication (Valerius et al. 1991). Subsequent studies, such as the ELITE and EPIC studies, have shown that inhaled tobramycin for 4 wk is also an effective antibiotic eradication therapy (Ratjen et al. 2010; Treggiari et al. 2011). In two small studies, no difference was identified between these two regimes with a comparable eradication rate (Taccetti et al. 2012; Proesmans 2013).

Early intervention is now a recommended part of most guidelines for the treatment of new or repeated infection with *P. aeruginosa* (Döring et al. 2012). It is less clear what the best treatment for failure to eradicate using either the colistin/ciprofloxacin or inhaled tobramycin regimes should be. A recent recommendation from a consensus group was that after two attempts at eradication using inhaled and oral therapy, intravenous antipseudomonal antibiotics should be considered. A clinical trial (Torpedo; www.torpedo-cf.org.uk/index.html) comparing colistin/ciprofloxacin and intravenous antibiotics is currently under way in the United Kingdom.

Treatment of Chronic *P. aeruginosa* Infection

In ~75% of people with CF, *P. aeruginosa* eventually cannot be eradicated *and* chronic infection ensues (Döring et al. 2012). When this occurs, the bacteria frequently have a mucoid phenotype and exist in a biofilm. This and other important adaptions reduce the effectiveness of innate host defense and make treatment with antibiotics less effective (Williams and Davies 2012). When chronic infection has developed, treatment with long-term aerosolized antibiotics is an important and effective intervention.

Tobramycin (TOBI, Bramitob) and aztreonam lysine (Cayston) are approved therapies by most regulators throughout the world. Both of these treatments have shown superiority over placebo for measurements of lung function, quality of life, and reduction in pulmonary exacerbations (Döring et al. 2012). Colistin is also used in chronic suppressive therapy in its nebulized form. Tobramycin and colistin have recently been developed into dry-powder inhaled antibiotics. Tobramycin inhalation powder (TobiPodhaler) is approved by the EMA and also is under review by the FDA, and colistin inhalation powder (Colobreathe) has been approved by the EMA for treatment (Konstan et al. 2011; Schuster et al. 2012). Both of these drugs have shown noninferiority for key clinical outcomes to tobramycin delivered by nebulizer, and have increased cough as a common side effect.

In general, inhaled antipseudomonal antibiotics decrease bacterial burden (measured in CFU/g of sputum), improve lung function and quality of life, and reduce the frequency of pulmonary exacerbations (Döring et al. 2012). Colistin is given every day, whereas tobramycin is licensed for administration as an alternate-month therapy. Many patients describe increased symptoms and reduced FEV_1 during the month off tobramycin. Therefore, many patients cycle between two different inhaled antibiotics. The optimal regime for aerosolized antibiotic therapy as long-term suppressive therapy has yet to be determined. It is unlikely that a randomized controlled trial will allow any further direct comparisons between regimes or answer the challenging question of whether continuous nebulized antibiotics using alternate regimes, for example, alternating tobramycin, aztreonam lysine, and colistin, would be superior to the current recommended month-off month-on regimes.

Several other inhaled antibiotic therapies against *P. aeruginosa* are in development. Two formulations of levofloxacin, ciprofloxacin as a dry powder inhaler, and ciprofloxacin and amikacin as liposomal formulations are in various stages of development and may become available in the near future (Döring et al. 2012).

Treating Exacerbations in Patients with Chronic *P. aeruginosa* Infection

Pulmonary exacerbations (PEs) are significant events in people with CF. These events are associated with an increase in pulmonary symptoms, systemic symptoms including weight loss, sometimes new clinical signs, physiological changes in lung function, and often evidence of local and systemic inflammation. Frequency of pulmonary exacerbations is associated with a more rapid decline in FEV_1 and reduced survival (Smyth and Elborn 2008; Döring et al. 2012). Prevention of pulmonary exacerbations is, therefore, a key intervention, and there is good evidence that long-term suppressive antibiotic therapy, concomitant human DNase (Pulmozyme), azithromycin, and inhaled hypertonic saline reduce the time to the next exacerbation in clinical trials, and, by extrapolation, probably reduce the frequency of these events in people with CF. Optimizing long-term treatment to reduce these events and the inflammatory injury impact they have on the lungs in people with CF is a critically important intervention (Stenbit and Flume 2011).

Optimizing the antibiotic therapy and duration of treatment is important in treating pulmonary exacerbations. There is some evidence that ~25% of pulmonary exacerbations result in a failure to return to baseline lung function and/or rapid relapse (Sanders et al. 2010; Stenbit and Flume 2011; Parkins et al. 2012). Selection of antibiotics for treatment of such exacerbations is largely empirical. Using current techniques to determine in vitro antimicrobial susceptibility does not appear to improve the outcomes (Döring et al. 2012). This is a particular problem with *P. aeruginosa* because with repeated antibiotic therapy, these organisms develop resistance. However, response to antibiotics does not appear to relate directly to in vitro susceptibility as currently determined in microbiology laboratories. The choice of initial antibiotics to treat a pulmonary exacerbation is often based on antibiotic susceptibility testing and is affected by patient tolerability. If clinical response is suboptimal, changes in antibiotic regimen are often made based on clin-

ical experience irrespective of susceptibility testing.

There is a good historical and current rationale for the use of a combination of an extended-action penicillin and an aminoglycoside. Conventionally, this therapy has been for 14 d, although in a recent registry-based study, 10 d may be a sufficient period of time for the majority of patients (Van Devanter et al. 2010). However, this approach has not been subjected to a prospective study.

ANTIBIOTIC TREATMENT FOR OTHER BACTERIAL PATHOGENS IN CF

S. maltophilia and *Achromobacter* Species

S. maltophilia and *Achromobacter* species are commonly isolated in CF sputum, although their prevalences are variable. Single centers have reported up to 25% prevalence of either organism. The clinical significance of these organisms is hard to determine, although it is clear that individual patients with either of these bacteria as their predominant pathogen can have progressive deterioration in lung function and pulmonary exacerbations. A range of antibiotics can be useful against these bacteria including tetracyclines, co-trimoxazole, colistin, and piperacillin/tazobactam. There are no eradication studies in the people infected with these bacteria, but it is logical to attempt to clear these organisms if possible. Inhaled colistin can be used for *S. maltophilia* infection, and combinations of the above antibiotics can be used for treatment of exacerbations (Döring et al. 2012).

B. cepacia Complex Strains

Treatment of *B. cepacia* complex strains is quite problematic. In general, members of this complex have intrinsic antibiotic resistance, and *Burkholderia cenocepacia*, which historically has been the most common *Burkholderia* in people with CF, is almost universally pan-resistant. There are no long-term antibiotics recommended for treatment of patients with chronic infection with *B. cepacia* complex strains, and a recent study of inhaled aztreonam lysine showed no benefit. Intravenous antibiotic treatment is based on several studies that would suggest that tetracycline, co-trimoxazole, chloramphenicol, colistin, ceftazidime, meropenem, and piperacillin/tazobactam combinations may have clinical efficacy (Döring et al. 2012).

Nontuberculous Mycobacterial Infection

Nontuberculous mycobacterial infection is a currently emerging area that is causing diagnostic and therapeutic challenges. Several nontuberculous mycobacteria have been isolated from people with CF, although the predominant groups are the *Mycobacterium avium* complex and *Mycobacterium abscessus* organisms. *M. abscessus* infection in CF and in other chronic lung conditions is associated with particular challenges (Döring et al. 2012). Treatment for both organisms requires long-term therapy with combinations of oral, intravenous and/or inhaled antimycobacterial agents. *Mycobacterium avium* complex organisms are typically treated with combinations of ethambutol, rifampin, macrolides, and inhaled amikacin. For *Mycobacterium abscessus*, combinations of amikacin, linezolid, macrolides, cefoxitin, tigicycline, and/ or meropenem are utilized. Currently, international guidelines are being developed to help further guide treatment of this difficult group of bacteria.

Other Bacterial Species

Many other bacterial species are being identified in the airway microbiota in people with CF. Treatment of organisms such as *Streptococcus* species and anaerobic bacteria such as *Prevotella*, *Veillonella*, *Rothia*, and *Actinomyces* species is not clear (Tunney et al. 2011; Zhao et al. 2012). There may be changes in how we perceive treatment with antibiotics over the next several years because second-generation sequencing provides more insight into the complex microbiota of the airways in people with CF. Antimicrobial therapy remains a key cornerstone for the modulation of infection and inflammation in CF.

THERAPIES DIRECTED AT THE INFLAMMATORY RESPONSE OF THE CF AIRWAY

Addressing the inflammatory response is warranted because it is the major process leading to lung destruction. Several steps in the pathophysiological cascade could be targeted. However, efficacy at any step may be beneficial. Early anti-inflammatory drug studies focused on systemic and inhaled corticosteroids, ibuprofen, and other nonsteroidal anti-inflammatory drugs (NSAIDs). Many anti-inflammatory drugs have been studied in CF since 1990. Although many have been translated from the bench to a clinical trial, precious few are recommended for clinical use (Table 2). The length of this review precludes a detailed discussion of all of these drugs. This review focuses on drugs that have undergone in-depth evaluation or more recent developments. Corticosteroids and ibuprofen are reviewed first, followed by cytokines and anticytokines, modulators of intracellular signaling, antioxidants, protease inhibitors, and other anti-inflammatory therapies.

CORTICOSTEROIDS

Corticosteroids were the first anti-inflammatory drugs studied in CF. They reduce the formation of mucus and edema; inhibit chemotaxis, adhesion, and activation of leukocytes; inhibit NF-κB activation; and interfere with the synthesis or actions of inflammatory mediators. Several studies evaluated corticosteroids as a chronic therapy for attenuating CF airway inflammation. Two 4-yr, double-blind, placebo-controlled studies of oral corticosteroids laid the foundation for future anti-inflammatory trials (Auerbach et al. 1985; Eigen et al. 1995). In these studies, alternate-day corticosteroids were associated with better lung function, improved weight gain, and fewer hospital admissions. In one trial, beneficial effects on lung function were observed primarily in patients infected with *P. aeruginosa* (Eigen et al. 1995). Adverse effects, including glucose intolerance, growth impairment, and cataract formation, limit the long-term practical application of alternate-day oral

Table 2. Anti-inflammatory therapies evaluated in CF

Therapy	Target
Evaluated only in preclinical studies	
Anti-ICAM-1	Leukocyte adhesion molecules
Anti-IL-8	Cytokines
Anti-IL-17	Cytokines
IL-10	Intracellular signaling and cytokines
Interferon-γ	Intracellular signaling
p38 Mitogen-activated protein kinase inhibitors	Intracellular signaling
Evaluated in clinical trials	
α1-Protease inhibitor	Proteases
Corticosteroids	Intracellular signaling, eiscosanoids, cytokines
CXCR2 antagonist	Neutrophil chemotaxis/activation
Cyclosporine-A	Intracellular signaling
DHA	Eicosanoid modulator
EPI-hNE4	Proteases
Glutathione	Oxidants
Hydroxychloroquine	Unknown
L-Arginine	Intracellular signaling
LTB$_4$ receptor antagonist (BIIL 284 BS)	Leukotriene receptors, eicosanoids
Methotrexate	Adhesion molecules, cytokines, transmethylation reactions, increases adenosine release
Monocyte/neutrophil elastase inhibitor	Proteases
Montelukast	Leukotriene receptors
N-Acetyl cysteine	Intracellular signaling, oxidants
Omega-3-fatty acids	Eicosanoids
Secretory leukoprotease inhibitor	Proteases
Simvastatin	Intracellular signaling
Thiazolidinediones/pioglitazone	Intracellular signaling
Vitamins C, E, and β-carotene	Oxidants
Vitamin D	Bacteria and cytokines
Therapies recommended for clinical use	
Antibiotics	Bacteria (inflammatory stimuli)
Azithromycin	Unknown
Dornase-alfa	DNA
Ibuprofen	Intracellular signaling, eicosanoids, cytokines

corticosteroids. In a follow-up study, growth deficits persisted after therapy discontinuation (Lai et al. 2000). In view of these results and other known toxicities of corticosteroids including osteopenia, osteoporosis, and skeletal muscle weakness, the long-term use of systemic corticosteroids for slowing lung function decline is not advocated (Flume et al. 2007).

Adverse effects associated with systemic corticosteroids prompted investigations of inhaled corticosteroids (ICS). The inhaled route might afford benefit with less risk than systemic administration, but systemic administration may be necessary to affect neutrophil migration. Consistent ICS use was associated with a slower rate of decline of lung function. However, ICS use also was associated with decreased weight and height for age, and increased use of insulin and oral hypoglycemic agents in one large observational study (Ren et al. 2008). In another observational study, ICSs were associated with slower FEV$_1$ decline in children with CF age

6–12 yr but not other age groups (De Boeck et al. 2011). A benefit of ICS has not been shown in randomized placebo-clinical trials. None of the trials of ICS in CF showed convincing effects on lung function or airway inflammatory markers (Ross et al. 2009). Small study populations and short observation periods hampered these trials. In a large, prospective, multicenter study, withdrawal of ICS for a 6-mo period was not associated with significant worsening of CF lung disease (Balfour-Lynn et al. 2006). Patients in whom steroids were discontinued did not have worsening of lung function, an increased need for oral or intravenous antibiotics, or a shorter time to pulmonary exacerbation. Thus, although inhaled steroids may be of benefit in patients with coexisting asthma, their use as an anti-inflammatory agent in CF has not been substantiated. Potential long-term complications of ICS have not been adequately evaluated in CF. Caution should be heeded when prescribing ICS to patients with CF. A Cystic Fibrosis Foundation expert panel advised against the long-term use of ICS in patients with CF older than 6 yr who did not have coexistent asthma or allergic bronchopulmonary aspergillosis (ABPA) (Flume et al. 2007).

IBUPROFEN

NSAIDs possess properties similar to corticosteroids but have fewer adverse effects. Ibuprofen has received the most attention in CF, largely because it has specific activity against neutrophils. Neutrophil chemotaxis to mucosal epithelium is significantly decreased with oral doses of ibuprofen that result in a peak plasma concentration >50 μg/mL (Konstan et al. 2003). In a 4-yr, double-blind, placebo-controlled clinical trial in the United States in patients with CF, twice-daily high-dose ibuprofen was associated with a slower rate of decline in pulmonary function measures, better preservation of body weight, fewer hospital admissions, and better Brasfield chest radiograph scores, with no significant adverse effects (Konstan et al. 1995). The beneficial effects were most pronounced in the youngest patients (5–13 yr old). A 2-yr placebo-controlled multicenter trial in Canada in

145 CF children also showed a beneficial effect of ibuprofen on lung function (Lands et al. 2007). Results from an analysis of observational data from the CF Foundation Patient Registry revealed that "real world" clinical use of ibuprofen in 1365 CF children treated on average for 4 yr reduced the annual rate of FEV_1 decline by 29% compared with those not treated with ibuprofen (Konstan et al. 2007).

Nonetheless, ibuprofen has not been widely adopted, largely because of the logistic challenges associated with the need to establish the dose in each patient with a 3-h pharmacokinetic test and to concerns related to adverse effects of the drug (Konstan 2008). Based on U.S. CF Foundation Patient Registry data, ibuprofen use is associated with gastrointestinal bleeding, but the occurrence is rare (annual incidence 0.37% vs. 0.14% in those not treated with ibuprofen [Konstan et al. 2007]). Concomitant use of antacids, proton pump inhibitors, or misoprostol (a PGE_1 analog) would likely limit this adverse event. Renal failure associated with ibuprofen therapy has been the subject of a few case reports, but CF Foundation Registry data suggest that the incidence of renal failure is not increased among CF patients treated with ibuprofen (Konstan et al. 2007). Thus, the benefits of this therapy appear to outweigh the associated risks. An expert panel recommended high-dose ibuprofen for CF patients with mild lung disease (Flume et al. 2007). A *Cochrane Review* also concluded that "high-dose ibuprofen can slow the progression of lung disease in people with CF, especially in children, and this suggests that strategies to modulate lung inflammation can be beneficial for people with CF" (Lands and Stanojevic 2007).

CYTOKINES AND ANTICYTOKINES

Since the first trial of corticosteroids, many new anti-inflammatory drugs have been introduced for the treatment of other diseases. Anti-inflammatory cytokines and antibodies to proinflammatory cytokines may have efficacy in CF. Most cytokines are pleiomorphic, and although some show a preponderance of proinflammatory effects, others show a preponderance of anti-inflammatory effects. IL-10 possesses many anti-inflammatory properties. It inhibits the production of many proinflammatory cytokines, inhibits NF-κB activation, induces neutrophil apoptosis, and decreases antigen presentation and T-cell stimulation. Because IL-10 terminates the inflammatory response, its deficiency, as shown in CF (Bonfield et al. 1999), might result in persistent inflammation even after stimulus removal. IL-10 deficiency worsens endobronchial inflammation in *P. aeruginosa*-infected mice (Chmiel et al. 2002b). Administration of IL-10 to infected mice showed beneficial effects on airway inflammation (Chmiel et al. 1999). A planned clinical trial of the drug was abandoned for reasons unrelated to CF, but this remains an interesting therapeutic option. Interferon-γ, another cytokine with anti-inflammatory effects, did not improve pulmonary function or alter sputum inflammatory mediators in a multicenter clinical trial despite its ability to restore counter-regulatory functions in CF cell models (Moss et al. 2005).

An alternative approach to administering anti-inflammatory cytokines would be to inhibit specific proinflammatory mediators. Antibodies to ICAM-1 and IL-8 have been evaluated in preclinical studies but never came to fruition in clinical trials. IL-17, a proinflammatory cytokine involved in TH-17 cell signaling, is elevated in CF lungs (McAllister et al. 2005; Dubin and Kolls 2007) and plays an important role in recruiting neutrophils to the airway in response to infectious stimuli. Anti-IL-17 antibodies reduced airway neutrophilia in mice exposed to lipopolysaccharide (LPS) (Ferretti et al. 2003). Clinical trials of anti-IL-17 in rheumatoid arthritis and psoriasis recently have been completed. Given the similarities between CF lung inflammation and these hyperinflammatory conditions, anti-IL-17 could be considered for CF.

EICOSANOID MODULATORS

LTB$_4$, a potent neutrophil chemoattractant, is present in high concentrations in the CF airway (Konstan et al. 1993). BIIL 284 BS (amelubant), a specific LTB$_4$ receptor antagonist, was studied in CF, but the trial was terminated early because of a statistically significant increase in pulmonary-related serious adverse events in adults receiving BIIL 284 BS (Konstan et al. 2005). One possible explanation is that LTB$_4$ has some other previously unrecognized beneficial effect in CF, and its inhibition is detrimental. A more plausible explanation might be that the inhibitory effect of BIIL 284 BS on the LTB$_4$ pathway was too potent, resulting in impaired antimicrobial defenses and increasing the risk of an exacerbation. Regardless, the results of this study show that care must be taken when selecting an anti-inflammatory agent for future clinical trials.

Deficiencies in some fatty acids may contribute to CF pulmonary inflammation. There is an increase in arachidonic acid and a decrease in docosahexaenoic acid (DHA) in *cftr*$^{-/-}$ mice (Freedman et al. 1999). Investigators have shown that oral administration of DHA to *cftr*$^{-/-}$ mice corrected the lipid imbalance and reversed the observed pathological manifestations (Freedman et al. 2002; Beharry et al. 2007). There have been a handful of studies of DHA supplementation in CF patients (Van Biervliet et al. 2008; Aldamiz-Echevarria et al. 2009). Unfortunately, the dose of DHA, study size, duration of treatment, and outcome measures vary widely. Few are placebo-controlled, and some do not show a clinical effect on lung function, possibly because of short observational periods. Despite these drawbacks, the cumulative data from these studies indicate that DHA oral supplementation may effectively increase serum and phospholipid concentrations of this essential fatty acid, and that this may have beneficial health effects, including improved pulmonary function. To validate this claim, larger, placebo-controlled trials with an adequate observation period are necessary.

MODULATORS OF INTRACELLULAR SIGNALING

Because cytokines contribute to the excessive inflammatory response in CF, it seems logical to target signaling pathways responsible for upregulating their production. Unfortunately, there is no consensus as to what these pathways are and how they may interact with the basic defect in CF. High-dose ibuprofen and IL-10 inhibit NF-κB activation, thereby down-regulating the inflammatory response at the transcriptional level. Other drugs that limit proinflammatory molecule transcription may be useful in CF. Another mechanism of inhibiting NF-κB activity occurs via up-regulation of peroxisome proliferator activating receptor (PPAR) with the thiazolidinediones or glitazones (Ruan et al. 2003; Vanden Berghe et al. 2003; Zingarelli et al. 2003). CF tissues appear to be deficient in PPAR (Ollero et al. 2004; Perez et al. 2008). Decreased PPAR expression leads to an imbalance between IκB and NF-κB and favors increased inflammation. Activation of PPAR may quell the CF inflammatory response. One of the mechanisms by which ibuprofen may work in CF is through the inhibition of NF-κB by activating PPARγ (Jaradat et al. 2001; Davis et al. 2003). Troglitazone and ciglitazone activate PPAR in primary CF airway epithelial cells and CF epithelial cell lines, and reduce production of proinflammatory mediators in response to *P. aeruginosa* (Perez et al. 2008). A 28-d clinical trial of pioglitazone did not show a beneficial effect on sputum inflammatory mediators (Konstan et al. 2009). However, this may have occurred because the dose of pioglitazone was not adequate to inhibit inflammation, the duration of the study was too short, the number of subjects was too small ($N = 20$), or sputum measures may not have been sensitive enough to detect subtle changes. Although this study did not show a beneficial effect, further evaluation of the glitazones in CF is warranted.

Multiple studies show decreased NO in exhaled air (eNO) from patients with CF (Balfour-Lynn et al. 1996; Grasemann et al. 1997). NO is a highly reactive molecule with important antimicrobial and anti-inflammatory properties. The loss of NO production from the airways may contribute to bacterial infection and overzealous inflammation (Kelley and Drumm 1998; Meng et al. 1998). One potential cause of decreased eNO in CF is up-regulation of RhoGTPase, a signaling molecule that reduces NOS2 expression (Kraynack et al. 2002). Up-regulation of RhoGTPase may also contribute to the inflammatory response by increasing IL-8 production. RhoGTPase can be inhibited by blocking 3-hydroxy-3-methylglutaryl-CoA reductase (HMG-CoAR) with the commercially available statins (e.g., simvastatin) (Kraynack et al. 2002). The statins have other anti-inflammatory effects including the ability to inhibit neutrophil migration, decrease proinflammatory cytokine production, and increase transcriptional activation of PPAR (Dunzendorfer et al. 1997; Rezaie-Majd et al. 2002; Zelvyte et al. 2002). A clinical trial of simvastatin showed a trend toward increased eNO, but no effect was seen on sputum inflammatory markers. This may be for the same reasons cited above for pioglitazone (Kraynack et al. 2008). Other agents that increase NO production have also been studied. Arginase activity is increased in blood and sputum of CF patients and may further decrease NO by degrading L-arginine, an NO substrate (Grasemann et al. 2005a). In a rodent model, L-arginine was associated with reduced tissue damage, decreased neutrophil recruitment, and reduced IL-1β (Hopkins et al. 2006). A small study of L-arginine in CF was associated with increased eNO (Grasemann et al. 2005b). L-Arginine has potential, but large prospective clinical trials have not been performed.

Synthetic triterpenoids are small-molecule derivatives of naturally occurring compounds that increase Nrf2 activity. Nrf2, a transcription factor active in respiratory epithelia and pivotal to mitigating the acute inflammatory response, is deficient in CF cells (Chen et al. 2008; Nichols et al. 2009). Small molecules like the synthetic triterpenoids that target proinflammatory signaling abnormalities in CF cells may be candidates for clinical trials. Preclinical studies show the ability of 2-cyano-3,12-dioxooleana-1,9-dien-28-oic acid (CDDO) to reduce the inflammatory response (Nichols et al. 2009). Much research is still needed to determine their utility, but early data suggest the theory that compounds affecting transcriptional regulators of inflammation (both pro- and anti-inflammatory) are promising therapeutic agents.

ANTIOXIDANTS

Neutrophils release oxygen radicals that contribute to local tissue damage and perpetuate inflammation. Elevated concentrations of oxidants have been detected in the airways of CF mice and patients (Hull et al. 1997; Velsor et al. 2001). The oxidant–antioxidant imbalance contributes to the exaggerated and damaging inflammatory response. Some biochemical abnormalities in CF can be reversed by antioxidants. N-Acetyl cysteine, initially developed as a mucolytic, is receiving renewed interest as an antioxidant that both inhibits H_2O_2 and increases glutathione. In a pilot study, oral N-acetyl cysteine was associated with increased glutathione levels in whole blood and decreased sputum neutrophils, IL-8, and elastase activity (Tirouvanziam et al. 2006). The study was of short duration and not powered to detect changes in pulmonary function or other clinical outcome measures. A large randomized double-blind placebo-controlled trial of N-acetyl cysteine is ongoing in patients with CF.

The oxidant–antioxidant imbalance may be exacerbated by abnormalities in CFTR. If glutathione is transported by CFTR (Linsdell and Hanrahan 1998), then CFTR abnormalities would decrease the concentrations of this important antioxidant on epithelial surfaces, thus leaving the airway vulnerable to even ordinary levels of oxidative stress. Because decreased lung concentrations of glutathione have been shown in CF mice and patients (Roum et al. 1993; Velsor et al. 2001), it seems logical to augment its concentration in the CF lung. Glutathione in epithelial lining fluid may be increased by twice-daily treatment with aerosolized glutathione. This treatment also reduced superoxide production by inflammatory cells (Roum et al. 1993). However, subjects treated with inhaled glutathione had no detectable change in BAL markers of oxidative stress (Bishop et al. 2005; Hartl 2005). Inhaled glutathione therapy for CF has attracted much attention, but its therapeutic benefit has not been adequately studied.

PROTEASE INHIBITORS

Early in life, neutrophils infiltrate the CF airway and release massive amounts of proteases that overwhelm antiproteases, including alpha-1-protease inhibitor (α_1-PI) and secretory leuko-cyte protease inhibitor (SLPI). The concentration of α_1-PI in BAL fluid from CF patients is several-fold higher than in healthy subjects, but even in mild patients there is a several hundred to several thousand fold excess of elastase and other neutrophil proteases, which vastly exceeds the capacity of the inhibitors (Birrer et al. 1994; Konstan et al. 1994; Chmiel et al. 2002a). The imbalance between the proteases and their inhibitors becomes worse during exacerbations and undoubtedly results in airway injury. Antiproteases have been under investigation in CF since 1990. Aerosol delivery of α_1-antitrypsin suppressed inflammatory markers including free neutrophil elastase, proinflammatory cytokines, and neutrophils (McElvaney et al. 1991; Berger et al. 1995; Bilton et al. 1999; Griese et al. 2007). Other inhibitors of neutrophil elastase studied in CF include rSLPI and the small-molecule drug EPI-hNE4 (McElvaney et al. 1993; Grimbert et al. 2003). Although these inhibitors showed positive effects in CF, further trials have not been conducted with these inhibitors. Data regarding antiprotease therapy suggest that antiprotease therapy possesses the potential to impact CF inflammation. Of the antiprotease therapies studied in CF, inhalation of plasma-derived α_1-antitrypsin appears closest to possible clinical use.

OTHER THERAPIES THAT IMPACT INFLAMMATION IN THE CF AIRWAY

Hydroxychloroquine, a dihydrofolate reductase inhibitor that increases intracellular pH and possesses anecdotal efficacy in some rare interstitial lung diseases, was evaluated in a small 28-d study in CF (Williams et al. 2008). The drug was well tolerated, but there was no change in sputum inflammatory markers (Williams et al. 2008). CXCR2 is an important receptor in neutrophil chemotaxis. Clinical and preclinical studies have evaluated the effects of CXCR2 antagonists in many lung diseases including COPD, asthma, and ozone-induced airway inflammation (Lazaar et al. 2011). A clinical trial in CF of SB-656933, a CXCR2 antagonist, was recently completed and shows some promise in modulating airway inflammation (Moss et al.

2012). Although chemotherapeutics have been studied in refractory asthma, there have been few studies in CF. Low-dose cyclosporin A decreased the need for systemic corticosteroids in one small case series (Bhal et al. 2001). In a small pilot study, methotrexate increased FEV_1 and decreased total serum immunoglobulins in five CF patients after 1 yr of treatment (Ballmann et al. 2003). However, in a more recent study, methotrexate was not well tolerated and was associated with an increased need for IV antibiotics (Oermann et al. 2007). Its routine use cannot be advocated. Because of toxicities, chemotherapeutics should be studied first in animal models. However, even if toxic, some agents may have a favorable risk/benefit profile.

CONCLUSION

CF lung disease begins early in infancy and continues unabated. Persistent infection and excessive inflammation are key contributors to the progression of CF lung disease. Studies showed that judicious use of antibiotics and anti-inflammatory drugs is beneficial. The development of both systemic and inhaled antibiotics has greatly improved survival and remains a cornerstone therapy. With the wider range of antibiotics now available, it is important that the most effective regimes for chronic therapy, which reduce pulmonary exacerbations, be determined. It is also critical that pulmonary exacerbations be identified early and treated with appropriate antibiotic combinations.

Anti-inflammatory drugs are another important weapon in the fight against CF. Although ICSs are the most frequently prescribed anti-inflammatory therapy in CF, their benefits, like their risks, have not been fully established. Thus far, high-dose chronic ibuprofen treatment is the best-proven anti-inflammatory drug, although its use in the CF community has been surprisingly low. Newer anti-inflammatory agents will probably come to clinical trial over the next few years. More studies of anti-inflammatory agents are necessary to fully elucidate their mechanisms of action, to determine which would provide the most benefit to CF patients, and to establish the optimal patient age and stage of lung disease for initiation of treatment.

Drugs designed to target abnormal CFTR may not fully ameliorate the relentless progression of CF lung disease in patients with established bronchiectasis. Therefore, vigorous research and development into anti-infective and anti-inflammatory therapeutics that address secondary disease pathology must continue. Improved antibacterial and anti-inflammatory therapies should lessen morbidity, improve quality of life, and prolong survival.

ACKNOWLEDGMENTS

Grant support from the National Institutes of Health (Grant P30-DK27651) and the U.S. Cystic Fibrosis Foundation is gratefully acknowledged. We gratefully acknowledge the editorial assistance of Robert C. Stern, M.D.

REFERENCES

Aldámiz-Echevarría L, Prieto JA, Andrade F, Elorz J, Sojo A, Lage S, Sanjurjo P, Vázquez C, Rodríguez-Soriano J. 2009. Persistence of essential fatty acid deficiency in cystic fibrosis despite nutritional therapy. *Pediatr Res* **66:** 585–589.

Armstrong DS, Grimwood K, Carzino R, Carlin JB, Olinsky A, Phelan PD. 1995. Lower respiratory infection and inflammation in infants with newly diagnosed cystic fibrosis. *BMJ* **310:** 1571–1572.

Armstrong DS, Grimwood K, Carlin JB, Armstrong DS, Grimwood K, Carlin JB, Carzino R, Gutièrrez JP, Hull J, Olinsky A, et al. 1997. Lower airway inflammation in infants and young children with cystic fibrosis. *Am J Respir Crit Care Med* **156:** 1197–1204.

Auerbach HS, Williams M, Kirkpatrick JA, Colten HR. 1985. Alternate-day prednisone reduces morbidity and improves pulmonary function in cystic fibrosis. *Lancet* **2:** 686–688.

Balfour-Lynn IM, Laverty A, Dinwiddie R. 1996. Reduced upper airway nitric oxide in cystic fibrosis. *Arch Dis Child* **75:** 319–322.

Balfour-Lynn IM, Lees B, Hall P, Phillips G, Khan M, Flather M, Elborn JS, CF WISE (Withdrawal of Inhaled Steroids Evaluation) Investigators. 2006. Multicenter randomized controlled trial of withdrawal of inhaled corticosteroids in cystic fibrosis. *Am J Respir Crit Care Med* **173:** 1356–1362.

Ballmann M, Junge S, von der Hardt H. 2003. Low-dose methotrexate for advanced pulmonary disease in patients with cystic fibrosis. *Respir Med* **97:** 498–500.

Balough K, McCubbin M, Weinberger M, Smits W, Ahrens R, Fick R. 1995. The relationship between infection and inflammation in the early stages of lung disease from cystic fibrosis. *Pediatr Pulmonol* **20**: 63–70.

Beharry S, Ackerley C, Corey M, Kent G, Heng YM, Christensen H, Luk C, Yantiss RK, Nasser IA, Zaman M, et al. 2007. Long-term docosahexaenoic acid therapy in a congenic murine model of cystic fibrosis. *Am J Physiol Gastrointest Liver Physiol* **292**: G839–G848.

Berger M, Konstan MW, Hilliard JB. 1995. Aerosolized prolastin (α_1-protease inhibitor) in CF. *Pediatr Pulmonol* **20**: 421.

Bhal GK, Maguire SA, Bowler IM. 2001. Use of cyclosporin A as a steroid sparing agent in cystic fibrosis. *Arch Dis Child* **84**: 89.

Bilton D, Elborn S, Conway S, Edgar J, Redmond A. 1999. Phase II trial to assess the clinical efficacy of transgenic α-1-antitrypsin (tg-hAAT) as an effective treatment of cystic fibrosis. *Pediatr Pulmonol* **28** (Suppl 19): 246.

Birrer P, McElvaney NG, Rüdeberg A, Sommer CW, Liechti-Gallati S, Kraemer R, Hubbard R, Crystal RG. 1994. Protease–antiprotease imbalance in the lungs of children with cystic fibrosis. *Am J Respir Crit Care Med* **150**: 207–213.

Bishop C, Hudson VM, Hilton SC, Wilde C. 2005. A pilot study of the effects of inhaled buffered reduced glutathione on the clinical status of patients with cystic fibrosis. *Chest* **127**: 308–317.

Bonfield TL, Panuska JR, Konstan MW, Hilliard KA, Hilliard JB, Ghnaim H, Berger M. 1995a. Inflammatory cytokines in cystic fibrosis lungs. *Am J Respir Crit Care Med* **152**: 2111–2118.

Bonfield TL, Konstan MW, Burfeind P, Panuska JR, Hilliard JB, Berger M. 1995b. Normal bronchial epithelial cells constitutively produce the anti-inflammatory cytokine interleukin-10, which is downregulated in cystic fibrosis. *Am J Respir Cell Mol Biol* **13**: 257–261.

Bonfield TL, Konstan MW, Berger M. 1999. Altered respiratory epithelial cell cytokine production in cystic fibrosis. *J Allergy Clin Immunol* **104**: 72–78.

Chen J, Kinter M, Shank S, Cotton C, Kelley TJ, Ziady AG. 2008. Dysfunction of Nrf-2 in CF epithelia leads to excess intracellular H_2O_2 and inflammatory cytokine production. *PLoS ONE* **3**: e3367.

Chmiel JF, Davis PB. 2003. State of the art: Why do the lungs of patients with cystic fibrosis become infected and why can't they clear the infection? *Respir Res* **4**: 8–21.

Chmiel JF, Konstan MW, Knesebeck JE, Hilliard JB, Bonfield TL, Dawson DV, Berger M. 1999. IL-10 attenuates excessive inflammation in chronic *Pseudomonas* infection in mice. *Am J Respir Crit Care Med* **160**: 2040–2047.

Chmiel JF, Konstan MW, Berger M. 2002a. The role of inflammation in the pathophysiology of CF lung disease. *Clin Review Allergy Immunol* **23**: 5–27.

Chmiel JF, Konstan MW, Saadane A, Krenicky JE, Lester Kirchner H, Berger M. 2002b. Prolonged inflammatory response to acute *Pseudomonas* challenge in interleukin-10 knockout mice. *Am J Respir Crit Care Med* **165**: 1176–1181.

Dasenbrook EC, Merlo CA, Diener-West M, Lechtzin N, Boyle MP. 2008. Persistent methicillin-resistant *Staphylo-coccus aureus* and rate of FEV1 decline in cystic fibrosis. *Am J Respir Crit Care Med* **178**: 814–821.

Dasenbrook EC, Checkley W, Merlo CA, Konstan MW, Lechtzin N, Boyle MP. 2010. Association between respiratory tract methicillin-resistant *Staphylococcus aureus* and survival in cystic fibrosis. *JAMA* **303**: 2386–2392.

Davis PB, Drumm M, Konstan MW. 1996. Cystic fibrosis. State of the art. *Am J Resp Crit Care Med* **154**: 1229–1256.

Davis PB, Gupta S, Eastman J, Konstan MW. 2003. Inhibition of proinflammatory cytokine production by PPARγ agonists in airway epithelial cells. *Pediatr Pulmonol* **36** (Suppl 25): 268–269.

De Boeck K, Vermeulen F, Wanyama S, Thomas M, Members of the Belgian CF Registry. 2011. Inhaled corticosteroids and lower lung function decline in young children with cystic fibrosis. *Eur Respir J* **37**: 1091–1095.

Döring G, Flume P, Heijerman H, Elborn JS, Consensus Study Group. 2012. Treatment of lung infection in patients with cystic fibrosis: Current and future strategies. *J Cyst Fibros* 2012 **11**: 461–479.

Dubin PJ, Kolls JK. 2007. IL-23 mediates inflammatory responses to mucoid *Pseudomonas aeruginosa* lung infection in mice. *Am J Physiol Lung Cell Mol Physiol* **292**: L519–L528.

Dubin PJ, McAllister F, Kolls JK. 2007. Is cystic fibrosis a TH17 disease? *Inflamm Res* **56**: 221–227.

Dunzendorfer S, Rothbucher D, Schratzberger P, Reinisch N, Kahler CM, Wiedermann CJ. 1997. Mevalonate-dependent inhibition of transendothelial migration and chemotaxis of human peripheral blood neutrophils by pravastatin. *Circ Res* **81**: 963–969.

Eigen H, Rosenstein BJ, FitzSimmons S, Schidlow DV. 1995. A multicenter study of alternate-day prednisone therapy in patients with cystic fibrosis. Cystic Fibrosis Foundation Prednisone Trial Group. *J Pediatr* **126**: 515–523.

Ferretti S, Bonneau O, Dubois GR, Jones CE, Trifilieff A. 2003. IL-17, produced by lymphocytes and neutrophils, is necessary for lipopolysaccharide-induced airway neutrophilia: IL-15 as a possible trigger. *J Immunol* **170**: 2106–2112.

Flume PA, O'Sullivan BP, Robinson KA, Goss CH, Mogayzel PJ Jr, Willey-Courand DB, Bujan J, Finder J, Lester M, Quittell L, et al. 2007. Cystic fibrosis pulmonary guidelines. *Am J Respir Crit Care Med* **176**: 957–969.

Freedman SD, Katz MH, Parker EM, Laposata M, Urman MY, Alvarez JG. 1999. A membrane lipid imbalance plays a role in the phenotypic expression of cystic fibrosis in $cftr^{-/-}$ mice. *Proc Natl Acad Sci* **96**: 13995–14000.

Freedman SD, Weinstein D, Blanco PG, Martinez-Clark P, Urman S, Zaman M, Morrow JD, Alvarez JG. 2002. Characterization of LPS-induced lung inflammation in $cftr^{-/-}$ mice and the effect of docosahexaenoic acid. *J Appl Physiol* **92**: 2169–2176.

Grasemann H, Michler E, Wallot M, Ratjen F. 1997. Decreased concentration of exhaled nitric oxide (NO) in patients with cystic fibrosis. *Pediatr Pulmonol* **24**: 173–177.

Grasemann H, Schwiertz R, Matthiesen S, Racke K, Ratjen F. 2005a. Increased arginase activity in cystic fibrosis airways. *Am J Respir Crit Care Med* **172**: 1523–1528.

Cite this article as *Cold Spring Harb Perspect Med* doi: 10.1101/cshperspect.a009779

Grasemann H, Grasemann C, Kurtz F, Tietze-Schillings G, Vester U, Ratjen F. 2005b. Oral L-arginine supplementation in cystic fibrosis patients: A placebo-controlled study. *Eur Respir J* **25:** 62–68.

Griese M, Latzin P, Kappler M, Weckerle K, Heinzlmaier T, Bernhardt T, Hartl D. 2007. α1-Antitrypsin inhalation reduces airway inflammation in cystic fibrosis patients. *Eur Respir J* **29:** 240–250.

Grimbert D, Vecellio L, Delépine P, Attucci S, Boissinot E, Poncin A, Gauthier F, Valat C, Saudubray F, Antonioz P, et al. 2003. Characteristics of EPI-hNE4 aerosol: A new elastase inhibitor for treatment of cystic fibrosis. *J Aerosol Med* **16:** 121–129.

Hartl D, Starosta V, Maier K, Beck-Speier I, Rebhan C, Becker BF, Latzin P, Fischer R, Ratjen F, Huber RM, et al. 2005. Inhaled glutathione decreases PGE2 and increases lymphocytes in cystic fibrosis lungs. *Free Radic Biol Med* **39:** 463–472.

Hopkins N, Gunning Y, O'Croinin DF, Laffey JG, McLoughlin P. 2006. Anti-inflammatory effect of augmented nitric oxide production in chronic lung infection. *J Pathol* **209:** 198–205.

Hull J, Vervaart P, Grimwood K, Phelan P. 1997. Pulmonary oxidative stress response in young children with cystic fibrosis. *Thorax* **52:** 557–560.

Jaradat MS, Wongsud B, Phornchirasilp S, Rangwala SM, Shams G, Sutton M, Romstedt KJ, Noonan DJ, Feller DR. 2001. Activation of peroxisome proliferators-activated receptor isoforms and inhibition of prostaglandin H$_2$ synthases by ibuprofen, naproxen, and indomethacin. *Biochem Pharmacol* **62:** 1587–1595.

Kelley TJ, Drumm ML. 1998. Inducible nitric oxide synthase expression is reduced in cystic fibrosis murine and human airway epithelial cells. *J Clin Invest* **102:** 1200–1207.

Khan TZ, Wagener JS, Bost T, Martinez J, Accurso FJ, Riches DW. 1995. Early pulmonary inflammation in infants with cystic fibrosis. *Am J Respir Crit Care Med* **151:** 1075–1082.

Kirchner KK, Wagener JS, Khan TZ, Copenhaver SC, Accurso FJ. 1996. Increased DNA levels in bronchoalveolar lavage fluid obtained from infants with cystic fibrosis. *Am J Respir Crit Care Med* **154:** 1426–1429.

Konstan MW. 2008. Ibuprofen therapy for cystic fibrosis lung disease: Revisited. *Curr Opin Pulm Med* **14:** 567–573.

Konstan MW, Berger M. 1997. Current understanding of the inflammatory process in cystic fibrosis—Onset and etiology. *Pediatr Pulmonol* **24:** 137–142.

Konstan MW, Walenga RW, Hilliard KA, Hilliard JB. 1993. Leukotriene B4 is markedly elevated in the epithelial lining fluid of patients with cystic fibrosis. *Am Rev Respir Dis* **148:** 896–901.

Konstan MW, Hilliard KA, Norvell TM, Berger M. 1994. Bronchoalveolar lavage findings in cystic fibrosis patients with stable, clinically mild lung disease suggest ongoing infection and inflammation. *Am J Respir Crit Care Med* **150:** 448–454.

Konstan MW, Byard PJ, Hoppel CL, Davis PB. 1995. Effect of high-dose ibuprofen in patients with cystic fibrosis. *N Engl J Med* **332:** 848–844.

Konstan MW, Krenicky JE, Finney MR, Kirchner HL, Hilliard KA, Hilliard JB, Davis PB, Hoppel CL. 2003. Effect of ibuprofen on neutrophil migration in vivo in cystic fibrosis. *J Pharmacol Exp Ther* **306:** 1086–1091.

Konstan MW, Doring G, Lands LC, Hilliard KA, Koker P, Bhattacharya S, Staab A, Hamilton AL. 2005. Results of a phase II clinical trial of BIIL 284 BS (a LTB$_4$ receptor antagonist) for the treatment of CF lung disease. *Pediatr Pulmonol* **40** (Suppl 28): 125–127.

Konstan MW, Schluchter MD, Xue W, Davis PB. 2007. Clinical use of ibuprofen is associated with slower FEV1 decline in children with cystic fibrosis. *Am J Respir Crit Care Med* **176:** 1084–1089.

Konstan M, Krenicky J, Hilliard K, Hilliard J. 2009. A pilot study evaluating the effect of pioglitazone, simvastatin, and ibuprofen on neutrophil migration in vivo in healthy subjects. *Pediatr Pulmonol* **44** (Suppl 32): 289–290.

Konstan MW, Geller DE, Minić P, Brockhaus F, Zhang J, Angyalosi G. 2011. Tobramycin inhalation powder for *P. aeruginosa* infection in cystic fibrosis: The EVOLVE trial. *Pediatr Pulmonol* **46:** 230–238.

Kraynack NC, Corey DA, Elmer HL, Kelley TJ. 2002. Mechanisms of NOS2 regulation by Rho GTPase signaling in airway epithelial cells. *Am J Physiol Lung Cell Mol Physiol* **283:** L604–L611.

Kraynack NC, Chmiel JF, Xue W, Schluchter MD, Gibson RL, Kelley TJ, Hilliard JB, Konstan MW. 2008. Effect of simvastatin on exhaled nitric oxide and inflammatory markers in sputum in patients with cystic fibrosis. *Pediatr Pulmonol* **43** (Suppl 31): 300.

Lai H-C, FitzSimmons SC, Allen DB, Kosorok MR, Rosenstein BJ, Campbell PW, Farrell PM. 2000. Risk of persistent growth impairment after alternate-day prednisone treatment in children with cystic fibrosis. *N Engl J Med* **342:** 851–859.

Lands L, Stanojevic S. 2007. Oral non-steroidal anti-inflammatory drug therapy for cystic fibrosis. *Cochrane Database Syst Rev* **17:** CD001505.

Lands LC, Milner R, Cantin AM, Manson D, Corey M. 2007. High-dose ibuprofen in cystic fibrosis: Canadian safety and effectiveness trial. *J Pediatr* **151:** 228–230.

Lazaar AL, Sweeney LE, MacDonald AJ, Alexis NE, Chen C, Tal-Singer R. 2011. SB-656933, a novel CXCR2 selective antagonist, inhibits ex vivo neutrophil activation and ozone-induced airway inflammation in humans. *Br J Clin Pharmacol* **72:** 282–293.

Linsdell P, Hanrahan JW. 1998. Glutathione permeability of CFTR. *Am J Physiol* **275** (Pt 1): C323–C326.

LiPuma J. 2010. The changing microbial epidemiology in CF. *Clin Microbiol Rev* **23:** 299–323.

LiPuma J. 2012. The new microbiology of cystic fibrosis: It takes a community. *Thorax* **67:** 851–852.

McAllister F, Henry A, Kreindler JL, Dubin PJ, Ulrich L, Steele C, Finder JD, Pilewski JM, Carreno BM, Goldman SJ, et al. 2005. Role of IL-17A, IL-17F, and the IL-17 receptor in regulating growth related oncogene-α and granulocyte colony-stimulating factor in bronchial epithelium: Implications for airway inflammation in cystic fibrosis. *J Immunol* **175:** 404–412.

McElvaney NG, Hubbard RC, Birrer P, Chernick MS, Caplan DB, Frank MM, Crystal RG. 1991. Aerosol α1-

antitrypsin treatment for cystic fibrosis. *Lancet* **337:** 392–394.

McElvaney NG, Doujaiji B, Moan MJ, Burnham MR, Wu MC, Crystal RG. 1993. Pharmacokinetics of recombinant secretory leukoprotease inhibitor aerosolized to normals and individuals with cystic fibrosis. *Am Rev Respir Dis* **148** (Pt 1): 1056–1060.

Meng QH, Springall DR, Bishop AE, Morgan K, Evans TJ, Habib S, Gruenert DC, Gyi KM, Hodson ME, Yacoub MH, et al. 1998. Lack of inducible nitric oxide synthase in bronchial epithelium: A possible mechanism of susceptibility to infection in cystic fibrosis. *J Pathol* **184:** 323–331.

Moss RB, Mayer-Hamblett N, Wagener J, Daines C, Hale K, Ahrens R, Gibson RL, Anderson P, Retsch-Bogart G, Nasr SZ, et al. 2005. Randomized, double-blind, placebo-controlled, dose-escalating study of aerosolized interferon γ-1b in patients with mild to moderate cystic fibrosis lung disease. *Pediatr Pulmonol* **39:** 209–218.

Moss RB, Mistry SJ, Konstan MW, Pilewski JM, Kerem E, Tal-Singer R, Lazaar AL, for the CF2110399 Investigators. 2012. Safety and early treatment effects of the CXCR2 antagonist SB-656933 in patients with cystic fibrosis. *J Cyst Fibros* **pii:** S1569-1993(12)00155-5

Muhlebach MS, PW Stewart, Leigh MW, Noah TL. 1999. Quantitation of inflammatory response to bacteria in young cystic fibrosis and control patients. *Am J Respir Crit Care Med* **160:** 186–191.

Nichols DP, Ziady AG, Shank SL, Eastman JF, Davis PB. 2009. The triterpenoid CDDO limits inflammation in preclinical models of cystic fibrosis lung disease. *Am J Physiol Lung Cell Mol Physiol* **297:** L828–L836.

Noah TL, Black HR, Cheng PW, Wood RE, Leigh MW. 1997. Nasal and bronchoalveolar lavage fluid cytokines in early cystic fibrosis. *J Infect Dis* **175:** 638–647.

Oermann CM, Katz M, Wheeler C, Cumming S. 2007. A pilot study evaluating the potential use of low-dose methotrexate as an anti-inflammatory therapy for cystic fibrosis lung disease. *Pediatr Pulmonol* **42** (Suppl 30): 292–293.

Ollero M, Junaidi O, Zaman MM, Tzameli I, Ferrando AA, Andersson C, Blanco PG, Bialecki E, Freedman SD. 2004. Decreased expression of peroxisome proliferators activated receptor γ in *cftr*$^{-/-}$ mice. *J Cell Physiol* **200:** 235–244.

Parkins MD, Rendall JC, Elborn JS. 2012. Incidence and risk factors for pulmonary exacerbation treatment failures in cystic fibrosis patients chronically infected with *Pseudomonas aeruginosa*. *Chest* **141:** 485–493.

Perez A, van Heeckeren AM, Nichols D, Gupta S, Eastman JF, Davis PB. 2008. Peroxisome proliferator–activated receptor-γ in cystic fibrosis lung epithelium. *Am J Physiol Lung Cell Mol Physiol* **295:** L303–L313.

Proesmans M. 2013. Comparison of two treatment regimens for eradication of *P. aeruginosa* infection in children with cystic fibrosis. *J Cyst Fibros* **12:** 29–34.

Ratjen F, Munck A, Kho P, Angyalosi G, ELITE Study Group. 2010. Treatment of early *Pseudomonas aeruginosa* infection in patients with cystic fibrosis: The ELITE trial. *Thorax* **65:** 286–291.

Ren CL, Pasta DJ, Rasouliyan L, Wagener JS, Konstan MW, Morgan WJ, Scientific Advisory Group the Investigators

and Coordinators of the Epidemiologic Study of Cystic Fibrosis. 2008. Relationship between inhaled corticosteroid therapy and rate of lung function decline in children with cystic fibrosis. *J Pediatr* **153:** 746–751.

Rezaie-Majd A, Maca T, Bucek RA, Valent P, Müller MR, Husslein P, Kashanipour A, Minar E, Baghestanian M. 2002. Simvastatin reduces expression of cytokines interleukin-6, interleukin-8, and monocyte chemoattractant protein-1 in circulating monocytes from hypercholesterolemic patients. *Arterioscler Thromb Vasc Biol* **22:** 1194–1199.

Ross KR, Chmiel JF, Konstan MW. 2009. The role of inhaled corticosteroids in the management of cystic fibrosis. *Paediatr Drugs* **11:** 101–113.

Roum JH, Buhl R, McElvaney NG, Borok Z, Crystal RG. 1993. Systemic deficiency of glutathione in cystic fibrosis. *J Appl Physiol* **75:** 2419–2424.

Ruan H, Pownall HJ, Lodish HF. 2003. Troglitazone antagonizes tumor necrosis factor-α-induced reprogramming of adipocyte gene expression by inhibiting the transcriptional regulatory functions of NF-κB. *J Biol Chem* **278:** 28181–28192.

Sanders DB, Bittner RC, Rosenfeld M, Hoffman LR, Redding GJ, Goss CH. 2010. Failure to recover to baseline pulmonary function after cystic fibrosis pulmonary exacerbation. *Am J Respir Crit Care Med* **182:** 627–632.

Schuster A, Haliburn C, Döring G, Goldman MH, for the Freedom Study Group. 2012. Safety, efficacy and convenience of colistimethate sodium dry powder for inhalation (Colobreathe DPI) in cystic fibrosis patients: A randomised study. *Thorax* **68:** 344–350.

Smyth A, Elborn JS. 2008. Exacerbations in cystic fibrosis: 3—Management. *Thorax* **63:** 180–184.

Smyth AR, Walters S. 2012. Prophylactic anti-staphylococcal antibiotics for cystic fibrosis. *Cochrane Database Syst Rev* **12:** CD001912. 23235585.

Stenbit AE, Flume PA. 2011. Pulmonary exacerbations in cystic fibrosis. *Curr Opin Pulm Med* **17:** 442–447.

Taccetti G, Bianchini E, Cariani L, Buzzetti R, Costantini D, Trevisan F. 2012. Early antibiotic treatment for *Pseudomonas aeruginosa* eradication in patients with cystic fibrosis: A randomised multicentre study comparing two different protocols. *Thorax* **67:** 853–859.

Tirouvanziam R, Conrad CK, Bottiglieri T, Herzenberg LA, Moss RB, Herzenberg LA. 2006. High-dose oral *N*-acetylcysteine, a glutathione prodrug, modulates inflammation in cystic fibrosis. *Proc Natl Acad Sci* **103:** 4628–4633.

Treggiari MM, Retsch-Bogart G, Mayer-Hamblett N, Khan U, Kulich M, Kronmal R, Williams J, Hiatt P, Gibson RL, Spencer T, et al. 2011. Comparative efficacy and safety of 4 randomized regimens to treat early *Pseudomonas aeruginosa* infection in children with cystic fibrosis. *Arch Pediatr Adolesc Med* **165:** 847–856.

Tunney MM, Field TR, Moriarty TF, Patrick S, Doering G, Muhlebach MS, Wolfgang MC, Boucher R, Gilpin DF, McDowell A, et al. 2008. Detection of anaerobic bacteria in high numbers in sputum from patients with cystic fibrosis. *Am J Respir Crit Care Med* **177:** 995–1001.

Tunney MM, Klem ER, Fodor AA, Gilpin DF, Moriarty TF, McGrath SJ, Muhlebach MS, Boucher RC, Cardwell C, Doering G, et al. 2011. Use of culture and molecular analysis to determine the effect of antibiotic treatment on microbial community diversity and abundance during exacerbation in patients with cystic fibrosis. *Thorax* **66:** 579–584.

Valerius NH, Koch C, Høiby N. 1991. Prevention of chronic *Pseudomonas aeruginosa* colonisation in cystic fibrosis by early treatment. *Lancet* **338:** 725–726.

Van Biervliet S, Devos M, Delhaye T, Van Biervliet JP, Robberecht E, Christophe A. 2008. Oral DHA supplementation in ΔF508 homozygous cystic fibrosis patients. *Prostaglandins Leukot Essent Fatty Acids* **78:** 109–115.

Vanden Berghe W, Vermeulen L, Delerive P, De Bosscher K, Staels B, Haegeman G. 2003. A paradigm for gene regulation: Inflammation, NF-κB, and PPAR. *Adv Exp Med Biol* **544:** 181–196.

VanDevanter DR, O'Riordan MA, Blumer JL, Konstan MW. 2010. Assessing time to pulmonary function benefit following antibiotic treatment of acute cystic fibrosis exacerbations. *Respir Res* **11:** 137.

Velsor LW, van Heeckeren A, Day BJ. 2001. Antioxidant imbalance in the lungs of cystic fibrosis transmembrane conductance regulator protein mutant mice. *Am J Physiol Lung Cell Mol Physiol* **281:** L31–L38.

Williams HD, Davies JC. 2012. Basic science for the chest physician: *Pseudomonas aeruginosa* and the cystic fibrosis airway. *Thorax* **67:** 465–467.

Williams B, Robinette M, Slovis B, Deretci V, Perkett E. 2008. Hydroxychloroquine—Pilot study of anti-inflammatory effects in cystic fibrosis. *Pediatr Pulmonol* **43** (Suppl 31): 314.

Zelvyte I, Dominaitiene R, Crisby M, Janciauskiene S. 2002. Modulation of inflammatory mediators and PPARγ and NFκB expression by pravastatin in response to lipoproteins in human monocytes in vitro. *Pharmacol Res* **45:** 147–154.

Zhao J, Schloss PD, Kalikin LM, Carmody LA, Foster BK, Petrosino JF, Cavalcoli JD, VanDevanter DR, Murray S, Li JZ, et al. 2012. Decade-long bacterial community dynamics in cystic fibrosis airways. *Proc Natl Acad Sci* **109:** 5809–5814.

Zingarelli B, Sheehan M, Hake PW, O'Connor M, Denenberg A, Cook JA. 2003. Peroxisome proliferator activator receptor-γ ligands, 15-deoxy-$\Delta^{12,14}$-prostaglandin J$_2$ and ciglitazone, reduce systemic inflammation in polymicrobial sepsis by modulation of signal transduction pathways. *J Immunol* **171:** 6827–6837.

New Pulmonary Therapies Directed at Targets Other than CFTR

Scott H. Donaldson[1] and Luis Galietta[2]

[1]Cystic Fibrosis Research and Treatment Center, University of North Carolina at Chapel Hill, Chapel Hill, North Carolina 27599

[2]Lab. di Genetica Molecolare, Instituto Giannina Gaslini, 16148 Genova, Italy

Correspondence: scott_donaldson@med.unc.edu

Our current understanding of the pathogenesis of cystic fibrosis (CF) lung disease stresses the importance of the physical and chemical properties of the airway surface liquid (ASL). In particular, the loss of cystic fibrosis transmembrane conductance regulator (CFTR) chloride channel function in CF reduces the volume and fluidity of the ASL, thus impairing mucociliary clearance and innate antimicrobial mechanisms. Besides direct approaches to restoring mutant CFTR function, alternative therapeutic strategies may also be considered to correct the basic defect of impaired salt and water transport. Such alternative strategies are focused on the restoration of mucociliary transport by (1) reducing sodium and fluid absorption by inhibiting the ENaC channel; (2) activating alternative chloride channels; and (3) increasing airway surface hydration with osmotic agents. Therapeutic approaches directed at targets other than CFTR are attractive because they are potentially useful to all patients irrespective of their genotype. Clinical trials are underway to test the efficacy of these approaches.

The study of the normal physiologic processes that contribute to innate lung defenses has allowed the identification of important defects involved in the pathogenesis of cystic fibrosis (CF) lung disease. There is substantial evidence that mucociliary clearance is perhaps the most important initial line of lung defense and that defective mucociliary clearance is responsible for the development and progression of the chronic lung disease that typifies CF. It is important, therefore, to review the elements that compose the mucus-clearance apparatus.

In the last 30 years, a series of studies revealed the importance of the airway-lining fluid for defense against microbial pathogens. This airway-surface liquid layer (ASL) is composed of two layers: a periciliary layer (PCL), containing water, low-molecular-weight ions, and a complex collection of secreted and cell-surface-tethered proteins; and an overlying mucus layer, primarily composed of secreted mucins. Acting in concert with secreted antimicrobial substances (e.g., lysozyme, lactoferrin, and defensins), the mucus layer efficiently traps and clears inhaled pathogens and particulates. The PCL, in turn, provides a lubricant layer, in which cilia can beat freely to propel the mucus layer toward the mouth. Proper regulation of ASL volume, therefore, facilitates cilia motion and ensures adequate hydration of its component layers. Both attributes are critical to the maintenance of mucus clearance.

Although fluid may be donated to the ASL compartment via the distal lung compartment and submucosal glands, the superficial epithelia play a key role in regulating ASL volume through its capacity to initiate both liquid absorption and secretion. The absorption of ASL is accomplished via active sodium transport, with the apical epithelial sodium channel (ENaC) mediating the rate-limiting step in this process. Liquid secretion across the superficial epithelium is conducted through cystic fibrosis transmembrane conductance regulator (CFTR) and calcium-activated chloride channels (CaCC). CFTR appears to play a central role in this process: it is the primary pathway for chloride secretion under basal conditions, and it also regulates the activity of ENaC via mechanisms that are still being elucidated. By regulating liquid absorption and secretion, CFTR ensures that the PCL remains at least 7 μm thick (approximating the height of cilia) and that the mucus layer is adequately hydrated for transport.

The importance of CFTR as a homeostatic regulator of ASL volume is demonstrated by the severe consequences observed in CF patients. CF epithelia hyperabsorb liquid because of dysregulation of ENaC and the absence of capacity to secrete liquid through CFTR (Stutts et al. 1995). As demonstrated by studies using primary cell cultures, these ion-transport abnormalities result in an abnormally shallow PCL, which in turn causes spontaneous mucus transport to stop (Matsui et al. 1998). Though it is certain that chloride secretion through CaCC can partially compensate for the loss of CFTR (Tarran et al. 2002), the system is vulnerable to insults (e.g., viral infection) that, over time, result in the accumulation of poorly clearing lung regions and mucus obstruction (Tarran et al. 2005). ASL dehydration and impaired mucus transport are believed to create a favorable environment for the initial acquisition and subsequent proliferation of bacteria in the airways. Developing therapies that correct these ion-transport abnormalities, therefore, may provide the means to prevent or delay the onset and progression of CF lung disease.

TARGETING CFTR

Over the past decade, many efforts have been aimed at developing strategies to recover mutant CFTR function. In this respect, it must be considered that CF mutations cause loss of CFTR function by different mechanisms, including impaired exit of the protein from the endoplasmic reticulum and/or defective channel gating. Therefore, CFTR-directed therapies must be tailored to specific mutations, or groups of mutations, sharing the same type of functional defect. For example, pharmacological agents called potentiators, able to stimulate CFTR channel opening, have been developed to treat patients carrying mutations that impair CFTR gating, such as G551D. The potentiator VX-770 (Vertex Pharmaceuticals) has been tested in phase II and III clinical trials with positive results (Accurso et al. 2010). Pharmacological correctors, aimed at the rescue of mutant CFTR (e.g., ΔF508-CFTR) from the endoplasmic reticulum and encouraging ribosomal read-through of abnormal stop mutations, are also undergoing clinical trials, as are combinations of correctors and potentiators. The approaches used to target CFTR directly are covered much more extensively elsewhere.

Whereas targeting CFTR might be the most rational and effective approach to fully restoring ion transport in CF, therapeutics aimed at different targets are also under active investigation. Although indirect, this approach has potential advantages when compared to CFTR-directed therapies. First, this approach utilizes receptors and ion channels that are known to be expressed at appropriate cellular locations and, therefore, do not require us to replace or restore CFTR at the apical membrane. Second, effective restoration of airway surface hydration by targeting proteins other than CFTR is expected to be beneficial for all CF patients, irrespective of their type of mutation. Fortunately, a number of potential therapeutic targets are available, which provide significant hope for developing novel, effective treatments for CF lung disease (Fig. 1).

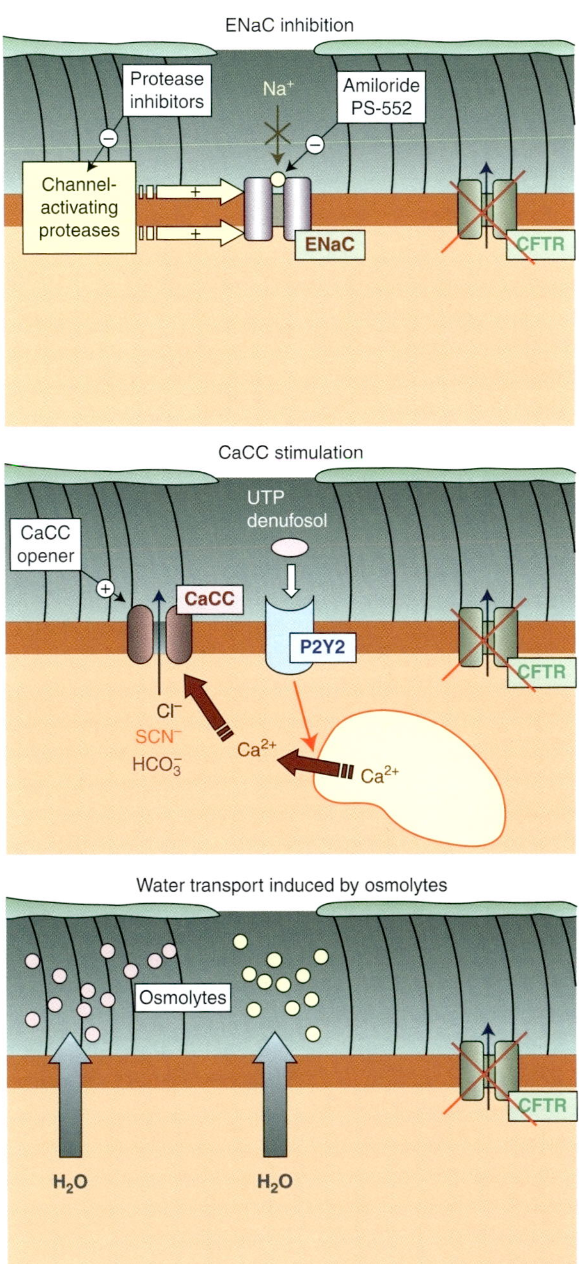

Figure 1. Three alternative strategies to correct the intrinsic airway surface liquid defect (ASL) in cystic fibrosis. (*Top*) Inhibition of sodium absorption through ENaC may be obtained by direct inhibition of the channel (e.g., with amiloride or analogs) or by preventing its activation mediated by proteases. (*Middle*) Stimulation of calcium-activated chloride channels (CaCCs) by direct activators (openers) or by inducing intracellular calcium elevation through a surface receptor (e.g., the purinergic P2Y2 receptor) is another potential way to promote ASL normalization in CF. (*Bottom*) Improved hydration of airway surfaces can be obtained by aerosolized osmolytes (e.g., hypertonic saline or mannitol).

TARGETING ENaC

One of the possible therapeutic targets outside of CFTR is ENaC. Sodium hyperabsorption through ENaC has been described as a primary pathophysiologic feature of CF epithelia. The data supporting this conclusion come from a variety of in vitro and in vivo systems, including nasal and lower-airway potential-difference measurements (Knowles et al. 1981) and short-circuit current measurements in excised and cultured tissues (Willumsen and Boucher 1991). Following these observations, a mouse model was created that, with overexpressed β-subunit of ENaC in its airways, was used to test the hypothesis that sodium hyperabsorption was sufficient to cause a CF-like lung disease (Mall et al. 2004). Indeed, these mice developed PCL depletion and mucus obstruction. Similarly, deletion of Nedd4 L, a ubiquitin ligase involved in down-regulation of ENaC, also led to the development of Na^+ hyperabsorption and a CF-like lung phenotype (Kimura et al. 2011). Identification of the mechanism(s) underlying sodium hyperabsorption in CF, however, has been a difficult and sometimes contentious problem. Recently, it appears that CFTR expression (wild type, but not ΔF508) prevents proteolytic activation of the α- and γ-ENaC subunits by endogenous proteases involved in the normal regulation of this channel (Gentzsch et al. 2010). Because ENaC and CFTR coimmunoprecipitate, this relationship could result from a direct physical association or from the formation of a regulatory complex. Additional disruption of ENaC regulation by endogenous proteases may come from changes in the ASL environment itself in CF (Myerburg et al. 2006; Tarran et al. 2006). It has been shown that ENaC activity responds to changes in the volume and/or composition of the overlying ASL (Myerburg et al. 2010). Dilution of the thin lining fluid on airways leads to a rapid upregulation of ENaC activity, whereas progressive ASL absorption is normally associated with reduced ENaC activity. It has become increasingly clear that, in CF, the balance between serine protease and antiprotease activity is disrupted, causing full proteolytic activation of ENaC, even under low ASL

volume conditions where down-regulation of ENaC activity would be expected and desirable.

Although altered levels of protease and antiprotease activities could account for these findings, other factors may also be involved. For example, SPLUNC1—a highly expressed, secreted protein—was found to bind to ENaC and protect it from proteolytic cleavage/activation (Garcia-Caballero et al. 2009; Rollins et al. 2010). While not possessing intrinsic antiprotease activity itself, alterations in the expression or binding of SPLUNC to ENaC could also result in constitutive activation of ENaC by endogenous proteases.

The evidence that sodium hyperabsorption contributes to ASL volume depletion in CF makes ENaC activity an attractive therapeutic target. Most simply, this could be accomplished through the use of inhaled ENaC blockers, such as amiloride (Knowles et al. 1990). The major challenge facing this approach is achieving the desired drug effectiveness and durability in the lung while avoiding systemic exposure that could lead to unwanted renal effects (i.e., hyperkalemia) (Knowles et al. 1992). One potential criticism of this therapeutic approach is that while slowing sodium/volume absorption might be desirable in CF, if adequate chloride/volume secretion does not accompany it, the desired improvements in ASL hydration and mucociliary clearance (MCC) might not be achieved. The combination of an ENaC inhibitor with a secretagogue (e.g., a hyperosmotic agent or CaCC stimulator), therefore, might be more beneficial.

A second approach to preventing sodium hyperabsorption is to block proteolytic activation of ENaC, which appears to be dysregulated in CF airways (Tarran et al. 2006; Myerburg et al. 2008). Peptide and small molecule inhibitors of relevant proteases have been proposed for this purpose, with supportive data from in vitro and animal studies (Bridges et al. 2001; Coote et al. 2009). However, testing the safety and efficacy of this approach in humans has yet to be accomplished. This approach is complicated by the fact that multiple endogenous proteases appear to be able to activate ENaC (e.g., prostasin, TMPRSS4, matriptase) under normal conditions (Donaldson et al. 2002; Vuagniaux et al. 2002; Garcia-Caballero et al. 2008), and activation by neutrophil elastase in the highly inflamed CF airway is also possible (Caldwell et al. 2005). It is likely that a broad-spectrum antiprotease, or knowledge that selective inactivation of a key protease in a signaling cascade would be effective, will be needed. Alternatively, given the recent data that CFTR and/or SPLUNC shield ENaC from proteolysis by several proteases expressed in cell culture systems, a therapeutic approach that leverages this knowledge could be conceived of as well.

CLINICAL STUDIES OF ENaC INHIBITORS

Clinical studies of inhaled amiloride, the prototype channel blocker, showed acute improvements in mucociliary clearance (Kohler et al. 1986) but little to no improvement in lung function or other clinically meaningful parameters (Knowles et al. 1990; Graham et al. 1993). Because amiloride has a short half-life in airways, likely limiting its efficacy, efforts have been made to develop more potent and longer-acting analogs. One blocker, PS552-02 (Parion Sciences), has entered phase II trials in CF (Hirsh et al. 2008). This compound is roughly 60-fold more potent than amiloride at blocking ENaC currents in cultured human airway cells, and it has also demonstrated a significantly increased ability to stimulate MCC in a sheep model. Importantly, whereas the effect of amiloride had waned by 4 h postdose in this model, PS552-02 continued to demonstrate an effect on clearance. Similar effects on MCC were demonstrated in healthy human volunteers (Donaldson et al. 2005), and preliminary reports of a multidose study in CF subjects did not reveal major safety concerns. Whether the regular use of this or other ENaC inhibitors currently being developed results in clinically important outcomes has not yet been demonstrated.

The concept that agents from different drug classes (e.g., osmotic agent or P2Y$_2$ agonist with an ENaC blocker) might yield additive or synergistic results is quite appealing. Although attractive, these approaches will require careful clinical testing. Interestingly, the combination of amiloride and hypertonic saline (HS) did not yield any additional improvements over HS alone, and in fact appeared to inhibit the beneficial effects of HS on MCC and lung function (Donaldson et al. 2006). Subsequent in vitro experiments suggested that off-target effects of amiloride in CF epithelia (i.e., reductions in osmotically driven water transport) may have prevented the expected ASL volume responses to HS that would lead to augmented MCC improvements. Similar in vitro studies with PS552-02 have not shown evidence of comparable off-target activities, and indeed pretreatment before HS administration increases the peak ASL volume response and extends the duration of ASL volume expansion (Hirsh et al. 2008). We await clinical trials to further test the concept that combination therapy may be useful in CF.

TARGETING ALTERNATIVE CHLORIDE CHANNELS

The molecular identity of CaCC has remained elusive for many years (Galietta 2009). However, three independent studies in 2008 identified TMEM16A, also known as anoctamin-1, as a membrane protein responsible for CaCC activity (Caputo et al. 2008; Schroeder et al. 2008; Yang et al. 2008). Expression of TMEM16A by transfection in null cells (i.e., devoid of endogenous activity of CaCCs) causes the appearance of CaCC-like membrane currents. Conversely, silencing TMEM16A expression in airway epithelial cells via RNA interference causes significant inhibition of calcium-dependent chloride secretion. These observations indicate that TMEM16A is an important component of CaCCs.

However, there are still several unanswered questions. First, it is not clear whether TMEM16A is directly activated by calcium or if another protein, such as calmodulin, represents the "calcium sensor" (Tian et al. 2011). It is also possible that activation by calcium involves a phosphorylation event, as indicated by some studies (Wagner et al. 1991; Xie et al. 1996). In other cases, phosphorylation seems to have an inhibitory role, leaving this issue unresolved (Wang and Kotlikoff 1997; Greenwood et al. 2001). Obviously, elucidation of CaCC regulation mechanisms is important if we are to devise pharmacological strategies that lead to stimulation of alternative chloride transport in CF cells. The recent identification of TMEM16A may help our understanding of the molecular mechanisms affecting CaCC function.

The TMEM16A discovery also opens the possibility of finding pharmacological activators that can directly open CaCC channels without altering cytosolic calcium. For this purpose, TMEM16A-expressing cells could be utilized in high-throughput screenings of large chemical libraries, as has been done to identify activators and inhibitors of the CFTR channel (Verkman and Galietta 2009). However, it is important to note that a recent study has questioned the role of TMEM16A as a CaCC protein in airway epithelia. According to this study, a selective TMEM16A inhibitor was able to strongly block calcium-dependent chloride secretion in salivary glands but had minor effects on bronchial epithelial cells (Namkung et al. 2011). This finding would suggest that another protein, not TMEM16A, is responsible for CaCC activity in airway epithelial cells. But this conclusion contrasts with the findings of other studies. For example, silencing TMEM16A in human bronchial epithelial cells caused a significant decrease in calcium-dependent chloride secretion (Caputo et al. 2008). Furthermore, chloride secretion and MCC were also significantly decreased in TMEM16A knockout mice (Ousingsawat et al. 2009; Rock et al. 2009). Additional studies are required to resolve these controversial findings.

Despite lacking knowledge of the molecular identify of CaCC, early studies indicated that airway epithelia possessed a chloride secretory pathway other than CFTR (Anderson and Welsh 1991; Mason et al. 1991). In these studies, increases in cytosolic calcium by different agonists caused the activation of chloride channels with biophysical and pharmacological properties distinct from those of CFTR. Experiments in CF epithelia provided the most conclusive demonstration that CaCCs and CFTR represent separate entities (Anderson and Welsh 1991; Knowles et al. 1991; Mason et al. 1991). Such

cells had CaCC-dependent membrane currents but no cAMP-activated chloride transport. In vitro measurements of the PCL height with a confocal microscope also demonstrated that CaCCs have the ability to influence airway surface hydration (Tarran et al. 2005). Stimulation of these channels by calcium agonists evokes a transient increase in PCL thickness and improvement in mucociliary transport in both normal and CF epithelia. This is a clear indication that stimulation of alternative chloride channels may, at least in part, compensate the CFTR defect.

How CaCC compensates for the lack of functional CFTR in CF is not completely established. In fact, it is not clear whether these two channels have fully overlapping physiological functions. In this respect, it has been shown that CFTR transports anions other than chloride that may have crucial physiological functions. For example, secretion of bicarbonate through CFTR (Smith and Welsh 1992; Illek et al. 1997) could be important in the expansion of mucins following exocytosis (Garcia et al. 2009) and in the regulation of PCL pH (Coakley et al. 2003). CFTR activity has also been directly linked to an antimicrobial function. Indeed, CFTR is permeable to thiocyanate, a physiological anion that is actively transported through the basolateral membrane by a sodium-coupled mechanism (Fragoso et al. 2004). When secreted onto the apical surface of the epithelium, thiocyanate is converted to hypothiocyanite by lactoperoxidase, an enzyme that also consumes hydrogen peroxide as the second substrate (Gerson et al. 2000). Hypothiocyanite has been reported to be an effective bactericidal molecule. Therefore, impaired thiocyanate transport through CFTR could generate a deficit in an important innate defense mechanism (Conner et al. 2007; Moskwa et al. 2007). Reassuringly, activation of CaCCs could provide a route not only for chloride secretion, but also for HCO_3^- and SCN^- transport. However, consideration of CaCC as a suitable drug target to circumvent the CFTR defect requires a better understanding of CaCC structure, regulation, and function.

In the absence of direct CaCC activators, strategies aiming at the stimulation of alternative chloride transport in the airway epithelium have been based on the mobilization of cytosolic calcium. One of the most effective ways to promote intracellular calcium increase in airway epithelial cells is by stimulation of purinergic receptors (Mason et al. 1991). Administration of ATP or UTP to the apical surface of the epithelium binds to the $P2Y_2$ nucleotide receptor and causes a rapid increase in cytosolic free calcium concentration and activation of CaCCs. The resulting increase in chloride secretion reaches a peak within a few seconds, though the largest part of this lasts for only 5–15 min (Mason et al. 1991; Galietta et al. 2002). The transient nature of this effect is likely multifactorial. First, ATP and UTP are rapidly hydrolyzed by ectoenzymes (Picher et al. 2004). Second, calcium mobilization is intrinsically transient because it is largely dependent on release from intracellular calcium stores (Mason et al. 1991). Third, desensitization mechanisms may affect purinergic receptors (Clarke et al. 1999) as well CaCCs themselves. Whether these factors impact the suitability of the purinergic receptor as a therapeutic target is unknown.

CLINICAL STUDIES UTILIZING CaCC PATHWAYS

To attempt to take advantage of the CaCC pathway in CF, inhaled $P2Y_2$ receptor agonists were developed, thus providing an opportunity to circumvent the CFTR-mediated chloride secretory defect. This approach has the additional potential benefits of increasing cilia beat frequency and mucus secretion. While exogenous UTP, a natural ligand for this receptor, was able to potently stimulate mucociliary clearance in vivo in non-CF subjects (Olivier et al. 1996; Noone et al. 1999), it was less efficacious as a monotherapy (i.e., without amiloride) in CF (Bennett et al. 1996). One logical explanation for this discrepancy was that UTP is rapidly degraded by the abundant extracellular nucleotidases found in CF secretions.

A longer acting agonist, denufosol, was subsequently synthesized to resist metabolism in the CF airway. Following the completion of a series of promising phase II studies (Deterding et al.

2005, 2007), two phase III studies were performed. In the first (TIGER-1), 352 patients were randomized to denufosol or placebo. As a function of the study design, these patients were relatively young (mean age 14.2 yr) and had mild lung disease (baseline FEV_1 92% of predicted). After 24 wk of therapy, a small but statistically significant improvement in FEV_1 (0.045 L; $p = 0.047$) was observed, without significant effects on other lung function parameters (FVC, FEF_{25-75}) or on the already low exacerbation frequency. Continued improvement in the raw FEV_1 values occurred during the subsequent 24-wk open label extension period (Accurso et al. 2011).

Following these results, a second phase III study (TIGER-2) that included a 48-wk placebo-controlled treatment period and 466 patients was conducted. These patients were similar to those studied in TIGER-1, with respect to their mild severity of lung disease and the intensity of baseline treatment regimen. Unfortunately, preliminary results from this study revealed no difference between denufosol- and placebo-treated subjects in lung function improvement, rate of lung function change over time, or time to first exacerbation frequency.

Why this program failed to achieve drug registration could relate to a number of factors. First, the adequacy of the drug itself (e.g., inadequate potency or durability in the CF lung) or appropriateness of the target (e.g., receptor/ channel already occupied/activated in vivo; receptor target or signaling pathway desensitizes) could be questioned. Second, inadequate delivery to a critical lung compartment (i.e., small airways); or trial design issues (i.e., targeting patients with very mild disease, where FEV_1 and exacerbation endpoints may be difficult to impact) could also be problematic. Because these issues will also impact the development of other inhaled agents that target non-CFTR ion channels, they deserve additional scrutiny and study.

Other approaches to the treatment of CF lung disease through modulation of ASL volume via non-CFTR mechanisms continue to be explored. Like denufosol, Moli-1901 (duramycin) is purported to stimulate calcium-activated chloride channels, although the mechanism underlying this effect is unclear. Small studies examining nasal potential-difference responses and safety after short-term inhalation use have been conducted and published (Zeitlin et al. 2004; Grasemann et al. 2007). A larger phase IIb dose-ranging study was subsequently initiated in Europe and completed in 2009, but to date no results are available. Clearly, while targeting CaCC holds promise, clinical benefits have not yet been achieved, and more needs to be learned in order to intelligently design effective therapies.

TARGETING ASL DEHYDRATION VIA OSMOTIC AGENTS

Perhaps the simplest approach to increasing ASL volume is through the use of osmotically active agents that draw water into the airway lumen across the water permeable airway epithelium (Tarran et al. 2007). HS is one currently available osmotic "hydrator" that has been shown to greatly reduce the frequency of pulmonary exacerbations and to improve lung function (Elkins et al. 2006). The ability of HS to produce these effects in infants and toddlers with CF is currently the subject of a large study spanning North America.

Mannitol, inhaled as a dry powder, is an alternative osmotic. Dry-powder mannitol has been shown to improve mucus rheologic properties of CF sputum (Daviskas et al. 2010) and to stimulate the rate of mucociliary clearance in CF and non-CF bronchiectasis patients (Daviskas et al. 1997, 1999, 2001, 2008; Robinson et al. 1999). As might be predicted, the effect of mannitol on both mucociliary clearance and spirometry outcomes was dose related, with doses ≥ 400 mg being most effective (Daviskas et al. 2008; Teper et al. 2011). A short-term, crossover, phase II study in 39 CF subjects revealed a 7% improvement from baseline in FEV_1 (121 mL), versus no change during the placebo-treatment period. A rapid fall in lung function back to baseline levels occurred during the 2-wk washout period that followed mannitol treatment (Jaques et al. 2008). Two large, long-term, phase III studies have subsequently been performed in a combined 643 patients from 11 countries. Re-

sults from these studies have not yet been published in a peer-reviewed journal, but have been reported at major scientific meetings (Bilton et al. 2009). During 6 mo of double-blinded treatment, mannitol led to an average 7.3% improvement in FEV_1 compared to baseline ($p < 0.001$), and a similar significant improvement when compared to placebo treatment ($p < 0.001$). In contrast to data from a small study by Minasian et al. (2010), patients using rhDNase with mannitol also showed lung function improvement (FEV_1 5.3%; $p < 0.001$), although this was numerically smaller than the FEV_1 change in those not using rhDNase (9.44% improvement from baseline). The effect of mannitol on pulmonary exacerbations appears to be smaller than that reported by Elkins et al. (2006) in their study of HS. One possible explanation for this difference could be the more aggressive use of other CF medications (including inhaled antibiotics, rhDNase, and azithromycin) during the mannitol studies, which in turn reduced baseline exacerbation rates and lessened the ability to see an impact on this important clinical outcome measure. Generally speaking, the safety profile of dry-powder mannitol appears to be good, with cough, bronchospasm, oropharyngeal irritation, and hemoptysis being the major reported adverse events. Because mannitol is delivered as a dry powder, the requirement for a nebulizer is obviated, and the treatment time associated with inhaling the required 10 capsules per treatment is ~ 5 min. On the basis of these findings, dry-powder mannitol (Bronchitol; Pharmaxis) was recently approved for use in Australia.

SUMMARY

A concerted effort to understand the epithelial ion transport processes that govern ASL formation in normal airways, and the pathophysiologic changes that are responsible for the onset and progression of CF lung disease, has identified several potential therapeutic targets. A multifaceted strategy for finding new CF treatments that leverages this knowledge provides the greatest opportunity to discover beneficial, complementary treatment approaches and yield significant reductions in morbidity and mortality. The challenge that we now face, and that will continue to grow, is learning how to maximize the benefits from multiple effective agents—rather than the previous challenge of having little to offer these patients.

REFERENCES

Accurso FJ, Rowe SM, Clancy JP, Boyle MP, Dunitz JM, Durie PR, Sagel SD, Hornick DB, Konstan MW, Donaldson SH, et al. 2010. Effect of VX-770 in persons with cystic fibrosis and the G551D-CFTR mutation. *N Engl J Med* **363:** 1991–2003.

Accurso FJ, Moss RB, Wilmott RW, Anbar RD, Schaberg AE, Durham TA, Ramsey BW, TIGER-1 Investigator Study Group. 2011. Denufosol tetrasodium in patients with cystic fibrosis and normal to mildly impaired lung function. *Am J Respir Crit Care Med* **183:** 627–634.

Anderson MP, Welsh MJ. 1991. Calcium and cAMP activate different chloride channels in the apical membrane of normal and cystic fibrosis epithelia. *Proc Natl Acad Sci* **88:** 6003–6007.

Bennett WD, Olivier KN, Zeman KL, Hohneker KW, Boucher RC, Knowles MR. 1996. Effect of uridine 5′-triphosphate plus amiloride on mucociliary clearance in adult cystic fibrosis. *Am J Respir Crit Care Med* **153:** 1796–1801.

Bilton D, Robinson P, Cooper P, Charlton B. 2009. Randomized, double blind, placebo-controlled phase iii study of inhaled dry powder mannitol (Bronchitol) in CF. *Pediatr Pulmonol* **44** (Suppl): A216.

Bridges RJ, Newton BB, Pilewski JM, Devor DC, Poll CT, Hall RL. 2001. Na$^+$ transport in normal and CF human bronchial epithelial cells is inhibited by BAY 39–9437. *Am J Physiol Lung Cell Mol Physiol* **281:** L16–L23.

Caldwell RA, Boucher RC, Stutts MJ. 2005. Neutrophil elastase activates near-silent epithelial Na$^+$ channels and increases airway epithelial Na$^+$ transport. *Am J Physiol Lung Cell Mol Physiol* **288:** L813–L819.

Caputo A, Caci E, Ferrera L, Pedemonte N, Barsanti C, Sondo E, Pfeffer U, Ravazzolo R, Zegarra-Moran O, Galietta LJ. 2008. TMEM16A, a membrane protein associated with calcium-dependent chloride channel activity. *Science* **322:** 590–594.

Clarke LL, Harline MC, Otero MA, Glover GG, Garrad RC, Krugh B, Walker NM, González FA, Turner JT, Weisman GA. 1999. Desensitization of P2Y2 receptor-activated transepithelial anion secretion. *Am J Physiol* **276:** C777–C787.

Coakley RD, Grubb BR, Paradiso AM, Gatzy JT, Johnson LG, Kreda SM, O'Neal WK, Boucher RC. 2003. Abnormal surface liquid pH regulation by cultured cystic fibrosis bronchial epithelium. *Proc Natl Acad Sci* **100:** 16083–16088.

Conner GE, Wijkstrom-Frei C, Randell SH, Fernandez VE, Salathe M. 2007. The lactoperoxidase system links anion transport to host defense in cystic fibrosis. *FEBS Lett* **581:** 271–278.

Coote K, Atherton-Watson HC, Sugar R, Young A, MacKenzie-Beevor A, Gosling M, Bhalay G, Bloomfield G, Dunstan A, Bridges RJ, et al. 2009. Camostat attenuates airway epithelial sodium channel function in vivo through the inhibition of a channel-activating protease. *J Pharmacol Exp Ther* **329:** 764–774.

Daviskas E, Anderson SD, Brannan JD, Chan HK, Eberl S, Bautovich G. 1997. Inhalation of dry-powder mannitol increases mucociliary clearance. *Eur Respir J* **10:** 2449–2454.

Daviskas E, Anderson SD, Eberl S, Chan HK, Bautovich G. 1999. Inhalation of dry powder mannitol improves clearance of mucus in patients with bronchiectasis. *Am J Respir Crit Care Med* **159:** 1843–1848.

Daviskas E, Anderson SD, Eberl S, Chan HK, Young IH. 2001. The 24-h effect of mannitol on the clearance of mucus in patients with bronchiectasis. *Chest* **119:** 414–421.

Daviskas E, Anderson SD, Eberl S, Young IH. 2008. Effect of increasing doses of mannitol on mucus clearance in patients with bronchiectasis. *Eur Respir J* **31:** 765–772.

Daviskas E, Anderson SD, Jaques A, Charlton B. 2010. Inhaled mannitol improves the hydration and surface properties of sputum in patients with cystic fibrosis. *Chest* **137:** 861–868.

Deterding R, Retsch-Bogart G, Milgram L, Gibson R, Daines C, Zeitlin PL, Milla C, Marshall B, Lavange L, Engels J, et al. 2005. Safety and tolerability of denufosol tetrasodium inhalation solution, a novel P2Y2 receptor agonist: Results of a phase 1/phase 2 multicenter study in mild to moderate cystic fibrosis. *Pediatr Pulmonol* **39:** 339–348.

Deterding RR, Lavange LM, Engels JM, Mathews DW, Coquillette SJ, Brody AS, Millard SP, Ramsey BW, Cystic Fibrosis Therapeutics Development Network Inspire 08–103 Working Group. 2007. Phase 2 randomized safety and efficacy trial of nebulized denufosol tetrasodium in cystic fibrosis. *Am J Respir Crit Care Med* **176:** 362–369.

Donaldson SH, Hirsh A, Li DC, Holloway G, Chao J, Boucher RC, Gabriel SE. 2002. Regulation of the epithelial sodium channel by serine proteases in human airways. *J Biol Chem* **277:** 8338–8345.

Donaldson S, Smith R, Doran J, DiMassimo B, Zeman K, Bennett B, Hurd H, Hopkins S. 2005. Safety, pharmacokinetics and effects on mucus clearance following administration of 552–02 to normal healthy volunteers. *Pediatr Pulmonol* **28** (Suppl): 218.

Donaldson SH, Bennett WD, Zeman KL, Knowles MR, Tarran R, Boucher RC. 2006. Mucus clearance and lung function in cystic fibrosis with hypertonic saline. *New Engl J Med* **354:** 241–250.

Elkins MR, Robinson M, Rose BR, Harbour C, Moriarty CP, Marks GB, Belousova EG, Xuan W, Bye PT, National Hypertonic Saline in Cystic Fibrosis (NHSCF) Study Group. 2006. A controlled trial of long-term inhaled hypertonic saline in patients with cystic fibrosis. *N Engl J Med* **354:** 229–240.

Fragoso MA, Fernandez V, Forteza R, Randell SH, Salathe M, Conner GE. 2004. Transcellular thiocyanate transport by human airway epithelia. *J Physiol* **561:** 183–194.

Galietta LJ. 2009. The TMEM16 protein family: A new class of chloride channels? *Biophys J* **97:** 3047–3053.

Galietta LJ, Pagesy P, Folli C, Caci E, Romio L, Costes B, Nicolis E, Cabrini G, Goossens M, Ravazzolo R, et al. 2002. IL-4 is a potent modulator of ion transport in the human bronchial epithelium in vitro. *J Immunol* **168:** 839–845.

Garcia MA, Yang N, Quinton PM. 2009. Normal mouse intestinal mucus release requires cystic fibrosis transmembrane regulator-dependent bicarbonate secretion. *J Clin Invest* **119:** 2613–2622.

Garcia-Caballero A, Dang Y, He H, Stutts MJ. 2008. ENaC proteolytic regulation by channel-activating protease 2. *J Gen Physiol* **132:** 521–535.

Garcia-Caballero A, Rasmussen JE, Gaillard E, Watson MJ, Olsen JC, Donaldson SH, Stutts MJ, Tarran R. 2009. SPLUNC1 regulates airway surface liquid volume by protecting ENaC from proteolytic cleavage. *Proc Natl Acad Sci* **106:** 11412–11417.

Gentzsch M, Dang H, Dang Y, Garcia-Caballero A, Suchindran H, Boucher RC, Stutts MJ. 2010. The cystic fibrosis transmembrane conductance regulator impedes proteolytic stimulation of the epithelial Na$^+$ channel. *J Biol Chem* **285:** 32227–32232.

Gerson C, Sabater J, Scuri M, Torbati A, Coffey R, Abraham JW, Lauredo I, Forteza R, Wanner A, Salathe M, et al. 2000. The lactoperoxidase system functions in bacterial clearance of airways. *Am J Respir Cell Mol Biol* **22:** 665–671.

Graham A, Hasani A, Alton EW, Martin GP, Marriott C, Hodson ME, Clarke SW, Geddes DM. 1993. No added benefit from nebulized amiloride in patients with cystic fibrosis. *Eur Respir J* **6:** 1243–1248.

Grasemann H, Stehling F, Brunar H, Widmann R, Laliberte TW, Molina L, Döring G, Ratjen F. 2007. Inhalation of Moli1901 in patients with cystic fibrosis. *Chest* **131:** 1461–1466.

Greenwood IA, Ledoux J, Leblanc N. 2001. Differential regulation of Ca^{2+}-activated Cl$^-$ currents in rabbit arterial and portal vein smooth muscle cells by Ca^{2+}-calmodulin-dependent kinase. *J Physiol* **534:** 395–408.

Hirsh AJ, Zhang J, Zamurs A, Fleegle J, Thelin WR, Caldwell RA, Sabater JR, Abraham WM, Donowitz M, Cha B, et al. 2008. Pharmacological properties of *N*-(3,5-diamino-6-chloropyrazine-2-carbonyl)-*N'*-4-[4-(2,3-dihydroxypropoxy) phenyl]butyl-guanidine methanesulfonate (552-02), a novel epithelial sodium channel blocker with potential clinical efficacy for cystic fibrosis lung disease. *J Pharmacol Exp Ther* **325:** 77–88.

Illek B, Yankaskas JR, Machen TE. 1997. cAMP and genistein stimulate HCO3- conductance through CFTR in human airway epithelia. *Am J Physiol* **272:** L752–L761.

Jaques A, Daviskas E, Turton JA, McKay K, Cooper P, Stirling RG, Robertson CF, Bye PT, Lesouëf PN, Shadbolt B, et al. 2008. Inhaled mannitol improves lung function in cystic fibrosis. *Chest* **133:** 1388–1396.

Kimura T, Kawabe H, Jiang C, Zhang W, Xiang YY, Lu C, Salter MW, Brose N, Lu WY, Rotin D. 2011. Deletion of the ubiquitin ligase Nedd4 L in lung epithelia causes cystic fibrosis-like disease. *Proc Natl Acad Sci* **108:** 3216–3221.

Knowles M, Gatzy J, Boucher R. 1981. Increased bioelectric potential difference across respiratory epithelia in cystic fibrosis. *N Engl J Med* **305:** 1489–1495.

Knowles MR, Church NL, Waltner WE, Yankaskas JR, Gilligan P, King M, Edwards LJ, Helms RW, Boucher RC. 1990. A pilot study of aerosolized amiloride for the treatment of lung disease in cystic fibrosis. *N Engl J Med* **322:** 1189–1194.

Knowles MR, Clarke LL, Boucher RC. 1991. Activation by extracellular nucleotides of chloride secretion in the airway epithelia of patients with cystic fibrosis [see comments]. *N Engl J Med* **325:** 533–538.

Knowles MR, Church NL, Waltner WE, et al. 1992. Amiloride in CF: Safety, pharmacokinetics, and efficacy in the treatment of pulmonary disease. In *Amiloride and its analogs: Unique cation transport inhibitors* (ed. Cragoe EJ, et al.), pp. 301–316. VCH, Chichester, U.K.

Kohler D, App E, Schmitz-Schumann M, Wurtemberger G, Matthys H. 1986. Inhalation of amiloride improves the mucociliary and the cough clearance in patients with cystic fibroses. *Eur Respir J Dis Suppl* **146:** 319–326.

Mall M, Grubb BR, Harkema JR, O'Neal WK, Boucher RC. 2004. Increased airway epithelial Na$^+$ absorption produces cystic fibrosis-like lung disease in mice. *Nat Med* **10:** 487–493.

Mason SJ, Paradiso AM, Boucher RC. 1991. Regulation of transepithelial ion transport and intracellular calcium by extracellular ATP in human normal and cystic fibrosis airway epithelium. *Br J Pharmacol* **103:** 1649–1656.

Matsui H, Grubb BR, Tarran R, Randell SH, Gatzy JT, Davis CW, Boucher RC. 1998. Evidence for periciliary liquid layer depletion, not abnormal ion composition, in the pathogenesis of cystic fibrosis airways disease. *Cell* **95:** 1005–1015.

Minasian C, Wallis C, Metcalfe C, Bush A. 2010. Comparison of inhaled mannitol, daily rhDNase and a combination of both in children with cystic fibrosis: A randomised trial. *Thorax* **65:** 51–56.

Moskwa P, Lorentzen D, Excoffon KJ, Zabner J, McCray PB Jr, Nauseef WM, Dupuy C, Bánfi B. 2007. A novel host defense system of airways is defective in cystic fibrosis. *Am J Respir Crit Care Med* **175:** 174–183.

Myerburg MM, Butterworth MB, McKenna EE, Peters KW, Frizzell RA, Kleyman TR, Pilewski JM. 2006. Airway surface liquid volume regulates ENaC by altering the serine protease-protease inhibitor balance: A mechanism for sodium hyperabsorption in cystic fibrosis. *J Biol Chem* **281:** 27942–27949.

Myerburg MM, McKenna EE, Luke CJ, Frizzell RA, Kleyman TR, Pilewski JM. 2008. Prostasin expression is regulated by airway surface liquid volume and is increased in cystic fibrosis. *Am J Physiol Lung Cell Mol Physiol* **294:** L932–L941.

Myerburg MM, Harvey PR, Heidrich EM, Pilewski JM, Butterworth MB. 2010. Acute regulation of the epithelial sodium channel in airway epithelia by proteases and trafficking. *Am J Res Cell Mol Biol* **43:** 712–719.

Namkung W, Phuan PW, Verkman AS. 2011. TMEM16A inhibitors reveal TMEM16A as a minor component of calcium-activated chloride channel conductance in airway and intestinal epithelial cells. *J Biol Chem* **286:** 2365–2374.

Noone PG, Bennett WD, Regnis JA, Zeman KL, Carson JL, King M, Boucher RC, Knowles MR. 1999. Effect of aerosolized uridine-5′-triphosphate on airway clearance with

Cite this article as *Cold Spring Harb Perspect Med* doi: 10.1101/cshperspect.a009787

cough in patients with primary ciliary dyskinesia. *Am J Respir Crit Care Med* **160:** 144–149.

Olivier KN, Bennett WD, Hohneker KW, Zeman KL, Edwards LJ, Boucher RC, Knowles MR. 1996. Acute safety and effects on mucociliary clearance of aerosolized uridine 5′-triphosphate $+/-$ amiloride in normal human adults. *Am J Respir Crit Care Med* **154:** 217–223.

Ousingsawat J, Martins JR, Schreiber R, Rock JR, Harfe BD, Kunzelmann K. 2009. Loss of TMEM16A causes a defect in epithelial Ca^{2+}-dependent chloride transport. *J Biol Chem* **284:** 28698–28703.

Picher M, Burch LH, Boucher RC. 2004. Metabolism of P2 receptor agonists in human airways: Implications for mucociliary clearance and cystic fibrosis. *J Biol Chem* **279:** 20234–20241.

Robinson M, Daviskas E, Eberl S, Baker J, Chan HK, Anderson SD, Bye PT. 1999. The effect of inhaled mannitol on bronchial mucus clearance in cystic fibrosis patients: A pilot study. *Eur Respir J* **14:** 678–685.

Rock JR, O'Neal WK, Gabriel SE, Randell SH, Harfe BD, Boucher RC, Grubb BR. 2009. Transmembrane protein 16A (TMEM16A) is a Ca^{2+}-regulated Cl^- secretory channel in mouse airways. *J Biol Chem* **284:** 14875–14880.

Rollins BM, Garcia-Caballero A, Stutts MJ, Tarran R. 2010. SPLUNC1 expression reduces surface levels of the epithelial sodium channel (ENaC) in *Xenopus laevis* oocytes. *Channels (Austin)* **4:** 255–259.

Schroeder BC, Cheng T, Jan YN, Jan LY. 2008. Expression cloning of TMEM16A as a calcium-activated chloride channel subunit. *Cell* **134:** 1019–1029.

Smith JJ, Welsh MJ. 1992. cAMP stimulates bicarbonate secretion across normal, but not cystic fibrosis airway epithelia. *J Clin Invest* **89:** 1148–1153.

Stutts MJ, Canessa CM, Olsen JC, Hamrick M, Cohn JA, Rossier BC, Boucher RC. 1995. CFTR as a cAMP-dependent regulator of sodium channels [see comments]. *Science* **269:** 847–850.

Tarran R, Loewen ME, Paradiso AM, Olsen JC, Gray MA, Argent BE, Boucher RC, Gabriel SE. 2002. Regulation of murine airway surface liquid volume by CFTR and Ca^{2+}-activated Cl^- conductances. *J Gen Physiol* **120:** 407–418.

Tarran R, Button B, Picher M, Paradiso AM, Ribeiro CM, Lazarowski ER, Zhang L, Collins PL, Pickles RJ, Fredberg JJ, et al. 2005. Normal and cystic fibrosis airway surface liquid homeostasis: The effects of phasic shear stress and viral infections. *J Biol Chem* **280:** 35751–35759.

Tarran R, Trout L, Donaldson SH, Boucher RC. 2006. Soluble mediators, not cilia, determine airway surface liquid volume in normal and cystic fibrosis superficial airway epithelia. *J Gen Physiol* **127:** 591–604.

Tarran R, Donaldson S, Boucher RC. 2007. Rationale for hypertonic saline therapy for cystic fibrosis lung disease. *Semin Respir Crit Care Med* **28:** 295–302.

Teper A, Jaques A, Charlton B. 2011. Inhaled mannitol in patients with cystic fibrosis: A randomised open-label dose response trial. *J Cystic Fibrosis* **10:** 1–8.

Tian Y, Kongsuphol P, Hug M, Ousingsawat J, Witzgall R, Schreiber R, Kunzelmann K. 2011. Calmodulin-dependent activation of the epithelial calcium-dependent chloride channel TMEM16A. *FASEB J* **25:** 1058–1068.

Verkman AS, Galietta LJ. 2009. Chloride channels as drug targets. *Nat Rev Drug Discov* **8:** 153–171.

Vuagniaux G, Vallet V, Jaeger NF, Hummler E, Rossier BC. 2002. Synergistic activation of ENaC by three membrane-bound channel-activating serine proteases (mCAP1, mCAP2, and mCAP3) and serum- and glucocorticoid-regulated kinase (Sgk1) in Xenopus Oocytes. *J Gen Physiol* **120:** 191–201.

Wagner JA, Cozens AL, Schulman H, Gruenert DC, Stryer L, Gardner P. 1991. Activation of chloride channels in normal and cystic fibrosis airway epithelial cells by multifunctional calcium/calmodulin-dependent protein kinase. *Nature* **349:** 793–796.

Wang YX, Kotlikoff MI. 1997. Inactivation of calcium-activated chloride channels in smooth muscle by calcium/calmodulin-dependent protein kinase. *Proc Natl Acad Sci* **94:** 14918–14923.

Willumsen NJ, Boucher RC. 1991. Transcellular sodium transport in cultured cystic fibrosis human nasal epithelium. *Am J Physiol* **261:** C332–C341.

Xie W, Kaetzel MA, Bruzik KS, Dedman JR, Shears SB, Nelson DJ. 1996. Inositol 3,4,5,6-tetrakisphosphate inhibits the calmodulin-dependent protein kinase II-activated chloride conductance in T84 colonic epithelial cells. *J Biol Chem* **271:** 14092–14097.

Yang YD, Cho H, Koo JY, Tak MH, Cho Y, Shim WS, Park SP, Lee J, Lee B, Kim BM, et al. 2008. TMEM16A confers receptor-activated calcium-dependent chloride conductance. *Nature* **455:** 1210–1215.

Zeitlin PL, Boyle MP, Guggino WB, Molina L. 2004. A phase I trial of intranasal Moli1901 for cystic fibrosis. *Chest* **125:** 143–149.

Asynchronous ventilation, mucus clearance
        promotion, 232
ATP
    gating. *See* Cystic fibrosis transmembrane
        conductance regulator
    mucus hydration regulation, 228–229
    release stimulation by lung mechanical stress, 230
Autogenic drainage (AD), mucus clearance
        promotion, 233

B

Bacterial infection. *See* Infection
Bicarbonate secretion. *See also* Anion secretion
    airway cells
        Calu-3 cell studies, 147
        chloride-dependent secretion, 149–150
        microelectrode and impedance analysis
            studies, 150–151
        modeling, 148–149
        transport overview, 146–147
    intestine in CFTR knockout mouse model, 279
    mucin unpacking role, 221
    overview, 145–146, 156
    pancreas defects in cystic fibrosis, 255–256
    pancreas
        transport overview, 155–156
        model, 164–166
    transporters
        AE2, 159
        apical membrane
            calcium-activated chloride channel,
                160–161, 258
            chloride/bicarbonate exchanger, 160
            cystic fibrosis transmembrane conductance
                regulator, 159–160, 256–258
            ENaC, 161
            potassium channels, 161
        aquaporins, 162
        carbonic anhydrase, 162
        NKCC1, 159
        potassium channels, 158–159, 258–259
        proton/potassium ATPase, 158, 258
        scaffold proteins, 162–164
        sodium/bicarbonate cotransporter, 158,
            257–258
        sodium/potassium-ATPase, 156–157
        sodium/proton exchanger, 157–158, 161–162,
            257–258
Bronchiectasis. *See* Disseminated bronchiectasis
Burkholderia cepacia, infection management, 308–309

C

CACC. *See* Calcium-activated chloride channel
Calcium-activated chloride channel (CACC)
    anion secretion, 130–131
    bicarbonate secretion role in pancreas, 160–161, 258
    therapeutic targeting in cystic fibrosis
        clinical trials, 327–328
        overview, 325–327
Calcium flux, mucin secretion regulation, 191
Carbonic anhydrase, bicarbonate secretion role in
        pancreas, 162
CBAVD. *See* Congenital bilateral absence of vas deferens
CD. *See* Celiac disease
Celiac disease (CD), cystic fibrosis comorbidity,
        274–275
CFLD. *See* Cystic fibrosis liver disease
CFMDB. *See* Cystic Fibrosis Mutation Database
CFRD. *See* Cystic fibrosis-related diabetes
CFTR. *See* Cystic fibrosis transmembrane conductance
        regulator
CFTR
    animal studies
        knockout, 9–10, 259–261
        sequence conservation, 10
    coding regions, 4
    haplotype mapping, 12
    history of study
        identification as cystic fibrosis gene, 1–3
        mutation analysis, 3
        structure, 3
    mutation. *See* Mutation, *CFTR*
    prospects for study, 11, 13
    segmental duplication, 3
    splicing
        overview, 6–7
        regulation in disease, 7–8
        trans-splicing, 8–9
    transcriptional regulation overview, 3, 5–7
Chloride/bicarbonate exchanger, bicarbonate secretion
        role in pancreas, 160, 256
Chloride secretion. *See* Anion secretion
Clostridium difficile, clinical features of infection, 273
Collateral ventilation, mucus clearance promotion, 232
Congenital bilateral absence of vas deferens (CBAVD),
        CFTR mutations, 22–23
CoPo-22, cystic fibrosis transmembrane conductance
        regulator effects, 293
Corr-4a, cystic fibrosis transmembrane conductance
        regulator correction, 290
Corticosteroids, anti-inflammatory therapy in cystic
        fibrosis, 309–311
Cough, mucus clearance promotion, 231
CPX, cystic fibrosis transmembrane conductance
        regulator correction, 295, 297
CREB, mucin expression regulation, 186, 188
Crohn's disease, cystic fibrosis comorbidity, 274
Cystic fibrosis liver disease (CFLD)
    gene modifier identification by candidate gene
        approach, 102–103

overview, 3, 17–18
Mycobacterium, nontuberculous infection
    management, 309

N

N-Acetylcysteine, anti-inflammatory therapy in cystic
    fibrosis, 314CXCR2 antagonists, 314
Nasal potential difference (NPD), pancreatitis in cystic
    fibrosis, 253
NBC. *See* Sodium/bicarbonate cotransporter
NBD. *See* Nucleotide-binding domain
NBS. *See* Newborn screening
Neuregulin, mucin expression regulation, 187
Neutrophil, airway infection and inflammation in cystic
    fibrosis, 304, 314
Newborn screening (NBS), cystic fibrosis, 25
NF-κB. *See* Nuclear factor-κB
NHE. *See* Sodium/proton exchanger
NHERF1, cystic fibrosis transmembrane conductance
    regulator assembly role, 135–136, 162
NHERF2, cystic fibrosis transmembrane conductance
    regulator assembly role, 135, 162
Nitric oxide (NO), airway production in cystic fibrosis, 313
NKCC1
    basolateral membrane, 129
    bicarbonate secretion role in pancreas, 159
NMR. *See* Nuclear magnetic resonance
NMRD. *See* Nonsense-mediated RNA decay
NO. *See* Nitric oxide
Nonsense-mediated RNA decay (NMRD), *CFTR*, 19
NPD. *See* Nasal potential difference
Nuclear factor-κB (NF-κB)
    inflammation therapeutic targeting, 313
    mucin expression regulation, 186
Nuclear magnetic resonance (NMR), NBD1 isotope
    exchange studies, 85–86
Nucleotide-binding domain (NBD)
    ABC transporters
        mechanochemistry, 53–54
        structural organization, 49, 51–53
    CFTR
        dimerization in gating, 34–35
        NBD1
            F508del mutation effects on structure and
                folding, 60–64, 82, 86, 89
            structure and function, 84–89
            variants, 56–60
        NBD2, 64–66
        R region in regulation, 60, 83–84

O

Oscillating positive expiratory pressure (OscPEP),
    mucus clearance promotion, 235
OscPEP. *See* Oscillating positive expiratory pressure

P

Pancreas
    anion secretion defects in cystic fibrosis, 138
    bicarbonate secretion
        model, 164–166
        overview, 155–156, 256
        regulation, 259
    cystic fibrosis effects on exocrine pancreas
        duodenal acidity effects, 254
        malnutrition, 275–276
        overview, 251–253
        pancreatitis. *See* Idiopathic chronic pancreatitis
        replacement therapy, 275
    cystic fibrosis transmembrane conductance regulator
        knockout animal models
        ferret, 261
        mouse, 259–261
        pig, 261
    therapeutic targets in cystic fibrosis, 262
    transporters
        AE2, 159
        apical membrane
            calcium-activated chloride channel,
                160–161, 258
            chloride/bicarbonate exchanger, 160
            cystic fibrosis transmembrane conductance
                regulator, 159–160, 256–258
            ENaC, 161
            potassium channels, 161
        aquaporins, 162
        carbonic anhydrase, 162
        NKCC1, 159
        potassium channels, 158–159, 258–259
        proton/potassium ATPase, 158, 258
        scaffold proteins, 162–164
        sodium/bicarbonate cotransporter, 158,
            257–258
        sodium/potassium-ATPase, 156–157
        sodium/proton exchanger, 157–158, 161–162,
            257–258
Pancreatitis. *See* Idiopathic chronic pancreatitis
Paraoxonase (PON), history of study in cystic fibrosis,
    1–2
PEP. *See* Positive expiratory pressure
PG-01, cystic fibrosis transmembrane conductance
    regulator potentiation, 290
PgP. *See* Multidrug resistance transporter
4-Phenylbutyrate, cystic fibrosis transmembrane
    conductance regulator correction, 295
PKA. *See* Protein kinase A
PON. *See* Paraoxonase
Positive expiratory pressure (PEP), mucus clearance
    promotion, 233–234
PPQ-102, cystic fibrosis transmembrane conductance
    regulator inhibition, 290
Premature termination codon (PTC), *CFTR*, 19